CLYMER® *Collection Series*

BSA
500 & 650cc UNIT TWINS • 1963-1972

NORTON
750 & 850cc COMMANDOS • 1969-1975

TRIUMPH
500, 650 & 750cc TWINS • 1963-1979

More information available at haynes.com
Phone: 805-498-6703

Haynes UK
Sparkford Nr Yeovil
Somerset BA22 7JJ England

Haynes North America, Inc
859 Lawrence Drive
Newbury Park
California 91320 USA

ISBN 10: 0-89287-604-2
ISBN-13: 978-0-89287-604-4
Library of Congress: 93-78509

Disclaimer

There are risks associated with automotive repairs. The ability to make repairs depends on the individual's skill, experience and proper tools. Individuals should act with due care and acknowledge and assume the risk of performing automotive repairs.

The purpose of this manual is to provide comprehensive, useful and accessible automotive repair information, to help you get the best value from your vehicle. However, this manual is not a substitute for a professional certified technician or mechanic.

This repair manual is produced by a third party and is not associated with an individual vehicle manufacturer. If there is any doubt or discrepancy between this manual and the owner's manual or the factory service manual, please refer to the factory service manual or seek assistance from a professional certified technician or mechanic.

Even though we have prepared this manual with extreme care and every attempt is made to ensure that the information in this manual is correct, neither the publisher nor the author can accept responsibility for loss, damage or injury caused by any errors in, or omissions from, the information given.

M330, 10S1, 14-640

Common spark plug conditions

NORMAL

Symptoms: Brown to grayish-tan color and slight electrode wear. Correct heat range for engine and operating conditions.
Recommendation: When new spark plugs are installed, replace with plugs of the same heat range.

WORN

Symptoms: Rounded electrodes with a small amount of deposits on the firing end. Normal color. Causes hard starting in damp or cold weather and poor fuel economy.
Recommendation: Plugs have been left in the engine too long. Replace with new plugs of the same heat range. Follow the recommended maintenance schedule.

CARBON DEPOSITS

Symptoms: Dry sooty deposits indicate a rich mixture or weak ignition. Causes misfiring, hard starting and hesitation.
Recommendation: Make sure the plug has the correct heat range. Check for a clogged air filter or problem in the fuel system or engine management system. Also check for ignition system problems.

ASH DEPOSITS

Symptoms: Light brown deposits encrusted on the side or center electrodes or both. Derived from oil and/or fuel additives. Excessive amounts may mask the spark, causing misfiring and hesitation during acceleration.
Recommendation: If excessive deposits accumulate over a short time or low mileage, install new valve guide seals to prevent seepage of oil into the combustion chambers. Also try changing gasoline brands.

OIL DEPOSITS

Symptoms: Oily coating caused by poor oil control. Oil is leaking past worn valve guides or piston rings into the combustion chamber. Causes hard starting, misfiring and hesitation.
Recommendation: Correct the mechanical condition with necessary repairs and install new plugs.

GAP BRIDGING

Symptoms: Combustion deposits lodge between the electrodes. Heavy deposits accumulate and bridge the electrode gap. The plug ceases to fire, resulting in a dead cylinder.
Recommendation: Locate the faulty plug and remove the deposits from between the electrodes.

TOO HOT

Symptoms: Blistered, white insulator, eroded electrode and absence of deposits. Results in shortened plug life.
Recommendation: Check for the correct plug heat range, over-advanced ignition timing, lean fuel mixture, intake manifold vacuum leaks, sticking valves and insufficient engine cooling.

PREIGNITION

Symptoms: Melted electrodes. Insulators are white, but may be dirty due to misfiring or flying debris in the combustion chamber. Can lead to engine damage.
Recommendation: Check for the correct plug heat range, over-advanced ignition timing, lean fuel mixture, insufficient engine cooling and lack of lubrication.

HIGH SPEED GLAZING

Symptoms: Insulator has yellowish, glazed appearance. Indicates that combustion chamber temperatures have risen suddenly during hard acceleration. Normal deposits melt to form a conductive coating. Causes misfiring at high speeds.
Recommendation: Install new plugs. Consider using a colder plug if driving habits warrant.

DETONATION

Symptoms: Insulators may be cracked or chipped. Improper gap setting techniques can also result in a fractured insulator tip. Can lead to piston damage.
Recommendation: Make sure the fuel anti-knock values meet engine requirements. Use care when setting the gaps on new plugs. Avoid lugging the engine.

MECHANICAL DAMAGE

Symptoms: May be caused by a foreign object in the combustion chamber or the piston striking an incorrect reach (too long) plug. Causes a dead cylinder and could result in piston damage.
Recommendation: Repair the mechanical damage. Remove the foreign object from the engine and/or install the correct reach plug.

CONTENTS

QUICK REFERENCE DATA

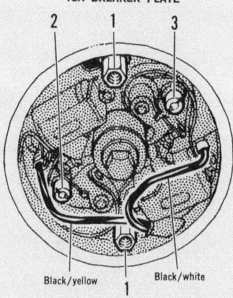

4CA BREAKER PLATE

Black/yellow

Black/white

1

1. Base plate nuts (to adjust ignition timing)
2. Breaker point gap adjuster (right cylinder)
3. Breaker point gap adjuster (left cylinder)

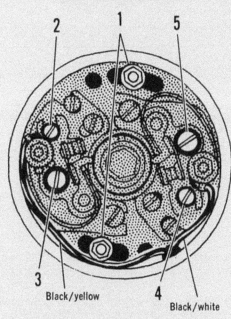

6CA BREAKER PLATE

Black/yellow

Black/white

1. Base plate nuts (to adjust ignition timing)
2. Breaker point gap adjuster (right cylinder)
3. Lock screw (right cylinder)
4. Breaker point gap adjuster (left cylinder)
5. Lock screw (right cylinder)

ENGINE TUNE-UP

Spark plug type	
All except A65T	Champion N-3
A65T only	Champion N-4
Spark plug gap	0.020-0.022 in. (0.5-0.6mm)
Breaker point gap	0.012-0.016 in. (0.3-0.4mm)
Valve clearance	
Intake	0.002 in. (0.05mm)
Exhaust	0.004 in. (0.10mm)
Ignition timing	
Piston position	0.304 in. (7.22mm) BTDC
Crankshaft position	34° BTDC

ELECTRICAL SYSTEM

Battery	Lucas	PUZ5A
Replacement lamps		
Headlight	Lucas 370	50/40 watt
Parking light	Lucas 989	6 watt
Stop/taillight	Lucas 380	21/6 watt
Speedometer light	Smiths	2.2 watt
Headlight main beam indicator bulb	Lucas 281	2 watt
Oil pressure indicator warning	Lucas 281	2 watt
Direction indicator warning	Lucas 281	2 watt
Direction indicator	Lucas 382	21 watt

RECOMMENDED LUBRICANTS AND CAPACITIES

	Capacity	Type
Engine oil	5 U.S. pints 4 Imp. pints	SAE 20W-50
Gearbox	16.5 U.S. oz. 13.8 Imp. oz. 490cc	EP 90 gear lube
Primary chaincase	4.7 U.S. oz. 4.0 Imp. oz. 140cc	SAE 10W-30
Front forks (each leg)	6.5 U.S. oz. 5.4 Imp. oz. 190cc	SAE 5W

TIRES

	Size	Pressure
Front	3.25x19	21 psi
Rear	4.00x18	22 psi

CHAIN ADJUSTMENTS

Primary chain play	⅛ to ¼ in.
Drive chain play	1¼ in.

CHAPTER ONE

GENERAL INFORMATION

This manual provides maintenance and repair information for 1963 and later BSA 650cc and 500cc motorcycles. Engines used for the various models have many similarities and share many components. The principal differences are in carburetion, ignition, and bore and stroke dimensions. Dimensions and capacities are expressed in English units, which are familiar to U.S. mechanics, as well as in metric units.

Procedures common to different models are combined to avoid duplication. Read the *Service Hints* section to make the work as easy and pleasant as possible.

MANUAL ORGANIZATION

This chapter, in addition to providing general information, also discusses equipment and tools useful both for preventive maintenance and troubleshooting.

Chapter Two explains all periodic lubrication and routine maintenance necessary to keep the motorcycle in proper running condition. Chapter Two also includes recommended tune-up procedures which eliminate the need to constantly consult chapters on the various sub-assemblies.

Chapter Three provides methods and suggestions for quick and accurate diagnosis and repair of problems. Troubleshooting procedures discuss typical symptoms and logical methods to pinpoint trouble spots.

Subsequent chapters describe specific systems such as the engine, transmission, and electrical system. Each chapter provides disassembly, repair, and assembly procedures in simple step-by-step form. If a repair is impractical for the owner-mechanic, it is so indicated. It is usually faster and less expensive to take such repairs to a dealer or competent repair shop.

Some of the procedures in this manual specify special tools. A well-equipped mechanic may find he can substitute similar tools already on hand or can fabricate his own.

The terms NOTE, CAUTION, and WARNING have specific meanings in this manual. A NOTE provides additional information to make a step or procedure easier or clearer. Disregarding a NOTE could cause inconvenience, but would not cause damage or personal injury.

A CAUTION emphasizes areas where equipment damage could result. Disregarding a CAUTION could cause permanent mechanical damage; however, personal injury is unlikely.

A WARNING emphasizes areas where personal injury or even death could result from negligence. Mechanical damage may also occur.

Throughout this manual, keep in mind 2 conventions. "Front" refers to the front of the bike. The front of any component such as the engine is at that end which faces toward the front of bike. The left and right side refer to a person sitting on the bike facing forward. For example, the clutch lever is on the left side. These rules are simple, but even experienced mechanics occasionally bcome disoriented.

SERVICE HINTS

Most of the service procedures covered are straightforward and can be performed by anyone reasonably handy with tools. It is suggested, however, that you consider your own capabilities carefully before attempting any operation involving major disassembly of the engine.

Some operations, for example, require the use of a press. It would be wiser to have these performed by a shop equipped for such work, rather than to try to do the job at home with makeshift equipment. Other procedures require precision measurements. Unless you have the skills and equipment required, have a qualified repair shop make the measurements.

Repairs go much faster and easier if the machine is clean before beginning work. There are special cleaners for washing the engine and related parts. Just brush or spray on the cleaning solution, let it stand, then rinse away with a garden hose. Clean all oily or greasy parts with cleaning solution as they are removed.

WARNING
Never use gasoline as a cleaning agent. It presents an extreme fire hazard. Be sure to work in a well-ventilated area while using cleaning solvent. Keep a fire extinguisher, rated for gasoline and electrical fires, handy at all times.

Special tools are required for some repair procedures. These may be purchased from a dealer (or borrowed if you are on good terms with the service department personnel), or fabricated by a mechanic or machinist, often at considerable savings.

Much of the labor charge for repairs made by dealers is for the removal and disassembly of other parts to reach the defective unit. It is frequently possible to perform the preliminary operations first and then take the defective unit to a dealer for repair.

Once having decided to tackle the job at home, read the entire section in this manual which pertains to it. Make sure to identify the proper section. Study the illustrations and text to gain a good idea of what is involved in completing the job satisfactorily. If special tools are necessary, make arrangements to get them before starting. It is frustrating and time consuming to get partly into a job and then be unable to complete it.

Simple wiring checks are easily made at home, but knowledge of electronics is almost a necessity for performing tests with complicated electronic test gear.

During disassembly of parts keep a few general cautions in mind. Force is rarely needed to get things apart. If parts are a tight fit, as a magneto on a crankshaft, there is usually a tool designed to separate them. Never use a screwdriver to pry apart parts with machined surfaces such as crankcase halves and valve covers. It will mar the surfaces and cause leaks.

Make diagrams wherever similar-appearing parts are found. For instance, case cover screws are often of various lengths. Do not rely on remembering where everything came from—mistakes are costly. There is also the possibility of being sidetracked and not returning to work for days or weeks—during which interval, carefully laid out parts may become disarranged.

Tag all similar internal parts for location and mark all mating parts for position. Record number and thickness of shims are they are removed. Small parts, such as several identical bolts, can be identified by placing in plastic sandwich bags and sealing and labeling them with masking tape.

Wiring should be tagged with masking tape and marked as each wire is removed. Again, do not rely on memory alone.

Disconnect the battery ground cable before working near electrical connections and before disconnecting wires. Never run the engine with the battery disconnected; the electrical system could be seriously damaged.

Protect finished surfaces from physical damage or corrosion. Keep gasoline and cleaning solvent off painted surfaces.

Frozen or very tight bolts and screws can often be loosened by soaking with penetrating oil, then sharply striking the bolt head a few times with a hammer and punch (or screwdriver for screws). Avoid heat unless absolutely necessary, because it may melt, warp, or remove the temper from parts.

Avoid flames or sparks while working near a battery being charged or flammable liquids such as cleaning fluid, brake fluid, or gasoline.

No parts, except those assembled with a press fit, require unusual force during assembly. If a part is hard to remove or install, find out why before proceeding.

Cover all openings after removing parts to keep dirt, small tools, and other foreign matter from falling in.

While assembling 2 parts, start all fasteners and then tighten evenly to avoid warpage.

Clutch plates, wiring connections, and brake shoes and drums should be kept clean and free of grease and oil during assembly.

While assembling parts, be sure that all shims and washers are replaced exactly as they were before disassembly.

Wherever a rotating part butts against a stationary part, look for a shim or washer. Use new gaskets if there is any doubt about the condition of old ones. Generally, apply gasket cement to one mating surface only so the parts may be easily disassembled in the future. A thin coat of oil on gaskets helps them seal effectively.

Heavy grease can be used to hold small parts in place if they tend to fall out during assembly.

However, keep grease and oil away from electrical components or brake shoes and drums.

High spots may be sanded off a piston dome with sandpaper, but emery colth and oil do a much more professional job.

Carburetors are best cleaned by soaking the disassembled parts in a commercial carburetor cleaner. Never soak gaskets and rubber parts in cleaner. Never use wire to clean jets and air passages; they are easily damaged. Use compressed air to blow out the carburetor only if the float has been removed first.

A baby bottle makes a good measuring device for adding oil to forks and transmissions. Obtain one which is graduated in ounces and cubic centimeters.

Take sufficient time to do the job right. Do not forget that a newly rebuilt motorcycle engine must be broken in the same as a new one. Keep rpm within the limits given in the owner's manual.

SAFETY HINTS

Professional motorcycle mechanics can work for years and never sustain a serious injury. Observing a few rules of common sense and safety permits many safe enjoyable hours servicing your own machine. You can hurt yourself or damage the bike if you ignore the following precautions.

1. Never use gasoline as a cleaning solvent.

2. Never smoke or use a torch in the vicinity of flammable liquids such as cleaning solvent in open containers.

3. Never smoke or use a torch in an area where batteries are being charged. Highly explosive hydrogen gas is formed during the charging process.

4. If welding or brazing is required on the machine, remove the fuel tank to a safe distance (at least 50 feet away). Welding on gas tanks requires special safety procedures and must be performed only by someone skilled in the process.

5. Use the proper sized wrenches to avoid damage to nuts and personal injury.

6. While loosening a tight or stuck nut, be guided by what would happen if the wrench should slip.

7. Keep the work area clean and uncluttered.

8. Wear safety goggles during all operations involving drilling, grinding, or use of a chisel.

9. Never use worn tools.

10. Keep a fire extinguisher handy and be sure it is rated for gasoline and electrical fires.

PARTS REPLACEMENT

When ordering parts, always order by engine and frame number. Write the numbers down and carry them in your wallet. Compare new parts to old before purchasing. If they are not alike, have the parts clerk explain the difference.

TOOLS

Tool Kit

Most new bikes are equipped with fairly complete tool kits. These tools are satisfactory for most small jobs and emergency roadside repairs.

Shop Tools

For proper servicing, an assortment of ordinary hand tools is needed. As a minimum, these include the following.

1. Combination wrenches
2. Socket wrenches
3. Plastic mallet
4. Small hammer
5. Snap ring pliers
6. Pliers
7. Phillips screwdrivers
8. Slot (common) screwdrivers
9. Feeler gauges
10. Spark plug gauge
11. Spark plug wrench
12. Dial indicator

Electrical system servicing requires a multimeter, ohmmeter, or other device for determining continuity, and a hydrometer for battery equipped machines.

Advanced tune-up and troubleshooting procedures require the additional tools listed below.

1. *Timing gauge* (**Figure 1**). Piston position may be determined by screwing this instrument into the spark plug hole. The tool shown costs approximately $20, and is available from larger dealers and mail order houses. Less expensive ones, which utilize a vernier scale instead of a dial indicator, are also available. They are satisfactory but are not quite so quick and easy to use.

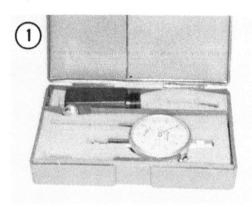

2. *Hydrometer* (**Figure 2**). This instrument measures a battery's state of charge and tells much about battery condition. Such an instrument is available at any auto parts store and through most larger mail order outlets. A satisfactory one costs approximately $3.

3. *Multimeter or VOM* (**Figure 3**). This instrument is invaluable for electrical system trouble-

shooting and service. A few of its functions may be duplicated by locally fabricated substitutes, but for the serious hobbyist, it is a must. Its uses are described in the applicable sections of this manual. Prices start at around $10 at hobbyist electronics stores and mail order outlets.

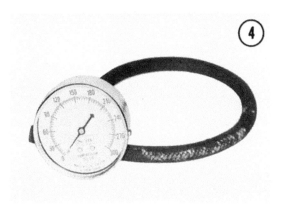

4. *Compression gauge* (**Figure 4**). An engine with low compression cannot be properly tuned and will not develop full power. A compression gauge measures engine cylinder pressure. The one shown has a flexible stem which enables it to reach cylinders where there is little clearance between the cylinder head and frame. Inexpensive ones start at approximately $3 and are available at auto accessory stores or by mail order from large catalog order firms.

5. *Impact driver* (**Figure 5**). This tool makes removal of engine cover screws easy and eliminates damaged screw slots. Good ones cost approximately $12 at larger hardware or motorcycle supply stores.

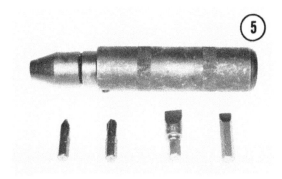

6. *Ignition gauge* (**Figure 6**). This tool measures point gap. It also has round wire gauges for measuring spark plug gap.

EXPENDABLE SUPPLIES

Required expendable supplies include grease, oil, gasket cement, wiping rags, cleaning solvent, and distilled water. Ask a BSA dealer for the special locking compounds, silicone lubricants, and commercial chain lube products. Solvent is available at most service stations. Distilled water for the battery may be obtained at almost any supermarket.

SERIAL NUMBERS

The model serial number must be available for registration and parts ordering. These numbers can be permanently recorded by placing a sheet of paper over the imprinted area and rubbing with the side of a pencil. Some motor vehicle registration offices will accept such evidence in lieu of inspecting the bike in person.

CHAPTER TWO

LUBRICATION AND MAINTENANCE

Regular maintenance is the best guarantee of a trouble-free motorcycle. An afternoon spent cleaning and adjusting can minimize costly mechanical problems and unexpected break-downs on the road.

This chapter describes all required preventive maintenance and procedures for engine tune-up. Anyone with average mechanical ability can perform the procedures.

MAINTENANCE AND LUBRICATION

Maintenance and lubrication intervals are shown in **Table 1** for bikes used on the highway and in **Table 2** for off-road use. Because of dust and dirt, off-road usage requires more frequent service for trouble-free performance.

Figure 1 shows lubrication points and gives the correct grade or type of lubricant required. Information in **Table 3** lists recommended lubricants. Selection need not be limited to these brands if comparable products are available.

In addition to the items listed in the pertinent table (Tables 1 or 2), the following parts should be carefully inspected after 24 months.

1. Brake cables
2. Brake light switches
3. Carburetor rubber dust caps
4. Fuel lines

TOOLS

The basic tools needed are listed in Chapter One. In addition, equipment required for a complete tune-up includes a static timing light, a strobe light, dwell tachometer, carburetor float gauge, and sets of flat and round feeler gauges.

LUBRICATION

Changing Engine Oil

Change oil and clean the reservoir and crankcase filters every 4,000 miles. If this is not done, the crankshaft centrifugal filter may become clogged by foreign particles in the oil.

1. Warm up the engine to operating temperature. The oil flows more freely when warm, and more impurities will remain in suspension immediately after circulating through the system.

2. Remove the reservoir drain plug and filter. In later models, this assembly is located at the base of the main frame down tube. Self-locking nuts secure the sump plate, and must be used when reassembling. Remove the sump plate, gaskets, and filter, allowing the oil to drain into

Table 1   LUBRICATION AND MAINTENANCE (STREET MODELS)

Intervals: Months or miles, whichever comes first

Service required	Months	2	4	6	8	12	16	18	20	24
	Miles	1,000	2,000	3,000	4,000	6,000	8,000	9,000	10,000	12,000
	Km	1,600	3,200	4,800	6,400	9,600	12,800	14,400	16,000	19,200
Engine										
Check level in oil-tank*										
Change engine oil			X							
Change oil filter element			X							
Clean oil screen filter			X							
Service spark plugs			X							
Service contact breaker points			X							
Oil point cam felt wick			X^1			X^2				
Adjust ignition timing			X							
Check ignition primary and secondary cables			X							
Adjust valve tappet clearances			X							
Adjust cam chain			X							
Service air cleaner			X							
Adjust carburetors**					X					
Check throttle valve operation					X					
Clutch										
Adjust clutch					X					
Battery										
Service battery			X							
Fuel System										
Clean fuel valve filter			X							
Check fuel tank and fuel lines					X					

(continued)

*Every 250 miles (400km)   **As required   1. Late models   2. Early models

2

Table 1 LUBRICATION AND MAINTENANCE (STREET MODELS) (continued)

Intervals: Months or miles, whichever comes first

Service required	Months	2	4	6	8	12	16	18	20	24
	Miles	1,000	2,000	3,000	4,000	6,000	8,000	9,000	10,000	12,000
	Km	1,600	3,200	4,800	6,400	9,600	12,800	14,400	16,000	19,200
Steering and Front Suspension										
Check steering head bearings										X
Check steering handle lock										X
Check handle bar holder					X					
Check front fork top plate						X				
Steering and Front Suspension										
Check front fork bottom case										X
Change front fork oil										X
Rear Suspension										
Grease swing arm pivot		X								
Check swing arm				X						
Check rear suspension mounting bolts						X				
Wheels and Brakes										
Check front and rear wheel spokes			X							
Check front and rear wheel rims and hubs									X	
Check front and rear wheels, bearings, and axles									X	
Check front and rear tires		X								
Check front brake caliper and pad linings						X				
Check front brake line		X								
Check brake fluid level		X								

(continued)

2

Table 1 LUBRICATION AND MAINTENANCE (STREET MODELS) (continued)

Intervals: Months or miles, whichever comes first

Service required	Months	2	4	6	8	12	16	18	20	24
	Miles	1,000	2,000	3,000	4,000	6,000	8,000	9,000	10,000	12,000
	Km	1,600	3,200	4,800	6,400	9,600	12,800	14,400	16,000	19,200
Wheels and Brakes (continued)										
Check and adjust brake pedal			X							
Check rear brake shoe linings							X			
Check rear brake stopper arm			X							
Lubricate control cables		X								
Chassis and Final Drive										
Check oil level in primary chaincase (chain oiler)*										
Check frame					X					
Check exhaust system					X					
Gearbox						X				
Service and adjust final drive chain		X								
Check final drive and driven sprockets		X								
Change oil in primary chaincase		X								
Lights and Accessories										
Check lights and switches			X							
Check horn			X							
Check speedometer and tachometer			X							

*Every 250 miles (400km)

Table 2   LUBRICATION AND MAINTENANCE (OFF-ROAD MODELS)

Intervals: Months or miles, whichever comes first

Service required	Months	2	4	6	8	12	16	18	20	24
	Miles	500	1,000	2,000	4,000	6,000	8,000	9,000	10,000	12,000
	Km	800	1,600	3,200	6,400	9,600	12,800	14,400	16,000	19,200
Engine										
Check level in oil tank*										
Change engine oil		X								
Change oil filter element		X								
Clean oil screen filter		X								
Service spark plugs		X								
Service contact breaker points			X							
Oil point cam felt wick		X				X				
Adjust ignition timing			X							
Check ignition primary and secondary cables			X							
Adjust valve tappet clearances			X							
Adjust cam chain			X							
Service air cleaner		X								
Adjust carburetors**										
Check throttle valve operation		X			X					
Clutch										
Adjust clutch					X					
Battery										
Service battery		X								

(continued)

*Every 100 miles (160km)   **As required

Table 2 LUBRICATION AND MAINTENANCE (OFF-ROAD MODELS) (continued)

Intervals: Months or miles, whichever comes first

Months	2	4	6	8	12	16	18	20	24
Service required — Miles	500	1,000	2,000	4,000	6,000	8,000	9,000	10,000	12,000
Km	800	1,600	3,200	6,400	9,600	12,800	14,400	16,000	19,200
Fuel System									
Clean fuel valve filter	X								
Check fuel tank and fuel lines	X								
Steering and Front Suspension									
Check steering head bearings	X								
Check steering handle lock									X
Check handle bar holder				X					
Check front fork top plate					X				
Check front fork bottom case									X
Change front fork oil					X				
Rear Suspension									
Grease swing arm pivot	X								
Check swing arm	X								
Check rear suspension mounting bolts	X								
Wheels and Brakes									
Check front and rear wheel spokes	X								
Check front and rear wheel rims and hubs	X								
Check front and rear wheels, bearings, and axles	X								
Check front and rear tires	X								

(continued)

Table 2   LUBRICATION AND MAINTENANCE   (OFF-ROAD MODELS)   (continued)

Intervals: Months or miles, whichever comes first

Service required	Months	2	4	6	8	12	16	18	20	24
	Miles	500	1,000	2,000	4,000	6,000	8,000	9,000	10,000	12,000
	Km	800	1,600	3,200	6,400	9,600	12,800	14,400	16,000	19,200
Wheels and Brakes (continued)										
Check and adjust brakes		X								
Check rear brake shoe linings							X			
Check rear brake stopper arm			X							
Lubricate control cables		X								
Chassis and Final Drive										
Check oil level in primary chaincase (chain oiler)*										
Check frame			X							
Check exhaust system					X					
Gearbox						X				
Service and adjust final drive chain		X								
Check final drive and driven sprockets		X								
Change oil in primary chaincase		X								

*Every 250 miles (400km)

NOTE: Off-road motorcycles are subjected to extremes of dust, heat, water and general abuse. Ideally all critical areas should be checked after each day of riding. Table 2 is a general guide for moderate use and represents the maximum limits of service.

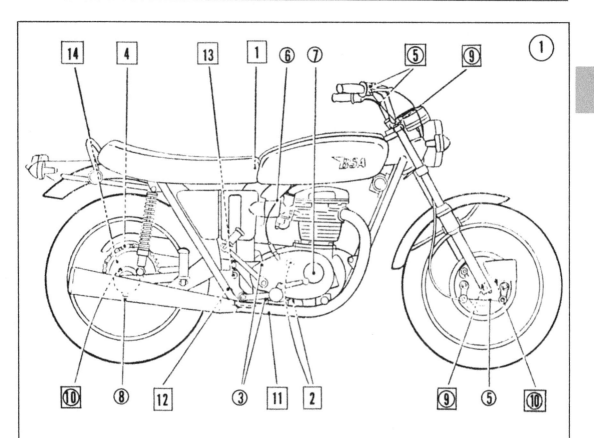

Figures within squares refer
to the left side of the machine.

Figures within circles refer
to the right side of the machine.

LUBRICATION POINTS

1. Oil reservoir 20W/50
2. Primary chaincase 10W/30
3. Gearbox filler and drain plug . . . . EP. 90
4. Rear chain . . . . Oil, grease, or chain lube
5. Exposed cables and joints 10W/30
6. Clutch cable . . . . Grease
7. Contact breaker and auto-advance . . 10W/30
8. Speedometer drive gearbox . . . . Grease
9. Front fork drain and filler plugs . . . 5W (190cc)
10. Clean and repack wheel bearings . . . Grease
11. Center stand and side stand pivots . . 10W/30
12. Brake pedal pivot 10W/30
13. Swing arm grease points . . . . . Grease
14. Rear brake cam spindle . . . . . Grease

Table 3   RECOMMENDED LUBRICANTS

Unit	Mobil	Castrol	Exxon	Shell	Texaco
Engine and primary chaincase	Mobiloil Super	Castrol GTX or Castrol XL 20/50	Uniflo	Shell Super Motor Oil	Havoline Motor Oil 20W/50
Gearbox	Mobilube GX 90	Castrol Hypoy	Exxon Gear Oil GX 90/140	Shell Spirax 90 EP	Multigear Lubricant EP 90
Front forks	Mobiloil Super	Castrolite	Uniflo	Shell Super Motor Oil	Havoline Motor Oil 10W/30
Wheel bearings, swing arm, and steering races	Mobilgrease MP or Mobilgrease Super	Castrol LM Grease	Exxon Multi-purpose Grease H	Shell Retinax A	Marfak All Purpose
Freeing rusted parts	Mobil Handy Oil	Castrol Penetrating Oil	Exxon Penetrating Oil	Shell Easing Oil	Graphited Penetrating Oil

NOTE: The above lubricants are recommended for all operating temperatures above 0°F (−18°C). Approval is given to lubricants marketed by companies other than those listed provided they meet the API Service MS Performance level.

a receptacle. See **Figure 2**. On earlier models, the drain plug with filter attached is located at the bottom of the oil tank (see **Figure 3**).

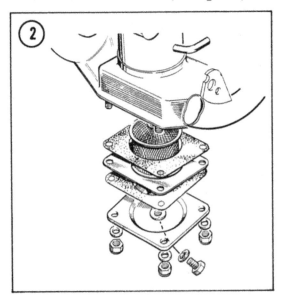

3. Remove the crankcase filter, nuts and spring washers (see **Figure 4**).

4. Wash both filters with solvent or kerosene using a stiff brush to clean thoroughly. Shake dry before reassembling.

5. The oil pump scavenge pipe is located in the crankcase filter cavity (D, Figure 4). This pipe has a one-way valve. Check the valve by lifting it off its seat. If the ball falls back into its seat when released, it is operating correctly.

6. Check the gaskets for the reservoir and crankcase filters to be sure they are in good condition. Replace if necessary.

7. The 3 oil lines are attached to the crankcase by a single connector. See **Figure 5**. Remove it by means of a single bolt, but do not disconnect the flexible lines. Allow oil to drain from the lines and replace the connector, making sure the O-rings are in place to prevent leakage.

8. Install the reservoir and crankcase filters. Tighten the nuts evenly to prevent distortion of the sump plates. Be sure everything is snugly in place with new gaskets, to prevent leakage.

9. Refill the reservoir with 5 pints of recommended oil. On later models, with a frame tube reservoir, the dipstick attached to the filter cap has oil level markings. Add oil until full.

10. Before starting the engine, add ½ pint of oil through the timing plug aperture at the front of the crankcase to prime the oil pump. With

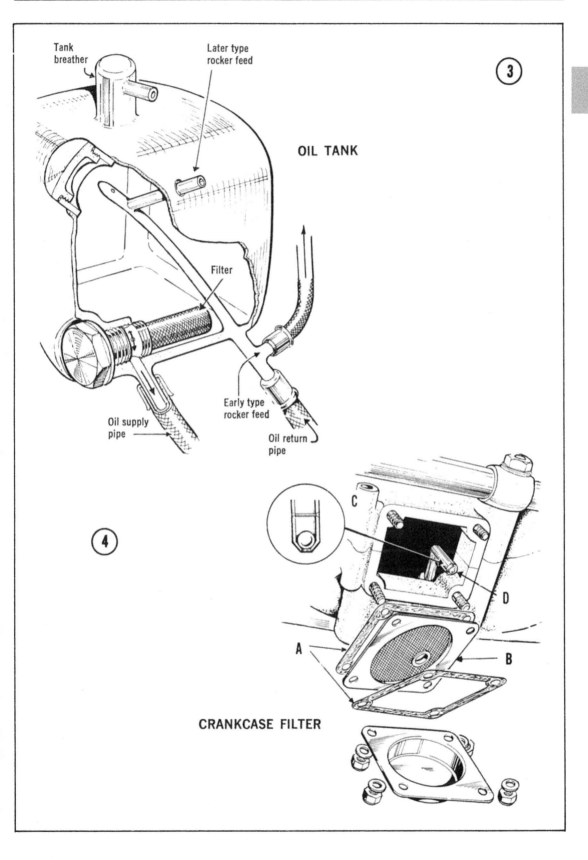

OIL TANK

Tank breather

Later type rocker feed

③

2

Filter

Oil supply pipe

Early type rocker feed

Oil return pipe

④

C

D

A

B

CRANKCASE FILTER

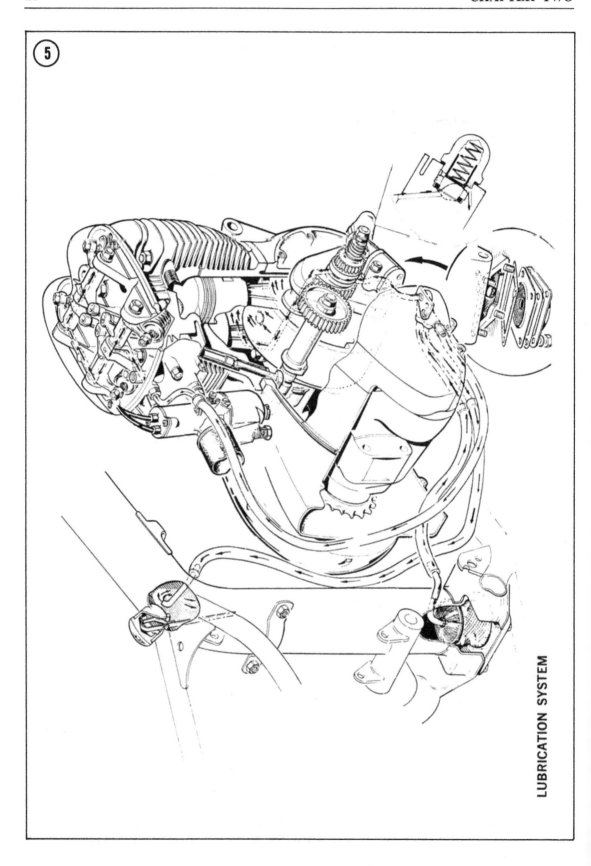

⑤

LUBRICATION SYSTEM

the ignition off, kick the starter lever through until oil flows from the return line below the filter cap in the reservoir. This will prevent excessive wear caused by oil starvation immediately after starting.

Oil Pressure

The lubrication system is designed to operate at a maximum pressure of 50 psi, despite variations in oil temperature or engine speed. This maximum is maintained by an oil pressure relief valve located on the right side of the crankcase (**Figure 6**). The valve is set to open automatically when the pressure reaches 50 psi, draining the surplus oil into the crankcase sump. The valve is pre-set at the factory and cannot be disassembled for adjustment. The only service required is the periodic cleaning of the gauge filter of the valve with a brush and solvent (**Figure 7**). To remove the valve, simply unscrew it. When replacing, be sure the washer is in good condition or there will be leakage around the valve.

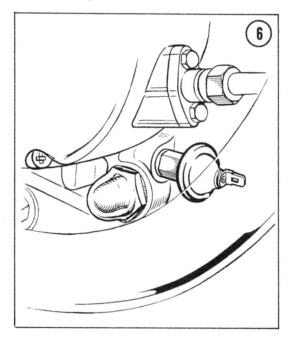

Wet Sumping

If large amounts of smoke are present in the exhaust when the engine is first started after standing for several hours, the cause may be

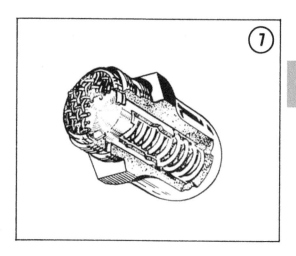

malfunction of either one or both of the non-return valves in the delivery and return oil lines. The valve in the return line is most often the culprit, allowing oil to drain from the reservoir into the crankcase where it builds up while the engine is stopped. If you have the problem, check both non-return valves immediately.

1. Check for improper seating of the ball valve in the scavenge side of the oil pump.

2. Check the ball valve seat. There may be an air leak in the crankcase oil scavenge pipe, oil pump to crankcase joint, or in an oil plug at the bottom of the engine. A crankcase casting may be too porous and seep oil. Malfunctions may be checked by performing the scavenge suction test given below. Additional causes for wet sumping could be blockage in the return oil pipe (caused by a misaligned oil junction block gasket), an oil pressure release valve piston in full bypass position because of a stuck piston or broken or missing spring, or a restriction in the oil tank vent pipe.

3. Perform the scavenge suction test using a vacuum gauge calibrated in inches (or mm) of mercury. Fit a length of BSA oil pipe to it and run the engine until it has warmed up. Remove the oil sump cap and screen and attach the vacuum gauge hose to the oil scavenge pipe. The gauge should read 18-26 inches (46-66 cm) of mercury with the engine at a fast idle. After the engine is stopped, the gauge needle should return slowly to zero.

4. If the test is not satisfactory, remove the oil pump, clean and inspect. Pay particular attention to the balls and ball seats. Reinstall the pump using a new gasket.

5. Check the crankcase scavenge tube for leaks and the crankcase itself for porosity by filling a pumper type oil can with light oil and then squirting into the pickup tube through a folded rag. Some back pressure should be felt after a few strokes of the pump.

Pressure Switch

The lubrication system pressure switch (H) is located on the right side of the crankcase next to the pressure relief valve (see Figure 6). When the oil pressure drops below 7 psi, the switch is actuated, illuminating a red warning light mounted on top of the headlight case. The light should be on when the ignition is switched on, before the engine is started.

If the light does not come on, the bulb is faulty and should be replaced. If the light does not go out a few seconds after the engine is started, oil pressure is too low or the switch is faulty. The switch is a sealed unit which cannot be adjusted or repaired. It must be replaced. If the switch is operating correctly and the red warning light stays on when the engine is running, turn off the ignition. Don't run the engine with low oil pressure. Serious damage could result, but at the very least there will be excessive wear.

1. Check the oil level in the reservoir. If it's low, fill to the proper level and start the engine. Do not replace the filler cap until you see the oil flowing steadily from the return line into the reservoir.

2. The oil must pass through a filter between the reservoir and engine. If the filter is clogged, insufficient oil will be delivered to the crankcase. Check the filter to be sure it's clean.

3. Badly worn crankshaft main bearings or big end rod bearings will cause a drop in oil pressure. If this is your problem, repair the lower end.

4. The oil pump may be either loosely mounted

or badly worn. If it's worn, it must be repaired as a unit. Do not overlook any components when inspecting the pump.

Transmission

The transmission is housed in a separate compartment within the engine crankcase. It has its own supply of special oil, which must be kept at the required level and changed every 4,000 miles. The correct level is shown by a mark on the dipstick of the filler cap (see **Figure 8**).

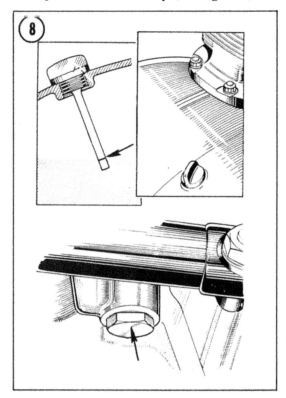

1. Ride the motorcycle until the oil is warm.

2. Place a container under the drain plug (see Figure 8).

3. Unscrew the filler cap to allow air into the transmission.

4. Remove the drain plug and drain the oil into the container.

5. The filler cap has a breather hole drilled in it. Be sure this hole is free from any obstruction to prevent a buildup of pressure in the transmission housing.

6. Replace the drain plug, making sure the O-ring is in good condition. If it isn't, it will leak.

7. Fill the transmission with oil to the mark on the dipstick. Capacity is approximately a pint.

8. Replace the filler cap snugly so it won't vibrate out.

Primary Chain

The primary chain and clutch are contained in a separate housing outside the crankcase. Since the oil supply lubricates the clutch as well as the chain, anti-friction compounds such as molybdenum sulphide or graphite must never be added to the oil, since they can cause the clutch to slip. The oil supply also lubricates the rear chain by a drip feed, so the level must be checked frequently to insure adequate lubrication. The oil should be changed every 4,000 miles.

1. Remove the inspection cap (see **Figure 9**).

2. Place a container under the chaincase and remove the drain screw.

3. Drain the oil into the container and replace the drain screw, making sure the sealing washer is in good condition to prevent leakage.

4. Remove the oil level screw from the chaincase.

5. Add oil through the inspection cap opening until overflow just appears through the oil level screw opening.

6. Replace the oil level screw, sealing washer, and the inspection cap. The chaincase holds less than a half pint of oil, so a frequent oil level check is necessary.

Secondary Chain

The lubrication of the rear chain is provided for by a non-adjustable drip feed at the rear of the primary chaincase. This causes the oil level to drop steadily, so proper lubrication of the secondary chain is dependent on maintaining proper oil level in the chaincase. In some cases, the drip feed may not supply adequate lubrication to the rear chain. Additional oil from a can, or the application of one of the commercial chain lubricants may be used to supplement the drip feed.

A special treatment every 2,000 miles will add considerably to the life of the rear chain. It's an easy chore, and it will save money.

1. Wash the chain thoroughly in kerosene and hang it up to drain.

2. Warm graphite grease in a tray until it is liquid.

WARNING
When heating grease, don't get is so hot that it will start a fire. Grease becomes flammable if it gets hot enough.

3. Immerse the chain in the grease, stirring and flexing it with a stick.

4. Hang the chain to drain and cool. Wipe off surplus grease.

5. When reinstalling the chain, be sure the spring clip is installed with its closed end toward the direction of chain travel, or facing forward on the top run of the chain (see **Figure 10**).

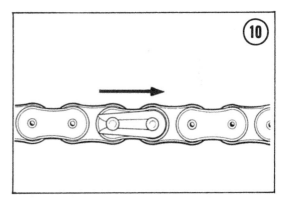

Contact Breaker

A few drops of oil on the bearings is all the lubrication required by the contact breaker. An oil seal is fitted between the contact breaker and the engine to prevent engine oil from penetrating into the contact breaker housing. If the seal has to be replaced, it must be inserted into the housing with its lip toward the inside of the engine.

1. The contact breaker cam is lubricated by a pair of felt pads (J, **Figure 11**). Apply 2 drops of engine oil to each pad every 2,000 miles.

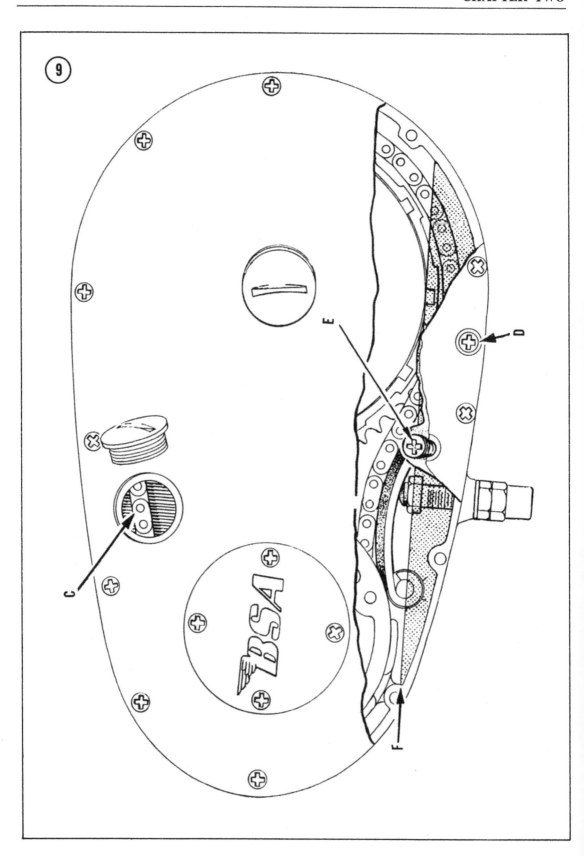

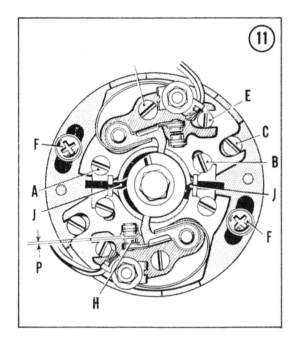

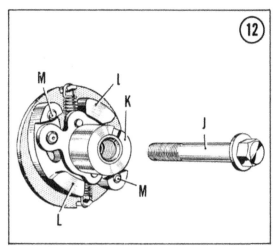

2. Rotate the engine slowly until the slot on the face of the cam is pointing up. Apply one drop of engine oil in the slot to lubricate the cam spindle.

3. Lubricating the auto-advance mechanism requires care, in order to keep the ignition properly timed. Draw a line across the contact breaker carrier plate and its housing, so it can be replaced in exactly the same position after lubrication.

4. Remove the 2 carrier plate screws (F, Figure 11) and remove the entire contact breaker assembly. The auto-advance mechanism is housed behind the contact breaker carrier plate.

5. The governor weights of the auto-advance mechanism (L, **Figure 12**), have 2 pivot points (M). Apply one drop of oil to each pivot point.

6. Check to be sure the governor weights move freely. Turn the auto-advance cam (K) until the governor weights are opened to the full advance position. When the cam is released, the springs should return the weights to the inner, full retard position.

7. Replace the contact breaker carrier plate, making sure the marks line up exactly. Tighten the screws (F, Figure 11). If the marks line up

correctly, the timing should not be changed after reassembly.

Steering Head

The steering head has tapered roller bearings fitted at the top and bottom. They are packed with grease at the factory before assembly, and cannot be lubricated while installed on the motorcycle. The bearings must be repacked with grease every 10,000 miles, an operation which requires the complete disassembly of the steering head. Refer to Chapter Seven.

Front Fork

The main function of the oil in the forks is to dampen the front suspension. For this reason, the correct grade and quantity of oil must be used, making sure the amount in each fork leg is exactly 190cc (or as marked on the cap nuts). The oil should be changed every 10,000 miles or at least once a year.

1. Remove the handlebar pillars (J, **Figure 13**), from the top yoke and lay the handlebars on the gas tank. Be sure to place a cloth between the handlebars and the tank to prevent scratches.

2. Disconnect the cable drive from the bottom of the speedometer and tachometer.

3. Unscrew the cap nut (A) from one fork tube at a time. This will release the mounting plates for the speedometer and tachometer.

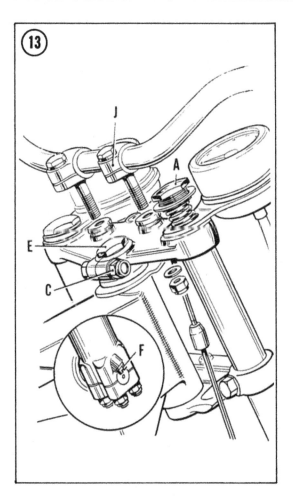

4. Remove the drain screws (F) at the bottom of each lower fork leg, draining the oil into a container. When the oil stops flowing, apply the front brake and pump the forks up and down to completely drain the oil.

5. Replace the drain screws and sealing washers. Make sure the washers are in good condition.

6. Pour 190cc's of oil into the top of each fork tube. Replace the speedometer and tachometer, locking them in place and sealing the fork tubes with the cap nuts.

7. Connect cable drives to the instruments and reassemble the handlebars.

8. A double lip oil seal is fitted inside the top of each lower fork leg. Leakage of oil onto the outside of the upper fork tubes indicates the seal is faulty and must be replaced. This operation is described in Chapter Seven.

Wheel Bearings

The front and rear wheel bearings are packed with grease before assembly, and cannot be lubricated on the motorcycle. They must be repacked with grease every 10,000 miles, which requires dismantling the hubs. This operation is described in Chapter Seven.

Grease Fittings

The BSA has a number of grease fittings (see Figure 1) which require one or 2 strokes of a hand grease gun every 2,000 miles.

Be careful not to use too much grease in lubricating the rear brake cam spindle. If grease gets on the brake linings, it will cause dangerously reduced braking.

The 2 swing arm pivot fittings require a high pressure grease gun. Grease should be applied until it appears at both ends of the pivot. If grease does not appear it means the passages are blocked and the bearings are not getting lubrication. This will cause excessive wear, resulting in a dangerous condition. In this case, the pivot assembly must be disassembled and cleaned. The procedure is described in Chapter Seven.

Control Cables

The inner wires of the control cables are treated with a molybdenum based grease to reduce friction and give long service.

The exposed portions of the inner wires at each end of the cable should be lubricated with grease every 2,000 miles. Grease is preferable to oil because it is more resistent to weather.

If lubrication of the entire inner wire is required, a suitable method is illustrated in **Figure 14**.

1. Unfasten the top of the cable from the control lever.

2. Tie a plastic bag tightly around the end of the armored cable covering. Suspend the top of the bag by a string, as shown.

3. Fill the bag with oil, allowing it to drain down

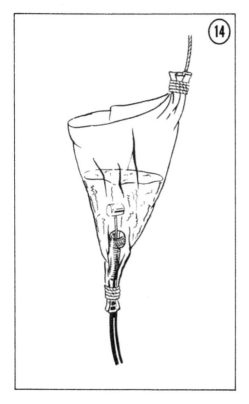

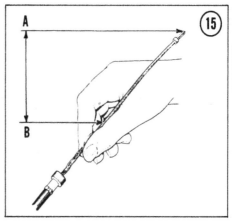

inside the cable until it seeps out of the lower end. Remove the bag and allow excess oil to drip out.

4. Reassemble the cables and control levers.

Speedometer and Tachometer

The cable drives of the speedometer and tachometer should be lubricated with grease every 2000 miles. It is important to grease these lightly, so excess can't work its way into the instrument and cause damage.

1. Disconnect the cable drive under the instrument. The inner cable can then be pulled out for cleaning and lubrication as shown in **Figure 15**.

2. Grease the inner cable lightly, but do not apply any grease to the top 6 inches of the wire (between A and B in **Figure 15**).

3. Reconnect the cables to the instruments. Proper lubrication is one of the commonly neglected maintenance operations. Reduction of wear and improved performance make the time and effort required worthwhile.

ENGINE TUNE-UP

Engine tune-up consists of several accurate and careful adjustments to obtain maximum engine performance. Since different systems in an engine interact to affect overall performance, a tune-up must be performed in the following order:

1. Valve clearance adjustment.

2. Compression test.

3. Ignition adjustment and timing.

4. Carburetor adjustment.

Perform an engine tune-up every 3000 miles or 6 months. **Table 4** summarizes tune-up specifications.

Table 4 ENGINE TUNE-UP

Spark plug	
Type	See Appendix
Gap	0.020-0.022 in. (0.5-0.6 mm)
Breaker point gap	0.012-0.016 in. (0.3-0.4 mm)
Valve clearance	0.008 in. (0.203 mm) intake
	0.010 in. (0.254 mm) exhaust
Ignition timing	See Appendix

Valve Clearance Adjustment

This series of simple mechanical adjustments must be performed while the engine is cold.

Valve clearance for the engine must be adjusted carefully. If the clearance is too small, the valves may be burned or distorted. Large clearance results in excessive noise. In either case, engine power is reduced.

To set valve clearance properly, the tappet must be seated on the base o fthe cam.

Set the clearance (E) of each valve with a feeler gauge (see **Figure 16**). The correct clearance is 0.008 in. for intake and 0.010 in. for exhaust.

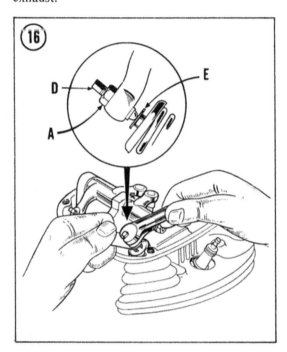

1. Set the left side exhaust valve clearance. To adjust the clearance, loosen the locknut (see Figure 16) and screw the adjuster (D) in or out as required. Always check the clearance again after the locknut (A) has been tightened. The right side exhaust valve spring must be fully compressed with the valve open.

2. Set the right side exhaust valve clearance. The left side exhaust valve spring must be fully compressed with the valve open.

3. Set the left side intake valve clearance. The right side intake valve spring must be fully compressed with the valve open.

4. Set the right side intake valve clearance. The left side intake valve spring must be fully compressed with the valve open.

5. Install the rocker cover with a new gasket.

6. If old spark plugs are used they should be cleaned and gapped. New plugs should be gapped.

Compression Readings

See Chapter Three under *Starting Difficulties*.

Spark Plugs

Spark plugs are available in various heat ranges, hotter or colder than the plugs originally installed at the factory.

Select plugs of a heat range designed for the loads and temperature conditions under which the engine will run. Use of incorrect heat range can cause seized pistons, scored cylinder walls, or damaged piston crowns.

In general, use a lower-numbered plug for low speeds, low loads, and low temperatures. Use a higher-numbered plug for high speeds, high engine loads, and high temperatures.

> NOTE: *Use the highest numbered plug that will not foul. In areas where seasonal temperature variations are great, the factory recommends a high-numbered plug for slower winter operation.*

The reach (length) of a plug is also important. A longer than normal plug could interfere with the piston, causing permanent damage. Refer to **Figures 17 and 18**.

1. Grasp the spark plug leads as near the plug as possible and pull them off.

2. Blow away any dirt which has accumulated in the spark plug wells.

> CAUTION
> *Dirt could fall into the cylinders when the plugs are removed, causing serious engine damage.*

3. Remove the spark plugs with a spark plug wrench.

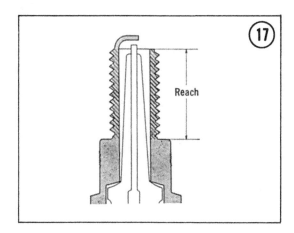

Reach

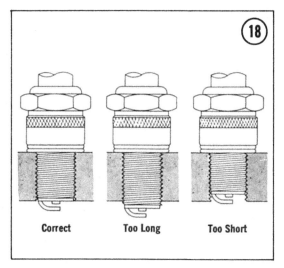

Correct   Too Long   Too Short

NOTE: *If the plugs are difficult to re-move, apply penetrating oil around base of plugs and let it soak in for 10-20 minutes.*

4. Inspect the spark plugs carefully. Look for broken porcelain, excessively eroded electrodes, and excessive carbon or oil fouling. Replace such plugs. If deposits are light, plugs may be cleaned in solvent with a wire brush or cleaned in a special spark plug sandblast cleaner.

5. Gap plugs to 0.020-0.022 in. (0.5-0.6mm) with a wire feeler gauge.

6. Install plugs with a new gasket. First, apply a small drop of oil to the threads. Tighten plugs finger-tight, then tighten with a spark plug wrench an additional ½ turn. If you must reuse an old gasket, tighten only an additional ⅛ turn.

NOTE: *Do not overtighten. This will only flatten the gasket and destroy its sealing ability.*

Much information about engine and spark plug performance can be determined by careful examination of the spark plugs. This information is only valid after performing the following steps.

1. Ride a short distance at full throttle in any gear.

2. Kill the engine and close the throttle simultaneously.

3. Pull in clutch and coast to a stop.

4. Remove the spark plugs and examine them. Compare to **Figure 19**.

Condenser (Capacitor)

The condenser (capacitor) is a sealed unit and requires no maintenance. Be sure connections are clean and tight.

The only proper test is to measure the resistance of the insulation with an ohmmeter. The value should be 5,000 ohms. A made-do test is to charge the capacitor by hooking the leads, or case and lead, to a battery. After a few seconds, touch the leads together, or lead to case, and check for a spark as shown in **Figure 20**. A damaged capacitor won't store electricity.

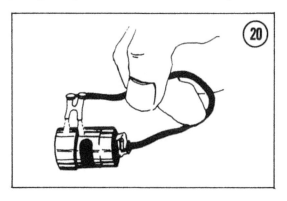

Most mechanics prefer to discard a condenser and replace it with a new one during engine tune-up.

On some 500cc models, the condenser is located under the gas tank and is protected by a rubber cover.

⑲

Normal plug appearance noted by the brown to grayish-tan deposits and light electrode wear. This plug indicates the correct plug heat range and proper air/fuel ratio.

Red, brown, yellow, and white coatings caused by fuel and oil additives. Such additives should not be used or damage will result.

Carbon fouling distinguished by dry, fluffy black carbon deposits which may be caused by an over-rich air/fuel mixture, excessive hand choking, clogged air filter, or excessive idling.

Shiny yellow glaze on insulator cone is caused when the powdery deposits from fuel and oil additives melt. Melting occurs during hard acceleration after prolonged idling. This glaze conducts electricity and shorts out the plug. Avoid the use of additives at all times.

Oil fouling indicated by wet, oily deposits caused by too much oil in the mix. A hotter plug tempo-rarily reduces oil deposits, but a plug that is too hot leads to preignition and possible engine damage.

Overheated plug indicated by burned or blistered insulator tip and badly worn electrodes. This condi-tion may be caused by preignition, cooling system defects, lean air/fuel ratios, low octane fuel, or over advanced ignition timing.

Spark plug condition photos courtesy of AC Spark Plug Division, General Motors Corporation.

Breaker Points

Check that the insulation between the breaker contacts and the contact breaker base is not defective. A short circuit will prevent the motor from running. To test for this condition, disconnect the wire or wires on the points, block the points open and measure insulation resistance between the movable point and a good ground, using the highest range on the ohmmeter. If there is any indication at all on the ohmmeter, the points are shorted.

Through normal use, points gradually pit and burn. See **Figure 21**. If this condition is not too serious, they can be dressed with a few strokes of a clean point file. Do not use emery cloth or sandpaper, as particles remain on the points and cause arcing and burning. If a few strokes of the file do not smooth the points completely, replace them.

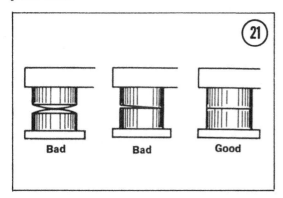

Bad	Bad	Good

If points are still serviceable after filing, remove all residue with lacquer thinner or special contact cleaner. Close the points on a piece of clean white paper such as a business card. Continue to pull the card through the closed points until no particles or discoloration remains on the card. Finally, rotate the engine and observe the points as they open and close. If they do not meet squarely, replace them.

Adjust point gap and ignition timing as described below.

Replacement

1. Disconnect the battery leads or remove the fuse from the holder near the battery.

2. Remove the contact breaker cover located on the right end of the exhaust camshaft.

3. Unscrew the cam center bolt and use a puller to remove the point cam.

4. If the contact breaker unit is to be removed, disconnect the coil leads and unhook the frame clips necessary to pass the leads through the crankcase and timing cover. Make a note of the degree figure stamped on the cam unit as an aid in static timing.

5. Place a drop of light oil on the pivot pins before reassembling the contact breaker unit.

6. Align the peg on the camshaft with the slot on the point cam and press it in place, tightening the bolt.

7. For the 6CA contact breaker plate used on early models (**Figure 22**), assemble the black/yellow leads to the rear. A few still earlier machines had the black/white leads to the rear, so it should be noted how they are routed before the unit is disassembled.

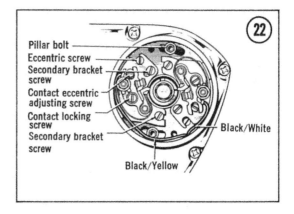

Pillar bolt
Eccentric screw
Secondary bracket screw
Contact eccentric adjusting screw
Contact locking screw
Secondary bracket screw
Black/White
Black/Yellow

8. To adjust the point gaps, rotate the crankshaft until the scribe mark on the breaker cam is aligned with one of the nylon point lifters. Set the gap to 0.015 in. (0.38mm), using a feeler gauge, by loosening the locking screw and turning the adjusting screw. Repeat the operation for the other set of points.

9. For the 4CA contact breaker plate used on later models (**Figure 23**), assemble the black/yellow leads to the rear with the pillar bolts in the centers of their adjusting slots.

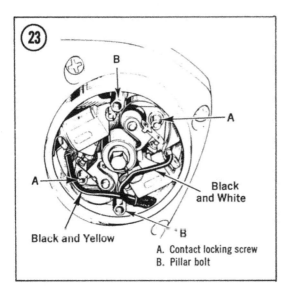

Black and White

Black and Yellow

B

A. Contact locking screw
B. Pillar bolt

10. To adjust the point gaps, loosen the sleeve nuts and set the gaps to 0.015 in. (0.38mm).

11. On earlier models, without a peg and slot on the cam unit and exhaust camshaft, the cam unit must be installed as follows. Position the base plate so the pillar bolts are in the center of their adjusting slots.

12. Remove the spark plugs and all 4 rocker caps. Turn the engine over using the rear wheel (with the transmission in 4th gear) until the right piston is on its compression stroke.

13. Use a dial indicator and set the engine so it is 1/32 in. (0.9mm) before top dead center; turn the contact breaker cam unit until the rear set of points is just about to open.

14. Tighten the center bolt at this point. Turn the cam unit clockwise to bring it to the point where the breaker points begin to open.

Ignition Timing Static

1. Remove both spark plugs.

2. Ignition timing varies with engine speed. It is retarded at low speeds and advanced at higher speeds above 3,000 rpm.

3. Timing is expressed as BTDC, which means before top dead center. This is expressed in number of crankshaft degrees BTDC when the spark plug fires.

4. The automatic advance unit does not give absolute accuracy at all speeds. There are slight variations between the retarded and advanced positions, so it is best to time the engine so it will operate most efficiently at high speeds, for best performance. If low running speeds or idle are somewhat ragged, this will have the least effect on performance.

5. The engine can be timed in the fully advanced position while it is not running by using a special service tool (B, **Figure 24**).

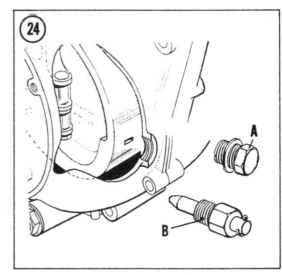

6. Remove the timing plug (A) in the front of the crankcase on the right side (A). Insert the service tool timing plunger in the aperature. With light finger pressure on the plunger, rotate the crankshaft until the plunger is located in the groove on the flywheel.

7. When the timing tool is in position, do not rotate the engine.

8. Check which piston is on its compression stroke. Both valves will be closed.

9. The ignition timing for the right piston is set with the upper contact breaker points, and for the left piston with the lower points.

10. Secure the backing plate of the contact breaker with the screws in their slots.

11. Place the right piston on its compression stroke and press the automatic advance unit into

its tapered sleeve on the idler gear. Turn the unit counterclockwise to open the governor weights. When the upper set of points just being to open, secure the automatic advance unit by tightening its center bolt. This results in fully advanced ignition; but when finger pressure is released on the advance unit, it will resume the retarded position.

12. Remove the service tool from the slot in the flywheel and rotate the engine until the upper set of points are fully opened. Chcek the gap. It should be 0.015 in. You can be sure the points are fully open by checking the dash marks on the contact breaker cam. It indicates the highest point of the cam which should be against the heel of the breaker points when the gap is checked.

13. To adjust the gap, loosen the screw (D, **Figure 25**) and turn the eccentric screw (E) until the gap is at 0.015 in. Tighten both screws and repeat the procedure for the lower set of points. If the gap requires a large amount of adjustment, the timing is probably not correct. Release the automatic advance unit (using special service tool No. 61-3816) and repeat the timing procedure already described.

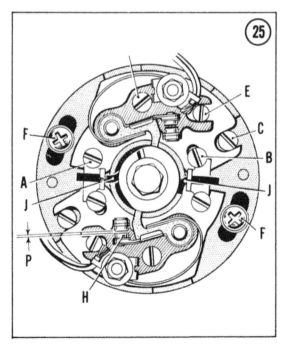

Ignition Timing (Fully Advanced)

1. Timing must always be set for the fully advanced position (34-36 degrees).

2. Place special service tool (No. 60-1859) in the timing groove on the flywheel again. The ignition has already been set approximately but the contact breaker points must be finely adjusted to give accurate ignition timing.

3. Lock the automatic advance unit in the fully advanced position. Remove the center bolt (A, **Figure 26**) without turning the unit. Fit another washer with a hole large enough to clear the cam bearing, so the washer can bear against the outer face of the cam.

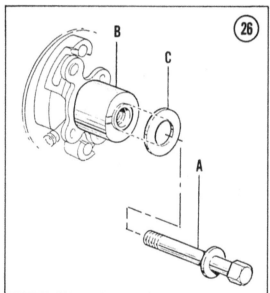

4. Rotate the cam as far as it will go in a counterclockwise direction. This fully advances the unit. Replace the center bolt and lock the unit in the fully advanced position.

5. Connect the battery and light bulb in a circuit with the contact breaker points (**Figure 27**). Attach one wire to the upper points spring (C, Figure 25) and the battery terminal. A second wire goes from the other terminal to the bulb. A third wire goes from the bulb to ground on the engine. The bulb will be lit when the points are closed and will go out as soon as they start to

open. The wire must be changed to the other set of points to adjust.

6. Be sure the timing plunger is located in the slot in the flywheel and the right piston is on its compression stroke. Loosen the 2 screws (A and B, Figure 25) and turn the eccentric pin (C, Figure 25) until the bulb just goes out. The points are now opening. Lock the 2 screws (A and B, Figure 25).

7. Remove the timing plunger from the flywheel and rotate the engine one revolution. Locate the timing plunger in the flywheel slot again and repeat the above operation for the lower set of points.

8. When the timing is finished, remove the timing plunger from the flywheel and replace the plug in the crankcase opening. Remove the large washer from the automatic advance unit and replace the standard washer on the center bolt. Failure to do this will result in the unit not working.

Ignition Timing (Strobe)

The strobe light can only be used to time a running engine, so the timing must be set at least approximately as described in the previous section for the engine to run.

1. Remove the inspection cover at the front of the primary case. The generator rotor is behind it. The rotor has 2 timing lines on opposite sides. The correct one to use is marked with a "2". At the bottom of the opening is a pointer projecting toward the rotor (see **Figure 28**).

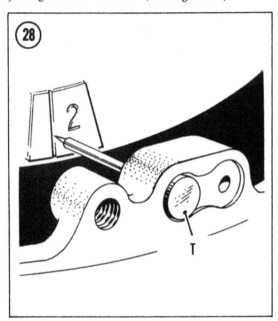

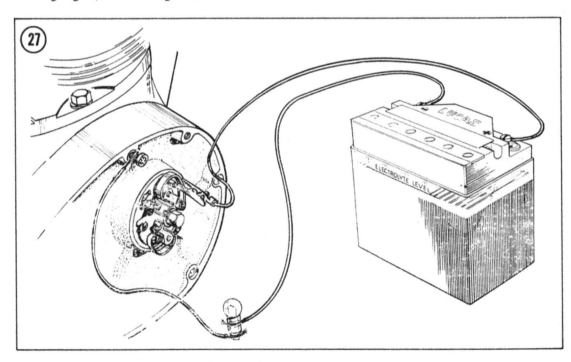

2. Connect the strobe light to an independent 12-volt battery outside the engine. The light voltage wire goes from the strobe light to the right spark plug (see **Figure 29**). Run the engine at a speed of at least 3,300 rpm to be sure the ignition is fully advanced. Shine the strobe light through the opening. If the timing is correct, the mark on the rotor will appear to be stationary, with the pointer pointing directly at it.

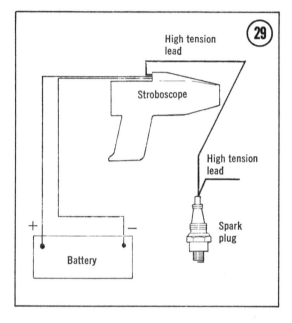

3. If timing is not correct, the upper set of points must be adjusted as previously described until it is right. Then change the wire to the left spark plug and check the lower set of points.

4. If a new primary chaincase is being installed during this overhaul, be sure to install the ignition pointer in the left of the 2 holes at the bottom of the inspection opening (T, Figure 28).

Carburetor Adjustment

Refer to Chapter Four for carburetor adjustment and synchronization.

CHAIN ADJUSTMENT AND INSPECTION

The first signs of inadequate chain lubrication are reddish brown deposits at the joints. This is rust and means more oil is required. Lubricate either the primary or secondary chain according to the details under *Lubrication* in this chapter.

Primary Chain Adjustment

1. The up and down movement in the center of the top run should be between ⅛ in. and ¼ in.

2. Remove the adjuster cap nut (F, **Figure 30**) under the chaincase.

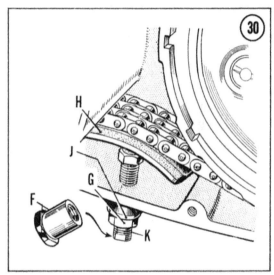

3. Loosen the adjuster locknut and screw in the adjuster screw until the tensioner takes up the excess slack.

4. Remove the filler cap and check slack of the top run of the chain.

5. Be sure the oil seal washer is in good condition (J, Figure 30). Tighten the adjuster locknut while holding the adjuster screw steady. Replace the cap nut and tighten.

Secondary Chain Adjustment

1. Loosen the nuts at both ends of the brake anchor strap. Loosen the axle nut (A, **Figure 31**) until the axle can move in the slotted ends of the swing arm forks.

2. Tighten the left side chain adjuster (D) until the chain has about 1¼ in. up and down movement in the center of the bottom run.

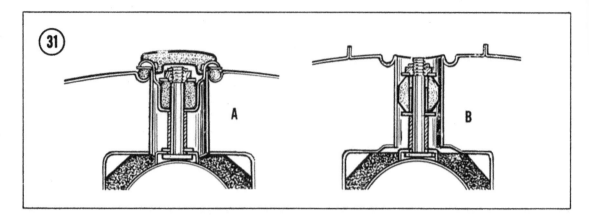

3. Tighten the right side chain adjuster an equal amount to the left, to align the axle. Check the chain slack again to be sure it hasn't changed.

4. Tighten the chain adjuster locknuts firmly. Tighten the axle nut and then the nuts at each end of the brake anchor strap.

5. Check wheel alignment as described in Chapter Seven.

Chain Inspection

1. A chain inevitably stretches after long use. When the stretch reaches ¼ inch per foot, the chain should be replaced.

2. Remove the chain and clean it thoroughly to free all bearings. Drive a nail into a bench and hook the end link of the chain over it.

3. Stretch the chain to its fullest length and measure. Compress the chain a few links at a time to its shortest length and measure length.

4. If the distance between the 2 measurements is ¼ inch per foot or more, replace the chain. See **Figure 32**.

5. Be sure the master link spring clip has its closed end facing forward (see **Figure 33**) on the top run of the chain when replacing.

Sprocket Inspection

1. The sprockets must be aligned to prevent excessive wear on the chain.
2. Remove the chain and check the inner faces of the inner plates. They should be lightly

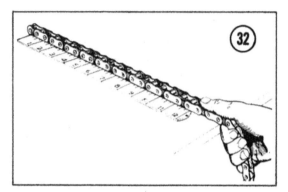

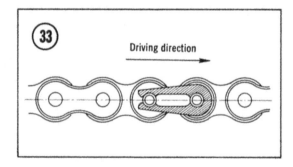

Driving direction

polished on both sides. If they show considerable wear on both sides, the sprockets are not aligned.

3. Worn sprockets will develop hooked teeth (**Figure 34**), instead of normal symmetrically pointed ones. A worn sprocket will rapidly wear the chain.

DECARBONIZATION

After an engine has been run for many hours, it will probably require the removal of carbon from the piston crown and cylinder head. The best way to detect this need is if the engine has

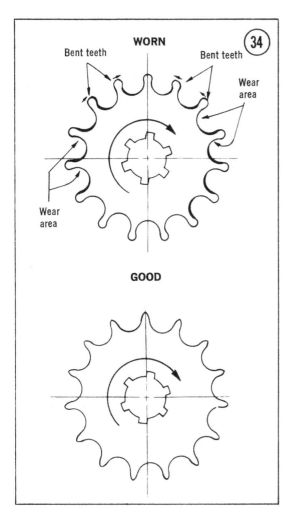

shown progressively worsening preignition or a gradual loss of power. Several new products are now being marketed to allow a simple approach to decarbonization without the need of dismantling the engine. These products will not be as thorough, but can be used periodically. The procedure for their use is as follows.

1. Start and warm the engine to normal operating temperature.

2. Slowly pour 10 ounces (284 grams) of cleaner into the carburetor with the engine running at a fast idle of 1,500 rpm.

3. Slowly increase the speed of the engine and kill it by dumping in the remaining liquid.

4. Let the engine stand for ½ hour with the liquid still in the cylinder.

5. Start the engine and run it at full throttle, under load, for at least 5 minutes to clear out the system and remove the last traces of fluid. If this fails to completely decarbonize the cylinder then you will have to remove the head as outlined in Chapter Six.

WINTER STORAGE

Several months of inactivity can cause serious problems and a general deterioration of bike condition. This is especially true in areas of weather extremes. During the winter months it is advisable to specially prepare the bike for lay-up.

Selecting a Storage Area

Most cyclists store their bikes in their home garage. Facilities suitable for long-term motorcycle storage are readily available for rent or lease in most areas. In selecting a building, consider the following points.

1. The storage area must be dry, free from dampness and excessive humidity. Heating is not necessary but the building should be well insulated to minimize extreme temperature variations.

2. Buildings with large window areas should be avoided, or such windows should be masked (also a good security measure) if direct sunlight can fall on the bike.

3. Buildings in industrial areas, where factories are liable to emit corrosive fumes, are not desirable, nor are facilities near bodies of salt water.

4. The area should be selected to minimize the possibility of loss by fire, theft, or vandalism. The area should be fully insured, perhaps with a package covering fire, theft, vandalism, weather, and liability. The advice of your insurance agent should be solicited on these matters. The building should be fireproof and items such as the security doors and windows, alarm facilities, and proximity of police should be considered.

Preparing for Storage

Careful pre-storage preparation will minimize deterioration and will ease restoring the bike to service in the spring. The following procedure is recommended.

1. Wash the bike completely making certain to remove any accumulation of road salt that may have collected during the first weeks of winter. Wax all painted and polished surfaces.

2. Run the engine 20-30 minutes to stabilize oil temperature. Drain oil regardless of mileage since last oil change and replace with normal quantity of fresh oil.

3. Remove battery and coat cable terminals with petroleum jelly. If there is evidence of acid spillage in the battery box, neutralize with baking soda, wash clean, and repaint. Batteries should be kept in an area where they will not freeze, and where they can be recharged every 2 weeks.

4. Drain all gasoline from fuel tank, settling bowl, and carburetor float bowls. Leave fuel cock on the RESERVE position.

5. Remove spark plugs and add a small quantity of oil to each cylinder. Turn the engine a few revolutions by hand. Install spark plugs.

6. To avoid ignition point oxidation, place a paper card, lightly saturated with silicone oil, between the points.

7. Check tire pressure. Move the bike to storage area and place on center stand. If preparation is performed in an area remote from the storage facility, the bike should be trucked, not ridden into storage.

Inspection During Storage

Try to inspect the bike weekly while it is in storage. Any deterioration should be corrected as soon as possible. For example, if corrosion of bright metal parts is observed, coat with a light film of grease or silicone spray.

Restoring to Service

A bike that has been properly prepared, and stored in a suitable area, requires only light maintenance to restore it to service. It is advisable, however, to perform a "spring tune-up."

1. Before removing the bike from the storage area, re-inflate tires to the correct pressure. Air loss during the storage period may have nearly flattened the tires, and moving the bike can cause damage to tires, tubes, or rims.

2. When the bike is brought to the work area, immediately install the battery (fully charged) and fill the fuel tank. (The fuel cock should be on the RESERVE position; do not move yet.)

3. Check the fuel system for leaks. Remove the carburetor float bowl or open the float bowl drain cock and allow several cups of fuel to pass through the system. Move the fuel cock slowly to the closed position, remove the settling bowl and empty any accumulated water.

4. Perform a normal tune-up as described earlier, adjust valve clearance, apply oil to camshaft, and while checking spark plugs add a few drops of oil to the cylinder. Be especially certain to de-grease ignition points if an oily card was used to inhibit oxidation during storage; use a non-petroleum solvent such as trichlorethylene or denatured alcohol.

5. Check safety items, i.e., lights, horn, etc., as oxidation of switch contacts and/or sockets during storage may make one or more of these critical devices inoperative.

6. Test ride and clean the motorcycle.

CHAPTER THREE

TROUBLESHOOTING

Diagnosing mechanical problems is relatively simple if orderly procedures are used and a few basic principles are kept in mind.

The troubleshooting procedures in this chapter analyze typical symptoms and show logical methods of isolating causes. These are not the only methods. There may be several ways to solve a problem, but only a systematic, methodical approach can guarantee success.

Never assume anything. Don't overlook the obvious. If while riding along, the bike suddenly quits, check the easiest, most accessible problem spots first. Is there gasoline in the tank? Is the gas petcock on? Has a spark plug wire fallen off? Check the ignition switch. Sometimes the weight of keys on a key ring may turn the ignition off suddenly.

If nothing obvious turns up in a cursory check, look a little further. Learning to recognize and describe symptoms will make repairs easier at home or by a mechanic at a shop. Saying that "it won't run" isn't the same as saying "it quit on the highway at high speed and wouldn't start," or that "it sat in my garage for 3 months and then wouldn't start."

Gather as many symptoms together as possible to aid in diagnosis. Note whether the engine lost power gradually or all at once, what color smoke (if any) came from the exhaust, and so on. Remember that the more complicated a machine is, the easier it is to troubleshoot because symptoms point to specific problems.

After the symptoms are defined, areas which could cause the problems are tested and analyzed. Guessing at the cause of a problem may provide the solution, but it can easily lead to frustration, wasted time, and a series of expensive, unnecessary replacement of parts.

Neither fancy equipment nor complicated test gear is needed to determine whether repairs can be attempted at home. A few simple checks could save a large repair bill and time lost while the bike sits in a dealer's service department. On the other hand, be realistic and don't attempt repairs beyond your abilities. Service departments tend to charge heavily for putting together a disassembled engine that may have been abused. Some won't even take on such a job—so use common sense; do not get in too deeply.

OPERATING REQUIREMENTS

An engine needs 3 basics to run properly: correct gas/air mixture, compression and a spark at the right time. If one or more are missing, the engine will not run. The electrical system is the weakest link of the three. More problems result from electrical breakdowns than from any other source. Keep that in mind before you begin tampering with carburetor adjustments and the like.

If a bike has been sitting for any length of time and refuses to start, check the battery for a charged condition first, and then look to the gasoline delivery system. This includes the tank, fuel petcocks, lines, and carburetors. Rust may have formed in the tank, obstructing fuel flow. Gasoline deposits may have gummed up carburetor jets and air passages. Gasoline tends to lose its potency after standing for long periods. Condensation may contaminate it with water. Drain old gas and try starting with a fresh tankful.

TROUBLESHOOTING INSTRUMENTS

Chapter One lists many of the instruments needed and detailed instructions on their use.

STARTING DIFFICULTIES

Check gas flow first. Remove the gas cap and look into the tank. If gas is present, pull off a fuel line at the carburetor and see if gas flows freely. If none comes out, the fuel tap may be shut off, blocked by rust or other foreign matter, or the fuel line may be stopped up or kinked. If the carburetor is getting usable fuel, inspect the electrical system next.

Check that the battery is charged by turning on the lights or by beeping the horn. Refer to your owner's manual for starting procedures with a dead battery. Have the battery recharged if necessary.

Pull off a spark plug cap, remove the spark plug, and reconnect the cap. Turn on the ignition. Lay the plug against the cylinder head so its base makes a good connection, and turn the engine over with the kickstarter (engine rotation of at least 500 rpm). A fat, blue spark should jump across the electrodes. If there is no spark, or only a weak one, there is electrical system trouble. Check for a defective plug by replacing

it with a known good one. Don't assume a plug is good just because it's new.

Once the plug has been cleared of malfunctioning but there's still no spark, start backtracking through the system. If the contact at the end of the spark plug wire can be exposed, it can be held about ⅛ inch from the head while the engine is turned over to check for a spark. Remember to hold the wire only by its insulation to avoid a nasty shock. If the plug wires are dirty, greasy, or wet, wrap a rag around them so you don't get shocked. If you do feel a shock or see sparks along the wire, clean or replace the wire and/or its connections.

If there's no spark, or only a weak spark, at the plug wire, look for loose connections at the coil and battery. If all seems in order there, check next for oily or dirty contact points. Clean points with electrical contact cleaner, or a strip of paper. On battery ignition models, with the ignition switch turned on, open and close the points manually with a screwdriver.

No spark at the points with this test indicates a failure in the ignition system. Refer to *Ignition System* section of this chapter for checkout procedures for the entire system and individual components. Refer to Chapter Two for checking and setting ignition timing.

Note that spark plugs of the incorrect heat range (too cold) may cause hard starting. Set gaps to specifications. If you have just ridden through a puddle or washed the bike and it won't start, dry off plugs and plug wires. Water may have entered the carburetor and fouled the fuel under these conditions, but wet plugs and wires are the more likely problem.

If a healthy spark occurs at the right time, and there is adequate gas flow to the carburetor, check the carburetor itself. Make sure all jets and air passages are clean. Check float level and adjust if necessary. Shake the float to check for gasoline inside it, and replace or repair as indicated. Check that the carburetors are mounted snugly and that no air is leaking past the manifold. Check for a clogged air filter.

Compression, or the lack of it, usually occurs

only in the case of older machines. Worn or broken pistons, rings, and cylinder bores could prevent starting. Generally a gradual power loss and harder starting will be readily apparent in this case.

Compression may be checked in the field by turning the kickstarter by hand and noting that adequate resistance is felt.

An accurate compression check gives a good idea of the condition of the basic working parts of the engine. To perform this test, you need a compression gauge. The motor should be warm.

1. Remove spark plug and clean out any dirt or grease.

2. Insert the tip of the gauge into the hole, making sure it is seated correctly.

3. Open the throttle all the way and make sure the choke on the carburetor is open.

4. Crank the engine several times and record the highest pressure reading on the gauge. Run the test on each of the cylinders. Normal compression value ranges are given in **Table 1**. If the reading obtained is not satisfactory, proceed to the next step.

5. Pour a tablespoon of motor oil into the cylinder and record the compression. If oil raises the compression significantly—10 psi in an old engine—the rings are worn and should be replaced.

Valve adjustments should be checked. Sticking, burned, or broken valves may hamper starting. As a last resort, check valve timing.

POOR IDLING

Poor idling may be caused by incorrect carburetor adjustment, incorrect timing, or ignition system defects. Check the gas cap vent for an obstruction.

MISFIRING

Misfiring can be caused by a weak spark or dirty plugs. Check for fuel contamination. Run the machine at night to check for spark leaks along plug wires and under the spark plug cap.

Table 1 COMPRESSION LIMITS

Maximum PSI	Minimum PSI	Maximum PSI	Minimum PSI
134	101	192	144
136	102	194	145
138	104	196	147
140	105	198	148
142	107	200	150
144	108	202	151
146	110	204	153
148	111	206	154
150	113	208	156
152	114	210	157
154	115	212	158
156	117	214	160
158	118	216	162
160	120	218	163
162	121	220	165
164	123	222	166
166	124	224	168
168	126	226	169
170	127	228	171
172	129	230	172
174	131	232	174
176	132	234	175
178	133	236	177
180	135	238	178
182	136	240	180
184	138	242	181
186	140	244	183
188	141	246	184
190	142	248	186
		250	187

Note: If the highest cylinder and lowest cylinder are not within a Maximum/Minimum range, valve or ring trouble is indicated.

WARNING
Do not run engine in closed garage. There is considerable danger of carbon monoxide poisoning.

If misfiring occurs only at certain throttle settings, refer to the fuel system chapter for the specific carburetor problem involved. Misfiring under heavy load, as when climbing hills or accelerating, is usually caused by bad spark plugs.

FLAT SPOTS

If the engine seems to die momentarily when the throttle is opened and then recovers, check for a dirty main jet in the carburetor, water in the fuel, or an excessively lean mixture.

POWER LOSS

Poor condition of rings, pistons, or cylinders will cause a lack of power and speed. Ignition timing should be checked.

OVERHEATING

If the engine seems to run too hot all the time, be sure you are not idling it for long periods. Air-cooled engines are not designed to operate at a standstill for any length of time. Heavy stop and go traffic or slow hill-climbing is hard on a motorcycle engine. Spark plugs of the wrong heat range can burn pistons. An excessively lean gas mixture may cause overheating. Check ignition timing. Don't ride in too high a gear. Broken or worn rings may permit compression gases to leak past them, heating heads and cylinders excessively. Check oil level and use the proper grade of lubricants.

ENGINE NOISES

Experience is needed to diagnose accurately in this area. Noises are hard to differentiate and harder yet to describe. Deep knocking noises usually mean main bearing failure. A slapping noise generally comes from loose pistons. A light knocking noise during acceleration may be a bad connecting rod bearing. Pinging should be corrected immediately or damage to pistons will result. Compression leaks at the head-cylinder joint will sound like a rapid on-and-off squeal.

PISTON SEIZURE

Piston seizure is caused by incorrect piston clearances when fitted, fitting rings with improper end gap, too thin an oil being used, incorrect spark plug heat range, or incorrect igni-

tion timing. Overheating from any cause may result in seizure.

EXCESSIVE VIBRATION

Excessive vibration may be caused by loose motor mounts, worn engine or transmission bearings, loose wheels, worn swing arm bushings, a generally poor running engine, broken or cracked frame, or one that has been damaged in a collision. See also *Poor Handling*.

CLUTCH

Clutch slip may be due to worn plates, improper adjustment, or glazed plates. A dragging clutch could result from damaged or bent plates, improper adjustment, or uneven clutch spring pressure.

All clutch problems, except adjustments or cable replacement, require removal to identify the cause and make repairs.

1. *Slippage*—This condition is most noticeable when accelerating in high gear at relatively low speed. To check slippage, drive at a steady speed in fourth or fifth gear. Without letting up on the accelerator, pull in the clutch long enough to let engine speed increase (one or two seconds). Then let the clutch out rapidly. If the clutch is good, engine speed will drop quickly or the bike will jump forward. If the clutch is slipping, engine speed will drop slowly and the bike will not jump forward.

Slippage results from insufficient clutch lever free play, worn friction plates, or weak springs. Riding the clutch can cause the disc surfaces to become glazed, resulting in slippage.

2. *Drag or failure to release*—This trouble usually causes difficult shifting and gear clash especially when downshifting. The cause may be excessive clutch lever free play, warped or bent plates, stretched clutch cable, or broken or loose disc linings.

3. *Chatter or grabbing*—Check for worn or misaligned steel plate and clutch friction plates.

TRANSMISSION

Transmission problems are usually indicated by one or more of the following symptoms:

a. Difficulty in shifting gears

b. Gear clash when downshifting

c. Slipping out of gear

d. Excessive noise in neutral

e. Excessive noise in gear

Transmission symptoms are sometimes hard to distinguish from clutch symptoms. Be sure the clutch is not causing the trouble before working on the transmission.

HANDLING

Poor handling may be caused by improper tire pressures, a damaged frame or swing arm, worn shocks or front forks, weak fork springs, a bent or broken steering stem, misaligned wheels, loose or missing spokes, worn tires, bent handlebars, worn wheel bearings, or dragging brakes.

BRAKES

Sticking brakes may be caused by broken or weak return springs, improper cable or rod adjustment, or dry pivot and cam bushings. Grabbing brakes may be caused by greasy linings which must be replaced. Brake grab may also be due to out-of-round drums or linings which have broken loose from the brake shoes. Glazed linings will cause loss of stopping power.

ELECTRICAL

Bulbs which continually burn out may be caused by excessive vibration, loose connections that permit sudden current surges, poor battery connections, installation of the wrong type bulb, or a faulty voltage regulator.

A dead battery or one which discharges quickly may be caused by a faulty alternator or rectifier. Check for loose or corroded terminals. Shorted battery cells or broken terminals will keep a battery from charging. Low water level will decease a battery's capacity. A battery left uncharged after installation will sulphate, rendering it useless.

A majority of light and horn or other electrical accessory problems are caused by loose or corroded ground connections. Check these first, and then substitute known good units for easier troubleshooting.

COIL IGNITION SYSTEM

The system has a contact breaker driven by the camshaft, with 2 sets of points connected to ignition coils. The capacitors are mounted on a rubber covered pack under the seat (**Figure 1**). The only maintenance for the capacitors and coils is to keep them clean and the terminals tightly connected.

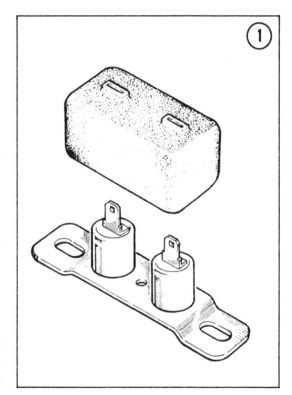

Low Voltage Test

1. Clean and adjust the contact breaker and spark plugs as described in Chapter Two.

2. Connect a 0-15 range DC voltmeter to the CB terminals of the contact breaker, with the red wire grounded.

3. Open the points. Turn on the ignition. Read the battery voltage on the voltmeter.

4. Close the points. The voltmeter reading should fall to zero.

5. Repeat the test on the other set of points. If the voltmeter does not react as described, there is trouble in the low voltage circuit.

Low Voltage Circuit Troubleshooting

1. Turn the ignition switch on. Insert a small piece of insulation between each set of points.

2. Disconnect the brown/blue lead from the diode center terminal to disconnect the zener diode.

3. Use the voltmeter and 2 test probes to check the low voltage circuit, beginning with battery.

4. Connect the voltmeter across the negative battery terminal and the frame ground. If the voltage reading is low or zero, the battery ground is poor or the battery has an inadequate charge.

5. Connct the voltmeter to the —VE (negative) terminal of each coil and the ground. No voltage reading means an open circuit between the battery and coil, or else poor switch connections.

6. Connect the voltmeter to the ignition switch input terminal and ground. No voltage reading means the brown/blue wire has a defective connection or the fuse has blown. Check the brown/blue wire connections at the rectifier.

7. Connect the voltmeter to the ignition switch output terminal and ground. No voltage reading means the ignition switch is defective and must be replaced. A voltage reading here but not at the ignition coil —VE terminals means the white wire is disconnected or loose.

8. Connect the voltmeter to the contact breaker terminal of each coil and ground. No voltage reading means the coil is damaged. Replace the coil.

9. Connect the voltmeter across each set of points with the insulation still in place. No voltage reading between the points and ground means a faulty connection or damaged internal

insulation in the capacitor. Replace the capacitor and test again.

10. Connect the zener diode brown/blue wire. Turn the ignition on. Connect the voltmeter to the diode center terminal and ground. No voltage reading means the zener diode is defective and must be replaced.

Ignition Coil

1. Disconnect the spark plug wires. Turn the ignition switch on and crank the engine until the points for the right cylinder are closed. This is the black/white wire from the ignition coil.

2. Hold the black/white wire from the ignition coil about ⅛ in. from the cylinder head. Open the points with a screwdriver. A strong spark should jump across to the cylinder head. If it doesn't, the ignition coil is defective.

3. Repeat the test for the other coil.

4. When testing, be sure the insulation on the wires is in good condition.

Ignition Coil Bench Testing

This test requires a special testing system (**Figure 2**) and should be performed by a dealer.

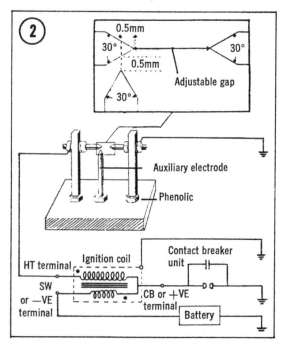

1. Connect the coil into the circuit as shown and set the adjustable gap to 8mm.

2. Run a single lobe contact breaker at 600 rpm. There should not be more than 5 percent missing at the spark gap in 15 seconds running.

3. Check the primary winding by connecting an ohmmeter across the low voltage terminals. The reading should be between 3.3 and 3.8 ohms.

Contact Breaker

1. Be sure the points are clean and properly adjusted as described in Chapter Two.

2. Turn the ignition switch on. Open contact points and connect the voltmeter across them. No voltage reading means the insulation in the capacitor is defective. Excessive arcing means the capacitor has a reduced capacity. Check by substituting a known good capacitor.

3. Pay careful attention to the lubrication procedure for the contact breaker described in Chapter Two. Be sure no oil or grease gets on the points.

4. The points can be ground if necessary. Remove as described in Chapter Two. Use a light emery cloth. After reassembly, the gap must be set again.

High Voltage Test

1. Check the coils and high voltage wires as described earlier.

2. If the spark plug wire has a good spark, the plug cap or plug itself may be damaged. Clean and gap the plugs, or replace them. If running is still ragged, replace the plug caps.

BATTERY CHARGING SYSTEM

The battery is charged by the alternator whose alternating current is converted into direct current by a rectifier (**Figure 3**). Excessive charge is absorbed by the zener diode. Be sure the ignition switch is always off when the engine is not running, to prevent heating of the coils and discharging the battery.

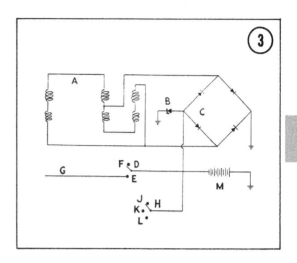

Rectifier DC Output

1. Check the battery for full charge.

2. Disconnect the brown/blue center wire at the rectifier.

3. Connect a DC ammeter of 0-15 amp capacity in series between the center wire of the rectifier and the brown/blue main wire. Run the engine at 3,000 rpm.

4. The minimum DC output to the battery should be 7.75 amps.

Alternator Output

1. Disconnect the 2 alternator cables above the engine. Run the engine at 3,000 rpm.

2. Connect an AC voltmeter of 0-15 volt capacity with a one ohm load resistor (see this chapter for resistor construction) in parallel between the green/white and green/yellow alternator wires.

3. If the voltage reading is 9.0 or more, the alternator is all right.

4. A low reading means: the wire insulation is defective; some turns of the coils are short circuited; the rotor is demagnetized.

5. A zero reading for any coil means it is disconnected or grounded.

6. A reading between a wire and ground means the coil windings or connections have become grounded.

7. In case of problems, always check the stator wires for damage before repairing or replacing the rotor.

Rectifier Testing (Mounted)

The rectifier is located below the electrical platform under the seat.

2. Remove the right side panel. Push the ignition coil up as much as possible to give access to the rectifier (see **Figure 4**). Loosen the rectifier nut at the platform to turn it for access to the wire connections.

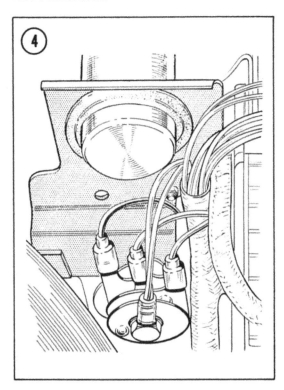

3. Disconnect the brown/white wire from the rectifier center terminal. Insulate the end of the wire to prevent a short circuit.

4. Connect a DC voltmeter with a one ohm load resistor in parallel to the rectifier center terminal and ground. Be sure the voltmeter negative terminal connects to the rectifier terminal and the voltmeter positive terminal to ground.

5. Run the engine at 3,000 rpm. The voltage reading should be 7.75 minimum.

6. If a lower, or zero, reading is obtained, the rectifier or the charging circuit wiring is faulty.

Rectifier Testing (Bench)

1. Disconnect the wires from the battery. Remove the rectifier.

2. Set up the bench test system (see **Figure 5**). Connect a 12-volt, 50-watt light bulb and 12-volt battery across rectifier terminals No. 1 and No. 2 (see **Figure 6**) for not more than 30 seconds, and repeat test with the wires reversed.

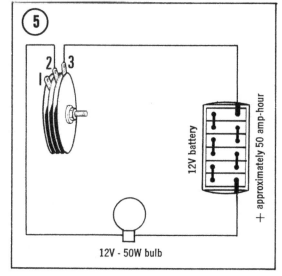

12V battery

+ approximately 50 amp-hour

12V - 50W bulb

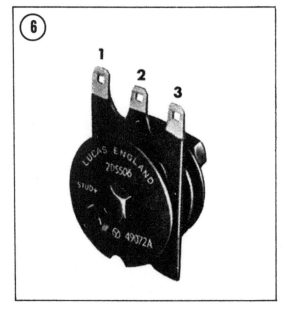

3. Repeat the test with the wires connected to the rectifier bolt and terminal No. 1, the bolt and terminal No. 3, and terminals No. 2 and No. 3 (see **Figure 7**).

4. The light bulb should come on in one direction only for each of the reverse connections made on each pair of contacts.

5. If the bulb does not light in either direction, or if it lights in both directions, replace the rectifier since it cannot be repaired.

Charging Circuit Testing

1. Be sure the battery has voltage and the fuse has not blown.

2. Connect a DC voltmeter with a one ohm (see next section) load resistor in parallel to the rectifier center terminal and ground. The meter should give a reading. If it doesn't, disconnect the green/white and green/yellow alternator wires from the engine snap connectors.

3. Connect a jumper wire across the brown/blue and green/yellow connections of the rectifier. Check the voltage at the snap connector. This test indicates whether the alternator wire is disconnected or the fuse has blown.

4. Repeat the test at the green/white rectifier wire.

5. If there is no voltage at the rectifier center terminal, the brown/blue wire is disconnected. Check the fuse.

6. Tighten the rectifier securing nut carefully using 2 wrenches (see **Figure 8**). If the plates are twisted, the internal connections will break.

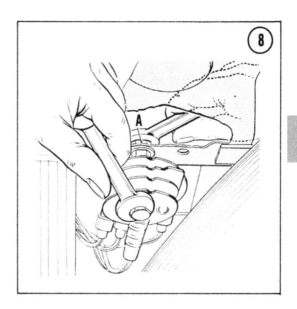

ONE OHM LOAD RESISTOR

This resistor can be made from 4 yards of 18 gauge standard wire.

1. Bend the wire into 2 equal lengths. Connect a heavy flexible wire to the folded end and the positive terminal of a battery.

2. Connect a DC voltmeter of 0-10 volt capacity to the battery terminals with a 0-10 amp ammeter between the battery negative terminal and the free ends of the wire resistor with an alligator clip.

3. Move the clip along the wires until the ammeter and voltmeter readings are numerically the same. The resistance is now one ohm. Cut the wires at this point. Twist the 2 ends together and wind the wire on an asbestos cylinder about 2 inches in diameter. Be sure each loop

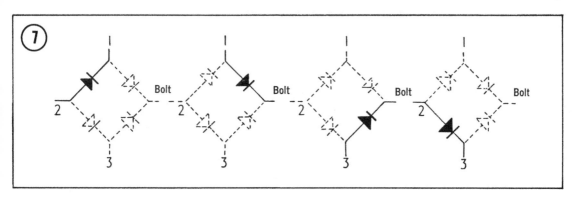

of the double wire does not touch the one next to it.

ZENER DIODE CHARGE CONTROL

Be sure a firm contact is maintained between the base of the diode (see **Figure 9**) and the air cleaner box, to prevent overheating. Be sure the ground connection is good. No maintenance is necessary to the diode unit.

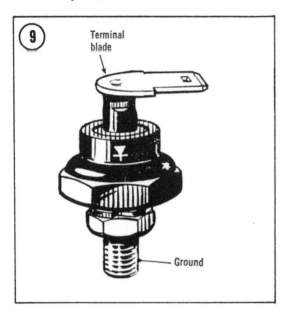

Zener Diode Location

The diode is mounted to the aluminum air cleaner box. To remove the diode, disconnect the brown/blue double connector from the diode and remove the mounting nut. Be careful not to tighten the nut too much when installing.

ZENER DIODE TESTING

1. Disconnect the cable from the diode. Connect an ammeter in series to the diode terminal and the cable. The ammeter positive wire connects to the diode terminal.

2. Connect a voltmeter to the zener diode and air cleaner box, with the positive wire to the air cleaner box.

3. Start the engine, gradually increasing the speed while reading both meters. There should be zero amps up to 12.75 volts. Increase engine speed until the reading is 2 amps. The voltage should then be from 13.5 to 15.5.

4. If the ammeter registers any current before the voltmeter reaches 12.75, replace the zener diode. If the voltage goes higher than 15.5 before 2 amps is reached, replace the zener diode.

TROUBLESHOOTING GUIDE

The following "quick reference" guide (see **Table 2**) summarizes the troubleshooting process. Use it to outline possible problem areas, then refer to the specific chapter or section involved.

Table 2   TROUBLESHOOTING GUIDE

Item	Problem or Cause	Things to Check
Loss of power	Poor compression	Piston rings and cylinder Head gasket Crankcase leaks
	Overheated engine	Lubricating oil supply Clogged cooling fins Ignition timing Slipping clutch Carbon in combustion chamber
	Improper mixture	Dirty air cleaner Restricted fuel flow Gas cap vent hole
	Miscellaneous	Dragging brakes Tight wheel bearings Defective chain Clogged exhaust system
Steering	Hard steering	Tire pressures Steering damper adjustment Steering stem head Steering head bearings Steering oil damper
	Pulls to one side	Unbalanced shock absorbers Drive chain adjustment Front/rear wheel alignment Unbalanced tires Defective swing arm Defective steering head Defective steering oil damper
	Shimmy	Drive chain adjustment Loose or missing spokes Deformed rims Worn wheel bearings Wheel balance
Gearshifting difficulties	Clutch	Adjustment Springs Friction plates Steel plates Oil quantity
	Transmission	Oil quantity Oil grade Return spring or pin Change lever or spring Drum position plate Change drum Change forks

(continued)

3

Table 2   **TROUBLESHOOTING GUIDE**   (continued)

Item	Problem or Cause	Things to Check
Brakes	Poor brakes	Worn linings Brake adjustment Oil or water on brake linings Loose linkage or cables
	Noisy brakes	Worn or scratched lining Scratched brake drums Dirt in brakes
	Unadjustable brakes	Worn linings Worn drums Worn brake cams

CHAPTER FOUR

FUEL SYSTEM

For proper operation, a gasoline engine must be supplied with fuel and air, mixed in a ratio of 15:1. A mixture in which there is an excess of fuel is said to be rich. A lean mixture is one which contains insufficient fuel. It is the function of the carburetor to supply the proper mixture to the engine under all operating conditions.

All BSA motorcycles are equipped with Amal carburetors of either monobloc (**Figure 1**) or concentric (**Figure 2**) type.

The complete fuel system includes the carburetor, air cleaner, and fuel tank.

CARBURETOR PRINCIPLES

Figure 3 is an exploded view of a typical Amal concentric carburetor. The essential functional parts are a float and float valve mechanism for maintaining a constant fuel level in the float bowl, a pilot system for supplying fuel at low speeds, a main fuel system which supplies the engine at medium and high speeds, and a tickler system, which supplies the very rich mixture needed to start a cold engine. The operation of each system is discussed in the following paragraphs.

Float Mechanism

Figure 4 illustrates a typical float mechanism. Proper operation of the carburetor is dependent on maintaining a constant fuel level in the carburetor bowl. As fuel is drawn from the float bowl, the float drops. When the float drops, the float moves away from its seat and allows fuel to flow past the valve and seat into the float bowl. As this occurs, the float is then raised, pressing the valve against its seat, thereby shutting off the flow of fuel. It can be seen from this discussion that a small piece of dirt can be trapped between the valve and seat (see **Figures 5 and 6**), preventing the valve from closing and allowing fuel to rise beyond the normal level, resulting in flooding.

Pilot System

Under idle or low speed conditions, at less than ⅛ throttle, the engine doesn't require much fuel or air and the throttle valve is almost fully closed. A separate pilot system is required for operation under such conditions. **Figure 7** illustrates the operation of the pilot system. Air is drawn through the pilot air inlet and controlled

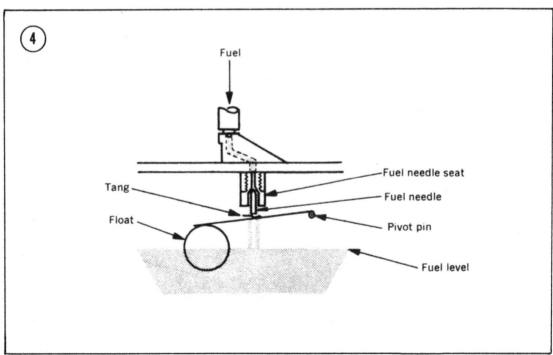

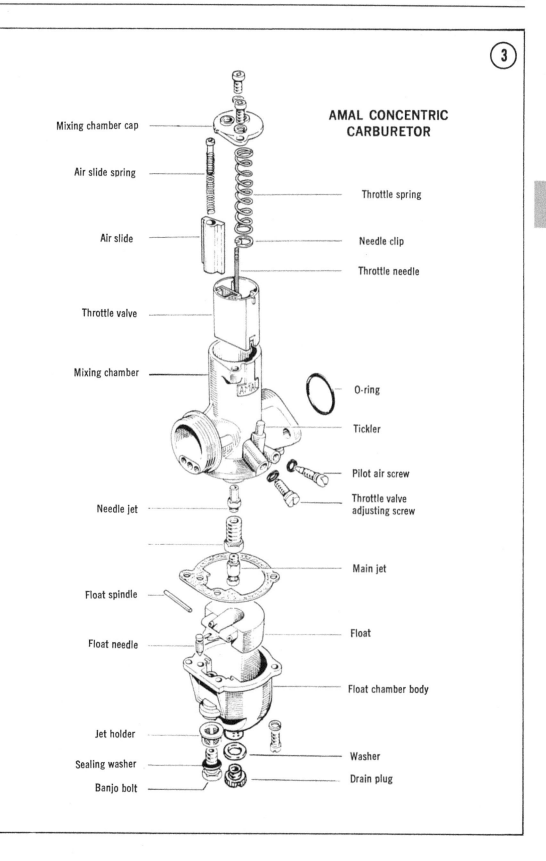

③

AMAL CONCENTRIC CARBURETOR

4

Mixing chamber cap

Air slide spring

Air slide

Throttle spring

Needle clip

Throttle needle

Throttle valve

Mixing chamber

O-ring

Tickler

Pilot air screw

Throttle valve adjusting screw

Needle jet

Main jet

Float spindle

Float

Float needle

Float chamber body

Jet holder

Washer

Sealing washer

Drain plug

Banjo bolt

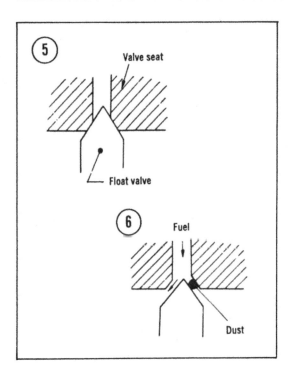

by the pilot air screw. The air is then mixed with fuel drawn through the pilot jet. The air/fuel mixture then travels from the pilot outlet into the main air passage, where it is further mixed with air prior to being drawn into the engine. The pilot air screw controls the idle mixture.

If proper idle and low speed mixture cannot be obtained within the normal adjustment range of the idle mixture screw, refer to **Table 1** for possible causes.

Table 1   IDLE MIXTURE

Too rich:	Clogged pilot air intake
	Clogged air passage
	Clogged air bleed opening
	Pilot jet loose
Too lean:	Obstructed pilot jet
	Obstructed jet outlet
	Worn throttle valve
	Carburetor mounting loose

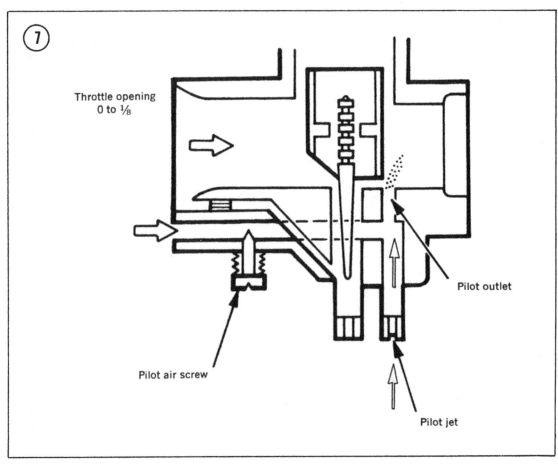

Main Fuel System

As the throttle is opened more (up to approximately ¼ open), the pilot circuit begins to supply less of the mixture to the engine as the main fuel system total, illustrated in **Figure 8**, begins to function. The main jet, the needle jet, the jet needle, and the air jet make up the main fuel circuit.

As the throttle valve opens more than approximately ⅛ of its travel, air is drawn through the main port, and passes under the throttle valve in the main bore. The velocity of the air stream results in reduced pressure around the jet needle. Fuel then passes through the main jet, past the needle jet and jet needle, and into the air stream where it is atomized and sent to the cylinder. As the throttle valve opens, more air flows through the carburetor. The jet needle, which is attached to the throttle slide, rises to permit more fuel to flow.

A portion of the air bled past the jet needle is mixed with the main air stream and atomized.

Air flow at small throttle openings is controlled primarily by the cutaway on the throttle slide.

As the throttle is opened still more, up to approximately ¾ open, the circuit draws air from 2 sources, as shown in **Figure 9**. The first source of air is through the venturi; the second source is through the air jet. Air passing through the venturi draws fuel through the needle jet. The jet needle is tapered, and therefore allows more fuel to pass. Air passing through the air jet passes to the needle jet to aid atomization of the fuel there.

Figure 10 illustrates the circuit at high speeds. The jet needle is withdrawn almost completely from the needle jet. Fuel flow is then controlled by the main jet. Air passing through the air jet continues to aid atomization of the fuel as described in the foregoing paragraphs.

Any dirt which collects in the main jet or in the needle jet obstructs fuel flow and causes a lean mixture. Any clogged air passages, such as

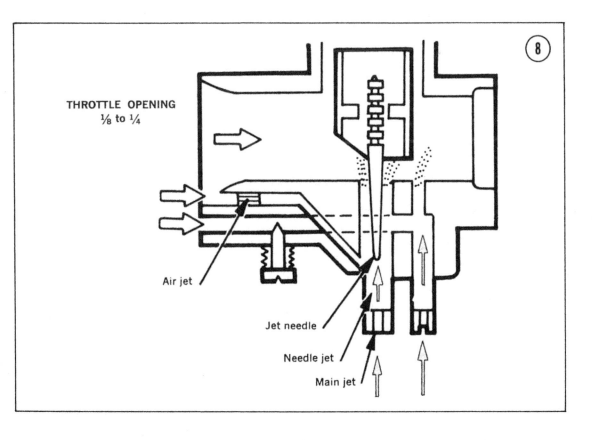

THROTTLE OPENING
⅛ to ¼

Air jet

Jet needle

Needle jet

Main jet

⑧

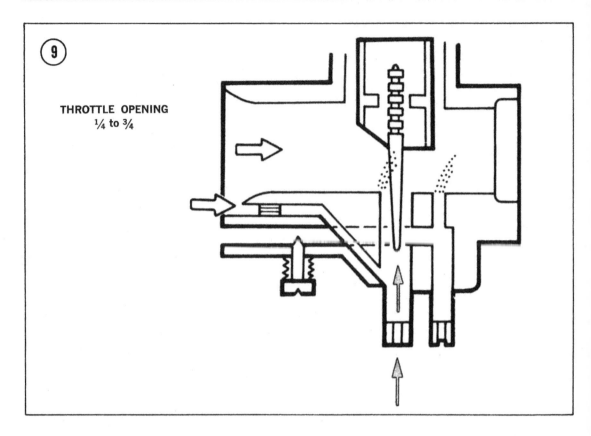

⑨ THROTTLE OPENING
¼ to ¾

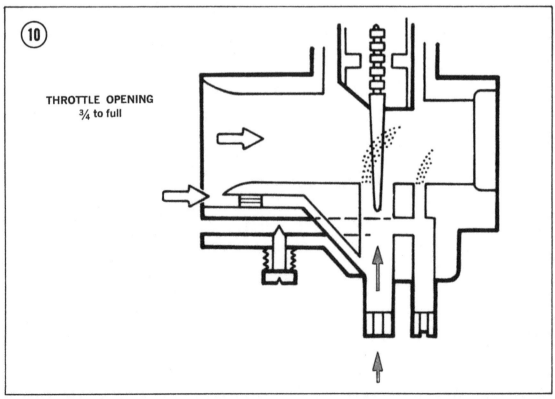

⑩ THROTTLE OPENING
¾ to full

the air bleed opening or air jet, may result in an overrich mixture. Other causes of a rich mixture are a worn needle jet, loose needle jet, or loose main jet. If the jet needle is worn, it should be replaced. However, it may be possible to effect a temporary repair by placing the jet needle clip in a higher groove.

Tickler System

A cold engine requires a mixture which is far richer than normal. A tickler system provides it. When the rider presses the tickler button, the float is forced downward, causing the float needle valve to open, and thereby allowing extra fuel to flow into the float chamber.

CARBURETOR OVERHAUL

There is no set rule regarding frequency of carburetor overhaul. A carburetor used primarily for street riding may go 5,000 miles (8,000 kilometers) without attention. If the bike is used off-road, the carburetor might need an overhaul

in less than 1,000 miles (1,600 km). Poor engine performance, hesitation, and little response to idle mixture adjustment are all symptoms of possible carburetor malfunctions. As a general rule, it is good practice to overhaul the carburetor each time you perform a routine decarbonization of the engine.

Carburetor Removal

1. Disconnect fuel lines at carburetor.

2. Remove the mixing chamber cap. See **Figure 11**.

3. Disconnect throttle cables at the slide.

4. Disconnect the flexible adapters between the carburetors and air cleaners.

5. Unbolt the carburetor at the manifold.

6. Remove the gaskets taking care not to damage them if they are to be reused.

Carburetor Installation

On single carburetor models, be sure that the faces between the manifold and cylinder head

4

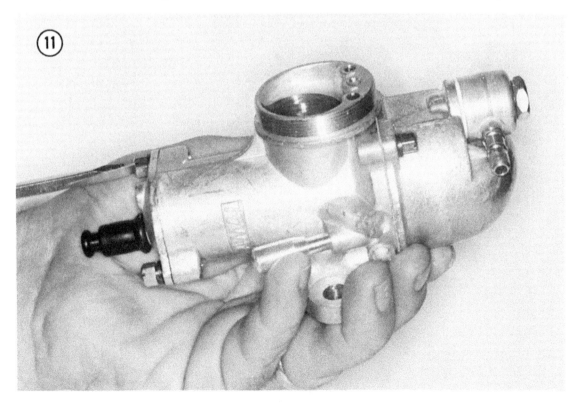

(11)

are flat and true. Use new gaskets and torque the nuts to 6 ft.-lb. The rest of the procedure is the same for all models.

1. Install the gasket over the carburetor studs on the head or manifold, depending on the model, and then install the insulating gasket.

2. Be sure the O-ring in the carburetor flange is in good condition.

3. Install the nuts, tightening them slowly and evenly to avoid distortion. Torque to 12 ft.-lb. Do not overtighten or air leaks can result.

4. Install the flexible adapters between the carburetors and the air cleaner. On twin carburetor models, the adapters are made to fit on either the left or right side, but not both. Install correctly with the mold joint line on top.

5. Install the control cables, attaching to the carburetor and the throttle grip and the frame clips.

6. Be sure the fuel filter and fiber washer at the carburetor end of the fuel lines are in good condition and attach the fuel lines to the carburetors.

AMAL CONCENTRIC

Disassembly/Assembly

1. Remove the mixing chamber cap (Figure 11). Note the position of each part as it is removed.

2. Withdraw the throttle valve (**Figure 12**). Don't lose the spring plate.

3. Remove the float bowl (**Figure 13**).

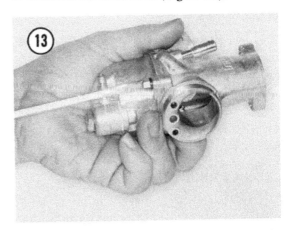

4. Remove the float and float needle together (**Figure 14**).

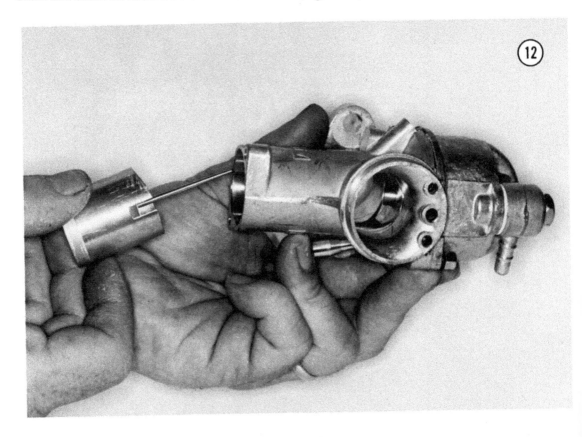

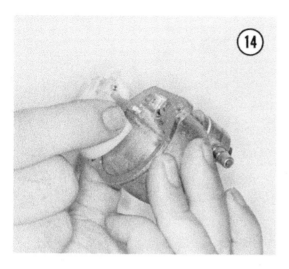

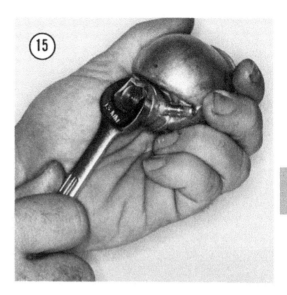

4

5. Remove the banjo bolt (**Figure 15**) and the fuel inlet fitting (**Figure 16**).

6. Carefully remove filter screen (**Figure 17**).

7. Remove the main jet (**Figure 18**).

8. Remove the jet holder (**Figure 19**). Unscrew the needle jet if necessary.

9. Remove the pilot jet (**Figure 20**).

10. Remove the mounting flange O-ring.

11. Remove the pilot air screw, throttle stop screw, and tickler assembly.

12. Reverse the disassembly procedure to reassemble the carburetor. Always use new gaskets for reassembly.

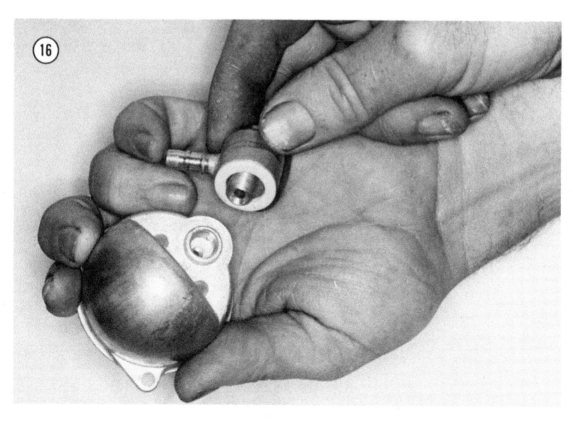

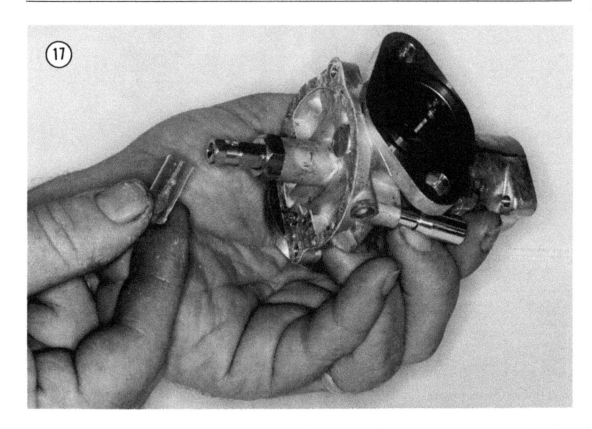

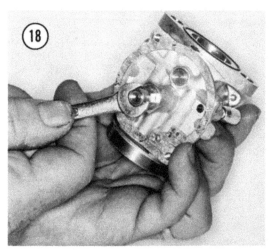

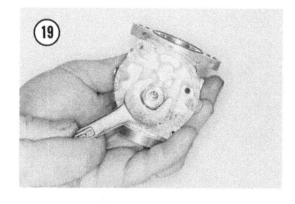

AMAL MONOBLOC

Disassembly/Assembly

Figure 21 is an exploded view of a typical Monobloc carburetor. Refer to this illustration during carburetor disassembly.

1. Remove the top ring nut (**Figure 22**).

2. Remove the throttle slide assembly. See **Figure 23**.

3. Note which jet needle groove the clip is in, then remove the jet needle clip and pull the jet needle from the throttle valve.

4. Remove the float chamber cover (**Figure 24**).

5. Remove the spacer from the float pivot shaft (**Figure 25**). Note the manner in which the float is installed, then pull it from the pivot shaft (**Figure 26**).

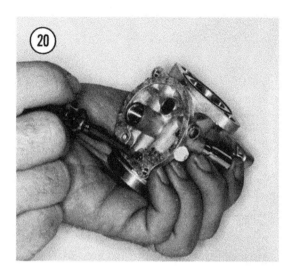

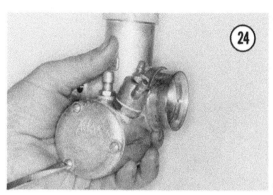

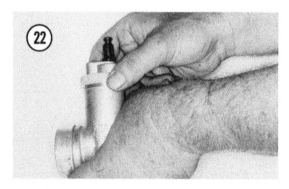

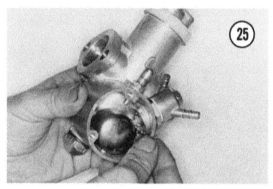

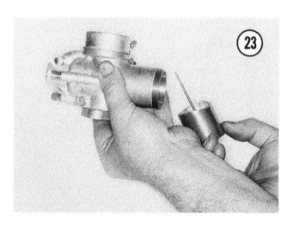

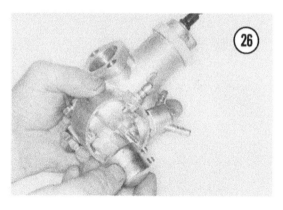

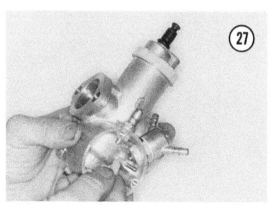

6. Remove the float needle (**Figure 27**), then the float needle seat.

7. Remove the main jet cover (**Figure 28**), then remove the main jet together with its gasket (**Figure 29**).

8. Remove the jet holder (**Figure 30**).

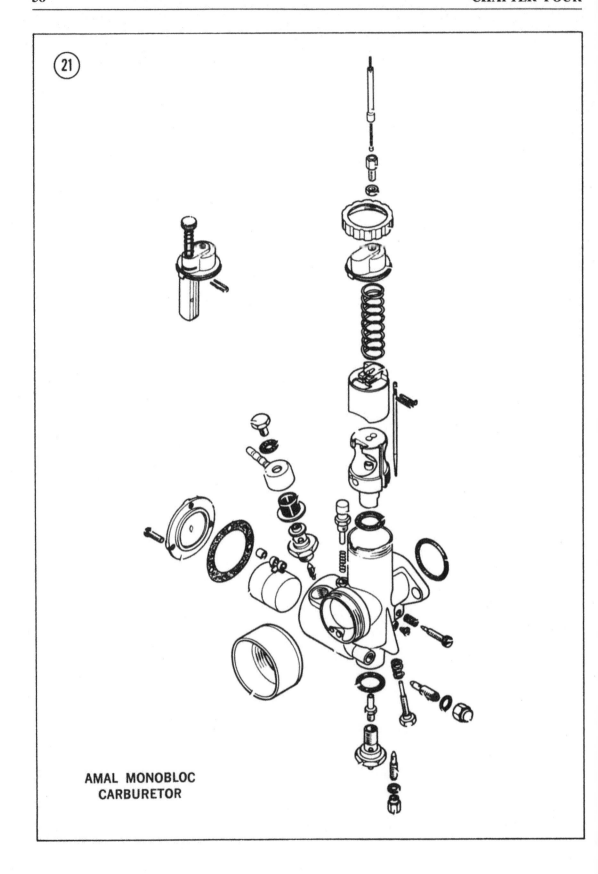

AMAL MONOBLOC
CARBURETOR

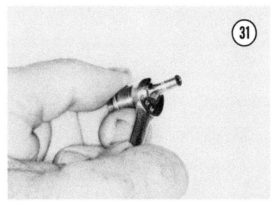

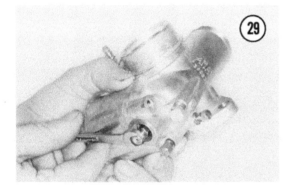

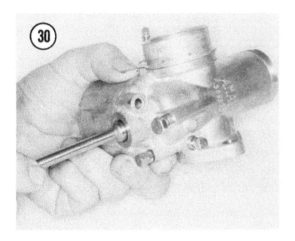

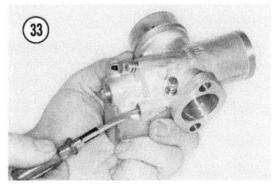

9. Separate the needle jet from the jet holder (**Figure 31**).

10. Remove the pilot jet cover (**Figure 32**), then the pilot jet (**Figure 33**).

11. Remove the banjo bolt, then the fuel inlet fitting (**Figure 34**).

12. Remove the fuel strainer (**Figure 35**), then the float needle seat.

13. Remove the pilot air screw, tickler button, and throttle stop screw.

14. Reverse the disassembly procedure to reassemble the carburetor. Always use new gaskets during assembly.

CARBURETOR ADJUSTMENT

Carburetor adjustment is not normally required except for occasional adjustment of idling

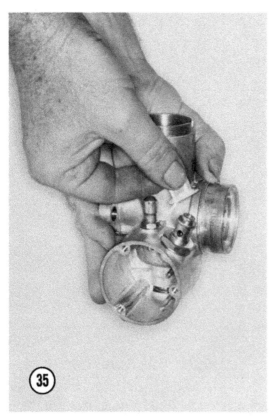

speed, or at time of carburetor overhaul. See the next section for synchronization of twin carburetors.

Float Level

The machine was delivered with the float level adjusted correctly. Floats on Amal carburetors do not have provision for adjustment.

Float Inspection

Shake the float to check for gasoline inside (**Figure 36**). If fuel leaks into the float, the float chamber fuel level will rise, resulting in an over-rich mixture. Replace the float if it is deformed or leaking.

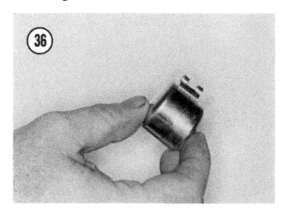

Replace the float valve if its seating end is scratched or worn. Press the float valve gently with your finger and make sure that the valve seats properly. If the float valve does not seat properly, fuel will overflow, causing a rich mixture and flooding the float chamber whenever the fuel petcock is open.

Component Parts

Clean all parts in carburetor cleaning solvent. Dry the parts with compressed air. Clean the jets and other delicate parts with compressed air after the float bowl has been removed. Never attempt to clean jets or passages by running a wire through them. To do so will cause damage and destroy their calibration. Do not use compressed air to clean an assembled carburetor, since the float and float valve can be damaged.

Speed Range Adjustments

The carburetor was designed to provide the proper mixture under all operating conditions. Little or no benefit will result from experimenting. However, unusual operating conditions such as sustained operation at high altitudes or unusually high or low temperatures may make modifications to the standard specifications desirable. The adjustments described in the following paragraphs should only be undertaken if the rider has definite reason to believe they are required. Make the tests and adjustments in the order specified.

Figure 37 illustrates typical carburetor components which may be changed to meet individual operating conditions. Shown left to right are the main jet, needle jet, jet needle and clip, and throttle valve.

Make a road test at full throttle for final determination of main jet size. To make such a test, operate the motorcycle at full throttle for at least 2 minutes, then shut the engine off, release the clutch, and stop the bike.

If at full throttle, the engine runs "heavily," the main jet is too large. If the engine runs better by closing the throttle slightly, the main jet is too small. The engine will run evenly at full throttle if the main jet is of the correct size.

After each such test, remove and examine the spark plug. The insulator should have a light tan color. If the insulator has black sooty deposits, the mixture is too rich. If there are signs of intense heat, such as a blistered white appearance, the mixture is too lean.

As a general rule, main jet size should be reduced approximately 5% for each 3,000 feet (900 meters) above sea level.

Table 2 lists symptoms caused by poor mixtures.

Adjust the pilot screw as follows.

1. Turn the pilot air screw in until it seats lightly, then back it out about 1½ turns.

2. Start the engine and warm it to normal operating temperature.

3. Turn the idle speed screw until the engine runs slow enough to falter.

4. Adjust the pilot screw to make the engine run smoothly.

5. Repeat Steps 3 and 4 to achieve the lowest stable idle speed.

Determine the proper throttle valve cutaway

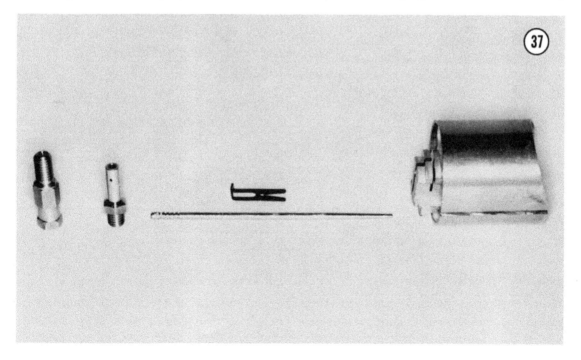

Table 2   CARBURETOR MIXTURE

Condition	Symptom
Rich Mixture	Rough idle Black exhaust smoke Hard starting, especially when hot "Blubbering" under acceleration Black deposits in exhaust pipe Gas-fouled spark plug Poor gas mileage Engine performs worse as it warms up
Lean mixture	Backfiring Rough idle Overheating Hesitation upon acceleration Engine speed varies at fixed throttle Loss of power White color on spark plug insulator Poor acceleration

Table 3   CARBURETOR ADJUSTMENTS

Throttle Opening	Adjustment	If too Rich	If too Lean
0 - ⅛	Air screw	Turn out	Turn in
⅛ - ¼	Throttle valve cutaway	Use larger cutaway	Use smaller cutaway
¼ - ¾	Jet needle	Raise clip	Lower clip
¾ - full	Main jet	Use smaller number	Use larger number

size. With the engine running at idle, open the throttle. If the engine does not accelerate smoothly from idle, turn the pilot air screw in (clockwise) slightly to enrich the mixture. If the condition still exists, return the air screw to its original position and replace the throttle valve with one which has a smaller cutaway. If engine operation is reduced by turning the air screw, replace the throttle valve with one having a larger cutaway.

6. For operation at ¼ to ¾ throttle opening, adjustment is made with the jet needle. Operate the engine at ½ throttle in a manner similar to that for full throttle tests described earlier. To enrich the mixture, place the jet needle clip in a lower groove. Conversely, placing the clip in a higher groove leans the mixture.

7. A summary of carburetor adjustments is given in **Table 3**.

Synchronization

1. First adjust throttle cable play to a minimum at each carburetor so that both throttle slides begin to move at the same time when the throttle is opened.

2. Start the engine and remove the left spark plug lead so the motor is running on the right cylinder only. Use dry gloves to avoid a shock.

3. Adjust the right carburetor, using the pilot air screw and the throttle stop screws so the motor is running at as smooth an idle as possible.

4. Replace the left spark plug lead and perform the same adjustments on the left carburetor.

5. Replace the right plug lead and unscrew both throttle stop screws the same amount until the engine idles just fast enough to keep it running with the throttle closed.

COMMON PROBLEMS

Throttle Cable

Adjust the throttle cable so the valve closes completely when the throttle grip is released. There should be minimal play in the throttle handle in the closed position. Be sure the throttle is not partially opened by turning the handlebars. This is a matter of correct routing of the cable.

Flooding

Flooding can be caused by wear on the float needle or by a punctured float. It is more commonly caused by debris in the gas tank entering the carburetor. If the float bowl filter is in good condition, the fuel lines may become plugged from the debris. If the condition is serious, the tank may have to be removed and completely cleaned.

Air Leaks

Uneven running, especially at idling speeds, may be caused by air leaks between the carburetor and the cylinder head. To check, apply oil around the gaskets and watch the exhaust with engine idling. If blue smoke appears, replace the gaskets between the carburetor and head, then check again. If leaks persist, check the O-ring of the carburetor body. If leaks still persist, check the carburetor flange with a steel straightedge. If it is warped, lap it on fine emery cloth placed over plate glass. On high mileage motorcycles, air leaks can be caused by a worn throttle valve or intake valve guides.

Pilot Air Screw

If the pilot air screw mixture is too rich or too lean, it will cause detonating in the exhaust at lower engine speeds as a result of unburned gas being ignited by the heat. If it occurs at higher speeds, it is an ignition problem.

Poor Gas Mileage

The most likely cause is a worn carburetor needle and needle jet, but be sure to check the entire fuel system, including the tank, for leaks. Poor riding techniques can also cause poor gas mileage when there is nothing wrong with the motorcycle.

Air Cleaner

The engine should be operated with carburetor air cleaners installed. The standard jet sizes are for running with air cleaners. If the cleaners are removed, the fuel/air mixture will be too lean for efficient combustion, and overheating the engine will result. If the engine runs better with the choke partly closed, the mixture is too lean. The needle must be raised or the main jet enlarged, or both.

Carburetor Choke

The choke should only be used when starting the engine cold. As soon as the engine is running, release the choke.

AIR FILTER

Many BSA motorcycles come equipped with paper filters. These can't be cleaned, but excess dirt should be shaken out at frequent intervals. See **Figure 38**.

Accessory houses have wet foam filters, made to fit any bike, which are more efficient and can be cleaned in kerosene. Allow the filter to air

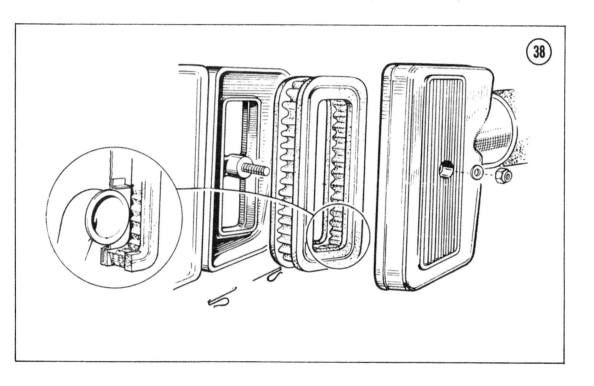

dry, dip in lightweight oil, squeeze out excess and insert in the stock housing.

Dirt riding will make it necessary to check the filter after every ride.

CAUTION

Even minute particles of dust can cause severe wear, so never run without a filter.

If a felt element is used, clean in kerosene and dry thoroughly. Replace any excessively dirty or damaged felt element.

FUEL TANK

Removal/Installation (2.5 Gallon)

1. Turn off the fuel petcocks and remove the fuel lines.

2. Remove the fuel tank filler cap and raise the seat.

3. Remove the 2 screws holding the center trim piece in place. Remove the trim, exposing the retaining bolt (**Figure 39**).

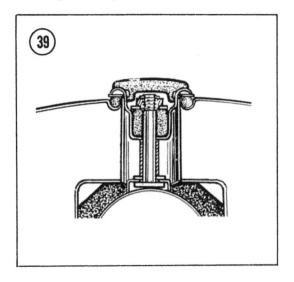

4. Remove the nut and lift off the tank, leaving the bolt in position on the frame.

5. When replacing the tank, be sure the bridge mounting rubbers and side pads are in place. Tighten the nut to expand the rubber bushing in its socket.

6. Connect the fuel lines.

Removal/Installation (4 Gallon)

1. Disconnect the fuel lines from the petcocks.

2. Remove the rubber grommet on top of the tank to expose the retainer bolt (**Figure 40**).

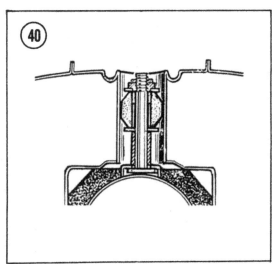

3. Remove the nut and washer, bead holder, and beading strips.

4. Lift off the tank, leaving the bolt in position on the frame.

5. When replacing the tank, be sure the bridge mounting rubbers and side pads are in place. Tighten the nut to expand the rubber bushing.

6. Reconnect the fuel lines.

CHAPTER FIVE

ELECTRICAL SYSTEM

The electrical system includes the battery, ignition system, charging system, lighting, and horn. Wiring diagrams are given at the end of this chapter. Specifications for service and replacement parts are given in the Appendix.

Any part of the electrical system may be repaired with a minimum of special tools by following the procedures described.

BATTERY

If two 6-volt batteries are connected in series for a 12-volt system, they must be connected as shown in **Figure 1**.

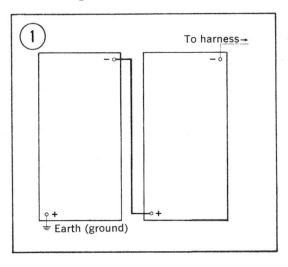

Battery construction is shown in **Figure 2**.

Be sure to check battery electrolyte level, especially during hot weather.

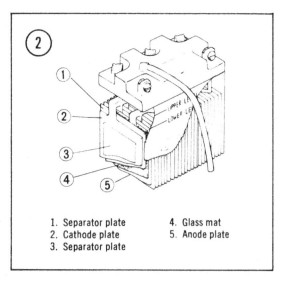

1. Separator plate
2. Cathode plate
3. Separator plate
4. Glass mat
5. Anode plate

Removal

1. Lift the seat and remove the battery retainer.

2. Disconnect the ground or positive (+) cable first, then the negative (-) cable.

3. Lift the battery from the carrier. Note that the battery vent tube is routed through the motorcycle and out behind the swing arm lug.

Safety Precautions

When working with batteries, use extreme care to avoid spilling or splashing the electrolyte. Electrolyte contains sulphuric acid which can destroy clothing and cause serious chemical burns. If any electrolyte is spilled or splashed on clothing, body, or other surfaces, neutralize it *immediately* with a solution of baking soda and water, then flush with plenty of clean water.

WARNING
Electrolyte splashed into the eyes is extremely dangerous. Safety glasses should always be worn when working with batteries. If electrolyte is splashed into the eye, call a physician immediately, force the eye open, and flood with cool, clean water for about 5 minutes.

While batteries are being charged, highly explosive hydrogen gas forms in each cell. Some of this gas escapes through the filler openings and may form an explosive atmosphere around the battery. This explosive atmosphere may exist for several hours. Sparks, open flame, or even a lighted cigarette can ignite this gas, causing an internal explosion and possible serious personal injury. The following precautions should be taken to prevent an explosion.

1. Do not smoke or permit open flame near any battery being charged or which has been recently charged.

2. Do not disconnect live circuits at battery terminals because a spark usually occurs when a live circuit is broken. Care must always be taken while connecting or disconnecting any battery charger; be sure its power switch is off before making or breaking connections. Poor connections are a common cause of electrical arcs which cause explosions.

Battery Inspection and Service

1. Measure the specific gravity of the battery electrolyte with a hydrometer. The specific gravity is calibrated on the hydrometer float stem. The reading is taken at the fluid surface

level with the float buoyant in the fluid (**Figure 3**).

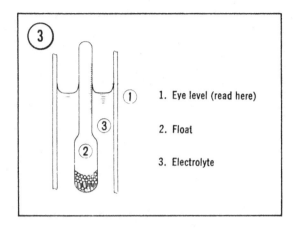

1. Eye level (read here)
2. Float
3. Electrolyte

2. If the reading is less than 1.20 with the temperature corrected to 68°F, recharge the battery. See **Figure 4** for a graph of specific gravity vs. residual capacity.

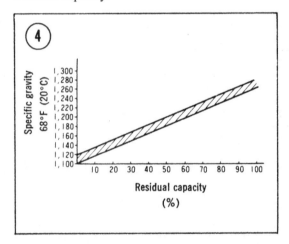

3. If any cell's electrolyte level is below the lower mark on the battery case, fill with distilled water to the upper mark.

4. Replace the battery if the case is cracked or damaged. Corrosion on the battery terminals causes leakage of current. Clean with a wire brush or with a solution of baking soda and water.

5. Check the battery terminal connections. If corrosion is present, the connection is poor. Clean the terminal and connector and coat with Vaseline and reinstall.

6. Vibration causes the corrosion of the battery plates to flake off forming a paste on the bottom (**Figure 5**). Replace the battery when the paste builds up considerably. Clear-cased batteries permit inspection.

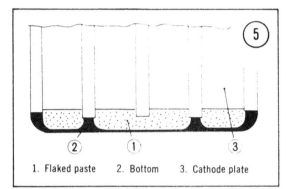

1. Flaked paste 2. Bottom 3. Cathode plate

Table 1 BATTERY CHARGE

Charging current rate	Maximum of 10% of ampere-hour rating.
Checking for full charge	① Specific gravity: 1.260-1.280 at 20°C (68°F) maintained constant for one hour.
	② 7.5-8.3V maintained constant at terminals for a 6 volt battery or 12.8-14.5V for a 12 volt battery, checked with voltmeter.
Charging duration	By this method, a battery with specific gravity of electrolyte below 1.220 at 20°C (68°F) will be fully charged in approximately 10-13 hours.

Battery Charging

Batteries are not designed for high charge or discharge rates. For this reason, it is recommended that a battery be charged at a rate not exceeding 10% of its ampere-hour capacity. That is, do not exceed 0.5 ampere charging rate for a 5 ampere-hour battery, or 1.5 amperes for a 15 ampere-hour battery. This charge rate should continue for 10-13 hours if the battery is completely discharged or until specific gravity of each cell is up to 1.260-1.280, corrected for temperature. If after prolonged charging, specific gravity of one or more cells does not come up to at least 1.230, the battery will not perform as well as it should, but it may continue to provide satisfactory service for a time.

Some temperature rise is normal as a battery is being charged. Do not allow the electrolyte temperature to exceed 110°F (43.3°C). Should temperature reach that figure, discontinue charging until the battery cools, then resume charging at a lower rate.

If possible, always slow-charge a battery (see **Table 1**). Quick-charging will shorten the battery service life. Use a quick-charge only if *absolutely* necessary.

After removing the battery, use the following procedure for charging.

1. Hook the battery to a charger by connecting the positive lead to the positive terminal on the battery and the negative lead to the negative terminal. To do otherwise could cause severe damage to the battery and result in injury if the battery explodes.

2. The electrolyte will begin bubbling, signifying that explosive hydrogen gas is being released. Make sure the area is adequately ventilated and that there are no open flames.

3. It will normally take at least 8 hours to bring the battery to a full charge. Test the electrolyte periodically with a hydrometer to see if the specific gravity is within the standard range of 1.26 to 1.28. If the reading remains constant for more than an hour, the battery is charged. See Table 1.

Installation

1. Wash the battery with water to remove spilled electrolyte. Coat the terminals with Vaseline or light grease before installing.

2. When replacing the battery, be careful to route the vent tube so that it is not crimped. Connect the negative terminal first, then the positive one. Don't overtighten the clamps.

3. Remeasure the specific gravity of the electrolyte with a bulb hydrometer, reading it as shown in Figure 3.

NEW BATTERY PREPARATION

1. Remove the vent hole sealing tape. Pour sulphuric acid into each cell to the colored line on the battery case (see **Figure 6**). Let set for an hour.

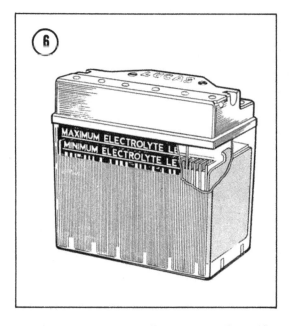

2. Charge the battery before installing. The electrolyte should be maintained with the periodic addition of distilled water.

WIRING DIAGRAMS

Wiring diagrams for the various models are given at the end of this chapter. Be sure that you follow the proper diagram when servicing the electrical system.

RECTIFIER (ALL MODELS)

The rectifier must be clean and dry at all times. When removing it for service, be sure that you hold the nut shown in **Figure 7** to avoid twisting the plates.

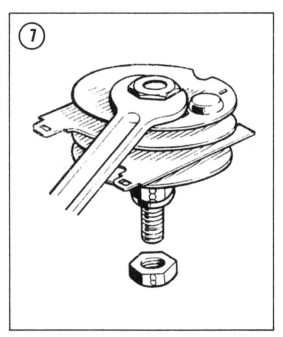

CAUTION
Never remove or loosen the nut holding the plates together as this breaks internal connections and destroys the rectifier.

ZENER DIODE CHARGE CONTROL

Be sure a firm contact is maintained between the base of the diode (see **Figure 8**) and the air cleaner box, to prevent overheating. Be sure the ground connection is good. No maintenance is necessary to the diode unit.

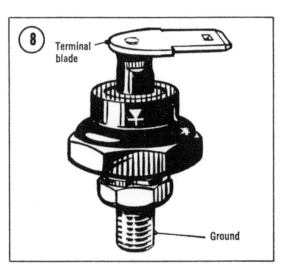

The diode is mounted to the aluminum air cleaner box. To remove the diode, disconnect the brown/blue double connector from the diode and remove the mounting nut. Be careful not to tighten the nut too much when installing.

ALTERNATOR
(ROTOR AND STATOR)

See *Alternator* section of Chapter Six for removal and installation.

SPARK PLUGS

See *Spark Plug* section of Chapter Two for removal, cleaning, and installation of plugs.

BREAKER POINTS

See *Breaker Point Service section* of Chapter Two for removal, installation, and setting of breaker points.

IGNITION COIL

Removal/Installation

Refer to **Figure 9** for an illustration of a typical coil.

1. Remove the gas tank. See Chapter Six.

2. Disconnect battery ground (red) lead.

3. Disconnect leads to coil.

4. Detach coil from mounting bracket and remove.

5. Install in reverse order of removal.

HORN ADJUSTMENT

Release the locknut on the back face of the horn and turn the screw (see **Figure 10**) until the loudest sound is obtained. Tighten the locknut while holding the screw.

HEADLIGHT

Replacement

1. Loosen the screw at the top of the light and remove the rim and reflector. Press in the bulb and turn counterclockwise to release from the cap. Replace entire lamp on sealed beam units.

2. The replacement bulb can only be fitted in the cap one way.

3. Reassemble in reverse sequence.

4. Adjustment nuts on the side of the headlight housing allow the light to be tilted up or down.

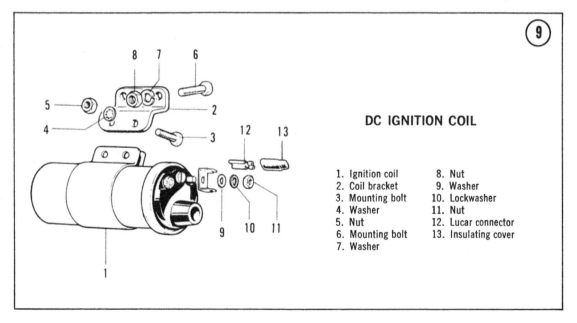

DC IGNITION COIL

1. Ignition coil	8. Nut
2. Coil bracket	9. Washer
3. Mounting bolt	10. Lockwasher
4. Washer	11. Nut
5. Nut	12. Lucar connector
6. Mounting bolt	13. Insulating cover
7. Washer	

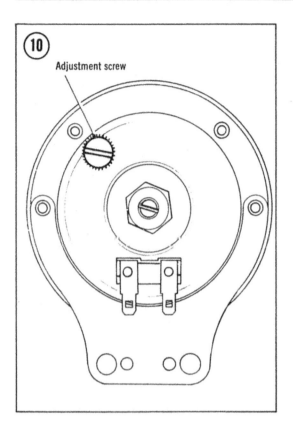

Adjustment

Proper headlight adjustment is essential to safe night riding. If the lights are set too low, the road will not be visible. If set too high, they will blind oncoming vehicles. Adjustment is very simple; proceed as follows.

1. Place the machine approximately 16 ft. from a white or light-colored wall. Refer to **Figure 11**.

2. Make sure the bike and wall are on level, parallel ground and that the machine is pointing directly ahead.

3. Measurements should be made with one rider sitting on the bike and both wheels on the ground.

4. Draw a cross on the wall equal in height to the center of the headlight.

5. Put on the high beam. The cross should be centered in the concentrated beam of light.

6. If the light does not correspond to the mark, loosen the bolts and adjust. Tighten the bolts and recheck positioning.

TURN SIGNAL LIGHTS

The lens is held by 2 screws. Remove them to replace the bulb. Be sure the ground clip in back makes good contact with the light reflector.

TAILLIGHT AND STOPLIGHT

The lens is held by 2 screws. Remove them to replace the bulb. The new bulb can only be fitted one way. Be sure the 2 electric wires and the ground are properly connected before replacing.

STOPLIGHT SWITCH

The front brake switch is not adjustable and requires no maintenance. The rear brake switch

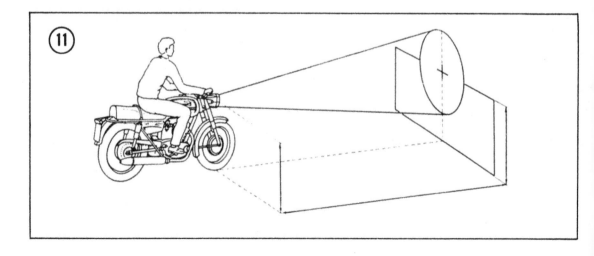

should have 1/32 in. movement of the pedal before the light comes on. Adjust by moving the switch up or down in its mounting.

FUSE

The fuse is in a housing in the brown/blue wire from the battery negative terminal. The housing is easily opened (see **Figure 12**) to replace the fuse. Be sure to locate any electrical problems before replacing the fuse, or it will just blow again. The fuse must not be more than 35 amp rating.

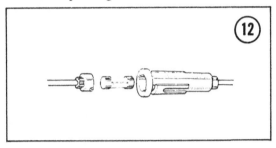

IGNITION SWITCH

1. Unscrew the bezel ring (C, **Figure 13**) to remove the side panel for access to the switch.

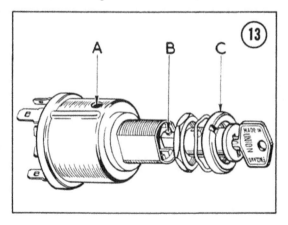

2. Check the 3 terminals in the switch for good connections.

3. Unscrew the locknut to remove the switch body from the bracket, if necessary. Be sure to disconnect the battery wires first.

4. A spring loaded plunger (B, Figure 13) holds the lock in the body. Press a pointed instrument through the small hole (A, Figure 13) to release the plunger and withdraw the lock.

WARNING LIGHTS

The 3 warning lights in the tip of the headlight indicate turn signals, headlight high beam, and low oil pressure. To change a bulb, remove the headlight rim and light unit for access to the bulb holders.

Left Handlebar Switches

These switches control the headlight on-off, headlight high-low beam, and the horn. The circuity is illustrated in **Figure 14**. To dismantle and reassemble the switch requires special equipment and should be left up to a dealer.

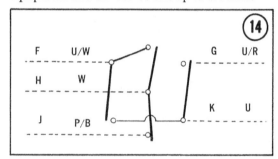

Right Handlebar Switches

These switches control the turn signals and the ignition kill button. The circuitry is illustrated in **Figure 15**. Special equipment is required for maintenance and should be left for a professional mechanic.

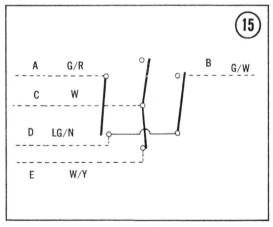

ALTERNATIVE CAPACITOR IGNITION SYSTEM

A kit is available, as part No. 00-4402, which allows the motorcycle to be run with or without the battery. Without the battery, the engine will readily start and run and the lights will work. When the engine is stopped, the lights will not work.

The kit capacitor has 2 terminals. The smaller is the positive ground and is marked with a spot of red paint. The larger double terminal is the negative.

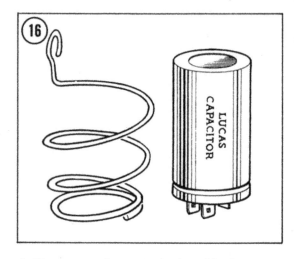

Installation

1. Push the capacitor down into the spring (see **Figure 16**) until the coil locates in the capacitor groove. Mount with the terminals down.

2. Connect the capacitor negative terminal and the zener diode to the rectifier brown/white center terminal. Connect the capacitor positive terminal to the rectifier center bolt ground (see **Figure 17**). The mounting spring is attached to any convenient point close to the battery.

3. If the battery is disconnected, remove the fuse from its holder and insert some insulation into the holder. This will prevent the battery from shorting to ground.

4. Never run the motorcycle with the zener diode disconnected, as excessive voltage can damage the electrical system.

5. If a capacitor is installed for emergencies only, and the battery left connected for lights with the engine off, disconnect the battery periodically to check the capacitor.

6. If the engine will not start with the battery disconnected, check the wiring between the capacitor and rectifier. If everything appears all right, install another capacitor to test starting.

7. If the engine will not start with the battery connected, disconnect the capacitor to test for a short circuit.

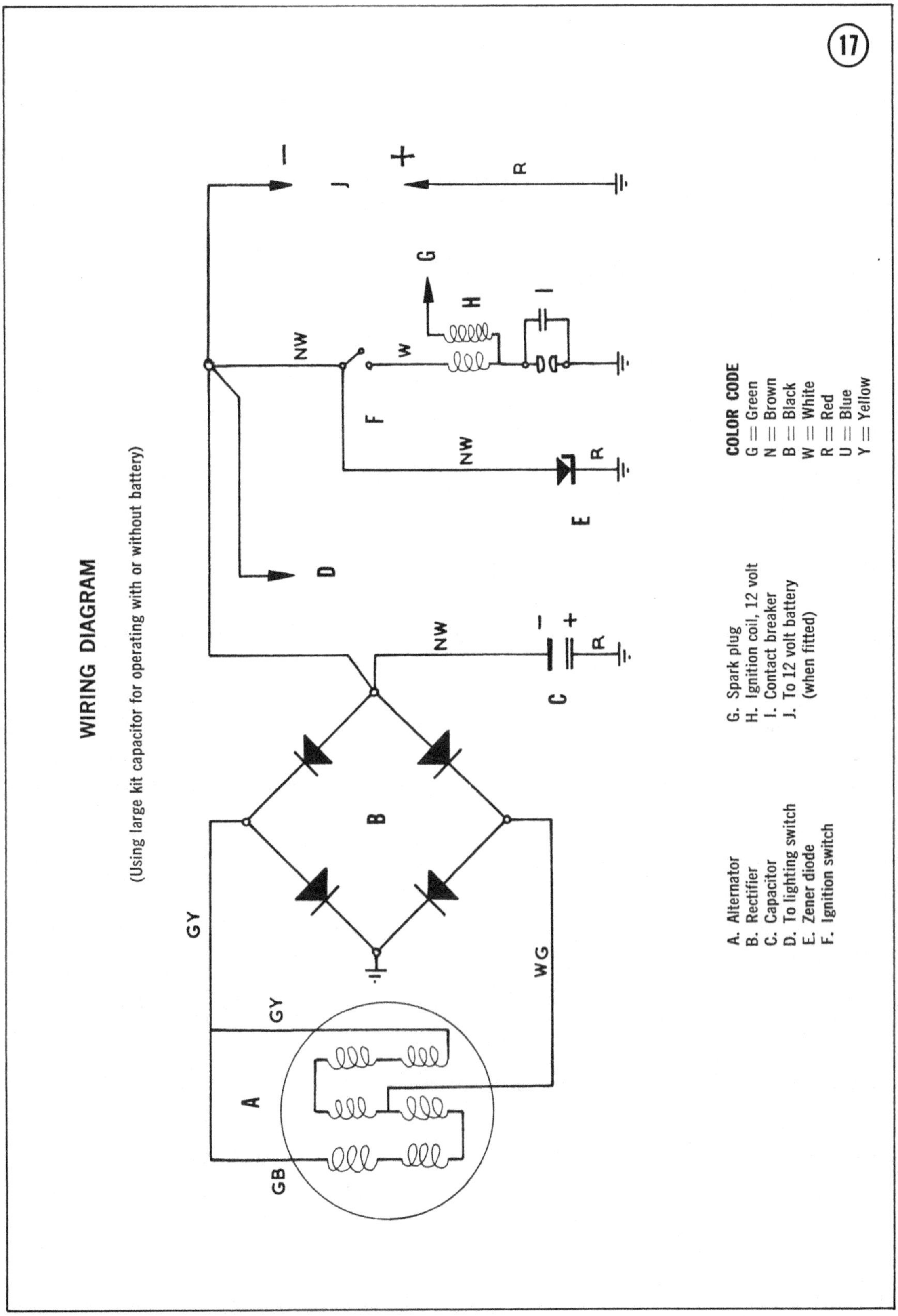
17
WIRING DIAGRAM
(Using large kit capacitor for operating with or without battery)
A
B
C
D
E
F
G
H
I
J
GY
GY
GB
WG
NW
NW
NW
W
R
R
R
−
+
−
+
A. Alternator
B. Rectifier
C. Capacitor
D. To lighting switch
E. Zener diode
F. Ignition switch
G. Spark plug
H. Ignition coil, 12 volt
I. Contact breaker
J. To 12 volt battery (when fitted)
COLOR CODE
G = Green
N = Brown
B = Black
W = White
R = Red
U = Blue
Y = Yellow

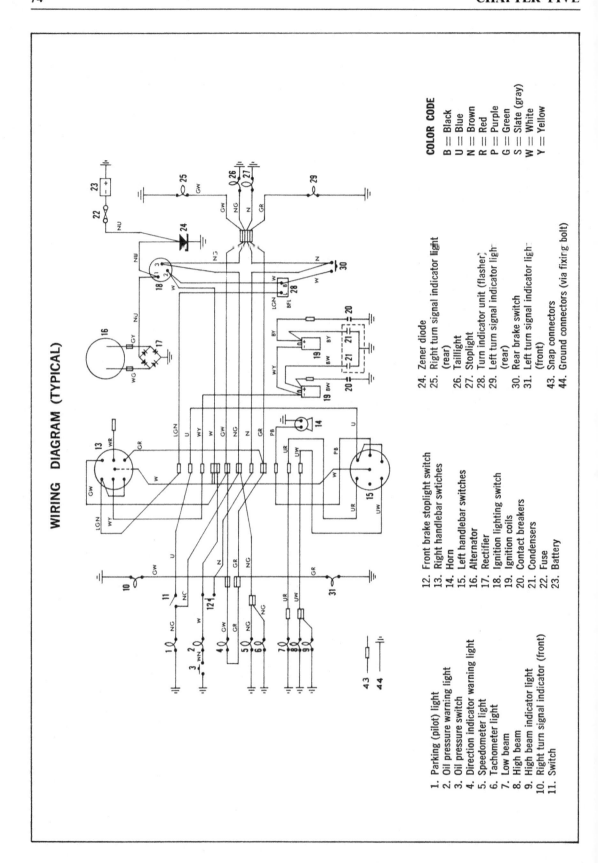

WIRING DIAGRAM (TYPICAL)

COLOR CODE

B = Black
U = Blue
N = Brown
R = Red
P = Purple
G = Green
S = Slate (gray)
W = White
Y = Yellow

1. Parking (pilot) light
2. Oil pressure warning light
3. Oil pressure switch
4. Direction indicator warning light
5. Speedometer light
6. Tachometer light
7. Low beam
8. High beam
9. High beam indicator light
10. Right turn signal indicator (front)
11. Switch
12. Front brake stoplight switch
13. Right handlebar swtiches
14. Horn
15. Left handlebar switches
16. Alternator
17. Rectifier
18. Ignition lighting switch
19. Ignition coils
20. Contact breakers
21. Condensers
22. Fuse
23. Battery
24. Zener diode
25. Right turn signal indicator light (rear)
26. Taillight
27. Stoplight
28. Turn indicator unit (flasher)
29. Left turn signal indicator ligh- (rear)
30. Rear brake switch
31. Left turn signal indicator ligh- (front)
43. Snap connectors
44. Ground connectors (via fixirg bolt)

CHAPTER SIX

ENGINE, TRANSMISSION, AND CLUTCH

The engine is a vertical twin with overhead valves. The transmission is contained in a separate housing within the crankcase. The crankcase and cylinder head are made of aluminum alloy, while the cylinders are cast iron. Power is transmitted from the crankshaft through a triple primary chain to the plate clutch, which contains a shock absorber. Power is then transmitted through a 4- or 5-speed transmission to the secondary chain which forms the final drive to the rear wheel.

ENGINE REMOVAL

Pay careful attention to any nuts or bolts that have come loose and were vibrating. Always replace these parts. Don't try to use them again. If any of the electrical wiring has had its insulation rubbed through until the wire is exposed, it should be replaced. Be sure to wrap new wire with several turns of tape at that spot to prevent it from happening again. A bare wire can cause a short circuit, resulting in a fire.

It is not necessary to remove the cylinder or engine during an ordinary top end overhaul. If the pistons, rings, or cylinder bores do not require repair, leave the cylinder in place.

1. Drain the oil from the reservoir, transmission, and primary crankcase, as described in Chapter Two. The oil feed and return lines can be left attached.

2. Remove the master link from the rear chain. Place the transmission in neutral and unwind the chain from both sprockets.

3. Remove the chain guard. See **Figure 1**. It is fastened in front by a bolt to the swing arm, and at the rear to the bottom of the shock absorbers.

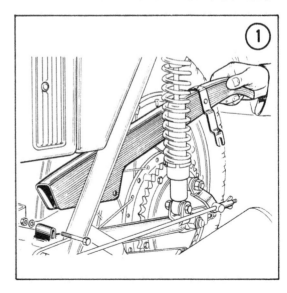

4. Remove the front bolt and loosen the lower shock absorber nut. The guard can then be pulled out to the rear.

5. Disconnect the generator wires at the snap connectors behind the engine. See **Figure 2**.

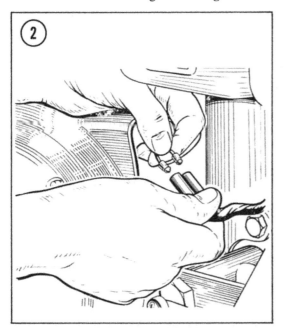

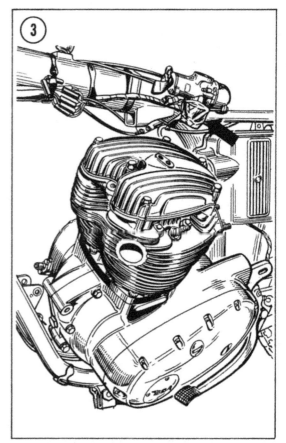

6. Follow the procedure for removing the gas tank in this chapter under *Gas Tank Removal/ Installation*.

7. Remove the engine brace from the main frame tube and the rocker cover.

8. Remove exhaust system as described in this chapter.

9. Remove the carburetors as described in Chapter Four. It is best to remove the carburetors completely, including the control cables. Otherwise the carburetors should be taped to the main frame tube (**Figure 3**) to avoid damage.

10. Remove the left foot peg. On later models it is attached by one nut and bolt (with a right-hand thread) with pegs that fit in slots in the frame to prevent it from turning. On earlier models, the bolt has a left-hand thread.

11. Loosen the rear brake pedal adjustment. Depress the pedal to allow the engine to be lifted. It isn't necessary to remove the pedal.

12. Disconnect the clutch cable at the handle-bar and coil it up. The other end can be left attached to the crankcase.

13. Disconnect the tachometer cable at the front of the timing case. Unscrew the cap and withdraw the cable from the driving spindle.

14. Disconnect the connector on the oil pressure switch and remove the switch to prevent damage.

15. Disconnect the contact breaker cables at their snap connectors.

16. Disconnect the oil line unions at the lower right side of the crankcase (**Figure 4**). Don't disconnect the oil lines from the union.

17. The rear frame engine plates are fastened by 5 bolts. Two of them go through the crankcase, while the other 3 fasten the plates to the frame. All bolts have self-locking nuts, so be sure not to lose them. Remove the right plate and 4 of the bolts. The lower crankcase bolt and

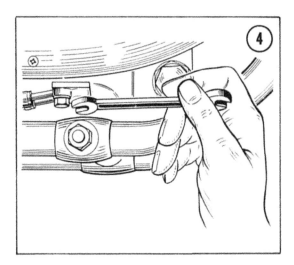

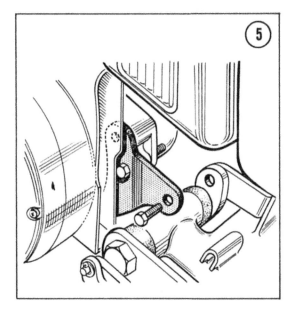

the left plate must be removed with the engine. The left plate can be removed later, after the chain cover plate has been removed (**Figure 5**).

18. The front engine mounting bolt should have been removed with the mufflers.

19. Remove the long bolt that passes through the frame and crankcase below the engine. Be sure not to lose any washers, and be careful of your fingers on this last operation. The engine will drop suddenly once the long bolt is removed.

20. Have someone help lift the engine out of the frame. One person on each side will simplify and ease the job. The shifter lever and kickstart lever

can be left attached. The latter makes a good handle for lifting.

21. From the right side, raise the front of the engine with a strong bar used as a lever (**Figure 6**) between the crankase and the frame. The front lug must clear the frame brackets and then be lowered on the left side of them.

22. Lift the rear of the engine by the kickstart lever.

23. From the right side, tilt the engine to the left and forward (**Figure 7**) and lift it out of the frame.

ENGINE INSTALLATION

1. Replace the transmission shift lever and the kickstarter. The primary cover should be to the left of the chain.

2. Attach the left side engine plate, leaving it loosely in place, held by the lower crankcase bolt. This bolt cannot be inserted properly if the chain cover is on the engine. Don't put the nut on this bolt until the engine is installed in its normal position in the frame.

3. The engine goes back into the frame from the left side. From the right side, use a lever (**Figure 8**) to raise the front of the crankcase and lower it into the frame mounting lugs. If another lever is used at the same time on the left side, the operation is much easier.

4. Replace the long bolt below the crankcase and right frame mounting plate. Replace the mounting plate bolts and tighten all nuts securely. The front engine mounting bolt goes on after the exhaust pipes.

5. Replace the rocker cover and the engine brace between the main frame tube and the rocker cover (**Figure 9**).

6. Replace the oil line union at its attachment under the crankcase. Replace the oil pressure switch and its electrical lead.

7. Attach the clutch cable to the lever on the handlebars. Be sure to route it through any frame clips provided.

6

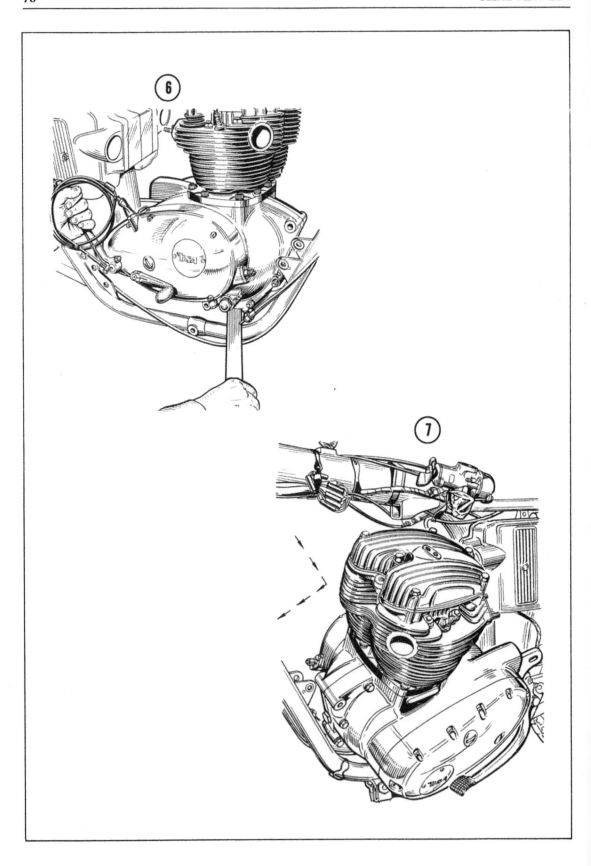

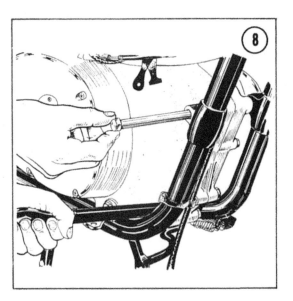

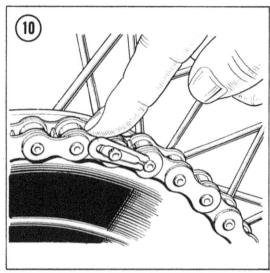

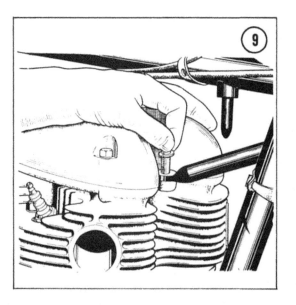

8. Connect the generator and contact breaker leads at their snap connectors. Be sure the cables are correctly matched by color.

9. Replace the rear chain, attaching the master link with the 2 end links in position on the rear sprocket. Be sure the closed end of the spring clip is toward the direction of chain travel (**Figure 10**). Replace the chain guard. Be sure the lower bolt on the shock absorber is tight.

10. The carburetor O-rings, gaskets, and insulators should be in good condition to be used again. Replace the carburetors as described earlier.

11. Connect the oil line to the rear of the cylinder head and install the left side foot peg. Readjust the rear brake pedal, making sure the stoplight comes on when the pedal is depressed.

12. Replace the exhaust system as described in this chapter. Remember to connect the interconnecting pipe before tightening everything.

13. Replace the gas tank as described in Chapter Four, connecting the fuel lines to the carburetors.

14. Before starting the engine, be very sure the reservoir, primary chaincase, and transmission have the proper quantity of oil. Add a half pint of oil through the timing plug opening to prime the crankcase. Kick the starter through with the ignition off until oil flows from the return line in the reservoir.

TOP END OVERHAUL

The necessity for a top end overhaul is usually a buildup of carbon on the piston crown, combustion chamber, ports, valves, or heads. Carbon is one product of combustion but is not harmful to the performance of the engine until deposits become excessive.

The presence of excessive carbon deposits in the engine is usually indicated by engine pinging while under load, power loss (especially on hills) and increased fuel consumption. The engine may also run hotter than usual and have pre-ignition which causes ragged running and poor power.

Before attempting a top end overhaul, you must have Whitworth-sized tools, and equipment. One of the most important requirements is a clean work bench and ample room to store parts as they are disassembled. Read all of Chapter One prior to starting work.

ROCKER COVER AND EXHAUST PIPE

Removal/Installation

1. Remove the engine brace from the main frame tube and rocker cover.

2. On the Lightning and Thunderbolt models, loosen the clip bolts on the right sleeve of the interconnecting exhaust pipe and slide the sleeve to the left. Remove the bolts holding the right exhaust pipe in place and tap it out of the cylinder head with a rubber mallet. Remove the pipe and repeat with the left side.

3. On the Firebird Scrambler model, remove the protective grille and all pipe connections to the frame. Loosen the clip bolts at the front of the mufflers and tap the pipes out of the cylinder heads, leaving the mufflers attached and in position.

4. Mufflers can be removed as a unit after disconnecting the mounting brackets.

5. Remove the 4 bolts and 2 nuts holding the rocker cover and loosen it with a soft mallet blow below the rim of the cover.

> NOTE: *On earlier models, the rocker was secured by 6 nuts.*

EXHAUST ROCKER ASSEMBLY

Removal/Installation

1. Loosen the 2 front exhaust valve rocker adjusting screws until the pushrods can be removed.

2. Cover the top of the pushrod tunnel with a clean rag to prevent debris from falling down inside the crankcase.

3. Remove the nut at the right front side holding the exhaust rocker shaft and tap the shaft through to the left, leaving the rockers in position. See **Figure 11**. Use a soft plastic mallet or similar tool to avoid damaging the end of the shaft.

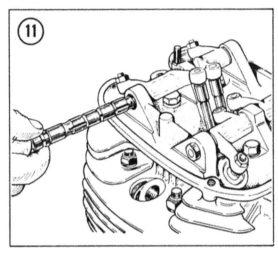

4. Note carefully the position of the spring and thrust washers and remove the rockers from between the shaft pillars (**Figure 12**). From left to right, the assembly should be: Thrust washer, Spring washer, Left side rocker, Thrust washer, Center post, Thrust rocker, Right side rocker, Spring washer, and Thrust washer.

> NOTE: *Spring washers are always fitted next to the rockers and never next to the shaft pillars.*

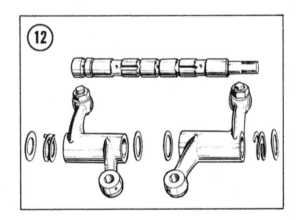

5. It is not necessary to remove the intake rocker assembly if the mechanism is operating correctly. The procedure will be the same.

6. Reverse procedure to install.

CYLINDER HEAD

Removal/Installation

The cylinder head is secured by 4 nuts on studs next to the spark plug holes, and one bolt in the center at the top of the pushrod tunnel. See **Figure 13**.

1. Disconnect the oil line at the rear of the cylinder head.

2. Remove the spark plug wires and the plugs.

3. Loosen the nuts and bolts a little at a time to avoid distorting the head. Start with the bolts below the rockers, then the center bolt, and finally the 4 nuts. When all are removed, break the head loose by tapping with a rubber mallet below the exhaust ports and lift it off. See **Figure 14**.

4. On single carburetor models, leave the intake manifold attached. The twin carburetor models do not have an intake manifold.

VALVE SPRINGS

Removal/Installation

1. Use a valve spring compressor to compress the springs until the split collets can be removed. If the compressor is tapped with a hammer on the spring side after the spring has been compressed, it will release the collets.

2. Remove the compressor and then the springs and collets. Wash all parts in solvent and place them on a numbered board to indicate their position in the head for assembly.

3. The springs may have partially collapsed from use. Check their length with a ruler (**Figure 15**) against the dimensions listed in the Appendix. If the springs have collapsed more than 1/16 in. or if they show any cracks, they should be replaced.

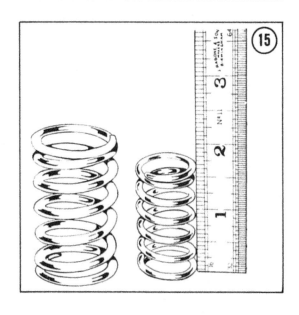

4. When reinstalling the springs in the head, the outer springs must have their closed coils placed downward into the spring cup.

5. Reverse to install.

PUSHRODS

Remove and examine the pushrod end cups for any wear or looseness. Roll the rods on a flat surface such as plate glass to see if they are bent. If the rod has any faults, it should be replaced.

VALVES AND GUIDES

Check the valves in the guides. There should not be excessive side play or carbon buildup on the part of the stem that slides in the guide. Carbon deposits can be removed by carefully scraping and very light use of fine emery cloth. If the stems are scored, it indicates a seizure has occurred and both the valve and guide should be replaced.

To avoid scoring the aluminum surface, immerse the head in hot water until it is warm. This will cause a slight expansion, making the guide easier to drive out. Drive the valve guide out with a drift. See **Figure 16**. Drive the new guide in with a drift while the head is still warm.

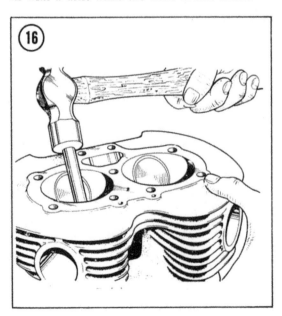

If the valve seats are only slightly pitted or discolored, they can be ground in by hand. If the seat is deeply pitted or the valve head is burned (**Figure 17**), the valve must be replaced and the new valve ground in.

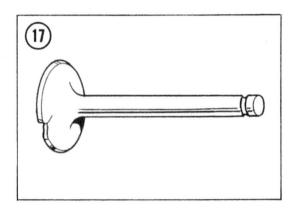

When a new valve guide has been installed, or a new valve, or both, the valve seat in the head must be recut to a 45 degree angle.

If regrinding has been repeated many times, the valve may become "pocketed," where the valve in its closed position in the valve seat has its head below the surface of the combustion chamber. See **Figure 18**. This impairs efficiency of the valve and interferes with the gas flow. A portion of the head around the valve must be removed to eliminate the pocket (see shaded area in Figure 18), using a special blending cutter. The valve seat can then be recut and the valve ground in again. This is a job best left for an experienced mechanic.

VALVE GRINDING

During a top end overhaul, all valves should be ground in. If the old valves are used again, be sure to return each to its own seat. Valve grinding should not be done until all carbon deposits have been removed from the combustion chamber.

1. Remove the carbon from the cylinder head intake and exhaust ports with a scraper or rotary file. Be sure not to let the tool slip against the valve seats. The resulting damage can make proper seating of the valve impossible.

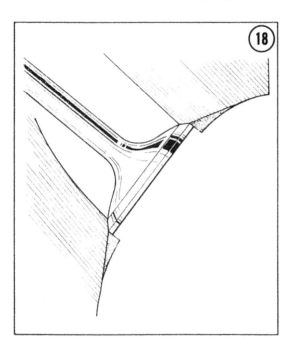

2. Final polishing of the decarboned areas is done with emery cloth dipped in solvent. Don't polish the valve seats.

3. Insert a light spring under the valve head so it can be raised easily to be rotated to a new position.

4. Spread a small quantity of fine grinding paste over the valve face.

5. Rotate the valve back and forth, maintaining a steady pressure. See **Figure 19**.

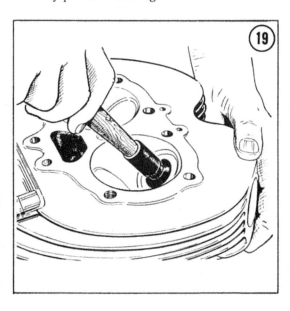

6. After a few strokes in one position, raise the valve and turn to a new position and repeat.

7. Remove the valve, clean off all paste and examine the valve face and seat. Repeat the grinding sequence until both the valve face and seat have a uniform matte finish with no visible flaws.

8. When the valve is ground in, carefully remove all traces of the grinding paste. Smear the valve stem with engine oil before reassembling the head.

9. Prolonged grinding will not produce the same results as recutting, so don't try it. You'll just wear out the valve. If recutting is necessary, have it done.

CYLINDER HEAD ASSEMBLY

While the old head gasket can be used again if no damage is evident, a new one is advisable. After taking the time to do a top end overhaul, the price is modest.

1. Fit the head gasket over the upper cylinder block face.

2. Loosen intake rocker adjusters completely.

3. Lower the head carefully over the 4 studs and loosely replace the nuts and washers.

4. The 5 cylinder head bolts are replaced next, with the shortest one fitting in the hole at the top of the rod tunnel. The bolt cannot be installed if the pushrods are sitting in the tunnel.

5. Tighten the nuts and bolts a little at a time, beginning with the center bolt at the top of the pushrod tunnel. Moving in diagonal order (**Figure 20**), tighten each nut and bolt gradually to insure even pressure across the cylinder head. This is most important to prevent distortion. Follow the sequence in Figure 20 until all are tightened to 33 ft.-lb.

6. Install the 2 short pushrods in the outer tappets at the bottom of the tunnel. The cupped ends of the rods fit under the intake or rear rocker ball pins.

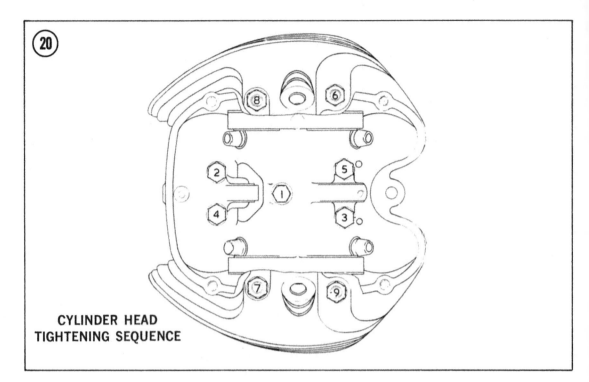

CYLINDER HEAD
TIGHTENING SEQUENCE

7. Reassemble the exhaust rocker assembly in the sequence shown in **Figure 21**. Be sure the spring washers are fitted against the rockers.

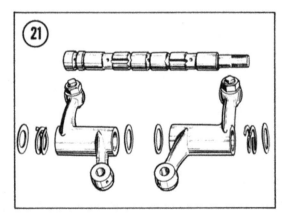

8. Install the 2 longer pushrods in the inner tappets at the bottom of the tunnel. The cupped ends of the rods fit under the exhaust or front rocker ball pins.

9. Squirt engine oil into the rocker oil supply opening at the rear of the head to fill the ducts with lubricant. Reconnect the oil line from the crankcase.

10. Adjust valve clearance, see Chapter Two.

CYLINDER

Cleaning/Inspection

1. Check for worn cylinder bores by placing fingers on top of the piston and pushing from side to side, watching for excessive movement. Worn piston rings are usually indicated by heavy oil consumption. If the valves are in good condition, poor compression can also mean worn rings. Loud piston slap when the engine is warm usually means a worn cylinder bore or possible seizure.

2. Measure cylinder bore with an internal diameter dial indicator. See **Figure 22**. This is done with the piston at the bottom of its stroke. See the Appendix for specifications.

3. Rotate the engine to bring the pistons to the top of their stroke.

4. Plug the pushrod tunnel with a clean rag.

5. Remove the carbon from the piston crowns with a scraper of some soft material. See **Figure 23**. Be sure the scraper is soft because the pistons are made of aluminum and are easily scored.

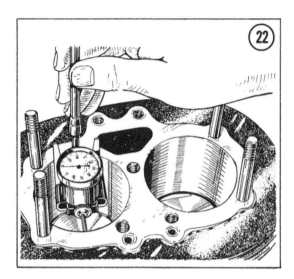

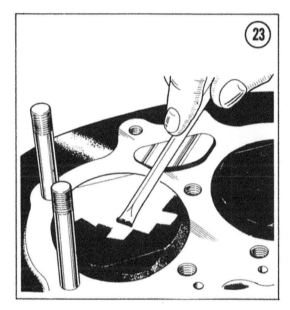

Leave a narrow ring of carbon around the edge of the piston crown and do not remove the ring of carbon at the top of the cylinder bore.

6. Rotate the engine until the pistons are at the bottom of their stroke again.

7. Carefully wipe all loose carbon from the cylinder walls.

8. Clean the joint faces of the cylinder block and head. Be careful not to score the faces with the scraper. Gas leakage could result. Be especially careful of the face on the head, as it is aluminum and softer than the cast iron cylinder.

Cylinder Removal

1. Remove the 8 nuts and rotate the engine until the pistons are at the bottom of their stroke.

2. Lift off the cylinder block carefully. See **Figure 24**. Before the pistons are free from the bores, get someone to hold them to avoid damage to the skirts when they fall free.

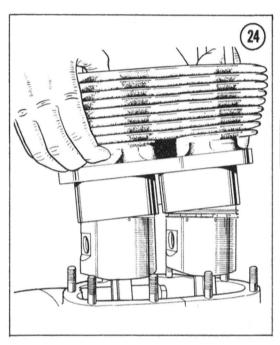

3. Stuff clean rags under the pistons to protect the skirts and prevent debris from falling into the crankcase.

Tappet Removal/Installation

1. Examine both ends for signs of wear or chipping, and be sure they move freely in the block. If the feet are scuffed, the tappets should be replaced.

2. Examine the camshaft for signs of the same types of wear.

3. Drive the tappet out of the block with a soft metal punch on the upper end.

4. Release the circlip from its groove.

5. Remove the tappet from the lower end of its guide.

6. To replace the tappet, insert it into the same guide from which it was removed.

7. Fit a new circlip from above.

Piston Removal/Installation

1. Remove one of the circlips holding the wrist pin in place. This can be done with a pointed instrument (see **Figure 25**) or circlip pliers. It is sometimes necessary to heat the piston by applying a hot electric iron to the crown for a short while, for easier removal of the circlip.

2. Push the wrist pin out and remove the piston from the connecting rod. Heat the piston if force is required.

3. Immediately mark the inside of the piston skirt to indicate which way it sat in the bore and which rod it came from for assembly.

4. Check the piston for high spots indicated by highly polished areas on the sides. These can be smoothed down with a fine file and emery cloth. Be careful not to remove too much material.

5. Use a new circlip and follow the procedure to install the cylinder.

Cylinder Bore Inspection

Examine the bores for signs of wear, scoring, or seizure. Maximum wear will occur at the top of the piston stroke, about ¼ in. from the top of the bore, across the thrust face (front to back of the bore). Bore wear from side to side and at the bottom is usually slight. If the original bore diameter is unknown, measure the difference between the maximum dimension across the thrust faces and at the bottom of the bores. If wear exceeds 0.005 in., reboring is necessary to the next oversize and requires a new oversize piston.

Cylinders of 0.020 in. and 0.040 in. above standard size are available. If the bore is not badly worn, but has deep score marks, it will also have to be rebored.

Piston Ring Inspection

1. The outside faces of the rings should have a smooth, shiny surface. Any discoloration means they must be replaced.

2. When released from the cylinders, the ends of the rings should have a gap of about 3/16 in. They should be free in their grooves, but with minimum side clearance.

3. If rings are stuck in their grooves, they must be carefully removed. Piston rings are brittle with limited flexibility.

4. Scrape all carbon from the ring grooves in the piston and from the back of the rings. An old broken piston ring makes a good tool for cleaning the grooves.

5. Check the ring gap by placing it in the bottom of the cylinder bore with the skirt end of the piston to keep the ring face parallel with the bore. Measure the gap with a feeler gauge. If the gap exceeds the figures listed in the Appendix by much, the ring should be replaced.

Piston Ring Replacement

1. Check the gap of new piston rings. Don't assume it must be right because it's a new part. The most common error is for the gap on a new ring to be too little, rather than too much. If it is less than the minimum given in the Appendix, file the ends until the gap is correct.

2. Piston rings are very brittle and can be easily broken by rough handling. Always check new rings to be sure they fit in their groove before installation. See **Figure 26**.

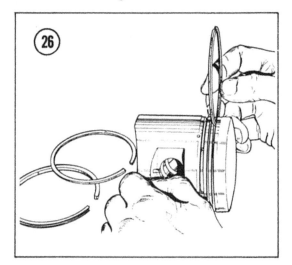

3. The proper sequence for reassembling piston rings is from bottom to top. The scraper ring on the bottom is installed first. The middle ring has a tapered face and is marked by the letter "T" on the face. The tapered edge is placed upward on the piston. The top ring is installed last.

4. Stagger ring gaps evenly around the piston and do not form a channel for gas to escape from the combustion chamber. Failure to do this will result in a loss of compression.

Wrist Pin Rod Bearing Replacement

The small end rod bearings do not usually wear much, but if badly worn, they will emit a high-pitched tapping sound when the engine is running.

1. The wrist pin should be a snug, sliding fit inside the bearing.

2. The old bearing is extracted and the new one installed at the same time. Be sure to line up the oil hole in the new bearing so it will correspond to the hole in the rod, before installing. Lack of lubrication for the bearing will cause serious damage.

3. Ream the new bearing to 0.7503-0.7506 in.

Cylinder Installation

1. Coat pistons completely with clean engine oil.

2. Space the piston ring gaps evenly around each piston at 120 degree intervals to restrict gas leakage through the gaps.

3. Use a ring compressor to compress the rings until they will slip inside the bores of the cylinder block. The tool must be loose enough to allow the rings to slide up as the block is lowered. A pair of hose clamps make good ring compressors.

4. Lower the cylinder block over the pistons carefully. When the rings have entered the bores, remove the compressors from the pistons. See **Figure 27**.

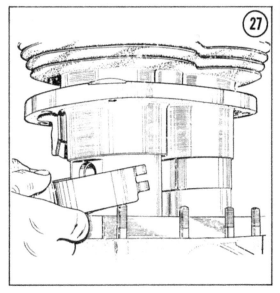

5. Remove the 2 pieces of hardwood and lower the block carefully onto the crankcase.

6. The 8 nuts holding the cylinder block must be tightened slowly and evenly to prevent distortion. Torque to 30 ft.-lb.

ALTERNATOR

Removal/Installation

The alternator rotor is connected to the crankshaft, while the starter is mounted by 3 studs in a housing in the crankcase.

1. The alternator must be removed before the clutch sprocket, primary chain, and crankshaft sprocket can be removed. The primary chain is endless, so the drive components must be removed as a unit and installed together.

2. Support the wrist pins through the connecting rod with hardwood pieces across the crankcase face. Straighten the lockwasher under the crankshaft nut and remove right-hand threaded nut.

3. The stator is fastened with self-locking nuts. Remove them and the stator.

4. Be careful with the stator. The electrical leads run from the outside face through a grommet at the rear of the case. It must be reassembled the same way to be sure the wires don't catch in any moving parts.

5. Remove the rotor from the crankshaft and remove the crankshaft key. See **Figure 28**.

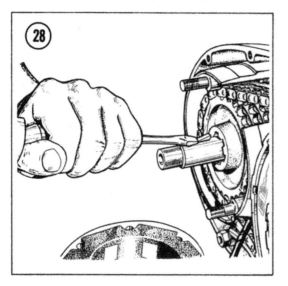

6. Release the clutch hub from the transmission. The main shaft sprocket, primary chain, and clutch sprocket can now be removed together.

7. Remove the key from the transmission main shaft.

8. Reverse the procedure to install.

CLUTCH

Operation

The clutch transmits power from the engine to the transmission by engaging the driving and driven plates. When the clutch lever is depressed, the plates disengage.

A cam thrust mechanism inside the right crankcase cover actuates the pushrods inside the hollow main shaft. The rod pushes the clutch pressure plate, compressing the springs and separating the driving and driven plates.

The cam thrust mechanism consists of 2 plates separated by 3 steel balls which sit in cups in the plates. See **Figure 29**. One plate is anchored to the crankcase cover while the other is attached to the clutch control cable. The cable rotates the plate, separating the 2 plates and operating the pushrod.

There must be a small amount of clearance between the pushrod, the camshaft mechanism, and the pressure plate.

Sprocket

Inspect the clutch sprocket carefully. The teeth should be even and regular. If any sign of wear is visible, the sprocket should be replaced.

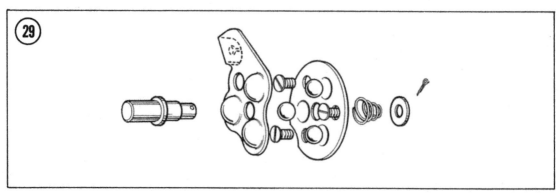

Reassemble the clutch center roller bearings and sprocket temporarily to check the sprocket bearing for tightness. If much play is evident, replace the bearings. The condition of the sprocket and center can also cause play. If so, they must also be replaced.

Dismantling

1. Drain the oil from the chaincase as described in Chapter Two.

2. Remove the screws holding the chaincase cover in place. Be sure not to lose the lockwashers. The screws are not all the same length, so remember which length goes where when reassembling.

3. To break the joint, tap the chaincase cover lightly with a soft mallet until it releases. The cover has a tubular dowel at the front, so tap at the rear and remove rear end first (**Figure 30**). Be careful not to bend the ignition timing pointer in the front of the case.

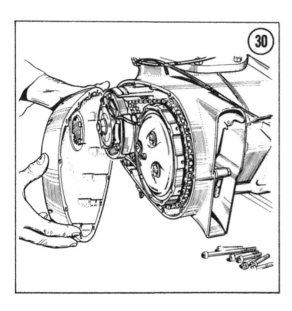

4. Remove the 3 spring retaining nuts on the face of the clutch. They have the normal right-hand thread, but they are difficult to remove because a projection on the nut engages with the spring. The spring may be destroyed in the removal and have to be replaced.

5. Remove the springs and cups. Remove the pressure plate and the remaining clutch plates. If these are the only items requiring repair, don't dismantle the clutch any farther.

6. To unscrew the clutch center nut, the clutch sprocket and the center must be locked together with a special service tool No. 61-3768. A suitable substitute can be made using parts from an old clutch. The wrist pins must be left in the connecting rods, with hardwood pieces between the pins and the crankcase face.

7. Unscrew the center nut. It has a self-locking nut with a right-hand thread, with a spacing collar behind it.

8. Remove the pushrod, but do not try to remove the clutch sprocket now.

9. Back off the adjusting screw for the chain tensioner until the tensioner blade no longer touches the chain.

Inspection

1. Examine the driving plates. See **Figure 31**. There are 6 driving plates and 6 driven plates, even though Figure 31 only shows one of each. The driving plates should be in good condition, with no friction material missing or damaged.

2. The driving plate tongues must slide freely and have no burrs or irregularities.

3. If the plates appear in good condition, measure the thickness of the plate and the friction material. The standard is 0.140 to 0.145 in. If wear is more than 0.030 in., the plates must be replaced.

4. The driven plates, Figure 31, are made of steel and should be flat and free from scoring. To test for flatness, lay each plate on a piece of plate glass. If it can be rocked, it is warped and must be replaced.

Transmission Shock Absorbers

The shock absorbers, in the clutch hub center, transmit power from the clutch sprocket through the clutch plates to the transmission main shaft. The shock absorber unit consists of 3 large

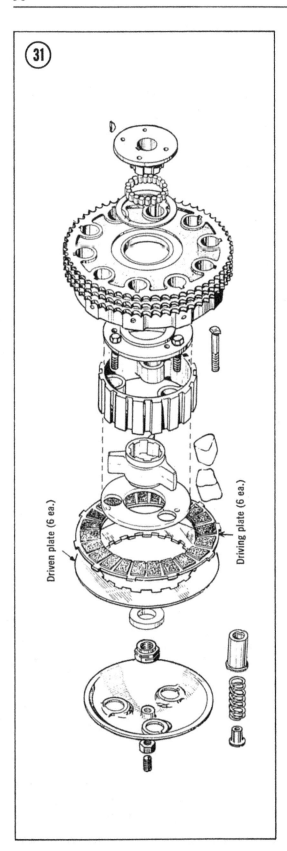

Driven plate (6 ea.)

Driving plate (6 ea.)

rubber pads engaging a 3-armed metal spider splined to the clutch center. The center is fastened to the transmission main shaft by a locking taper and key. There are also 3 smaller rubber rebound pads. The effect is to reduce the variations in engine torque at low rpm, reducing the strain on the transmission.

The shock absorber is a closed unit, with the bolts holding it together riveted over after assembly. To inspect it, the bolts must be drilled away. New bolts must then be used when reassembling, and the ends riveted over.

On earlier models, the shock absorber can be inspected by removing the 3 countersunk head screws next to the clutch spring cups and removing the plate. The units are interchangeable, but as a complete unit only.

The rubbers should be firm and springy to the touch. If they are deteriorating, they must be replaced. To remove the rubbers the smaller rebound ones come out first. To fit new rubbers in, liquid soap will help. Do not use oil or grease which will damage the rubber.

Final Drive Sprocket
Removal/Installation

1. The final drive sprocket is located inside the primary chaincase on the left side. In order to gain access to it, remove the clutch.

2. Remove the 6 screws holding the circular plate at the rear of the primary chaincase. See **Figure 32**.

3. Check that the joint of the plate has an oil seal which is in good condition.

4. Check the back of the plate for oil leakage. If any shows, the seal must be replaced. Be sure it is fitted with the lip toward the inside of the chaincase and not toward the final drive sprocket. Check that the seal lip has not scored the shaft upon which it bears.

A felt washer fits between the sprocket nut and the oil seal. It is important that this washer be in good condition, because it prevents road debris from damaging the seal. If the washer is worn, replace it.

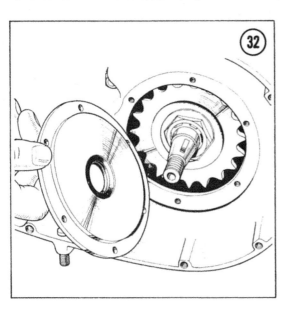

5. Remove the final drive sprocket with a puller. Be careful that the teeth are not damaged. If no puller is available, wrap a length of chain around the sprocket and secure the ends with a bolt. Flatten the lockwasher and remove the nut. Then ease the sprocket off its splined shaft slowly. Once it is free, it comes off easily.

6. Check the back of the sprocket area for signs of oil leakage. If any are present, the oil seal from the transmission is faulty and must be replaced. Be sure the sprocket boss is not worn or damaged. Never reuse a part that can cause oil leakage.

7. Oil the sprocket boss with engine oil to prevent damage to the transmission oil seal.

8. Press the sprocket back onto its splined shaft. The gearbox sprocket nut has a right-hand thread and must be torqued to 100 ft.-lb. Bend the lockwasher over one of the flat edges of the nut so it cannot turn.

Primary Drive Assembly

1. Replace the cover over the final drive sprocket at the rear of the primary chaincase. Use gasket sealer on one side of the paper gasket.

2. Grease the roller track. Assemble the rollers and slide the clutch sprocket over them. See **Figure 33**.

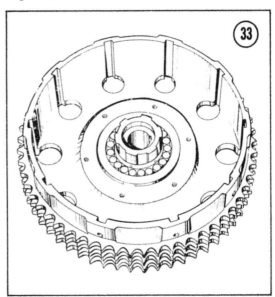

3. Replace the thick driving rubbers (**Figure 34**) on the right side of the center vanes and the thin rebound rubbers on the left side.

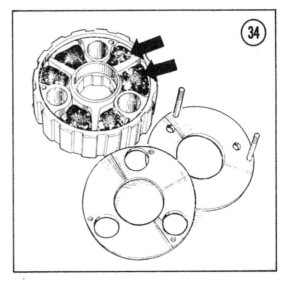

4. Use new bolts and rivet the ends. On earlier models, the plates are fastened by 3 screws secured by Loctite.

5. Insert the spring bolts from the rear and slide the shock absorber assembly into the clutch sprocket.

6. Install the clutch sprocket, primary chain, and crankshaft sprocket temporarily in place. See **Figure 35**. Be sure the chain between the 2 sprockets is properly aligned. Shims of 0.010 in. and 0.015 in. can be purchased for alignment if necessary.

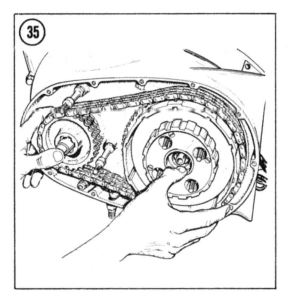

7. Remove the 2 sprockets and chain.

8. Thread the stator wires through the guide tube in back of the chaincase and slide the rubber sleeve over the guide. Clip the wires to the side of the generator housing.

9. Replace the key in the transmission main shaft.

10. Pass the stator plate through the primary chain between the crankshaft sprocket and clutch sprocket.

11. Slide the sprockets and chain into place.

12. Place the clutch spacing collar on the shaft with its chamfered bore facing out and secure with the shaft locking nut. Torque to 65 ft.-lb.

13. Place the Woodruff key in the crankshaft and slide the ignition rotor on. See **Figure 36**. Screw on the nut and washer. Torque to 60 ft.-lb. Be sure to bend the lockwasher over one of the flat edges of the nut.

14. Slide one driving plate with the bonded friction material into the clutch sprocket.

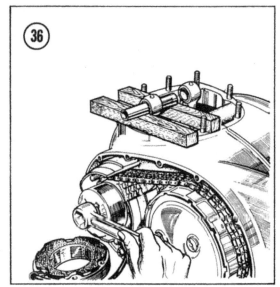

15. Place one driven plate on the clutch center. Repeat this sequence until all 6 of each plate are installed.

16. Fit the clutch pushrod into the transmission main shaft and then install the pressure plate, spring cups, springs, and nuts. Be sure the spring nuts are tightened evenly so the plates are parallel.

17. Install the stator on the 3 studs with the leads at the 2 o'clock position. Tighten the self-locking nuts to 15 ft.-lb.

18. The gap between the rotor and stator must be even all the way around. Check with a feeler gauge. The gap between the stator poles and the rotor should be 0.008 in.

19. Install the chain tensioner (**Figure 37**) and adjust until the top run of the chain has $1/8$-$1/4$ in. play by turning the screw and then tightening the locknut and the cap nut. The washer prevents oil leakage.

20. Smear gasket cement on both faces of the chaincase and install a new gasket, tightening the cover screws evenly to avoid distortion of the cover. Some of the screws are of different lengths.

21. Check the oil level and drain screws to be sure they are snug.

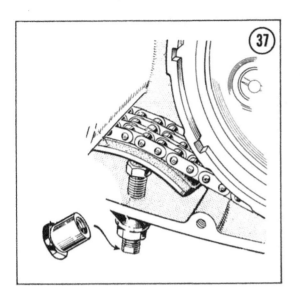

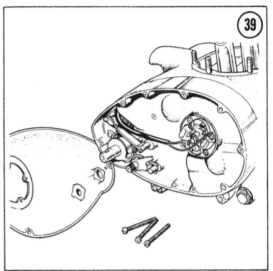

RIGHT CRANKCASE COVER

Removal/Installation

1. Remove the gearshift lever and kickstarter.

2. Remove the 6 screws holding the outer cover and take off the cover. See **Figure 38**. The clutch operating mechanism, kickstarter shaft, gearshift lever shaft, and contact breaker assembly are all behind the cover. See **Figure 39**.

3. Leave the clutch cable attached to the operating mechanism unless it requires service.

4. Take off the starter lever spring and anchor plate together.

5. Check the positions of the plate tongue and the spindle flats. Do not remove the screw (A, **Figure 40**) which holds the inner end of the spring.

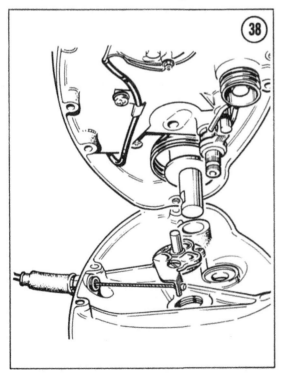

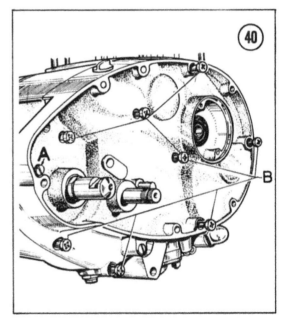

6. Remove the nut holding the gearshift return spring and cup.

7. Loosen the set screw in the body of the return lever and pull it off its shaft.

8. The clutch operating mechanism is attached to the outer plate and does not require service.

9. Reverse the procedure to install.

INNER CRANKCASE COVER

Removal

1. Straighten the tabs on the lockwashers and remove the 2 bolts holding the tachometer cable in place.

2. Remove the adapter, gasket, and O-ring oil seal. Be sure these are in good condition if they are to be used again.

3. Remove the driving spindle of the cable from the crankcase. When reassembling, the spindle tongue must engage with the driving slot in the oil pump spindle.

4. Remove the screws holding the inner cover. The screws are of different lengths, so note which one goes in each hole for reassembly. Do not remove the slotted screw.

5. Tap the cover gently with a soft mallet around the edges to break it loose. Remove the cover, leaving the gearshift quadrant and starter spindle in place.

6. The starter spindle may come off with the cover, but it is easily removed. Depress the 2 spring-loaded plungers on the gearshift quadrant until they clear the cam plate and it can be removed.

Installation

1. Replace the kickstarter and gearshift spindles and then replace the inner timing cover using a new gasket, as described earlier. See Figure 40.

2. Replace the gearshift lever, making sure it fits on the spindle. Put the return spring loosely over its stud with the legs of the spring on each side of the gearshift lever. See Figure 38.

3. Replace the spring cup and washer, rotating the cup until the gearshift lever can move the correct distance in both directions. Tighten the self-locking nut.

4. Tighten the small screw in the gearshift lever.

5. Replace the kickstarter spring with the hook on its inner arm on the screw head (see Figure 40). Place the stop plate on the flat of the spindle. Wind the spring until its outer arm is anchored by the tongue.

6. Replace the tachometer cable as described earlier.

7. Replace the ignition automatic advance assembly loosely (**Figure 41**), along with the contact breaker plate. The engine will have to be timed later as described in Chapter Two. Feed the wires through the grommet at the rear of the cover.

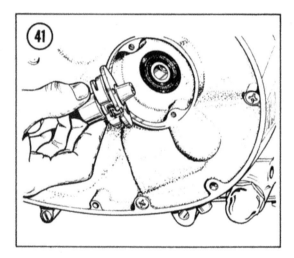

8. Replace the primary drive assembly as described earlier.

9. Reassemble the upper end of the engine as described in the earlier section on *Top End Overhaul*.

CONTACT BREAKER

Operation

The contact breaker assembly sits in its own housing inside the case cover on the right side of the engine. To gain access, remove the 2 screws holding the circular cover on the timing case cover.

The automatic advance mechanism is located behind the carrier plate. It has governor weights (**Figure 42**) which are controlled by springs attached to the cam. It is fastened to the tapered hole in the idler gear by a central bolt.

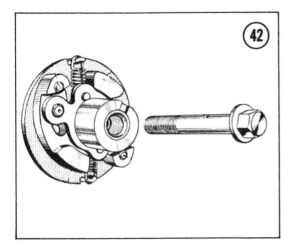

Engine oil is prevented from reaching the assembly by a seal in back of the housing. If the seal has to be replaced, the lip must point toward the inside of the engine.

For starting or running the engine slowly, the springs hold the governor weights in the inner or retarded position. It makes starting easier and prevents the starter lever from kicking back sharply.

As engine speed increases, centrifugal force forces the governor weights outward against the spring tension, rotating the cam to the advanced ignition position.

Removal/Installation

Removal of the contact breaker is simple. If it is done correctly, the ignition timing will not have to be readjusted.

1. Make a mark across the carrier plate and housing so it can be reinstalled in exactly the same position.

2. Remove the 2 screws (**Figure 43**) and remove the assembly as a unit.

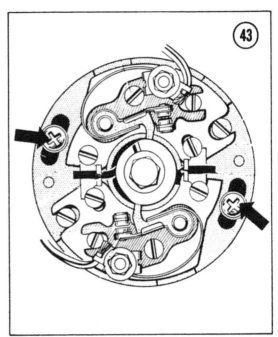

3. Do not remove the automatic advance assembly unless absolutely necessary. The timing must always be reset afterward.

4. To remove the automatic advance, unscrew the center bolt.

5. Screw a puller in until resistance is felt. Tap the end of the tool with a hammer to release the unit and pull it out.

6. The contact breaker has 2 sets of points (**Figure 44**). Remove the nut and washer holding the spring and wire on the terminal post and lift off the moving contacts.

7. Remove the screws to remove the stationary contacts.

8. Follow the same sequence in reverse when assembling. Don't forget the insulating bushing and fiber washer with the moving contact. Otherwise, a short circuit will occur.

9. Place the fiber washer over the terminal post before replacing the spring pivot. Install the lead connection, with the head of the insulating bushing above the connection.

Adjusting Point Gap

See Chapter Two under *Breaker Points*.

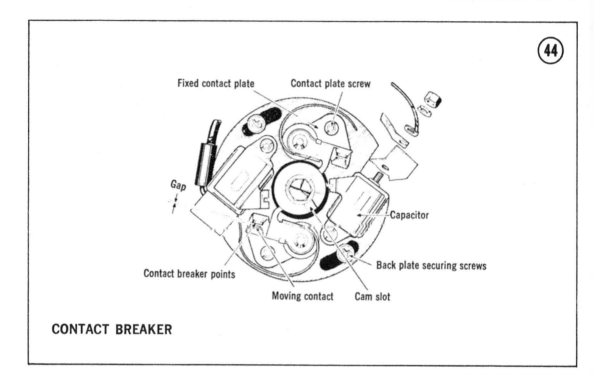

Fixed contact plate   Contact plate screw

Gap

Capacitor

Back plate securing screws

Contact breaker points

Moving contact   Cam slot

CONTACT BREAKER

TIMING GEAR AND OIL PUMP

BSA motorcycles use dry sump lubrication. Oil is contained in a separate reservoir outside the engine. In later models, the main frame tube serves as the reservoir (**Figure 45**). Earlier models are fitted with an oil tank. A gear pump draws oil from the reservoir for lubrication of the load bearing surfaces in the crankcase and cylinder walls. The oil then drains into the sump in the bottom of the crankcase, where another gear pump returns it to the reservoir.

Part of this return flow is tapped to provide the lubrication for the valve assembly in the cylinder heads. Oil enters the rear of the cylinder head and is metered by a split pin (**Figure 46**) to ducts which carry it to the necessary parts. Since the return pump has a greater capacity than the supply pump, the sump remains dry at all times. Advantages of this system are that the oil is constantly circulated through a series of filters which clean it and is cooled in the reservoir before being returned to the engine.

The oil pump, timing gear, starter ratchet assembly, and the transmission outer cover are exposed as outlined in the previous section.

Removal

1. Place the hardwood pieces between the wrist pins in the connecting rods and the crankcase face. Straighten the tab on the lockwasher and remove the crankshaft nut and the worm behind it. Both the nut and worm have left-hand threads, so they must be unscrewed in a clockwise direction.

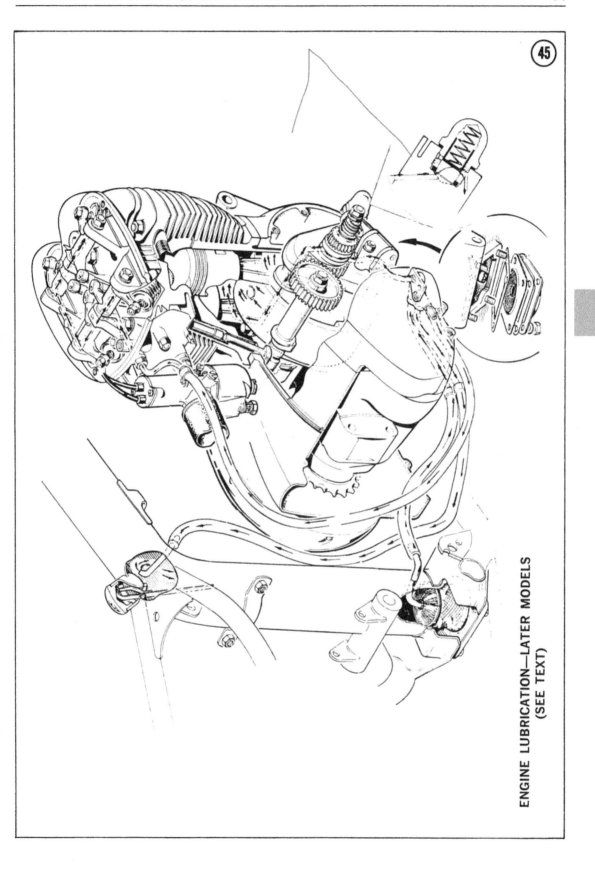

ENGINE LUBRICATION—LATER MODELS
(SEE TEXT)

6

2. Remove the plain washer between the worm and the crankshaft pinion.

3. Remove the 3 self-locking nuts and the oil pump. This will release the non-return ball valve and spring behind the pump (**Figure 47**).

4. Do not dismantle oil pump unless necessary.

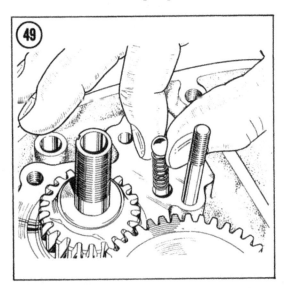

Installation

1. Be sure the mating faces of the crankcase and oil pump are clean.

2. Put a new gasket on the pump and fit the non-return valve and spring into the crankcase oil duct.

3. Put the plain washer against the pinion face and install the oil pump with the worm drive gears meshed together.

4. Slide the pump carefully until the worm is fully engaged with its thread. The pump spindle will rotate as the worm is screwed onto its shaft.

5. Be sure the valve is in the countersunk hole on the pump face and then tighten the 3 nuts evenly and slowly. Torque to 10 ft.-lb.

6. Tighten the worm gear, replace the lockwasher with a new one, and replace the crankshaft nut. Torque the worm gear and nut to 35 ft.-lb. Bend the tab on the lockwasher over one of the edges of the nut.

Timing Gear Removal/Installation

The timing gears have marks on their faces next to the teeth. These assist in proper assembly, so note their positions before removing. Removal of the camshaft nut also requires the wrist pins to be supported by hardwood pieces over the crankcase face.

1. Straighten the tab on the washer and unscrew the nut.

2. Extract the camshaft gear.

3. Remove the Woodruff key.

4. Remove the idler gear.

5. Remove the crankshaft nut and oil pump worm gear as previously described.

6. Remove the crankshaft pinion (**Figure 48**).

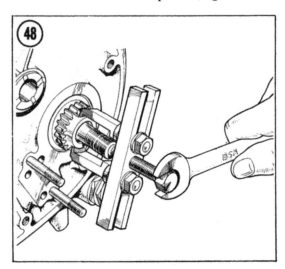

7. When replacing timing gears, the marks on the camshaft gear and the crankshaft gear cannot be changed. These 2 gears must be installed first, with the idler gear last and the timing marks carefully matched. See **Figure 49**.

TRANSMISSION

Operation

To understand how the transmission operates, you will have to refer to figures included and understand some of the terms used. See **Figures 50-54**.

1. The cam plate is shaped like a fan with notches at one end and slots at the other.

2. The large plunger is at the notched end of the cam plate and locates the gear positions.

3. The selector plungers are at the other end of the cam plate. They are attached to the selector quadrant and engage the slots in the plate.

4. The shifting forks are shown as spots in the wavy lines of the cam tracks at the large end of the cam plate. These are actually the rollers that move the shifting forks on their shaft, engaging or disengaging pairs of pinions.

5. Sliding gears move on the shafts, controlled by the shifting forks.

6. The transmission must be in the neutral position before starting the engine. See Figure 50. The large plunger locates the cam plate at the second or neutral notch. The selector quadrant plungers at the other end of the plate are compressed and ready to operate.

7. When the gearshift lever is pushed down, the selector plunger engages the cam plate slot to move it to first gear position. The large plunger moves to the first gear notch. The layshaft shifting fork will mesh the layshaft sliding gear with the layshaft first gear. See Figure 51. In this figure, the selector plunger has already engaged the second slot in the cam plate, ready to move it from first to second gear position.

8. To shift to second gear, the gearshift lever is pushed up. The cam plate moves in the opposite direction, causing the layshaft shifting fork to move the sliding gear in the opposite direction, meshing with second gear. See Figure 52. In this figure, the selector plungers have engaged 2 slots in the cam plate, ready to move second to first or neutral or back again. The large plunger has moved from the first gear notch past neutral to the second gear notch.

9. The gearshift is pushed up to shift from second to third. Both shifting forks move. One moves the layshaft sliding gear to neutral position while the other meshes the main shaft sliding gear with the main shaft third gear. See Figure 53. The large plunger moves to the third gear notch. The selector plungers are ready to move the cam plate either way.

10. The gearshift is pushed up again for fourth gear. See Figure 54. The main shaft shifting fork moves the sliding gear the opposite way to mesh with the sleeve pinion. The selector plungers are ready to move the cam plate back to the third gear position now. The large plunger has moved to the fourth gear notch.

Disassembly

It is not necessary to remove the gear cluster in order to gain access to the connecting rods or flywheel assembly. The gear cluster can be removed without removing the timing gears, but the primary drive and clutch must be removed first, as described earlier in this chapter. All transmissions are essentially the same as the 4-speed in design. Additional gears are added to the clusters.

1. Remove the 5 nuts and spring washers holding the transmission end cover.

2. Break the joint loose carefully with a soft mallet, because the transmission cover and case are dowelled together.

3. Remove the cover, gear cluster, cam plate, and selector forks as a unit.

4. Remove the fulcrum pin.

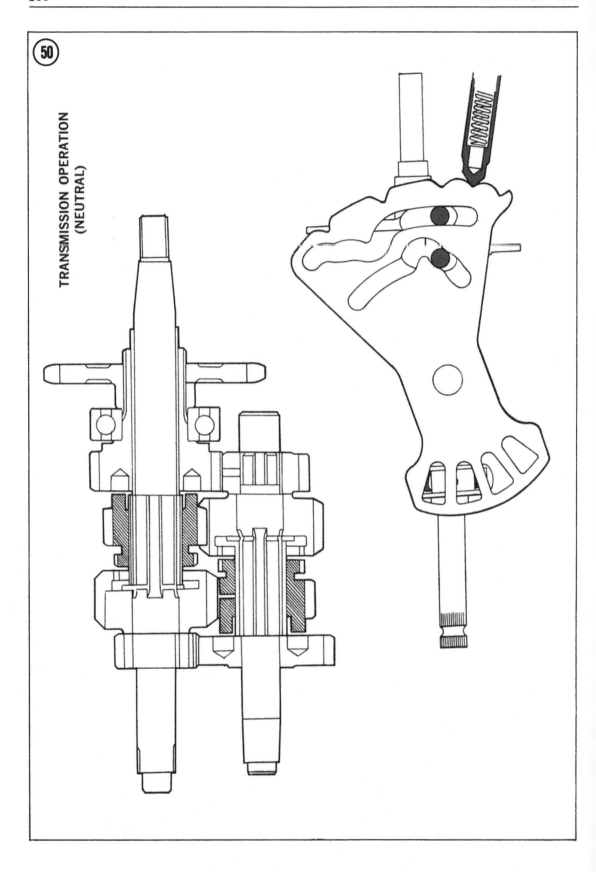

TRANSMISSION OPERATION
(NEUTRAL)

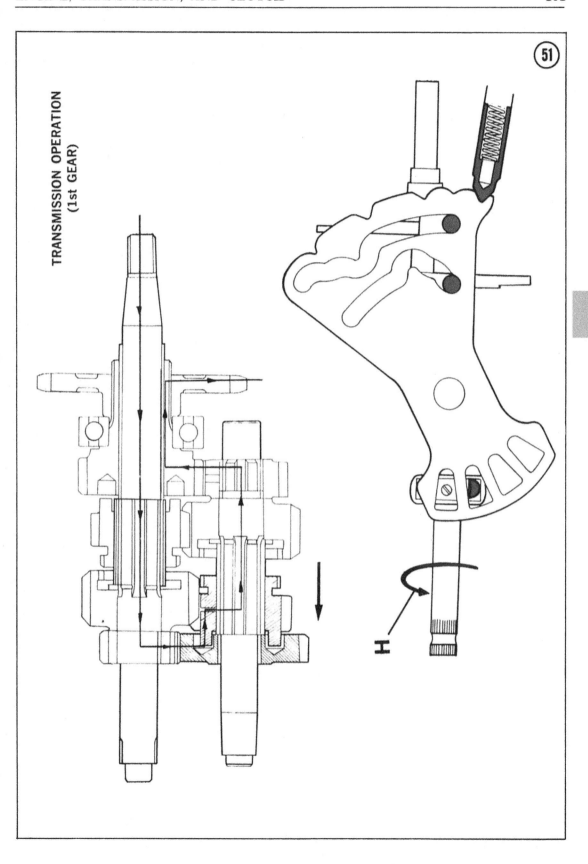

TRANSMISSION OPERATION
(1st GEAR)

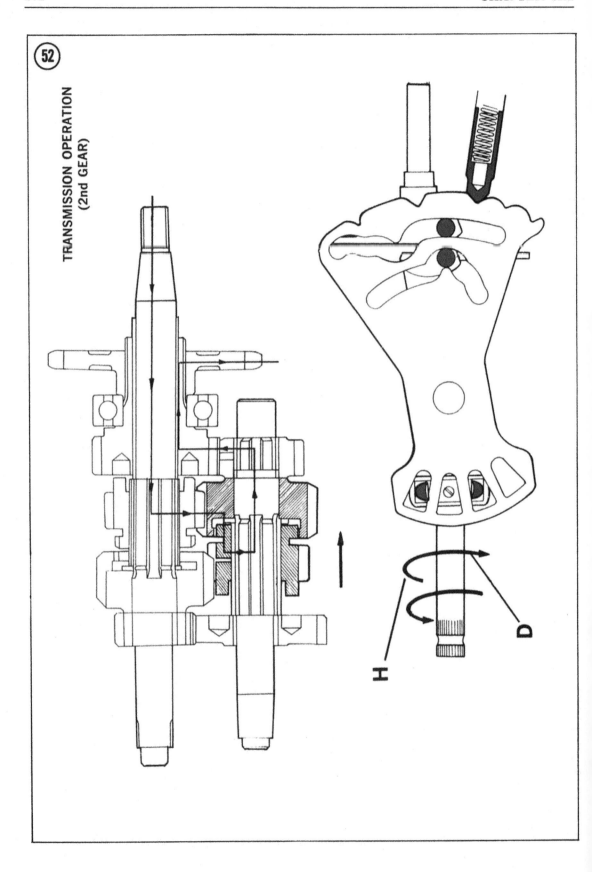

TRANSMISSION OPERATION
(2nd GEAR)

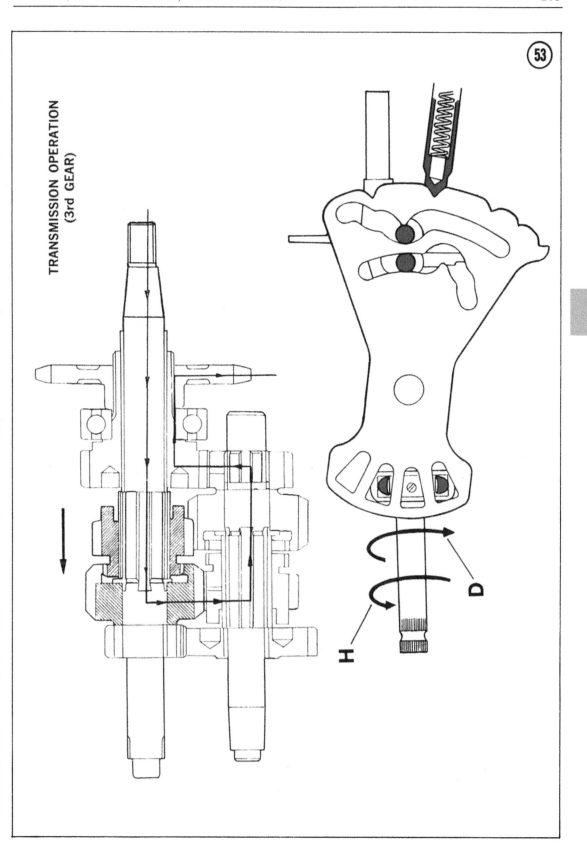

TRANSMISSION OPERATION
(3rd GEAR)

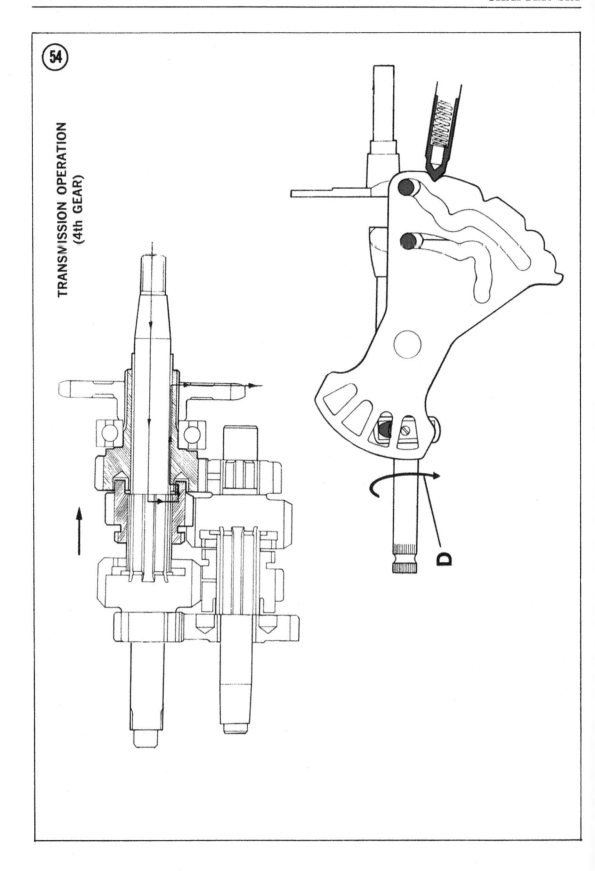

5. Remove the cam plate, selector forks, and shaft.

6. Extract the layshaft and remove the sliding third gear pinion and the first gear pinion. See **Figure 55**.

7. Hold the main shaft in a vise with soft metal clamps to protect it. Straighten the tab washer on the ratchet nut and remove the nut.

8. Remove the ratchet assembly (**Figure 56**).

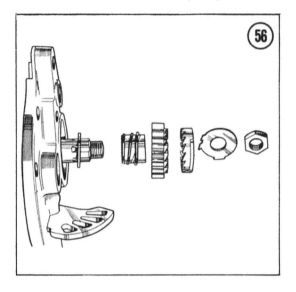

9. Drive the main shaft and gears out of the bearing with a lead or copper hammer so you do not damage the threads.

10. Remove the sliding gears. The main shaft now has first and third gears and the layshaft second and fourth. The smaller gear on both shafts is pressed on, holding the larger gear in place with a thrust washer between it and the end of the splines.

11. Examine the gears for any sign of wear or damage to the teeth. If any is present, the gear should be replaced by a professional mechanic.

Bearings

1. Examine the bearings and bushings for wear. Be sure to check the bronze bushings on the layshaft first gear and the main shaft fourth gear. See **Figure 57**. The layshaft rides in roller needle bearings on each end. One bearing is housed in the transmission case and the other in the end cover. To remove, they must be driven out with a drift.

2. The main shaft rides on ball bearings on each end. The transmission sprocket must be removed, as described earlier in this chapter, before the left end bearing can be removed.

3. Drive the top pinion gear through the bearing into the transmission.

4. Remove the oil seal and circlip, then press out the bearing.

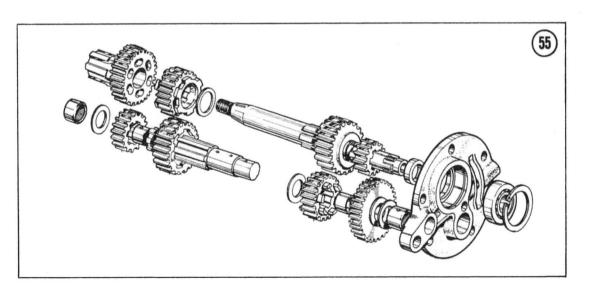

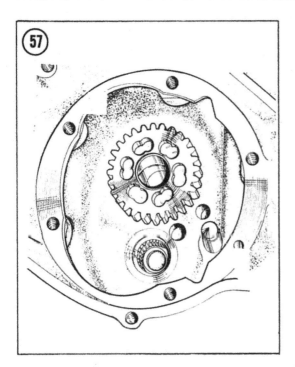

5. When reassembling, use a new oil seal with its lip inside the transmission. Always heat the aluminum case, including the transmission case, before removing a bearing or bushing. Press out the old item and press in the new while the case is still warm.

6. Remove the circlip and the right main shaft bearing from inside the end cover.

7. Check the cam plate plunger (Figure 57) in the left side of the transmission case. It should be free in its housing and the claws of the shifting forks should be in good condition, not chipped or worn, and should slide freely. The shifters are secured by a small plate and screw.

Assembly

Only good or new parts should be used for assembly, including gears, bearings, bushings, and seals.

1. Clean all the old cement off the faces of the case and end cover.

2. Put the main shaft into its bearing in the end cover. Be sure the spacer is between the bearing and the first gear. See Figure 55.

3. Use a vise with soft metal clamps to hold the shaft. Replace in order, the washer, bronze bushing, spring, ratchet pinion, ratchet, tab washer, and nut. See Figure 56. Torque the nut to 60 ft.-lb.

4. Temporarily assemble the layshaft in the transmission to be sure the thrust washer at each end gives proper clearance. Replace the first pinion gear and thrust washers and end plate. The movement at the end of the shart should be just perceptible. Replacement washers are available in thicknesses of 0.120-0.122 in. and 0.127-0.129 in.

5. After checking layshaft clearance, assemble the gear cluster on the end cover (**Figure 58**).

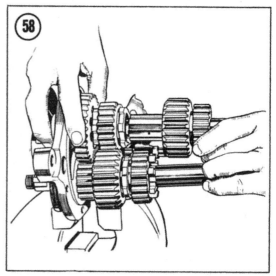

6. The main shaft second gear and the layshaft sliding third gear are interchangeable with standard ratio transmissions, but the shifting forks are different. With the forks on the shafts and both rollers below the shaft (**Figure 59**), the layshaft fork is on the left and the main shaft fork on the right.

7. Put one of the standard thrust washers on the inside face of the cover.

8. Install the layshaft first gear with its plain face against the washer.

9. Install the layshaft with its sliding gear and

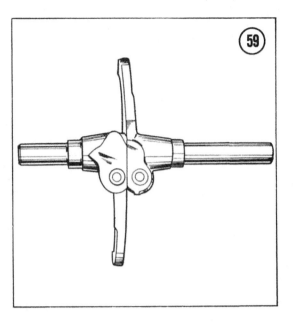

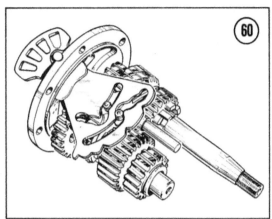

second and fourth gears. Be sure the sliding gear is installed on the shaft with its oil hole between 2 teeth aligned with the hole in the shaft.

10. Engage the shifting fork on the layshaft with its sliding third gear. Add the main shaft sliding gear and engage its shifting fork.

11. Slip the cam plate through the slot in the end cover, with the long end of the outer track on the bottom.

12. Adjust the sliding pinions until the selector rollers are in their tracks.

13. Install the shifting fork spindle and the cam plate fulcrum pin.

14. Press the main shaft fourth gear into its bearing and install it. Install the transmission sprocket and locknut and tab washer as described earlier in this chapter.

15. Be sure the cam plunger and spring are installed in the left side of the transmission case. Adjust the cam plate and sliding pinions until the shifting rollers are in the neutral position. See **Figure 60**.

16. Put the other thrust washer on the end of the layshaft and slide the gear cluster assembly into the transmission case with cement on both faces.

17. Revolve the shafts gently until the fourth gear and pinion mesh. When the end cover is seated, tighten the nuts evenly to 20 ft.-lb.

18. Install the gear selector quadrant spindle into the bearing in the end plate and engage the quadrant plungers in the cam plate (**Figure 61**). A thin strip of metal, as shown in the figure, will help this operation.

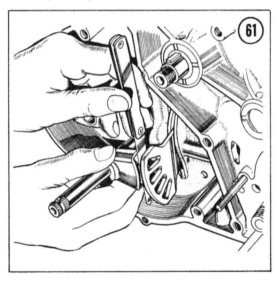

19. Install the small end of the kickstarter quadrant in the steel bushing on the inside of the cover plate before replacing the inner timing cover.

CRANKCASE

Before splitting the cases, the timing gear and primary drive must be removed as described

6

earlier. While the cases can be split without removing the gear cluster, it is always better to remove them for a thorough examination during a major overhaul.

Crankcase Splitting

1. See **Figure 62**. Remove the 2 nuts from the lower front (B) and upper front (C) of the crankcase. Remove the 3 stud nuts (A).

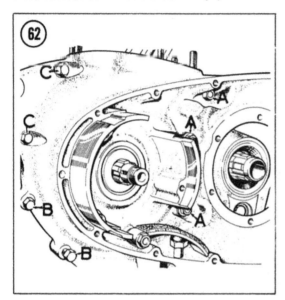

2. Remove the nut and washer from the bridge across the mouth of the case.

3. Remove the 4 nuts holding the crankcase sump and remove the cover, gaskets, and filter. See **Figure 63**.

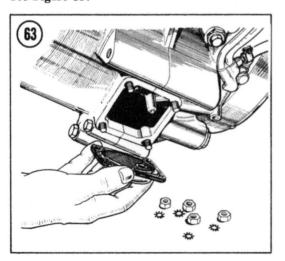

4. The oil return scavenge pipe can be left in place, but be careful not to damage it.

5. Remove the Woodruff keys which may be still in place in the shafts.

6. Tap the crankcase gently with a soft mallet to break the joint and remove the right half. Do not pry between the faces with any kind of sharp edge.

7. The left side crankshaft bearing will remain on the shaft.

Camshaft Removal

When the camshaft is removed, the breather valve and spring may remain attached to it or attached to the left half of the case. See **Figure 64**. The breather valve is a rotary disc, with the rotating half driven by the end of the camshaft and the stationary half held by a peg below the left side camshaft bearing. The bearing must be removed to get at the stationary half.

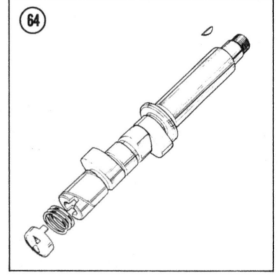

Examine the camshaft for signs of wear or chipping. If the lobes are worn, the valves will not open completely, causing poor performance. In this case, the camshaft should be replaced.

Remove camshaft from the right side of the case. See *Crankcase Assembly* for installation.

Crankshaft Removal

When removing the connecting rods from the crankshaft (**Figure 65**), note that the rods and big end caps have been marked with serial numbers. They must be assembled with the numbered faces together. For proper lubrication, each big end assembly must be put back on the crankshaft journal from which it was removed. The same bolts must be used in the same positions on the same rod. See **Figure 66**.

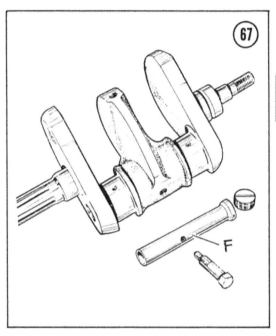

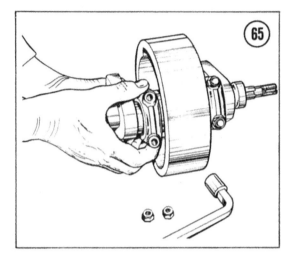

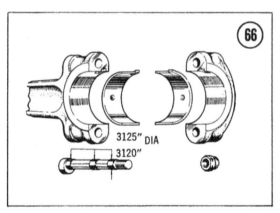

If the crankshaft bearings are reground, be sure to use the correct sizes to get the proper clearance with the undersize big end bearing shells.

Crankshaft Grinding

If the wear of the crankpins or right side journal exceeds 0.002 in., or if the surfaces have

been scored for any reason, they must be reground. Worn bearings will develop a knocking sound and the engine will run rough. Refer such work to a mechanic.

Crankshaft Centrifugal Filter

1. Clean the centrifugal filter while the crankshaft is out of the case. The filter (**Figure 67**) is a tubular sleeve that fits inside the crank journals.

2. Remove the end plug in the right side web and the flywheel bolt next to the crank journals. Remove the filter.

3. Clean all crankshaft oil passages with solvent and blow out with high pressure air.

4. When replacing the filter, install the flywheel bolt loosely to locate it. Install the filter and then the plug. Both the bolt and the plug should be secured with Loctite. Torque the flywheel bolt to 45 ft.-lb.

Crankshaft/Flywheel Balancing

The crankshaft and flywheel assembly (**Figure 68**) are balanced at the factory. They should not require rebalancing when installing new pistons or rods. If balancing is required, it must be done by a skilled mechanic with precision machinery.

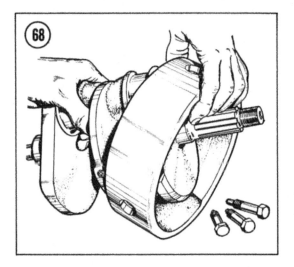

Bearing, Bushing, and Oil Seal Replacement

1. Examine all bushings and bearings and replace any that are worn.

2. Check the journal bearings for roughness indicating damaged balls, rollers, or tracks.

3. Bearings and bushings can be pushed out, but the crankcase must always be heated first and firmly supported. Replace the bearing or bushing while the case is warm (**Figure 69**).

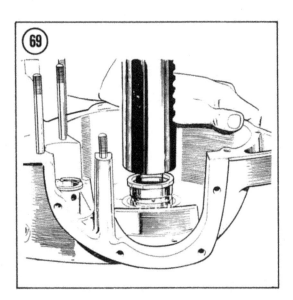

4. When the crankshaft is removed, the inner race and rollers will come out with it. To remove the outer race, the crankcase must be heated and the race released from its housing.

5. Rmove the camshaft bushings from the left case. The replacement bushings must be reamed to specifications given in the Appendix.

6. The crankcase halves must be bolted together after the bushings have been changed to ream the bushings. Ream to the required size, separate the case halves and very carefully remove all metal filings from inside with high pressure air.

7. Oil seals must always be handled carefully to avoid damaging the knife edge lip. Press into the housing squarely with the lip toward the part that is to be sealed.

If any part has been scored by an oil seal, it is useless to replace the seal and not the part. See **Figure 70**. Always replace both seal and part or oil leakage is sure to result.

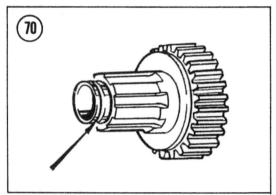

8. Check that all oil passages are clean and clear and the oil pump non-return valve in the crankcase sump is free to move in its seat.

Crankshaft Assembly

1. The maximum end float on the end of the crankshaft must not exceed 0.003 in. The end float is controlled by shims (**Figure 71**) fitted between the inner race of the left roller bearing and the crankshaft web. If the same crankshaft is being used again, reuse the same shims.

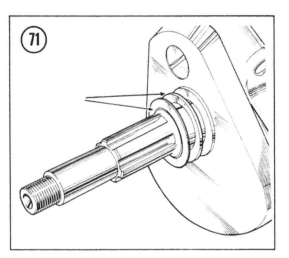

2. If the crankshaft has been ground or a new one is installed, it must be bolted tightly into the cases and the end float checked.

3. New crankcases and shafts are marked to make shimming easy. The case has a 3 digit number marked across the front mounting lug; the shaft number is on the flywheel. Subtract the number on the shaft from the number on the case to get the difference in measurement. Add shims to allow 0.0015 in. to 0.003 in. clearance. As an example, if the case measurement is 0.810 in. and the shaft is 0.791 in., the difference is 0.019 in. By adding 0.010 in. and two 0.003 in. shims, end float is reduced to 0.003 in.

4. As assembly progresses, coat all bearing surfaces with engine oil. Do *not* use additives such as STP or Stud.

5. Check the rod big ends for roundness before installing on the crankshaft. Bolt the rods and caps together without the bearing shells. If the big end eye is more than 0.0004 in. out-of-round, the rod and cap should be replaced.

6. Install the bearing shells in their original positions in the rod and cap, if the same ones are being used again. Do not scrape them with anything.

7. Attach each rod and cap to its original crank journal, making sure the marks correspond and the rod is installed correctly. The rod with a

small oil hole on each side of the web goes on the left journal (**Figure 72**). Install new self-locking nuts and use a torque wrench (**Figure 73**) to torque to 22 ft.-lb.

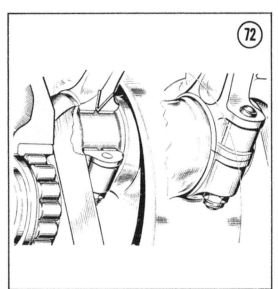

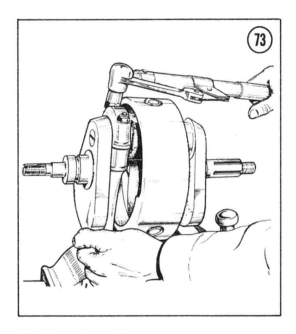

Crankcase Assembly

1. Get a block measuring about 8 x 8 x 6 inches with a hole large enough to accept the left end of the crankshaft.

2. Carefully clean all joint faces of joint cement. Be sure not to damage the soft aluminum faces.

3. Place the right side of the crank into the hole in the block. With the shims in position, place the left half of the case onto the left side of the crank, making sure the bearing fits squarely.

4. Place the left side of the crank in the hole in the block (**Figure 74**) and put the rotary breather valve into the camshaft bearing with the driving tongues up. Put on the spring and then the camshaft, making sure the tongues on the disc engage with the driving slot in the camshaft.

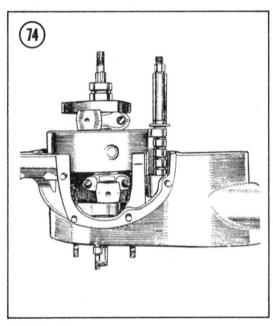

5. Put the thrust washer on the right side of the crankshaft with the oil grooves next to the crank web.

6. Smear both faces of the crankcase halves with cement and place the right half in position on top of the left which is face up. Bolt the 2 halves together. There are 4 nuts on studs, 3 inside the primary case and one on the bridge across the mouth, and 4 bolts on the front of the case. Tighten these carefully and evenly.

7. Check to be sure the crankshaft rotates freely (**Figure 75**). If it doesn't, the trouble must be found and corrected now.

8. Don't worry about the camshaft end float at this point. The pinion will correct it when installed.

9. Replace the crankcase filter and sump plate, using new gaskets with cement on both sides.

10. Replace the key in the right side of the crankshaft and replace the crankshaft gear. Be sure the timing mark is on the outside.

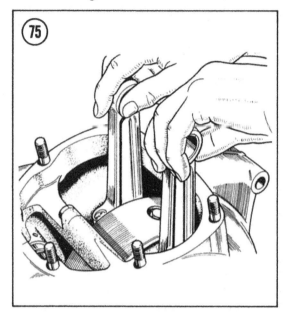

11. Replace the crankshaft spacer and then the oil pump and its driving worm together, as described earlier in this chapter. Remember the worm gear has a left-hand thread.

12. Replace the lockwasher and crankshaft nut, which also has a left-hand thread. Torque to the recommended settings quoted earlier. Bend the lockwasher over one of the flat edges of the nut.

13. Replace the key in the camshaft and then the gear with its timing mark on the outside face which is recessed. Remember to bend the washer tab over the nut edge after tightening. See **Figure 76**.

14. Replace the idler gear so its timing marks match those of the camshaft and camshaft gears (**Figure 77**).

15. Replace the gear cluster if it has been removed, as described earlier.

16. Other procedures to completely assemble engine are given in the various removal/installation procedures.

CHAPTER SEVEN

CHASSIS

Complete dismantling is seldom necessary unless the frame has been damaged in an accident or through rough use in competition. Most critical assemblies can be removed for repair or left on the machine for adjustment.

Figures 1 and 2 are exploded views of early and late chassis for 650cc models. BSA 500 chassis component service is similar, and in most cases identical, to the procedures for servicing the 650cc models. Diagrams showing the differences between the models are included whenever needed. Refer to **Figures 3 and 4** for BSA 500 chassis construction details.

Note that U.N.F. (Unified) threads have been introduced on chassis fittings. All nuts, bolts, and threaded components should be checked for proper thread type before they are replaced.

WHEELS

Removal (Front)

1. Disconnect the front brake stoplight wires at their snap connectors.

2. Loosen the brake cable at the handlebar and remove it.

3. Remove the cable from the brake levers on the front hub (A, **Figure 5**). Loosen the retaining nut (B, Figure 5).

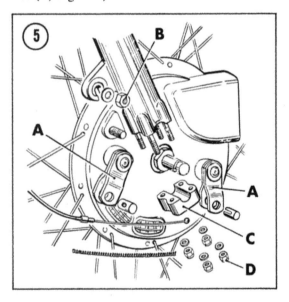

4. Remove the fork end caps (C, Figure 5) by unscrewing 4 nuts (D, Figure 5) on each fork leg.

5. Remove the brake anchor strap from the slotted ear on the fork leg and remove the wheel from the forks.

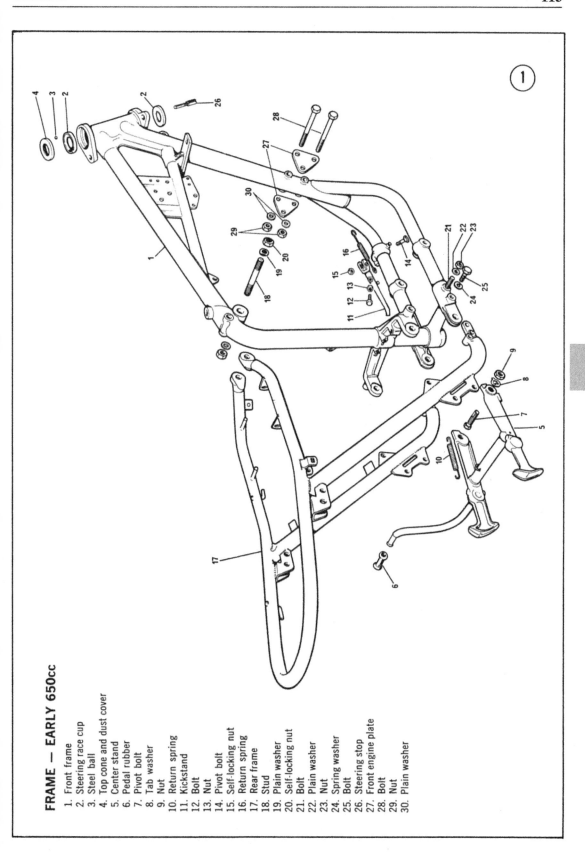

FRAME — EARLY 650cc

1. Front frame
2. Steering race cup
3. Steel ball
4. Top cone and dust cover
5. Center stand
6. Pedal rubber
7. Pivot bolt
8. Tab washer
9. Nut
10. Return spring
11. Kickstand
12. Bolt
13. Nut
14. Pivot bolt
15. Self-locking nut
16. Return spring
17. Rear frame
18. Stud
19. Plain washer
20. Self-locking nut
21. Bolt
22. Plain washer
23. Nut
24. Spring washer
25. Bolt
26. Steering stop
27. Front engine plate
28. Bolt
29. Nut
30. Plain washer

7

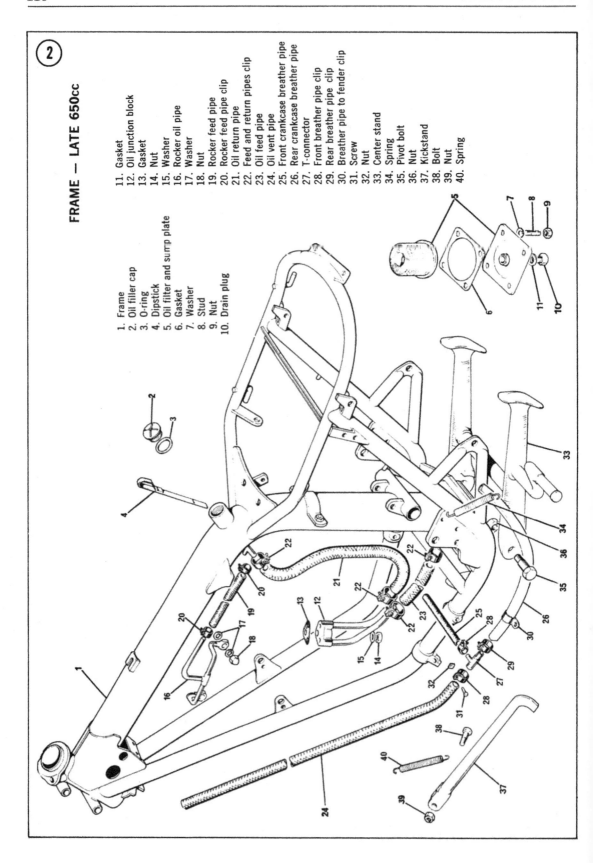

FRAME – LATE 650cc

1. Frame
2. Oil filler cap
3. O-ring
4. Dipstick
5. Oil filter and sump plate
6. Gasket
7. Washer
8. Stud
9. Nut
10. Drain plug
11. Gasket
12. Oil junction block
13. Gasket
14. Nut
15. Washer
16. Rocker oil pipe
17. Washer
18. Nut
19. Rocker feed pipe
20. Rocker feed pipe clip
21. Oil return pipe
22. Feed and return pipes clip
23. Oil feed pipe
24. Oil vent pipe
25. Front crankcase breather pipe
26. Rear crankcase breather pipe
27. T-connector
28. Front breather pipe clip
29. Rear breather pipe clip
30. Breather pipe to fender clip
31. Screw
32. Nut
33. Center stand
34. Spring
35. Pivot bolt
36. Nut
37. Kickstand
38. Bolt
39. Nut
40. Spring

FRAME—500cc

1. Frame
2. Oil filler cap
3. O-ring
4. Oil filter
5. Sump plate
6. Gasket
7. Washer
8. Stud
9. Nut
10. Drain plug
11. Gasket
12. Oil junction block
13. Gasket
14. Nut
15. Washer
16. Rocker oil pipe
17. Washer
18. Nut
19. Rocker feed pipe
20. Rocker feed pipe clip
21. Oil return pipe
22. Return pipe clip
23. Feed pipe clip
24. Oil feed pipe
25. Oil vent pipe
26. Front crankcase breather pipe
27. Rear crankcase breather pipe
28. T-connector
29. Front breather pipe clip
30. Rear breather pipe clip
31. Breather pipe to fender clip
32. Screw
33. Nut
34. Center stand
35. Spring
36. Pivot bolt
37. Nut
38. Kickstand
39. Bolt
40. Nut
41. Spring

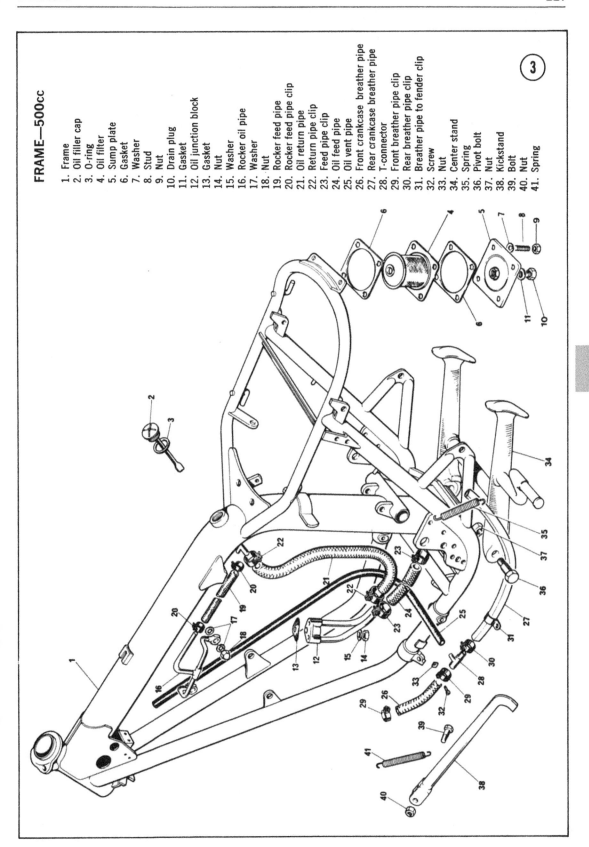

7

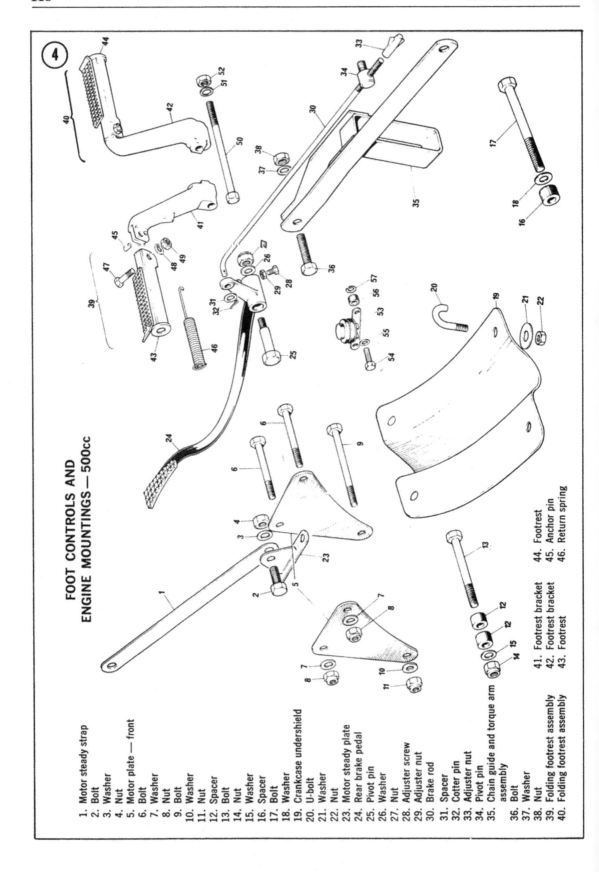

FOOT CONTROLS AND ENGINE MOUNTINGS — 500cc

1. Motor steady strap
2. Bolt
3. Washer
4. Nut
5. Motor plate — front
6. Bolt
7. Washer
8. Nut
9. Bolt
10. Washer
11. Nut
12. Spacer
13. Bolt
14. Nut
15. Washer
16. Spacer
17. Bolt
18. Washer
19. Crankcase undershield
20. U-bolt
21. Washer
22. Nut
23. Motor steady plate
24. Rear brake pedal
25. Pivot pin
26. Washer
27. Nut
28. Adjuster screw
29. Adjuster nut
30. Brake rod
31. Spacer
32. Cotter pin
33. Adjuster nut
34. Pivot pin
35. Chain guide and torque arm assembly
36. Bolt
37. Washer
38. Nut
39. Folding footrest assembly
40. Folding footrest assembly
41. Footrest bracket
42. Footrest bracket
43. Footrest
44. Footrest
45. Anchor pin
46. Return spring

Installation (Front)

1. Fit the brake anchor strap in the slotted ear of the fork leg. Fit the axle into the fork legs and replace the end caps. Do not fully tighten the 4 nuts on each end cap now.

2. Tighten the brake anchor plate nut (B, Figure 5). The axle grooves will align the wheel during this assembly. Then tighten the end cap nuts.

3. Replace the brake cable and stoplight switch connections.

Removal (Rear)

Refer to **Figure 6** for this procedure.

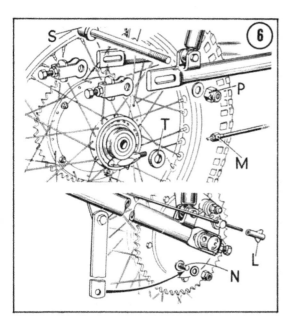

1. Disconnect the speedometer cable from its drive (see M, Figure 6). Unscrew the wing adjuster (L) from the brake rod. Remove the chain from the rear sprocket but not from the transmission sprocket.

2. Remove the anchor strap from the backing plate at the pivot (N). Loosen the front bolt on the anchor strap and leave it attached to the swing arm.

3. Remove the axle nut (D) on the right side and pull out the axle (S) from the left side. When

the spacer (T) is free, lower the wheel to the ground and remove it.

Installation (Rear)

Refer to **Figure 7** for this procedure.

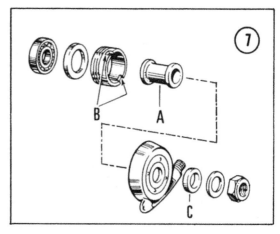

1. Install the speedometer driving ring (left-hand thread).

2. Install the speedometer drive on the hub with its spacer (A) and locate it in the slots in the driving ring (B).

3. Install the left side chain adjuster and insert the axle into the left swing arm fork. Lift the wheel into position and insert the axle through the hub. Put the spacer (C) on the axle and then the right side chain adjuster.

4. Insert the axle through the right swing arm fork. The speedometer cable should be parallel with the swing arm. Tighten the axle nut loosely and the brake anchor strap. Install the rear chain.

5. Adjust the rear chain to the correct tension as described in Chapter Two and tighten the axle nut. Reconnect the brake rod and check the stoplight adjustment.

BRAKE HUB

Disassembly/Assembly (Front)

1. Remove the nut holding the backing plate to the axle.

2. Remove the shoes with the backing plate.

3. Remove the bearing retainer (see **Figure 8**) which has a left-hand thread. Turn clockwise to remove.

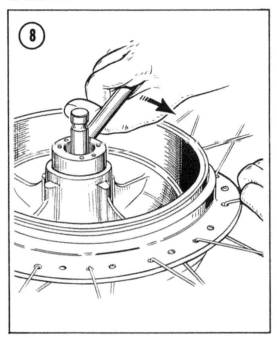

4. Insert the axle through the left side bearing and use it to drive out the right side bearing. Use a soft mallet or a piece of wood to protect the end of the axle.

5. Remove the bearing and spacer from the axle. Then remove the circlip.

6. Insert the axle in the left side bearing and drive out from the right side. This will remove the outer grease retainer, leaving the inner grease retainer in the hub.

7. Grease the new bearings. Apply pressure to the outer races only when inserting into the hub. Begin assembly at the right side of the hub. See **Figure 9**.

8. Install the grease retainer with its dish inward. Install the bearing as far as the circlip will allow. Install the bearing retainer and tighten the outer race.

9. Insert the axle and spacer through the bearing from the left side until the spacer is seated against the inner race.

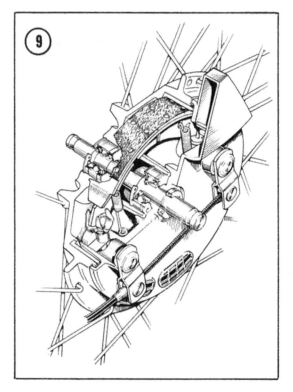

10. Fit the left side bearing and the outer grease retainer, and then the circlip in its groove.

Disassembly/Assembly (Rear)

1. Remove the speedometer drive from its ring, which will release the inner spacer.

2. Remove the backing plate and shoes from the hub (see **Figure 10**).

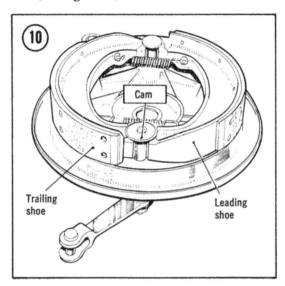

3. Use a special service tool (No. 61-3694) to remove the bearing retainer (**Figure 11**) which has a right-hand thread.

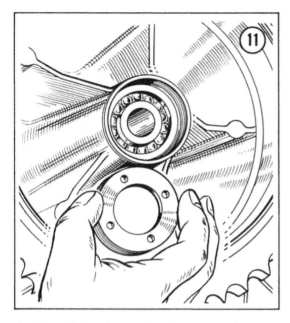

4. Use a drift of 0.78 in. diameter and a pilot of 0.62 in. diameter to drive out the center sleeve from the left side. This will remove the right side bearing, but not its grease retainer. Use the sleeve and drift to drive out the left side bearing (**Figure 12**).

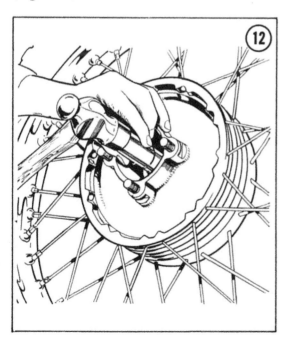

5. Grease the new bearings. Fit a bearing on the sleeve. Apply pressure to the inner race until the bearing seats on the center sleeve shoulder. Place the ring in the left side with the flat face inward and insert the bearing and sleeve. Apply pressure to the outer race until it is seated. Tighten the bearing retainer.

6. Be sure the grease retainer is in position on the right side. Insert the bearing in the hub. Apply pressure to the inner race until it is seated.

7. Install the spacer and speedometer drive on its ring.

BRAKES

Inspection

1. Remove the backing plate and brake shoes as described in the previous section.

2. The cam should have a small amount of float on its shaft to evenly apply pressure on the shoes.

3. The brake lining rivets must be below the surface of the lining (**Figure 13**) to prevent scoring the drum.

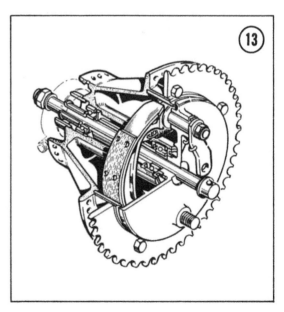

Shoe Replacement

1. Remove the nut holding the brake backing plate on the axle. The backing plate and brake shoes come out of the hub as a unit.

2. Pry the brake shoes up and out at the pivot points to remove from the backing plate. The brakes are fitted with pads against the pivot cam block. Be sure the pads are in place when re-assembling.

3. When reassembling, be sure the short part of the return spring is next to the adjusting screws (see **Figure 14**).

4. Remove the adjusting cam, cage, tappet, and O-ring. Remove the camshaft. Clean all parts in solvent. Be sure the camshaft and tappet move freely in the cam pivot block. Apply grease lightly to the operating cam, but be sure the O-ring is in good condition. Grease on the linings can be dangerous, causing a lack of braking effort.

5. Check the anchor plate for cracks and replace if necessary.

6. Check the brake return springs for fatigue or damage.

7. Clean and lightly grease the brake cam and shoe contacts. Do not overgrease or the linings will become contaminated and will have to be replaced.

8. Check the brake drums for scoring or an out-of-round condition. Repair or replace them if necessary.

9. If the brake linings have worn to a point that they are level with the rivets, they must be replaced.

10. Set the adjustment cam to the fully con-tracted position and install the brake shoes in reverse order of disassembly.

SPROCKET

1. Check the teeth for signs of wear. If serious, replace the sprocket. A worn sprocket will rapidly wear the chain.

2. The new sprocket must be squarely placed with its mating faces clean. Five self-locking nuts secure it.

WHEEL INSPECTION

1. Check spoke tension by grasping 2 spokes and pressing together. If they move noticeably, they are too loose.

2. Tighten the spoke by turning the nipple. It is very easy to get the wheel out-of-round when tightening spokes. To avoid this, proceed slowly and carefully.

3. Each spoke has an opposing one on the oppo-site side of the hub. Tightening only one will

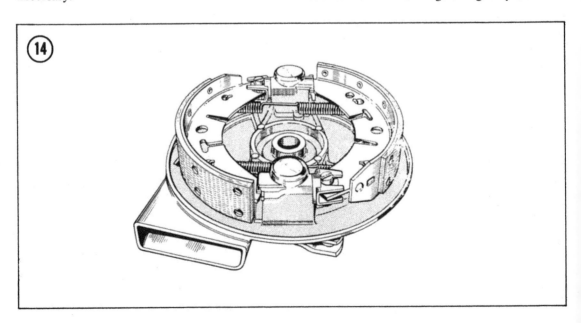

tend to distort the center line of the wheel rim, so its opposite should be tightened also.

4. Work slowly and carefully around the wheel, tightening each spoke a little at a time. If done correctly, it can be accomplished without seriously distorting the wheel.

WHEEL BALANCING

1. The wheel must revolve freely. For the back wheel, remove the chain.

2. Rotate the wheel and let it coast to a stop. Mark the top of the tire and repeat several times. If it stops on the same place, add weight to the marked spot.

3. Weights are available to attach to the spokes (**Figure 15**). The correct weight must be added until the wheel revolves freely, stopping at a different point each time.

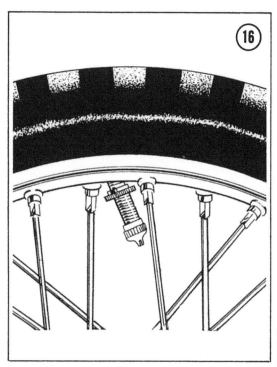

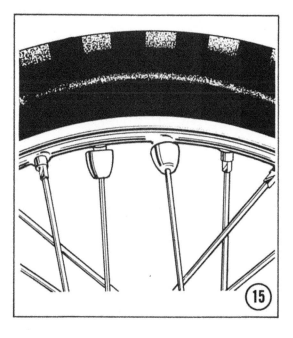

TIRE RIM LOCKS

1. An underinflated tire will creep on the rim, taking the tube with it. This will cause the valve to be pulled from the tube (**Figure 16**).

2. Two rim locks on the rear wheel 120 degrees on each side of the valve will prevent this.

3. Remove the tire and tube. Drill the rim between 2 spoke nipples to accommodate the rim lock bolt. Remove any burrs from the rim.

4. Install the rim locks loosely and install the tire so it covers the inner dish of the rim lock (**Figure 17**). Inflate the tire and then tighten the nuts on the rim lock bolts. The wheel must be balanced afterward.

WHEEL ALIGNMENT

1. Make a straightedge of wood about 80 in. long, with a step in the middle (D, **Figure 18**) for the difference in size between the 2 tires.

2. Lay the straightedge on blocks about 6 in. high and place it against the tires. If the wheels are aligned, the straightedge will touch each wheel at 2 points (B, Figure 18).

3. If the wheels are not aligned (A and C, Figure 18), the rear wheel must be aligned to correct it. The chain adjuster is moved in the direction indicated by the arrows in Figure 18.

4. If the frame has been badly bent in an accident, the wheels may not align using this method.

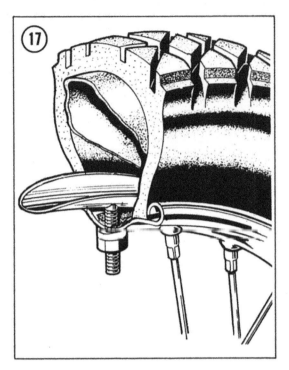

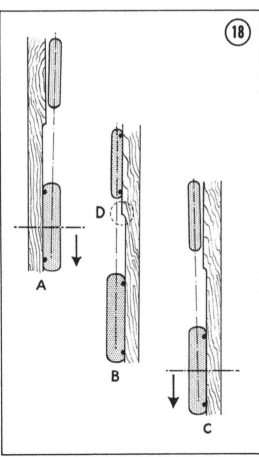

A frame expert must repair it, or else replace the frame.

TIRE CHANGING AND REPAIR

Removal

1. Remove the valve core to deflate the tire.

2. Press the entire bead on both sides of the tire into the center of the rim.

3. Lubricate the beads with soapy water.

4. Insert the tire iron under the bead next to the valve. Force the bead on the opposite side of the tire into the center of the rim and pry the bead over the rim with the tire iron (**Figure 19**).

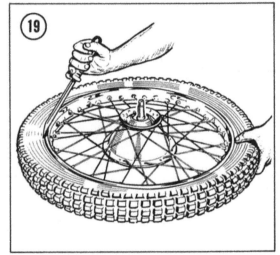

5. Insert a second tire iron next to the first to hold the bead over the rim. Work around the tire with the first tire iron, prying the bead over the rim (**Figure 20**). Be careful not to pinch the inner tube with the tire irons.

6. Remove the valve from the hole in the rim and remove the tube from the tire. Lift out and lay aside.

7. Stand the tire upright. Insert a tire iron between the second bead and the side of the rim that the first bead was pryed over (**Figure 21**). Force the bead on the opposite side from the tire iron into the center of the rim. Pry the second bead off the rim, working around as with the first.

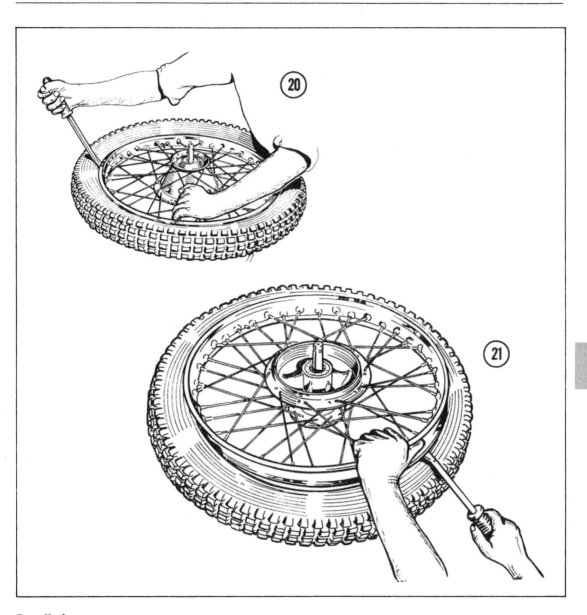

Installation

1. Carefully check the tire for any change, especially inside.

2. A new tire may have balancing rubbers inside. These are not patches and should not be disturbed. A white spot near the bead indicates a lighter point on the tire. This should be placed next to the valve or midway between the 2 rim locks, if so equipped.

3. Check that the spoke ends do not protrude through the nipples into the center of the rim to puncture the tube. File off any protruding ends.

4. Be sure the rim rubber tape is in place with the rough side toward the rim.

5. Put the core in the tube valve. Put the tube on the tire and inflate just enough to round it out. Too much air will make installing the tire difficult, and too little will increase the chances of pinching the tube with the tire irons.

6. Lubricate the tire beads and rim with soapy water. Pull the tube partly out of the tire at the valve. Squeeze the beads together to hold the tube and insert the valve into the hole in the rim (**Figure 22**). The lower bead should go into

the center of the rim with the upper bead outside it.

7. Press the lower bead into the rim center on each side of the valve, working around the tire in both directions (**Figure 23**). Use a tire iron for the last few inches of bead (**Figure 24**).

8. Press the upper bead into the rim opposite the valve. Pry the bead into the rim on both sides of the initial point with a tire iron, working around the rim to the valve (**Figure 25**).

9. Wriggle the valve to be sure the tube is not trapped under the bead. Set the valve squarely in its hole before screwing on the valve nut to hold it against the rim.

10. Check the bead on both sides of the tire for even fit around the rim. Inflate the tire slowly to seat the beads in the rim. It may be necessary to bounce the tire to complete the seating. Inflate to the required pressure. Tighten the rim locks if they are installed. Balance the wheel as described.

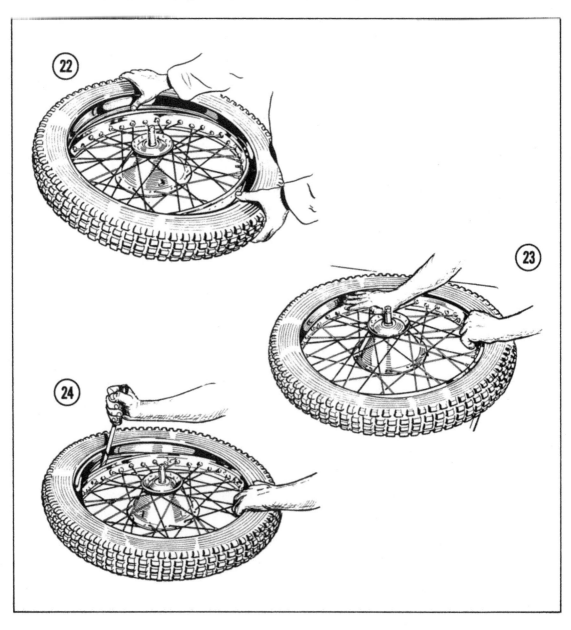

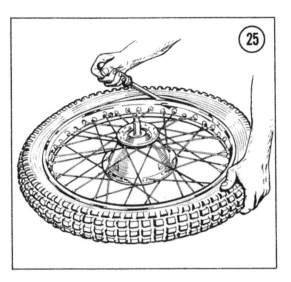

TIRE MAINTENANCE

Tire pressure is taken when cold. Check the pressure weekly. The tire pressure quoted in the *Appendix* are for a rider weighing 154 lb. Add one pound of tire pressure to the front tire for every 28 lb. of rider weight above 154 lb. Add one pound of air pressure to the rear tire for every 14 lb. of rider weight above 154 lb. For sustained high speed riding, the recommended pressures should be increased 5 lb. If a passenger is riding, the weight on each wheel should be taken and inflation followed according to **Table 1** below.

Table 1   WEIGHT AND TIRE PRESSURE

Tire Size	Weight	Pressure
3.25x19	240	16
3.25x19	295	20
3.25x19	350	24
3.25x19	405	28
3.25x19	465	32
4.00x18	345	16
4.00x18	410	20
4.00x18	470	24
4.00x18	530	28
4.00x18	595	32

SIDECAR ALIGNMENT

1. Align the front and rear motorcycle wheels as described earlier. Use the stepped straightedge (Figure 18) again. Place a second straightedge with no step against the sidecar tire (**Figure 26**). Be sure the machine is on a flat surface.

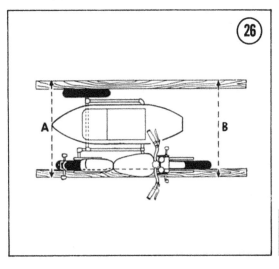

2. Straighten the front wheel to touch the straightedge at 2 points (B, Figure 18). Measure the distance between the straightedges (A and B, Figure 26). The distance in front (at B) should be ⅜-¾ in. less than at the rear (at A). This toe-in is necessary to make steering manageable at high speed. Adjustment is made at the front sidecar attachment.

3. Lean the motorcycle one inch vertically away from the sidecar. Hang a plumb weight from the handlebar and measure the distance from the center line of the motorcycle at top and bottom (**Figure 27**). Adjust at the 2 upper sidecar attachments until the top distance (C, Figure 27) is at least one inch greater than the bottom (D, Figure 27). Lean-out makes steering easier on cambered roads.

HANDLEBARS

1. Disconnect all wires at the junction farthest from the handlebars. Almost all wires are color coded (see Chapter Five for specifics) but old age may have caused them to fade beyond

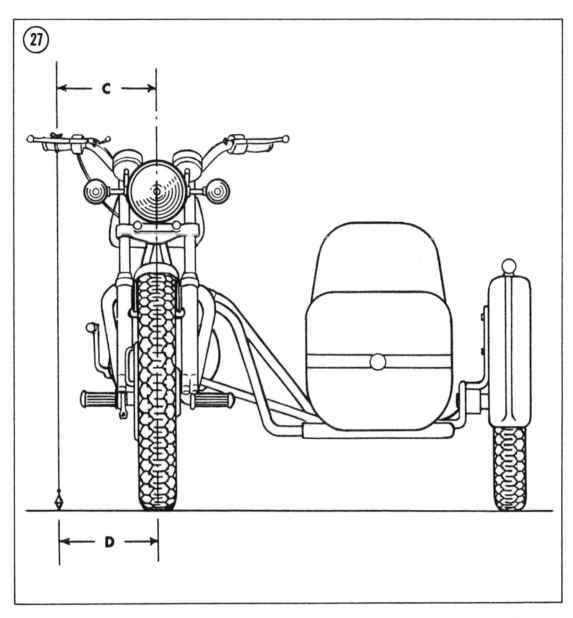

recognition. If the colors aren't discernible, mark them with masking tape and a felt tip pen to simplify reassembly.

2. Pull on the clutch cable, loosen the adjusting screw, then remove from the lever.

3. Loosen the front brake adjuster, then remove the cable end from the lever. The cable may be left with the front wheel assembly.

4. Remove the starter lever, choke, or compression release cables (if the bike is so equipped) using procedures in the following section.

5. Remove the screws on the body of the throttle twistgrip and rotate the housing until the cable can be moved.

6. Remove the handlebar clamp bolts, then lift off the handlebar.

7. Replace the handlebar or straighten if bent.

8. Check the cables and housing and lubricate with graphite.

9. Reverse the previous steps for reassembly. The clutch, front brake, and throttle cables should be adjusted to include a little play before

actuating. The throttle, for instance, should not be set up without any play or the bike will accelerate when the handlebars are turned.

THROTTLE CABLE

Replacement (Single Carburetor)

1. Open the throttle grip all the way. Close it while pulling on the cable to release the cable covering from its slot in the grip assembly.

2. Remove the 2 screws and take off the top half of the throttle grip mounting. Remove the cable nipple from the grip and remove the cable.

3. Install the nipple of the new cable in the bottom half of the mounting.

4. Replace the top of the mounting and fasten the screws.

5. Do not replace the cable covering in its slot yet.

6. Remove the gas tank as described in Chapter Two, and remove the carburetor.

7. Remove the carburetor top with the throttle valve assembly. Check which notch the carburetor needle clip is in. Compress the throttle spring and remove the needle clip.

8. Keep the spring compressed and push down on the cable to release the nipple from the valve (**Figure 28**).

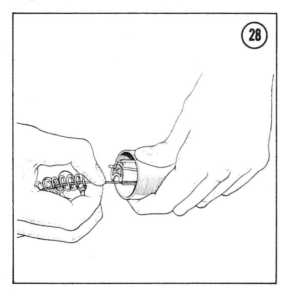

9. Slide the new cable through the top cap, spring, and needle clip. Compress the spring and insert the cable nipple in the valve. Replace the needle with the clip in the correct notch.

10. Slide the valve assembly into the carburetor body. Be sure the needle enters the jet squarely. Check the valve operation before tightening the cap screws. Replace the carburetor.

11. Route the cable with the frame clips and replace the cable covering in its notch in the throttle grip. Adjust the throttle grip.

12. Replace the gas tank as described in Chapter Two.

Replacement (Twin Carburetor)

1. Proceed as with the single carburetor models, except do not remove the carburetor. The twin cables join in a junction box to become a single cable.

2. Careful adjustment of the cables is necessary after replacing to keep the engine in tune.

3. If only the long cable from the throttle grip is being replaced, remove the cap from the junction box to expose the nipple (**Figure 29**).

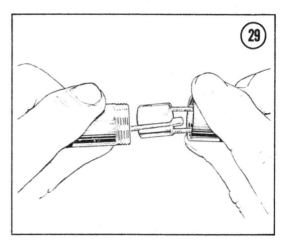

4. Replace the short cables to the carburetor as described earlier. The junction box has 2 nipple connections at the other end for connection to the carburetors.

5. After reassembly, carefully synchronize the carburetors as described in Chapter Three.

CHOKE CABLE

Replacement

1. Open the choke and then close it, pulling on the cable to release the covering from its notch. Remove the nipple from the choke control.

2. Remove the gas tank as described in Chapter Two and remove the carburetor. Remove the carburetor valve assembly as described above.

3. Slip the air slide up out of the throttle valve and compress the spring to release the cable nipple.

4. Slide the new cable through the top cap, spring guide tube, and spring. Compress the spring and replace the nipple in the air slide.

5. Replace the air slide in the throttle valve and proceed with reassembly as described above.

FRONT BRAKE CABLE

Removal/Installation

1. Disconnect the stoplight switch wires. Remove the handbrake lever pivot pin and nut.

2. Remove the lever from the bracket and remove the cable nipple. The adjuster is removed from the bracket with the cable.

3. Disconnect the cable end and covering from their attachments on the brake.

4. Lubricate the new inner cable before installing. Be sure the return spring is fitted on the front brake hub. If a rubber cover is fitted, replace it with a new spring.

5. Replace the pivot pin in the lever with the countersink facing forward to accommodate the cable nipple.

6. Carefully adjust the cable after replacement.

CLUTCH CABLE

Removal/Installation

1. Loosen the handlebar clutch cable adjustment. Loosen the crankcase adjustment (B, **Figure 30**) after loosening the locknut (A, Fig-

ure 30). A rubber sleeve (C, Figure 30) covers the adjuster and nut.

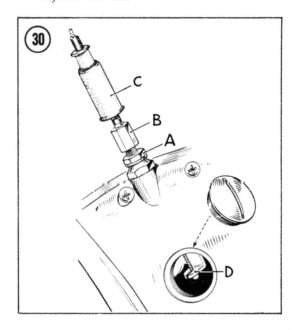

2. Remove the cable nipple from the slotted lever (D, Figure 30) which is reached through the inspection opening.

3. Unscrew the crankcase adjustment and remove the cable.

4. Remove the cable from the clutch lever at the handlebar. There should be plenty of slack now.

5. Replace the new cable in reverse sequence. The gas tank must be removed to reach the cable clip on the frame. Adjust the new cable carefully as described in Chapter Two.

STEERING HEAD BEARING ADJUSTMENT

1. Place a box under the engine to lift the front wheel off the ground.

2. Grab the lower fork legs and flex them to test for play in the head bearings (**Figure 31**).

3. To adjust, loosen the steering head pinch bolt (C, **Figure 32**).

4. Turn the adjusting nut (E, Figure 32) clockwise to tighten the bearings or counterclockwise to loosen.

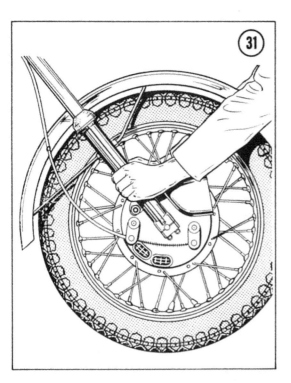

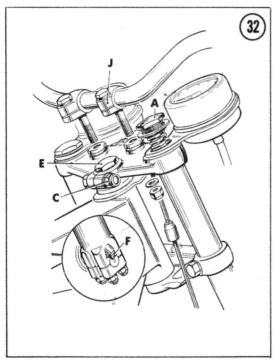

7. Test by moving the steering from lock to lock. The movement should be smooth and even. If it is jerky, replace the head bearings.

Bearing Replacement

1. Disconnect the cables from the base of the speedometer and tachometer.

2. Disconnect the headlight brackets from the triple clamps. Be careful with the rubber bushings.

3. Leave the headlight connected to its wiring, but support it carefully.

4. Remove the front brake cable as described under *Front Brake Cable* in this chapter.

5. Remove the front wheel as described under *Front Wheel* in this chapter.

7. Remove the front fender.

8. If a steering damper is installed, remove the split pin from the lower end of the rod. Remove the anchor bolt from the frame. Unscrew the rod and knob and lay the assembly aside.

9. Remove the handlebars as described under *Handlebars* in this chapter.

10. Loosen the pinch bolt (C, Figure 32) and remove the adjusting nut (E, Figure 32). Remove the cap nuts (A, Figure 32). Lay aside the speedometer and tachometer.

11. Tap upward on the top triple clamp with a soft mallet to release it from the fork tubes.

12. Pull the forks, lower triple clamp, and steering head down, out of the frame. Tap lightly on the top of the steering column with a soft mallet to release it.

13. Pry the lower bearing race from the column with a flat bladed lever.

14. Carefully drive the outer bearing races from inside the frame steering head by tapping all around it. Too much tilting will score the inside of the steering head.

15. Be sure the bearing rings are in place in the steering head (B, **Figure 33**). Drive the outer

5. The correct adjustment allows just a trace of play in the bearings. Do not tighten so that all play is eliminated.

6. Tighten the pinch bolt (C, Figure 32).

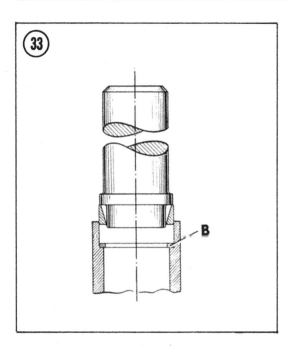

bearing races squarely into the steering head with a shouldered drift (Figure 33).

16. Install a new inner bearing race on the bottom of the steering column. Grease the bearing rollers and install the column in the steering head.

17. Grease the rollers of the upper inner bearing race and fit it over the column. Replace the dust cover and top triple clamp.

18. Tighten the adjusting nut (E, Figure 32) as described earlier and tighten the pinch bolt (C, Figure 32).

19. Replace the front fender and front wheel.

Fork Spring Replacement

1. Remove the handlebars and fork tube cap nuts as described earlier.

2. Pull out the springs and replace them.

3. Reassemble in reverse order.

Fork Damper Disassembly

1. Remove the headlight rim and light from the shell.

2. Disconnect the snap connectors and remove the headlight brackets.

3. Remove the handlebars as described under *Handlebar* in this Chapter.

4. Remove the front wheel and the front fender as described in this chapter.

5. Remove the speedometer and tachometer and fork cap nuts as described earlier.

6. Remove the drain screw as described in Chapter Two and drain the oil.

7. Remove the fork springs. Insert special service tool (No. 61-6113) in the top of the fork tube (**Figure 34**).

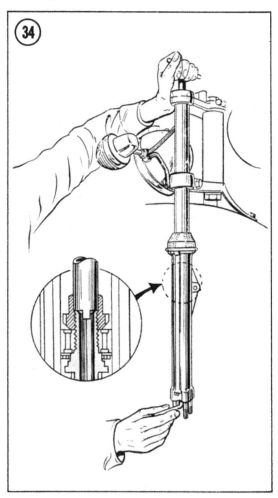

8. Hold the tool and remove the socket screw (**Figure 35**) from the lower fork leg.

9. Remove the lower fork leg from the tube. Carefully unscrew the aluminum end plug holding the damper assembly in the fork tube.

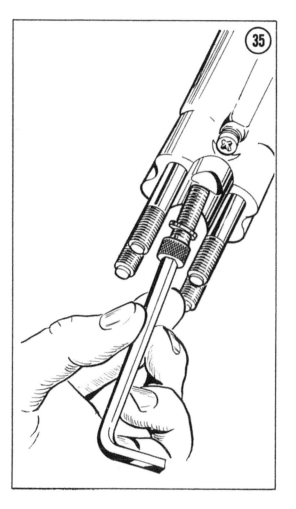

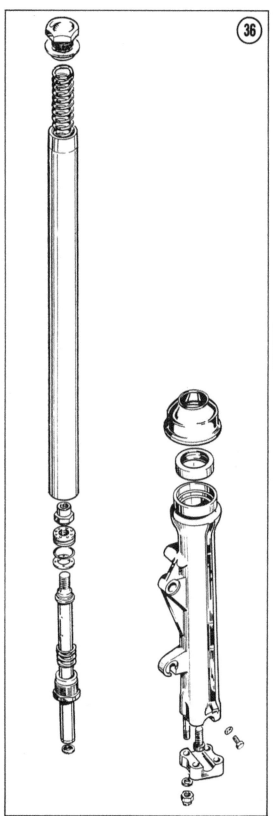

10. Remove the damper assembly and recoil spring (**Figure 36**). Remove the valve retaining nut and valve and washer. Check all parts for wear before replacing.

11. If necessary, remove the fork tubes from the triple clamps as described earlier.

Oil Seal Replacement

1. One oil seal is an O-ring around the damper valve. Inspect carefully.

2. Remove the rubber wiper from the top of the lower fork leg.

3. A double lipped oil seal is pressed into the top of the leg.

4. Hold the fork leg in a soft-jawed vise. Remove the seal with a small chisel (**Figure 37**) or

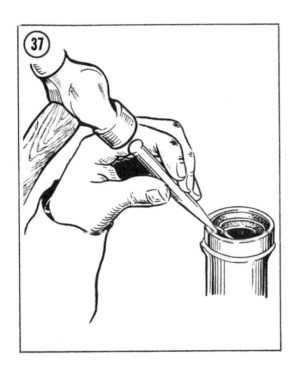

screwdriver. Once it is bent inward, it can be removed from the leg. Be careful not to damage the aluminum leg.

Fork Leg Assembly

1. Replace the fork tubes in the triple clamps and adjust the steering head as described earlier. Tighten the fork tube cap nuts to locate the tubes in the triple clamps and then remove the caps again.

2. Install new oil seals in the lower fork legs. Be sure they're square.

3. Reassemble the damper assembly as shown in Figure 36. After installing the shuttle washer, secure the damper valve with its plain face against the retaining nut. Install a new O-ring around the damper valve.

4. Install the recoil spring and slide the assembly up into the fork tube. Tighten the end plug. Slide the rubber wiper up over the end of the tube.

5. Install a new oil seal washer in the bottom of the lower fork leg. Lubricate the lower end of the fork tube and the oil seal in the lower fork leg with the correct fork oil. Carefully slide the

fork leg up over the fork tube. Be sure you don't damage the lips of the oil seal.

6. Clean the threads of the socket screw and apply a drop of Loctite. Hold the damper assembly in place with special service tool No. 61-6113 (Figure 34) and tighten the socket screw (Figure 35).

7. Fit the rubber wipers over the tops of the fork legs. Install the main springs, cap nuts, speedometer, tachometer, headlight, and handlebars as described earlier. Check all bolts and nuts on the fork assembly for tightness. Install the front fender and wheel as described in this chapter.

Alignment

Accurately checking the straightness of the fork tubes and trueness of the triple clamps requires special equipment, but a reasonably accurate check can be made by rolling the fork tubes on a sheet of plate glass (**Figure 38**). If the tube is bent more than 3/16 in., replace it. With straight tubes, check the forks for alignment as follows:

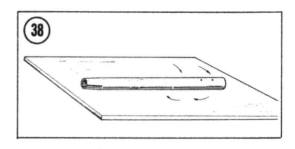

1. Install the tubes in the bottom triple clamp and tighten the pinch bolts.

2. Lay the tubes on a flat table to see if they are parallel. If they are not (**Figure 39**), the triple clamp is bent. If the bend is slight, it can be straightened.

3. Put the top of one tube above the triple clamp in a soft jawed vise. Grab the bottom of the other tube and bend the triple clamp as required. If more leverage is required, slide a larger diameter pipe over the lower part of the tube, but be very careful not to scratch it.

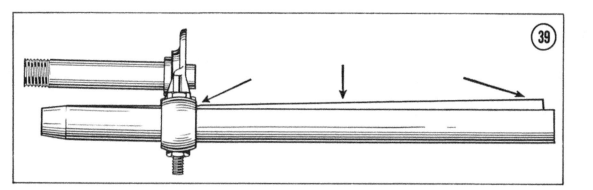

4. When the tubes are parallel in one plane, check the other plane between them. When both planes are parallel, install the top triple clamp and cap nuts to see if the tubes and steering column are parallel.

5. If they are not parallel, the column will be offset in the triple clamp (**Figure 40**). If the error is slight, you can try to straighten the triple clamp. If it's considerable, replace the item. Straightening a bend permanently weakens the metal.

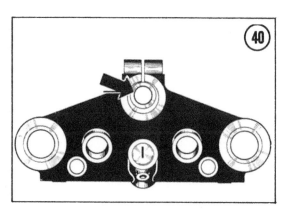

SWING ARM

Removal

1. Remove the footbrake pedal.

2. Remove the chain guard which is attached to the front of the swing arm and to the lower shock absorber mounting bolt.

3. Remove the rear chain.

4. Remove the rear wheel as described in this chapter.

5. Remove the shock absorbers as described later in this chapter.

6. Remove the self-locking nut from the end of the swing arm pivot rod and remove the rod and its components.

Bushing Replacement

1. There are 2 bushings in each half of the swing arm pivot mounting.

2. Remove bushings using a tool (**Figure 41**) made with a bolt, nuts and washers.

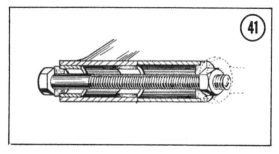

3. Replace the bushings with the same tool, pressing in from each side until the bushing end is even with the end of the housing.

4. New bushings are machined to the correct size for installation. Do not ream.

Assembly

1. Clean all parts with solvent and coat all bearing surfaces with grease.

2. Reassemble the swing arm pivot assembly (**Figure 42**). Insert the pivot rod into the frame from the left side. The thinner of the 2 spacing collars fits inside the frame plate on the right

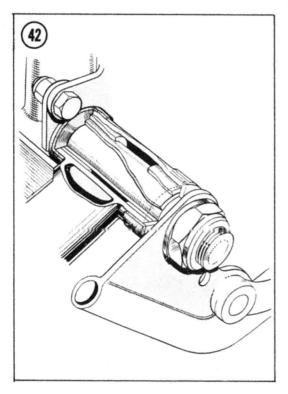

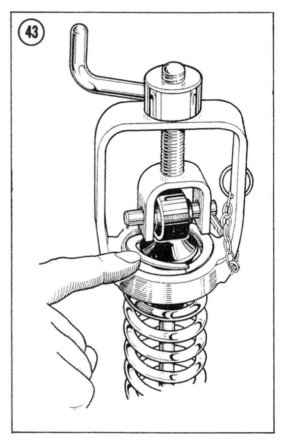

side, and the thicker collar inside the left frame plate.

3. Place the flexible sleeves over the working faces to keep out dirt and debris. Lubricate with grease at the nipples.

SHOCK ABSORBERS

The only service possible is the replacement of the springs.

1. Put a block of wood between the tire and rear fender and remove the upper and lower shock mounting bolts.

2. Compress the spring (**Figure 43**) and remove the split rings holding it.

3. The shock absorbers have 3 spring load positions (**Figure 44**). They must be in the lightest position (A) before dismantling.

4. The rubber bushings at each end can be driven out and new ones installed with liquid soap, if necessary.

5. Measure both shocks between the mounting eyes at both ends. If one is shorter than the other,

the spring has settled from age and both springs should be replaced.

6. Reverse to assemble.

SEAT

The seat is hinged, with a catch on the left side. Raise the seat and unscrew the front hinge. The seat can then be removed without loosening the rear hinge.

SIDE PANELS

1. Unscrew the master switch assembly.

2. Remove the air cleaner cover.

3. Loosen the 2 bolts from inside the air cleaner box (**Figure 45**) and remove the side panel with its front edge first, and then lift up from its peg on the frame.

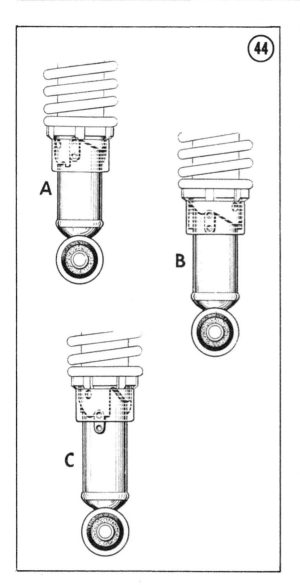

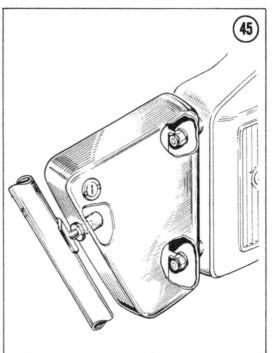

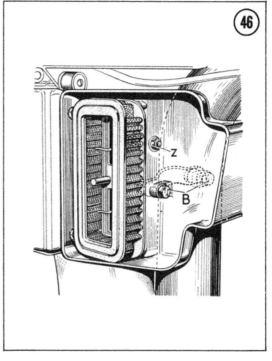

4. When installing the side cover, remember the grommet on the frame peg.

5. The left side is removed in the same way, except there is no master switch.

AIR CLEANER BOX

1. Remove the gas tank as described in Chapter Six.

2. Remove the "Lucar" connection (Z, **Figure 46**) inside the air cleaner box.

3. Do not split the 2 halves of the box body. Ease it out of the frame as one piece.

4. Unscrew the clamps and remove the flexible connections to the carburetors.

5. Remove the rubber outer covers and then the cleaner elements.

6. Wash the air cleaner in solvent and dry before reinstalling.

7. Remove the bolts (B, Figure 46) holding the box to the frame tube and remove the box from its mounting.

8. When installing, the flexible connections between the carburetor and the air cleaner box are different for the right and left side. Be sure the mold line is on top.

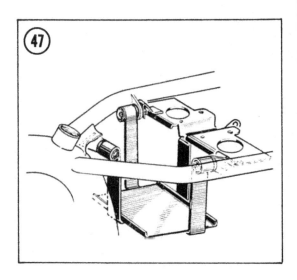

REAR FENDER

Removal/Installation

1. Remove the rear wheel as described in this chapter.

2. Unhook the taillight and turn signal wires at the snap connectors.

3. Remove the rear light assembly from fender.

4. Remove the 8 bolts holding the rear fender. All bolts are accessible from inside the fender. Remove the oil breather tube from its clip.

BATTERY AND MOUNTING

Removal

1. Disconnect the battery wires. Unhook the strap over the battery and lift it out.

2. When it is replaced, be sure the rubber tray and protective pads are in position in the mounting.

3. Remove the 3 screws holding the battery mounting to the frame (**Figure 47**).

Electrical Platform

1. Disconnect all wires, but leave them in position.

2. Remove one bolt and 2 screws holding the platform and remove it.

3. Always remove the battery before removing the electrical platform.

Tool Box

The tool box is fastened under the seat by one bolt which is accessible under the rear fender.

Center Stand

1. Remove the 2 bolts at the pivot and remove the stand.

2. To install, use a screwdriver to attach the spring to the frame (**Figure 48**). This must be done with the stand in the raised position.

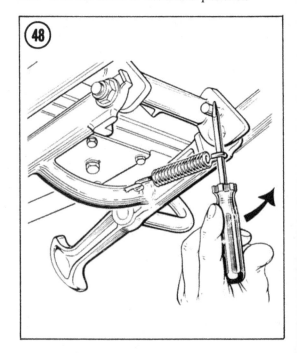

A P P E N D I X

SPECIFICATIONS

This appendix contains specifications for the BSA models covered by this book. Since there are differences between the various models, be sure to consult the correct table for your motorcycle.

8

Note: Specifications given apply to both 500 and 650cc models unless otherwise indicated.

General

Bore x stroke
   500   65.5 x 74.0mm
   650   75.0 x 74.0mm
Displacement
   500   499cc
   650   654cc
Compression ratio
   500   9.0:1
   650   9.0:1

Piston

Material   Aluminum alloy
Compression ratio   9.0:1
Clearance
   Bottom of skirt   0.0039-0.0054 in. (0.0990-0.137mm)
   Top of skirt   0.0094-0.0054 in. (0.2387-0.2768mm)
   (both measured on major axis)
Wrist pin hole diameter   0.7500-0.7502 in. (19.05-19.055mm)
Oversizes   +0.020 to +0.040 in. (+0.5 to +1.0mm)

Piston Rings

Material   Cast-iron
Compression rings (lower compression
   ring is tapered to same dimensions)
Width   0.0615-0.0625 in. (1.562-1.587mm)
   Thickness (radial)   0.114-0.121 in. (3.00-3.07mm)
   Clearance in groove   0.001-0.003 in. (0.025-0.076mm)
   Fitted gap   0.008-0.013 in. (0.203-0.330mm)
Oil control ring
   Width   0.124-0.125 in. (3.15-3.17mm)
   Thickness (radial)   0.114-0.121 in. (3.00-3.07mm)
   Clearance in groove   0.001-0.003 in. (0.025-0.076mm)
   Fitted gap   0.008-0.013 in. (0.203-0.330mm)

Wrist Pins

Material   Nickel-chrome high tensile steel
Diameter   0.7500-0.7502 in. (19.05-19.055mm)
Length   2.368-2.373 in. (60.15-60.27mm)

Small-end Bushings

Material   Phosphor-bronze
Outside diameter (before fitting)   0.8775-0.8785 in. (22.289-22.314mm)
Length   0.940-0.950 in. (23.88-24.13mm)
Finished bore (fitted)   0.7503-0.7506 in. (19.0576-19.0652mm)
Interference fit in rod   0.002-0.004 in. (0.0508-0.1016mm)

(continued)

Connecting Rods

Length between centers	6.0 in. (152.394mm)
Big-end bearing type	Vandervell D2 Bimetal
Rod side clearance	0.018-0.024 in. (0.457-0.605mm)
Bearing diametrical clearance	0.001-0.0025 in. (0.0254-0.0635mm)
Small-end bore diameter	0.8745-0.8755 in. (22.212-22.237mm)

Crankshaft

Type	One-piece forging with bolt-on flywheel
Main bearing (drive-side)	Hoffman RM.11L (roller)
Bore	1.125 in. (28.574mm) (nominal)
Outer diameter	2.812 in. (71.435mm) (nominal)
Width	0.812 in. (20.637mm) (nominal)
Main bushing (gear-side)	
Inner diameter (fitted)	1.4995-1.4990 in. (38.077-38.075mm)
Outer diameter (before fitting)	1.6300-1.6285 in. (41.402-41.364mm)
Width	0.940-0.960 in. (23.876-24.384mm)
Crankpin diameter	1.6865-1.6870 in. (42.837-42.849mm)
Minimum regrind	−0.010 in. (−0.254mm)
Second regrind	−0.020 in. (−0.508mm)
Third regrind	−0.030 in. (−0.762mm)
Gear-side journal regrind (2 times only)	−0.010 in. (−0.254mm)
	−0.020 in. (−0.508mm)
Drive-side shaft size	1.1251-1.1249 in. (28.5775-28.5725mm)
Crankshaft end-float	0.0015-0.003 in. (0.038-0.076mm)
Stroke	2.9134 in. (74.00mm)

Oil Pump

Pump body material	Cast-iron
Pump type	Double gear
Pump drive ratio	$\frac{1}{3}$ engine speed (1:3 ratio)
Pump non-return valve spring free length	0.8125 in. (20.637mm)
Pump non-return valve ball, size	$\frac{1}{4}$ in. diameter (6.35mm)
Relief valve blow-off pressure	50 lb. psi (3.51 kg/cm^2)
Oil pressure switch	Smith's PS.5300/1/07

Valves

Seat angle (inclusive)	90°
Head diameter	
Inlet	1.595-1.600 in. (40.51-40.64mm)
Exhaust	1.407-1.412 in. (35.74-35.86mm)
Stem diameter	
Inlet	0.3095-0.3100 in. (7.86-7.87mm)
Exhaust	0.3090-0.3095 in. (7.85-7.86mm)

Valve Guides

Bore diameter (inlet and exhaust)	0.312-0.313 in. (7.9248-7.950mm)
Outside diameter (inlet and exhaust)	0.5005-0.5010 in. (12.713-12.725mm)
Length (inlet and exhaust)	1.96-1.97 in. (49.78-50.04mm)
Clearance on valve stem	
Inlet	0.002-0.0035 in. (0.051-0.089mm)
Exhaust	0.0025-0.004 in. (0.0635-0.102mm)

(continued)

Valve Springs
  Free length
    Inner $1\frac{7}{16}$ in. (36.51mm)
    Outer $1\frac{3}{4}$ in. (44.45mm)
  Fitted length
    Inner   1.277 in. (32.44mm)
    Outer   1.37 in. (34.8mm)

Valve Timing
  Tappets set to 0.015 in. (0.381mm) for
    checking purposes only
    Intake opens   51° before top dead center
    Intake closes   68° after bottom dead center
    Exhaust opens   78° before bottom dead center
    Exhaust closes   37° after top dead center

Timing Gear
  Crankshaft pinion
    Number of teeth   22
    Fit on shaft   ±0.0005 in. (±0.0127mm)
  Camshaft pinion
    Number of teeth   44
    Interference fit   0.0000-0.001 in. (0.0254mm)
  Idler pinion
    Number of teeth   44
    Spindle dimensions, both ends   0.6875-0.6870 in. (17.449-17.463mm)
    Bushing dimensions, both (inside) 0.6880-0.6885 in. (17.475-17.488mm)
    Bushing dimensions, both (outside) 0.939-0.940 in. (23.851-23.876mm)
    Spindle working clearance   0.0005-0.0015 in. (0.0127-0.0381mm)

Tappet Clearance (cold)
  Intake   0.008 in. (0.203mm)
  Exhaust   0.010 in. (0.254mm)

Ignition Timing (standard ignition system)
  Piston position (BTDC) fully advanced   0.304 in. (7.22mm)
  Crankshaft position (BTDC) fully advanced 34°
  Contact breaker gap setting   0.015 in. (0.38mm)

Camshaft
  Journal diameter
    Left   0.810-0.8105 in. (20.574-20.586mm)
    Right   0.8735-0.874 in. (22.188-22.2mm)
  End float   Nil (spring-loaded)
  Cam lift (zero tappet clearance)   0.375 in. (9.52mm) (exhaust)
                                    0.385 in. (9.78mm) (inlet)
  Base circle diameter   0.812 in. (20.624mm)

Camshaft Bearing Bushings
  Bore diameter, fitted (left side)   0.8115-0.8125 in. (20.612-20.637mm)
  Outside diameter (left side)   0.906-0.907 in. (23.012-23.037mm)

(continued)

Camshaft Bearing Bushings (continued)

Interference fit in case (left side)	0.0018-0.0033 in. (0.046-0.084mm)
Bore diameter, fitted (right side)	0.875-0.876 in. (22.225-22.25mm)
Outside diameter (right side)	1.065-1.066 in. (27.051-27.076mm)
Interference fit in case (right side)	0.002-0.004 in. (0.0508-0.1016mm)

Cylinder Barrel

Material	Cast iron
Bore, standard	
500	2.576-2.579 in. (65.430-65.506mm)
650	2.952-2.953 in. (74.983-75.006mm)
Tappet bore size	0.3745-0.3750 in. (9.512-9.525mm)
Oversizes	+0.020 to +0.040 in. (+0.5 to +1.0mm)

Tappets

Tip radius	1.250 in. (31.75mm)
Tappet diameter	0.3740-0.3735 in. (9.5-9.49mm)
Clearance in barrel	0.0005-0.0015 in. (0.0127-0.0381mm)

Cylinder Head

Material	Aluminum alloy
Inlet port size (at valve)	$1\frac{1}{2}$ in. (38.1mm)
Exhaust port size (at valve)	$1\frac{5}{16}$ in. (33.34mm)
Valve seatings	Cast-iron (cast-in)

Intake Manifold Balance Pipe

Length	4 in. (101.6mm)
Bore	$\frac{9}{32}$ in. diameter (7mm)

Carburetor (concentric)

Type	Amal R.930/70 (right hand)
	Amal L.930/71 (left-hand)
Main jet	200
Needle jet size	0.106 in.
Needle position	1
Throttle valve	3
Nominal choke size	30mm
Throttle slide return spring	3 in. (76.2mm) (free length)
Air slide return spring	$3\frac{1}{4}$ in. (82.5mm) (free length)
Air cleaner type	Cloth filter

Clutch

Type	Multi-plate with built-in cush drive
Number of plates	
Driving (bonded)	6
Driven (plain)	6
Driving plate segments	
Number	288
Overall thickness	0.140-0.145 in. (3.56-3.68mm)

8

(continued)

Clutch (continued)

Clutch springs	3
Free length	1$\frac{13}{16}$ in. (46mm)
Working coils	7$\frac{1}{2}$ in.
Spring rate	121 lbs. per inch
Clutch sprocket	
Number of teeth	58
Bore diameter	1.8746-1.8736 in. (47.615-47.590mm)
Clutch hub bearing diameter	1.3740-1.3730 in. (34.9-34.874mm)
Clutch roller diameter (20)	0.2495-0.250 in. (6.337-6.35mm)
Clutch roller length	0.231-0.236 in. (5.867-5.99mm)

Primary Chain

Type	$\frac{3}{8}$ in. triple roller (80 links)

Clutch Operating Rod

Length	11$\frac{1}{8}$ in. (282.5mm)
Diameter	$\frac{7}{32}$ in. (5.5mm)

Gearbox

Internal ratios (standard)	
First	2.51:1
Second	1.60:1
Third	1.144:1
Fourth	1:1
Overall ratios (standard)	
First	12.23:1
Second	7.79:1
Third	5.57:1
Fourth	4.87:1

Gear Detail

Main shaft top gear	
Bushing diameter (fitted)	0.813-0.814 in. (20.6502-20.6756mm)
Bushing length (2)	1$\frac{3}{8}$ in. (35mm)
Bushing protrusion	$\frac{1}{2}$ in. (12.7mm)
Working clearance	0.0027-0.0042 in. (0.0685-0.1066mm)
Layshaft first gear	
Bushing diameter (fitted)	0.7495-0.7505 in. (19.0373-19.0627mm)
Working clearance	0.001 in. maximum (0.0127mm maximum)
Gearbox shafts	
Main shaft (left end) diameter	0.8098-0.8103 in. (20.568-20.581mm)
Main shaft (right end) diameter	0.7495-0.7499 in. (19.037-19.057mm)
Length	10$\frac{5}{8}$ in. (269.875mm)
Layshaft (left end) diameter	0.7495-0.750 in. (19.037-19.05mm)
Layshaft (right end) diameter	0.7495-0.750 in. (19.037-19.05mm)
Length	6$\frac{5}{8}$ in. (168.3mm)

Gearbox Bearings

Main shaft top gear bearing	2$\frac{1}{2}$ × 1$\frac{1}{4}$ × $\frac{5}{8}$ in. ball joint
Main shaft bearing right-side	$\frac{3}{4}$ × 1$\frac{7}{8}$ × $\frac{9}{16}$ in. ball joint
Layshaft bearing left-side	1 × $\frac{3}{4}$ × $\frac{3}{4}$ in. needle roller
Layshaft bearing right-side	1 × $\frac{3}{4}$ × $\frac{3}{4}$ in. needle roller

(continued)

Kickstarter Ratchet

Pinion bore diameter	0.937-0.938 in. (23.8-23.82mm)
Bushing (outside diameter)	0.933-0.935 in. (23.7-23.75mm)
Bushing (inside diameter)	0.750-0.751 in. (19.05-19.075mm)
Outside working clearance	0.002-0.005 in. (0.0508-0.127mm)
Inner working clearance	0.0001-0.0015 in. (0.0254-0.0381mm)
Ratchet spring free length	½ in. (12.7mm)

Gear Selector Quadrant

Plunger diameter	0.3352-0.3362 in. (8.514-8.539mm)
Housing diameter	0.3427-0.3437 in. (8.7045-8.73mm)
Working clearance	0.0065-0.0085 in. (0.165-0.216mm)

Cam Plate Plunger

Plunger diameter	0.4335-0.4325 in. (11.011-10.985mm)
Housing diameter	0.437-0.4375 in. (11.0998-11.1125mm)
Working clearance	0.0050-0.0035 in. (0.127-0.089mm)
Spring free length	2¼ in. (57.15mm)

FRAME AND FITTINGS

Steering Head

Taper roller bearings	Timken LM.11949L

Swing Arm

Bushing type	Glacier (butt joint)
Bushing bore	1 in. (25.4mm) (nominal)
Bushing outside diameter	1⅛ in. (28.57mm) (nominal)
Sleeve outside diameter	0.9984-0.9972 in. (25.359-25.328mm)
Sleeve bore	0.628-0.626 in. (15.951-15.900mm)
Housing diameter	1.1250-1.1260 in. (28.575-28.600mm)
Spindle diameter	0.625-0.619 in. (15.875-15.723mm)

Rear Shock Absorbers

Type	Coil-spring hydraulically damped
Springs	
Free length	9.48 in. (241mm)
Fitted length	8.44 in. (214mm)
Rate	88 lbs. per inch

Front Forks

Type	Coil-spring, hydraulically damped
Springs	
Free length	19.80 in. (503mm)
Fitted length	18.5 in. (470mm)
Rate	25 lbs. per inch
Number of coils	68
Shaft diameter	1.3610-1.3605 in. (34.569-34.557mm)
Outer member bore diameter	1.3634-1.363 in. (34.646-34.620mm)

8

(continued)

WHEELS, BRAKES AND TIRES

Wheels

Rim size and type
    Front   WM2 $\times$ 19
    Rear   WM3 $\times$ 18

Wheel Bearings

Front (left- and right-side)   Hoffmann 120
Front wheel bearing size   20 $\times$ 47 $\times$ 14mm
Rear (left- and right-side)   Hoffmann 120
Rear wheel bearing size   20 $\times$ 47 $\times$ 14mm
Spindle diameter   0.625-0.624 in. (15.875-15.849mm)

Chain

Chain size   ⅝ in. $\times$ ⅜ in. $\times$ 110 links

Brakes

Front (two leading shoes)
    Diameter   8 in. (203.2mm)
    Lining area, sq. in.   23¼ (150cm²)
Rear
    Diameter   7 in. (177.8mm)
    Width   1⅛ in. (28.6mm)
    Lining thickness   3/16 in. (4.7mm)
    Lining area, sq. in.   15.5 (100cm²)

Tires

Size
    Front   3.25 $\times$ 19 in. Dunlop
    Rear   4.00 $\times$ 18 in. Dunlop
Pressure
    Front   21 lbs. per sq. in.
    Rear   22 lbs. per sq. in.

ELECTRICAL EQUIPMENT (12-volt)

Alternator, type	Lucas RM.21	(54021167)
Zener diode	Lucas ZD.715	(54048288)
Rectifier	Lucas 2DS.506	(54048008)
Coils (2)	Lucas 17.M12	(45221)
Contact breaker	Lucas 6CA	
Battery	Lucas PUZ5A	(54027029)
Horn	6H	(70216)
Bulbs		
Headlight	Lucas 370	50/40 watt
Parking light	Lucas 989	6 watt
Stop/taillight	Lucas 380	21/6 watt
Speedometer light	Smiths	2.2 watt
Headlight main beam indicator bulb	Lucas	2 watt
Oil pressure indicator warning	Lucas	2 watt
Direction indicator warning	Lucas	2 watt
Direction indicator	Lucas	21 watt
Flasher unit	Lucas 8FL	(35048)
Handlebar switches	Lucas 169.SA	(39595 L.H.)
		(39596 R.H.)

(continued)

ELECTRICAL EQUIPMENT (12-volt) (continued)

Spark Plugs

Type	Champion N3
Gap setting	0.025 in. (0.63mm)
Thread size	14mm diameter × ¾ in. (19mm) reach

Capacities

Fuel tank	3 gals. (11.5 liters)
Fuel tank (special order only)	4.8 gals. (18 liters)
Oil	6 pints (3 liters)
Gearbox	1 pint (490 cm³)
Primary drive (approximate)	⅓ pint (140 cm³)
Front forks (each leg)	⅖ pint (190 cm³)

Basic Dimensions

Wheelbase	56 in. (142.2 cm)
Overall length	87½ in. (222 cm)
Overall width	33 in. (84 cm)
Overall height	43 in. (109 cm)
Ground clearance (unladen)	7½ in. (19 cm)

A65T THUNDERBOLT

Carburetor (concentric)

Type	Amal R.928/17 (concentric)
Main jet	230
Needle jet size	0.106 in.
Needle position	1st notch
Throttle valve	3½
Nominal choke size	28mm
Air cleaner type	Cloth filter

Spark Plugs

Type	Champion N4

Inlet Manifold

Carburetor port size	1⅛ in. (28.58mm)
Cylinder head port size	1¹⁄₁₆ in. (26.99mm)

Capacities

Fuel tank	4 gallons (18 liters)

A65FS FIREBIRD SCRAMBLER

Carburetor (concentric)

Type	Amal R.930/72 (right-hand, concentric)
	Amal L.930/73 (left-hand, concentric)
Main jet	220
Needle jet	0.106 in.
Needle position	1st notch
Throttle valve	3
Nominal choke size	30mm
Air cleaner type	Cloth filter

Spark Plugs

Type	Champion N3
Gap setting	0.025 in. (0.63mm)
Thread size	14mm diameter × ¾ in. (19mm) reach

8

INDEX

9

CLYMER® *Collection Series*

BSA
500 & 650cc UNIT TWINS • 1963-1972

NORTON
750 & 850cc COMMANDOS • 1969-1975

TRIUMPH
500, 650 & 750cc TWINS • 1963-1979

1
2
3
4
5
6
7
8
9
10
11
12

CONTENTS

QUICK REFERENCE DATA

BREAKER POINTS

BREAKER PLATE SCREWS

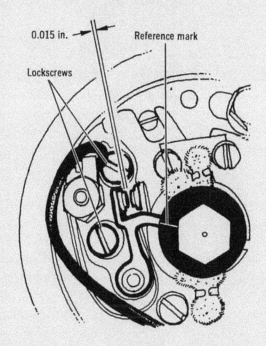

0.015 in.

Reference mark

Lockscrews

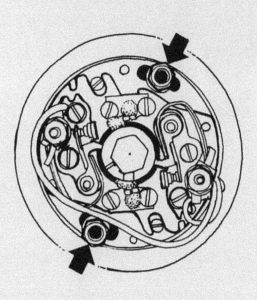

TUNE-UP SPECIFICATIONS

Cylinder head fastener torque	5/16 in. = 20 ft.-lb. (2.75 mkg)
	3/8 in. = 30 ft.-lb. (3.67 mkg)
Valve clearance	
Standard 750/850	Inlet = 0.006 in. (0.15mm)
	Exhaust = 0.008 in. (0.2mm)
Combat 750	Inlet = 0.008 in. (0.2mm)
	Exhaust = 0.010 in. (0.25mm)
Spark plug	
Type	Champion N7Y (standard engines)
	Champion N6Y (Combat engines and high-speed)
Gap	0.023-0.028 in. (0.59-0.71mm)
Contact breaker gap	0.014-0.016 in. (0.35-0.4mm)
Timing (fully advanced position)	28° BTDC
Idle speed	1,000 rpm

CARBURETOR ADJUSTMENT

Stop screw   Air screw

TIMING MARK ALIGNMENT

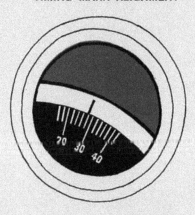

FLUIDS

	Type	Quantity
Engine oil	SAE 50W (API SE)	3 U.S. qt. (5 Imp. pt.; 2.8 liter)
Primary drive oil	SAE 20W-50	7 oz. (200cc)
Transmission oil	SAE 90 EP	14.5 U.S. oz. (0.75 Imp. pt.; 420cc)
Fork oil	SAE 10W 30 (also see Table 3, Chapter Two)	5 oz. (150cc) each leg
Swing arm pivot	SAE 140	As required
Brake fluid	Lockheed 329 hydraulic fluid	As required
Fuel	Premium	As required

TIRES

Type	Dunlop K-81*
Size	
Front	4.10 x 19 in.
Rear	4.10 x 18 in.
Pressure	
Solo	24 psi (front); 26 psi (rear)
With passenger	26 psi (front); 28 psi (rear)
With passenger and luggage (100 lb.)	28 psi (front); 32 psi (rear)

* The best all-around performance of any tire available for the Commando.

ADJUSTMENTS

	Inch	mm
Cam chain movement (maximum)	3/16	4.8
Primary chain free play	3/8	10
Clutch cable free play	3/16-1/4	5-6
Throttle cable free play	1/8	3

CLYMER® *Collection Series*

NORTON

750 & 850cc COMMANDOS • 1969-1975

CHAPTER ONE

GENERAL INFORMATION

This handbook covers all service operations for late-model Norton twins—from changing a spark plug to overhauling an engine. The information applies specifically to the following models:

750 Commando — Standard

750 Commando — Combat

850 Commando Mark II and Mark IIA

850 Commando Mark III
(Electric Start)

In addition, the majority of the engine and transmission service procedures, as well as those for the front and rear suspension units and drum types brakes, are applicable to the earlier 750 G15 CS and the 750 Atlas. To further aid servicing of these early models, an engine exploded view is included in addition to electrical system schematics.

A WORD OF CAUTION

The large-displacement Norton twin is a high-performance motorcycle designed for high-speed touring and open-class production racing. As such, it represents a total design approach; engine performance, chassis geometry, and component selection are all carefully integrated to provide not only high performance but safe

performance as well. Alterations to or substitution of suspension and brake components in an attempt to improve a late model or update an early one are not recommended. Changes in wheelbase, trail, caster, center of gravity, etc., resulting from uninformed chassis modification can render the motorcycle unsafe.

PARTS ORDERING

To prevent inadvertent installation of an incorrect part, always state your motorcycle's serial number when ordering replacement components. In this way your dealer can determine the correct part. Your motorcycle will be safer and more enjoyable as a result.

SERVICE HINTS

Most of the service procedures described can be performed by anyone reasonably handy with tools. However, carefully consider your own capabilities before attempting any operation which involves major disassembly of the engine or transmission.

Some operations, for example, require the use of a press. It would be wiser to have them performed by a shop equipped for such work, rather than to try the job yourself with makeshift equipment. Other procedures require precision measurements. Unless you have the skills and equipment to make them, it would be better to have a motorcycle shop do the work.

Repairs are faster and easier if the motorcycle is clean before you begin work. There are special cleaners for washing the engine and related parts. Just brush or spray on the solution, let it stand, then rinse it away with a garden hose. Clean all oily or greasy parts with cleaning solvent as you remove them. *Never use gasoline as a cleaning agent.* It presents an extreme fire hazard. Always work in a well-ventilated area when using cleaning solvent. Keep a fire extinguisher, rated for gasoline fires, handy just in case of emergency.

Special tools are required for some service procedures. Some of these may be purchased through Norton dealers. If you are on good terms with the dealer's service department, you may be able to borrow his.

Much of the labor charge for repairs made by dealers is for removal and disassembly of other parts to reach the defective one. It is frequently possible to do all this yourself, then take the affected subassembly to the dealer for repair.

Once you decide to tackle a job yourself, read the entire section in this handbook pertaining to it. Study the illustrations and the text until you have a thorough idea of what's involved. If special tools are required, make arrangements to get them before you begin work. It's frustrating to get part way into a job and then discover that you are unable to complete it.

TOOLS

To properly service your motorcycle, you will need an assortment of ordinary hand tools. As a minimum, these include:

1. Combination wrenches
2. Socket wrenches
3. Plastic mallet
4. Small hammer
5. Snap ring pliers
6. Phillips screwdrivers
7. Slot screwdrivers
8. Impact driver
9. Pliers
10. Feeler gauges
11. Spark plug gauge
12. Spark plug wrench
13. Dial indicator
14. Drift

An original equipment tool kit, like the one shown in **Figure 1**, is available through most Norton dealers and is suitable for most minor servicing.

Electrical system servicing requires a voltmeter, ohmmeter or other device for determining continuity, and a hydrometer for the battery.

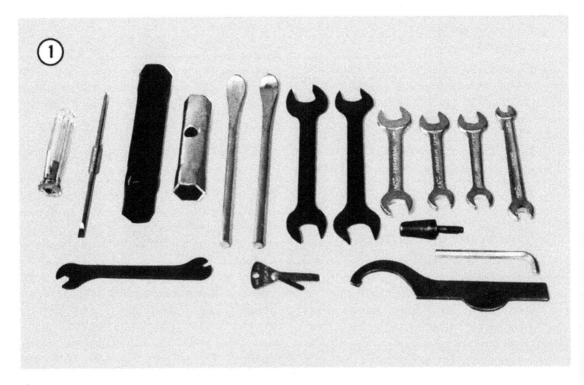

There are only 3 essential special tools required for working on the Norton — clutch spring compressor, Figure 69, Chapter Four; engine sprocket puller, Figure 73, Chapter Four; and a C wrench for removing and tightening the exhaust headpipe nuts.

EXPENDABLE SUPPLIES

Certain expendable supplies are required. These include grease, oil, gasket cement, liquid fastener-locking compound, rags, and cleaning solvent. These items are available at most motorcycle shops and auto supply stores. Distilled water for batteries is available at supermarkets.

SAFETY FIRST

A professional mechanic can work for years and never sustain a serious injury. If you observe a few rules of common sense and safety, you can enjoy many hours safely servicing your own machine. You can also hurt yourself or damage your motorcycle if you ignore these rules.

1. Never use gasoline as a cleaning solvent.

2. Never smoke or use a torch around flammable liquids, such as cleaning solvent.

3. Never smoke or use a torch in areas where batteries are being charged. Highly explosive hydrogen gas is formed during the charging process. And never arc the terminals of a battery to see if it has a charge; the sparks can ignite the explosive hydrogen as easily as would an open flame.

4. If welding or brazing is required on the motorcycle, remove the fuel tank and set it a safe distance away—at least 50 feet.

5. Always use the correct size wrench for turning nuts and bolts, and when a nut is tight, think for a moment what would happen to your hand if the wrench were to slip.

6. Keep your work area clean and uncluttered.

7. Wear safety goggles in all operations involving drilling, grinding, the use of a chisel, or an air hose.

8. Don't use worn tools.

9. Keep a fire extinguisher handy. Be sure it is rated for gasoline and electrical fires.

CHAPTER TWO

PERIODIC MAINTENANCE AND LUBRICATION

Regular maintenance is the best guarantee of a trouble-free, long-lasting motorcycle. An afternoon spent now, cleaning and adjusting, can prevent costly mechanical problems in the future and unexpected breakdowns on the road.

The procedures presented in this chapter can be easily carried out by anyone with average mechanical skills. The operations are presented step-by-step and if they are followed, it is difficult to go wrong.

SERVICE INTERVALS

The services and intervals shown in **Table 1** are recommended by the factory. Strict adherence to these recommendations will go a long way in ensuring long service life from your Norton.

For convenience in maintaining your motorcycle, most of the services shown in the table are described in this chapter. However, some procedures which require more than minor disassembly or adjustment are covered elsewhere in this book as indicated.

TIRE PRESSURE

Tire pressures should be checked and adjusted to accommodate rider and luggage weight. A simple, accurate gauge can be purchased for a few dollars and should be carried in the motorcycle tool kit. The appropriate tire pressures are shown in **Table 2**.

BATTERY ELECTROLYTE LEVEL

The battery is the heart of the electrical system. It should be checked and serviced as indicated. The majority of electrical system troubles can be attributed to neglect of this vital component.

The electrolyte level may be checked with the battery installed. However, it's necessary to remove the left cover plate (**Figure 1**). The electrolyte level should be maintained between the 2 marks embossed on the battery case (**Figure 2**). If the electrolyte level is low, it's a good idea to remove the battery from the motorcycle so that it can be thoroughly serviced and checked.

1. On 750 and 850 Mark II models, slide the metal loops off the hold-down bar and remove the bar (**Figure 3**). On 850 Mark III models, disconnect the battery strap buckle from the hook on the battery carrier (**Figure 4**).

2. Disconnect the electrical leads from the battery terminals—first the positive (ground) and then the negative (**Figure 5**).

3. Disconnect the vent pipe and lift the battery out of the holder.

Table 1   SERVICE INTERVALS

Weekly	Check tire pressure
Every 2 weeks	Check battery electrolyte level
Every 250 miles	Check engine oil tank level
Every 1,000 miles	Check primary chaincase oil level Adjust rear chain Lubricate all control cables with oil Adjust both brakes (optional disc brake is non-adjustable) Check disc brake fluid level Examine disc brake pads for wear
Every 2,500-3,000* miles	Check timing and adjust contact breaker points Clean spark plugs and set gaps Change primary chaincase oil Check clutch adjustment Check primary chain adjustment Change engine oil Lubricate and adjust rear chain Check gearbox oil level Check front and rear rubber engine mountings for side-play Change oil in forks Grease rear brake pedal pivot Check isolastic mountings for free-play
Every 5,000-6,000* miles	Change gearbox oil Replace oil filter element Clean contact breaker points Lubricate contact breaker cam felt and auto advance unit Grease brake expander pivots (one stroke of grease gun) Check and adjust valve rocker clearances Check and adjust camshaft chain Fit new air filter element Check and oil swinging arm bushings
Every 10,000-12,000* miles	Repack wheel bearings (including the rear wheel sprocket bearing) with grease Dismantle and clean both carburetors and check for wear

*Longer intervals correspond to factory-recommended intervals for 850 Mark II and Mark III models.

Table 2   TIRE PRESSURES

Load		Pressure*
Rider only (approx. 170 lb.)	Front	24 psi (1.7 Kg/sq. cm)
	Rear	26 psi (1.8 Kg/sq. cm)
Rider and passenger (approx. 340 lb.)	Front	26 psi (1.8 Kg/sq. cm)
	Rear	28 psi (1.969 Kg/sq. cm)
Rider and passenger plus 100 lb. luggage (approx. 440 lb.)	Front	28 psi (1.969 Kg/sq. cm)
	Rear	32 psi (2.250 Kg/sq. cm)

*Dunlop only—Avon min. 26 psi front and rear.

2

CAUTION
Be careful not to spill battery electrolyte on painted or polished surfaces. The liquid is highly corrosive and will damage the finish.

4. Remove the caps from the battery cells and add distilled water to correct the level. *Never add electrolyte (acid) to correct the level.*

5. After the level has been corrected and the battery allowed to stand for a few minutes, check the specific gravity of the electrolyte in each cell with a hydrometer. Follow the hydrometer manufacturer's instructions for reading the instrument. The specific gravity should be 1.270-1.290 at a temperature of 60°F. If it is substantially less, charge the battery at a rate of one ampere for at least 4 hours or until the specific gravity is at an acceptable level.

WARNING
During charging, highly explosive hydrogen gas is released from the battery. The battery should be charged only in a well-ventilated area, and open flames and lighted cigarettes should be kept

away. Never check the charge of the battery by arcing across the terminals; the resulting spark can ignite the hydrogen gas.

6. Connect the positive charger lead to the positive battery terminal and the negative charger lead to the negative battery terminal.

> NOTE: *The battery may be connected to a charger using the electrical outlet on the right side of the motorcycle* (**Figure 6**). *The plug should be wired as shown in* **Figure 7**. *The battery leads on the motorcycle must be connected.*

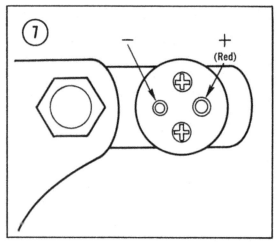

7. Remove the vent caps from the battery, set the charger at 12 volts, and switch it on. If the output of the charger is variable, it's best to select a low setting—1½ to 2 amps.

8. After the battery has been charged for about 8 hours, turn off the charger, disconnect the leads, and check the specific gravity. It should be within the limits specified above. If it is, and if it remains stable after one hour, the battery is charged.

9. Clean the battery terminals and case, and reinstall it in the motorcycle, reversing the removal steps. Coat the terminals with Vaseline or silicone spray to retard decomposition of the terminal material.

ENGINE OIL

Checking Level

Engine oil level is checked with the dipstick mounted in the tank filler cap (**Figure 8**).

1. Start the engine and allow it to run for a couple of minutes so that the excess oil will be returned from the crankcase to the oil tank.

2. Remove the seat and unscrew the filler cap from the oil tank. Visually check to see that the oil is circulating in the tank. Replace the filler cap to prevent oil from being splashed out.

3. Shut off the engine and allow the oil to settle in the tank. Then, check the level with the dipstick. The oil level should be above the "L" mark but not above the "H" mark (**Figure 9**); if the level is above the "H" mark, oil will overflow into the air filter. If necessary, add the recommended grade of oil to correct the level. Install the filler cap and tighten it securely.

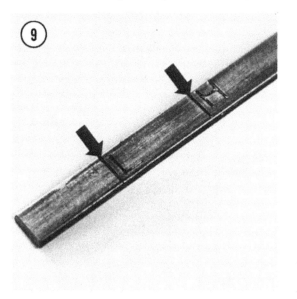

Changing

The factory recommended oil change interval is 2,500-3,000 miles. This assumes that the motorcycle is operated in moderate climates. In extremely cold climates, oil should be changed every 30 days. The time interval is more important than the use interval because acids formed by gasoline and water vapor from condensation will contaminate the oil even if the motorcycle is not run for several months. Also, if the motorcycle is operated under dusty conditions the oil will get dirty more quickly and should be changed more frequently than recommended.

Use only a detergent oil with an API rating of SE or better. The quality rating is stamped on the top of the can. Try always to use the same brand of oil. Oil additives are not recommended.

A 50W oil is recommended for use at ambient temperatures above 0°F (—18°C).

1. Start the engine and run it until it is warm. Then shut it off.

2. Remove the seat and the right side cover (**Figure 10**).

3. Remove the drain plug from the oil tank (**Figure 11**). On early models, remove the bolt which attaches the oil filter union to the tank.

4. Remove crankcase drain plug (with 1½ in. socket) and sump filter if fitted (**Figure 12**). Dismantle the sump filter (**Figure 13**) and clean it thoroughly in solvent. Clean the magnetic drain plug. Remove the filter from the oil tank (**Figure 14**) and thoroughly clean it with solvent.

5. Remove and discard the oil filter element (5,000-6,000-mile interval). See **Figure 15**.

6. When the oil has drained, install a new filter element and a new clamp and reinstall the drain

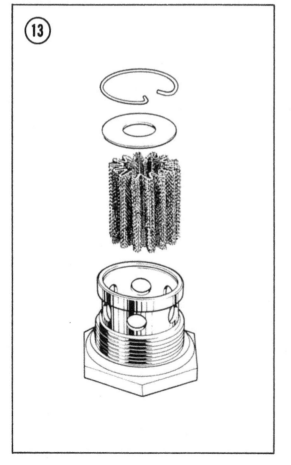

plugs and tank and sump filters. Fill the oil tank with fresh oil to the level described earlier (3 qts.—U.S.; 5 pts. Imp.—U.K.; 2.8 liters). Start the engine and run it at a moderate, steady speed. Visually check the oil circulation in the tank. Because the sump was drained earlier, the scavenge side of the oil pump will not immediately begin to pump oil back to the tank. Shut the engine off and allow the oil to settle for a couple of minutes. Recheck the level and correct it if necessary.

7. Reinstall the side cover and the seat.

PRIMARY CHAINCASE

The oil level in the primary chaincase should be checked every 1,000 miles and changed every 2,500 miles.

Checking Level

1. With the motorcycle sitting level, remove the level plug from the chaincase (**Figure 16**). If the oil level is correct, a small amount of oil should run out of the hole.

2. On 750 and 850 Mark II models, if oil does not run out, drain and refill the primary case as described below; it is important that the case contain exactly 7 ounces (200cc) of oil. On 850 Mark III models, if oil does not run out, remove the filler cap and slowly add oil (SAE 20W-50) until it begins to run out of the level hole.

3. Reinstall the level plug and the filler cap. See Figure 16.

Changing

1. Remove the left footrest. Place a drip pan beneath the chaincase and remove the center bolt. Tap along the edges of the chaincase with a soft mallet to break the seal and allow the oil to drain.

2. Remove the chaincase cover and clean it and the inner case with solvent.

3. Reinstall the chaincase cover and remove the filler cap and level plug.

4. Pour 7 ounces of clean SAE 20W-50 oil into the chaincase through the filler opening. Wait a moment until the oil ceases to run out of the level hole and then install the plug and cap. Reinstall the footrest.

TRANSMISSION

The transmission oil should be checked every 2,500 miles and should be changed every 5,000 miles.

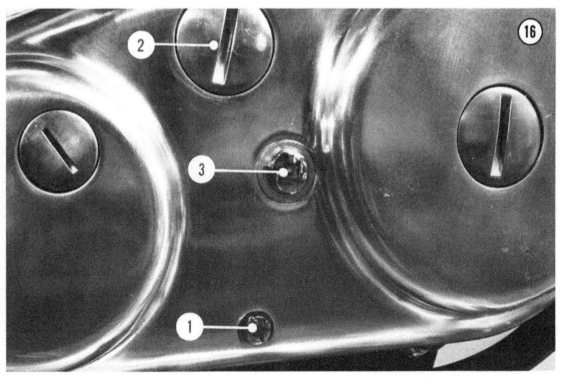

1. Level plug   2. Fill plug   3. Center bolt (except Mk III)

Checking Level

1. Remove the level plug from the transmission (**Figure 17**). If the oil level is correct, a small amount of oil should run out of the hole.

2. If oil does not run out, remove the filler cap (**Figure 18**) and slowly add oil (SAE 90 EP) until it just begins to run out of the level hole.

3. Reinstall the level plug and the filler cap.

Changing

1. Ride the motorcycle for several miles to warm up the oil in the transmission.

2. Place a drip pan beneath the transmission and remove the filler cap and drain plug (**Figure 19**).

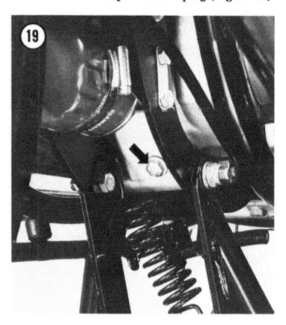

3. After the oil is drained, reinstall the drain plug and fill the transmission with 14.5 ounces of oil (U.S.); 0.75 pts. Imp. (U.K.); 420cc. Wait a few minutes until the oil has drained down into the transmission. Then remove the level plug and permit the excess oil to drain out.

4. Reinstall the level plug and the filler cap.

SWINGING ARM

Oil in the swinging arm pivot of 750 and 850 Mark II models should be checked every 5,000 miles and added to if necessary. The oiled wicks used in 850 Mark III models are good for the life of the swinging arm pivot bushings and need be replaced only when the bushings are replaced.

1. Remove the spindle locating bolt from the top of the swinging arm pivot (**Figure 20**).

2. Use a grease gun filled with SAE 140 oil. Inject oil into the pivot tube through the lube fitting (**Figure 21**) until it begins to run out of the spindle locating bolt hole. If a grease gun is not available, the lube fitting can be removed and oil added with a pumper-type oil can.

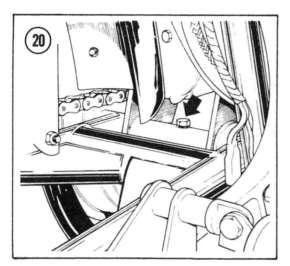

3. Reinstall the spindle locating bolt and the lube fitting (if removed).

FRONT FORK

The damping oil in the front fork should be changed every 2,500 miles or at any time excessive bouncing of the front end indicates a low oil level. There is no practical way of checking and correcting the level; each fork leg must contain exactly 150cc (5 ounces) of damping oil if the front supension is to operate correctly.

If after the oil has been changed the front suspension continues to bounce or "hobby-horse," a major service may be required. In such case, refer to Chapter Ten.

1. Turn the front end to left lock, place a drip pan beneath the left fork leg, and remove the drain screw (**Figure 22**).

2. When the oil has ceased to drain, lock the front brake and depress the front end several times to expel residual oil. Reinstall drain plug.

3. Turn the front end to right lock and drain the right leg in the same manner as for the left.

4. Place a clean shop towel over the gas tank to protect it.

5. Unscrew the fill plugs (1-5/16 in. socket) from the tops of the fork legs (**Figure 23**). Wrap a shop rag around each of the instruments and tape them in place to prevent damage to them and the gas tank.

6. Lift the front wheel to expose the springs and support it with a block of wood (**Figure 24**).

7. Unscrew the filler plugs from the tops of the damper rods (**Figure 25**).

8. Remove the block from beneath the wheel and fully extend the forks.

9. Fill each fork leg with 150cc (5 ounces) of any of the oils shown in **Table 3**. See **Figure 26**.

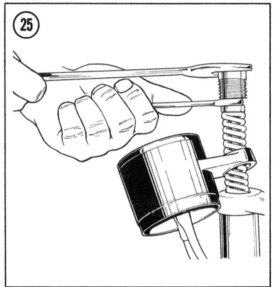

Table 3   FORK OIL

Brand	Type
Castrol	Castrolite 10W 30
BP	BP Super Visco-Static 10W 40
Shell	Shell Super Motor Oil
Mobil	Mobiloil Super
ESSO	Uniflo
Texaco	Havoline Motor Oil 10W 30
Duckham's	Duckham's Q5500
Sun Oil	Sunoco Special Motor Oil 20W 50

NOTE: *Each leg will probably not accept all of the oil at first. Cover the top of the fork tube with your hand and pump the slider slowly up and down to distribute the oil and then add more oil until the entire 150cc has been poured in.*

10. Raise and block the wheel once again. Screw the locknuts all the way on to the bottom thread (**Figure 27**). Slide the instrument mounts over the springs and screw on the filler caps and lock them with the nuts (**Figure 28**).

11. Remove the block and lower the front wheel. Screw in the top caps and tighten them securely. Reinstall the handlebars.

HYDRAULIC DISC BRAKE

The hydraulic fluid level in the disc brake master cylinder should be checked every 1,000 miles and the brake pads should be checked for wear. Bleeding the hydraulic system, servicing the master cylinder, caliper, and disc, and replacing brake pads are covered in Chapter Eight.

Fluid Level

1. Clean the reservoir cap thoroughly with a dry rag and unscrew it. Remove the flexible bellows and set it in the upturned cap to prevent it from getting dirty (**Figure 29**).

2. The fluid level in both brake reservoirs should be ½ in. (13mm) from the top edge (**Figures 30 and 31**). If necessary, correct the level by adding any DOT 3 approved hydraulic brake fluid.

3. Reinstall the bellows with the closed end down. Do not pour hydraulic fluid into the bellows. Check the vent hole in the top of the cap to make sure it is open and screw the cap on tightly.

CAUTION
Be careful not to spill the brake fluid on painted surfaces or bring it in contact with plastic or rubber components.

Brake Pad Wear

Inspect the brake pads for excessive or uneven wear, scoring, and oil or grease imbedded in

the friction material. If any of these conditions exist, replace the pads, as a set, as described in Chapter Eight.

REAR CHAIN

The rear drive chain should be checked and adjusted every 1,000 miles and removed, cleaned, and lubricated every 2,500 miles.

Adjustment

1. Loosen the axle nuts (**Figure 32**) and the chain adjuster locknuts.

2. Screw the adjusters either in or out as required, in equal amounts. The free movement of the chain, midway between the sprockets, should be ¾-1 in. (19-25mm) with a rider seated on the motorcycle (**Figure 33**). Rotate the rear wheel to move the chain to another position and recheck the adjustment; chains rarely wear or stretch evenly and as a result the free-play will not remain constant over the entire chain. At its tightest point it should have no less than ¾ in. free-play and at its loosest the free-play should not be greater than one inch. If the chain cannot be adjusted within these limits it is excessively worn and stretched and should be replaced.

3. When the adjustment is correct, sight along the chain from the rear sprocket to see that it is correctly aligned. It should leave the top of the rear sprocket in a straight line. If it is cocked to one side or the other, the rear wheel is incorrectly aligned and must be corrected by turning the adjusters counter to one another until the chain and sprocket are correctly aligned. When

3/4 –1'

the alignment is correct, readjust the free-play as described above and tighten the adjuster locknuts and axle nuts securely.

Cleaning and Lubrication

The rear chain is lubricated by a restricter type oiler connected to the engine oil tank. Every 2,500 miles, remove, thoroughly clean, and lubricate the chain.

1. Disconnect the master link and remove the chain from the motorcycle. If a piece of old chain is available, connect it to the motorcycle's chain before it is removed and leave the old chain around the gearbox sprocket to make chain replacement easier.

2. Inspect the chain for excessive wear and stretch. Scribe 2 marks 12½ in. (317mm) apart on a flat surface. With the chain compressed to its minimum free length, the marks should line

up with pivot pins exactly 20 links apart (**Figure 34**). When the chain is extended, the 20 links should not exceed 12¾ in. (324mm).

3. Immerse the chain in a pan of cleaning solvent and allow it to soak for about a half hour.

4. Scrub the rollers and side plates with a stiff brush and rinse the chain in clean solvent to carry away loosened grit. Hang up the chain and allow it to dry thoroughly.

5. Lubricate the chain with a good grade of chain lubricant carefully following the manufacturer's instructions. As an alternative, lubricate by soaking in a pan of heated all-purpose grease such as Castrol Graphited Grease, Shell Retinex A or DC, Mobilgrease MP, or Marfax All-Purpose Grease. Heating permits the grease to penetrate the rollers and pins, but extreme care must be taken.

WARNING

If the grease is heated excessively it may reach its flash point, resulting in a dangerous and difficult to extinguish fire. NEVER heat the grease with an open flame or on a hot plate. Heat it only by placing the grease pan in a larger pan containing about an inch of boiling water, and only after the water has been removed from the heat source.

After the chain has soaked in the grease for about a half hour, remove it from the pan and wipe off all the excess grease with a clean rag.

6. Reinstall the chain on the motorcycle. Use a new clip on the master link and install it so that the closed end of the clip faces the direction of chain travel (**Figure 35**).

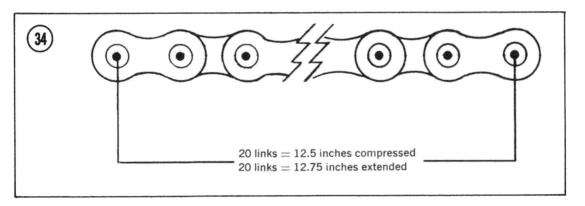

20 links = 12.5 inches compressed
20 links = 12.75 inches extended

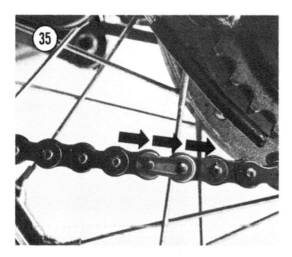

7. Adjust the free-play of the chain as described earlier in this section.

CONTROL CABLES

The control cables, with the exception of the Teflon lined throttle cable on 850 Mark III models, should be lubricated every 1,000 miles. In addition, all cables should be adjusted at this interval. The cables should be inspected periodically for fraying and the cable sheaths should be checked for chafing. The cables are expendable items and should be replaced when found to be faulty.

Lubrication

The control cables can be lubricated either with oil or with any one of the popular commercial cable lubricants and a cable lubricator. The first method requires more time and the complete lubrication of the entire cable is less certain.

Oil Method

1. Disconnect the cable from the control (throttle grip, clutch lever, etc.).
2. Make a cone from a stiff piece of paper and tape it to the end of the cable sheath (**Figure 36**).
3. Hold the cable upright and pour a small quantity of light oil into the cone. Work the cable in and out of the sheath for several minutes to help the oil migrate down between cable and sheath.

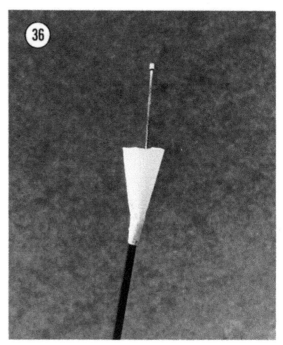

4. Remove the cone, reconnect the cable, and adjust it as described below.

Lubricator Method

1. Disconnect the cable from the control.
2. Attach the lubricator in accordance with the manufacturer's instructions (**Figure 37**).

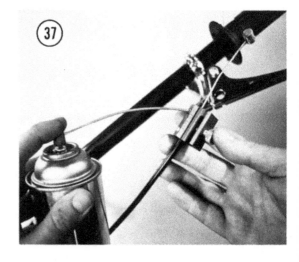

3. Insert the lubricant can nozzle in the lubricator, press the button on the can, and hold it down until the lubricant begins to flow out of the other end of the cable.

NOTE: *On throttle and choke cables,
remove the carburetor cap so the lubricant will not run into the carburetor.*

4. Remove the lubricator, reconnect the cable
to the control, and adjust it as described below.

Throttle Cable Adjustment

Throttle cable adjustment is described under
Tune-up at the end of this chapter.

Clutch Cable Adjustment

There should be 3/16-1/4 in. (4.6-6.3mm)
free play in the clutch cable sheath (**Figure 38**).
If this free play cannot be obtained with the
adjuster on the clutch control lever, screw the
adjuster all the way in and remove the clutch
adjuster cover from the primary chaincase
(**Figure 39**) and the inspection cover from the
transmission (**Figure 40**). Check to see if there
is any movement in the clutch operating lever
(**Figure 41**). If there is, the basic adjustment is
all right and it should be possible to adjust the
free play at the hand control. If there isn't any
movement, loosen the locknut (**Figure 42**) and
slowly turn the adjuster screw counterclockwise
until there is movement in the operating lever.
Turn the adjuster screw clockwise until you feel
it touch the clutch operating rod, then back it
out *one full turn*. Hold the screw to prevent it
from turning further and tighten the locknut.
Turn the cable adjuster at the hand control out
until the free play is correct, as shown above.

BRAKE ADJUSTMENT

Front Brake (Drum Type)

The front brake cable should be adjusted so
that there is a minimum of brake lever movement required to actuate the brake, but it must
not be so closely adjusted that the brake shoes
contact the drum with the lever relaxed. The
primary adjustment should be made with the
control lever adjuster. Because of normal brake
wear, this adjustment will eventually be "used
up." It is then necessary to loosen the control
lever adjuster and screw out the adjuster on the
brake backing plate (**Figure 43**) until the control

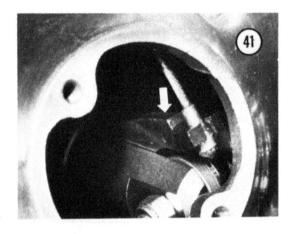

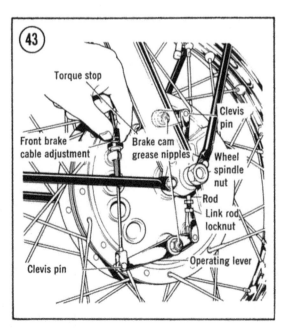

lever adjuster can be used once again for fine adjustment. After the backing plate adjuster has been extended, be sure to tighten the locknut.

Adjustment of the connecting link between the brake cams is described in Chapter Eight.

Rear Brake (Drum Type)

When the brake is fully applied the brake arm should be near perpendicular to the cable (**Figure 44**). Adjust the rear brake cable if necessary. After adjustment, release the brake pedal and spin the rear wheel to make sure the shoes do not contact the drum when the pedal is relaxed.

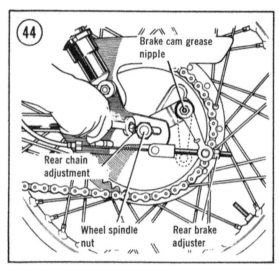

If the brake lever moves past vertical when the brake is applied, it is likely that the brake lining is worn and should be replaced. Refer to Chapter Nine.

GREASE FITTINGS

Grease fittings on 750 and 850 Mark II models are located on the brake cam pivots (2 cams on the drum type front brake, Figure 43), rear brake pedal pivot (**Figure 45**), and swinging arm pivot (**Figure 46**). On 850 Mark III models there is a single grease fitting on the rear brake pedal pivot (**Figure 47**). These points should be lubricated at the intervals shown in Table 1. For fittings on brake components, one stroke of a hand-type grease gun is sufficient. Lubrication of the swinging arm pivot was described earlier.

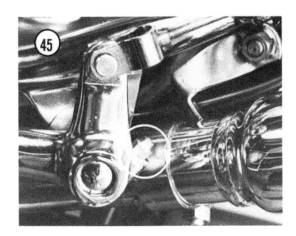

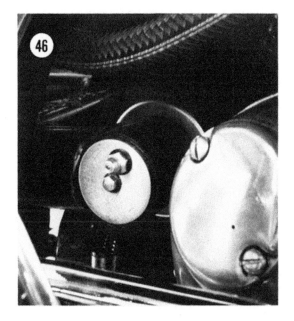

AIR FILTER

The carburetor air filter element on 750 and 850 Mark II models (**Figure 48**) should be replaced every 5,000 miles or sooner if it appears to be excessively clogged. The service life of the filter can be extended by cleaning the element. Large particles can be dislodged by tapping the filter on a work bench or other flat surface, and fine particles can be removed by blowing them out with low-pressure air applied to the inside of the element.

The carburetor air filter element on 850 Mark III models is an oil-wetted type that can be used almost indefinitely, provided it is cleaned and oiled periodically and the foam is not torn or deteriorated. The element should be serviced every 3,000 miles, or more frequently if the motorcycle is operated in extremely dusty conditions. Unscrew the 3 bolts which hold the filter cover in place (**Figure 49**) and slide the element out of the air box, taking care not to snag the foam on the carburetor drain plugs. Wash the filter thoroughly in clean gasoline, wring it out, and allow it to dry.

WARNING
Be extremely careful; clean the element out of doors and keep the gasoline away from all flame and potential

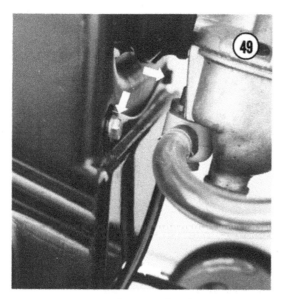

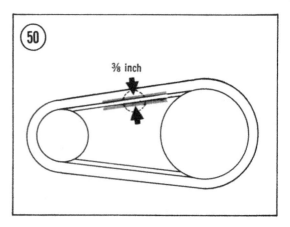

sources of sparks. Pour the dirty gaso-
line into a sealable container (not
glass) and discard it in the trash.

After the cleaned element has dried, saturate
it with clean engine oil, wring out the excess,
and install the element in the air box.

WHEEL BEARINGS

The wheel bearings should be cleaned and
repacked with grease every 10,000 miles or
after submergence or prolonged running in deep
water. The correct service procedures are presented in Chapters Eight and Nine.

PRIMARY CHAIN ADJUSTMENT

The primary chain on all models except the
850 Mark III which is fitted with a hydraulic
tensioner, must be adjusted each time the primary drive assembly is disturbed, and checked
every 25,000 miles and adjusted if necessary.
Total up-and-down movement of the chain
should be ⅜ in. (9.5mm). See **Figure 50**.

1. Remove the oil fill/inspection cap from the
top center of the primary chaincase and check
the movement of the chain with your finger.
Rotate the crankshaft and chain with the kickstarter and check the movement in several locations. At the tightest position, the movement
should be ⅜ in. If it is greater or less than this,
proceed with the next steps.

2. Loosen the top gearbox bolt (do not turn
the captive nut on the left side) and the bottom
nut (**Figure 51**).

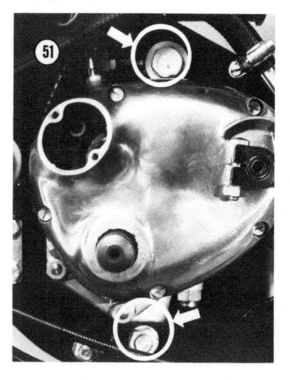

3. Loosen the front adjuster nut several turns
(**Figure 52**). Tighten the rear adjuster nut until
the chain is tight, then back it off until the movement of the chain is correct, while at the same
time tightening the front adjuster nut.

TUNE-UP

A complete tune-up should be performed
every 2,500 miles (650cc and 750cc models)

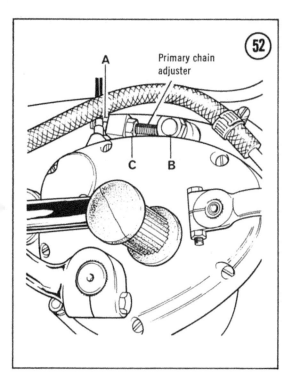

A Primary chain adjuster

C B

or 3,000 miles (850cc Mark II and Mark III models) of normal riding. More frequent tune-ups may be required if the motorcycle is ridden primarily in stop-and-go traffic.

The expendable ignition parts (spark plugs, points, and condenser) should be routinely replaced at every other tune-up or when the point contacts or spark plug electrodes show signs of erosion. In addition, this is a good time to replace the air filter element. Have the new parts on hand before you begin.

Because different systems in an engine interact, the procedures should be done in the following order:

    a. Tighten cylinder head bolts

    b. Adjust valve clearances

    c. Work on ignition system

    d. Adjust carburetors

Cylinder Head Bolts

1. Raise or remove the seat.

2. Shut off the fuel taps, disconnect the lines, and remove the fuel tank. On all original equipment Commando tanks, the front is mounted on 2 studs beneath the tank and held with self-

locking nuts. The rear mount on some models uses a rubber loop which passes beneath the top frame tube, while others are held in place with a metal cross bar.

3. Tighten the nuts and bolts in the sequence shown in **Figure 53** to 20 ft.-lb. (2.75 mkg) for the smaller (5/16 in.) bolts and 30 ft.-lb. (3.68 mkg) for the larger (⅜ in.) nuts and bolts. The fuel tank can be left off at this time.

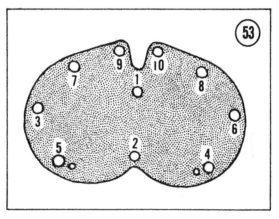

Valve Clearance Adjustment

Valve clearance adjustment must be made with the engine cold. For standard 750cc engines and all 850cc engines, the inlet clearance is 0.006 in. (0.15mm) and the exhaust clearance is 0.008 in. (0.2mm); for the 750cc Combat engine, the inlet clearance is 0.008 in. (0.2mm) and the exhaust clearance is 0.010 in. (0.25mm).

1. Remove the rocker covers and unscrew the spark plugs from the cylinder head.

2. Rotate the crankshaft with the kickstarter until the left inlet valve is completely open (**Figure 54**). Check the clearance of the right inlet valve and rocker. When the clearance is correct, there will be a slight resistance on the feeler gauge when it is inserted and withdrawn.

3. To correct the clearance, back off the locknut and screw the adjuster out far enough to insert the feeler gauge with no resistance. Screw in the adjuster until a slight resistance can be felt on the gauge. Hold the adjuster to prevent it from turning further and tighten the locknut. Then, recheck the clearance to make sure the adjuster did not turn after the correct clearance was achieved.

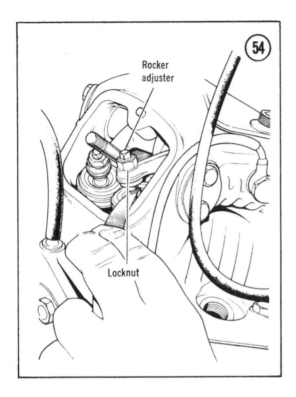

4. Rotate the crankshaft to open the right intake valve completely and check and adjust the clearance of the left inlet valve and rocker. Then, check and adjust the exhaust valves and rockers in the same manner.

5. When all the clearances have been checked and adjusted, install the rocker covers, making sure the gaskets are aligned and in good condition. Reinstall the fuel tank.

Ignition System

Two similar contact breaker assemblies are used on Norton Commandos. They are described in Chapter Seven along with instructions for removal, inspection, and installation. When the breaker assembly has been replaced, or determined to be serviceable, adjust as described below.

1. Remove the spark plugs from the cylinder head and rotate the crankshaft with the kickstarter to line up the mark on the point cam with the lifter heel on one of the contacts (**Figure 55**). This is the maximum contact opening.

2. Measure the contact gap with a flat feeler gauge (**Figure 56**). It should be 0.015 in.

(0.38mm). If it is not, loosen the breaker lockscrews (**Figure 57**) and move the plate in or out until the gap is correct. Tighten the lockscrews and recheck the gap to make sure the plate did not move when the screws were retightened.

3. Rotate the crankshaft to line up the cam mark with the other lifter heel and adjust this contact set in the same manner as above.

4. Examine the spark plugs and compare their condition to **Figure 58**. Their condition is an indication of engine condition and can warn of developing trouble. If the plugs are in good condition and can be reused, clean them with a wire brush and blow out scale and deposits with compressed air.

WARNING
*Hold the electrode end of the plug
away from you when blowing the scale
out with air.*

NOTE: *If the static timing procedure
referenced below is used, do not install
the spark plugs; it will be easier to rotate the crankshaft if there is no compression in the cylinders.*

5. Set the spark plug electrode gap at 0.028 in. (0.71mm) by bending the side electrode. For normal use, and for all models except the 750 Combat, Champion N7Y spark plugs are recommended; however, equivalent plugs of different manufacture can be used if their heat range corresponds to that of the N7Y. Plugs that are too cold can cause hard starting. If they are too hot, they can cause preignition which may result in engine damage.

Champion N6Y spark plugs are recommended for use in 750 Combat engines and for sustained high-speed operation in other engines.

6. Screw the plugs into the cylinder head using new gaskets to ensure a good seal. Tighten them firmly but not so tight that there is risk of damaging the threads in the head. Reconnect the high-tension leads.

7. Refer to *Ignition Timing*, Chapter Seven, and accurately time the ignition. Two methods can be used—static timing and dynamic timing (with the use of a strobe). The static method is accurate enough for most situations, but strobe timing is essential for complete timing accuracy required for maximum performance. If static timing is used, install the spark plugs and connect the high-tension leads as described above after the timing has been set.

Carburetion

For stock engines, standard factory jetting is good for most situations. Engine modifications that affect flow or cylinder volume, as well as prolonged operation at extremes of altitude or humidity, require changes in jetting and needle

SPARK PLUG CONDITION

(58)

NORMAL
• Identified by light tan or gray deposits on the firing tip.
• Can be cleaned.

GAP BRIDGED
• Identified by deposit buildup closing gap between electrodes.
• Caused by oil or carbon fouling. If deposits are not excessive, the plug can be cleaned.

OIL FOULED
• Identified by wet black deposits on the insulator shell bore electrodes.
• Caused by excessive oil entering combustion chamber through worn rings and pistons, excessive clearance between valve guides and stems, or worn or loose bearings. Can be cleaned. If engine is not repaired, use a hotter plug.

CARBON FOULED
• Identified by black, dry fluffy carbon deposits on insulator tips, exposed shell surfaces and electrodes.
• Caused by too cold a plug, weak ignition, dirty air cleaner, too rich a fuel mixture, or excessive idling. Can be cleaned.

LEAD FOULED
• Identified by dark gray, black, yellow, or tan deposits or a fused glazed coating on the insulator tip.
• Caused by highly leaded gasoline. Can be cleaned.

WORN
• Identified by severely eroded or worn electrodes.
• Caused by normal wear. Should be replaced.

FUSED SPOT DEPOSIT
• Identified by melted or spotty deposits resembling bubbles or blisters.
• Caused by sudden acceleration. Can be cleaned.

OVERHEATING
• Identified by a white or light gray insulator with small black or gray brown spots and with bluish-burnt appearance of electrodes.
• Caused by engine overheating, wrong type of fuel, loose spark plugs, too hot a plug, or incorrect ignition timing. Replace the plug.

PREIGNITION
• Identified by melted electrodes and possibly blistered insulator. Metallic deposits on insulator indicate engine damage.
• Caused by wrong type of fuel, incorrect ignition timing or advance, too hot a plug, burned valves, or engine overheating. Replace the plug.

position. These are described in detail in Chapter Eleven, *Performance Improvement*. The checks and adjustments that follow assume that jetting and needle position are correct.

1. Loosen the throttle stop screws in both carburetors until the throttle slides are completely closed (**Figure 59**).

2. Loosen the locknuts on the cable adjusters (**Figure 60**) and turn the adjusters either in or out until there is no perceptible end play in the cable sheaths when the slides are closed. Also, when the throttle grip is turned, the throttle slides must begin to lift together. When the throttle slides are synchronized, hold the adjusters to prevent them from turning and tighten the locknuts.

3. Start the engine, allow it to warm up for about a minute, and turn the throttle so that it idles slightly faster than normal. Turn each of the throttle stop screws in slowly until they hold the slides in this position with the throttle released.

4. With insulated pliers or screwdriver, pull the spark plug lead off one of the spark plugs. Screw in the pilot air screw (**Figure 61**) on the carburetor on the opposite cylinder until the engine begins to falter (too lean). Then screw

it out—the engine will begin to run smoothly again—until the engine falters once again (too rich). Slowly screw it in until the engine begins to run smoothly. Usually, the best position for the screw is midway between the too-lean and the too-rich idling positions (about 1 to 1½ turns out).

5. Reconnect the spark plug lead on the opposite cylinder, disconnect the lead on the cylinder just adjusted, and adjust the other cylinder in the same manner.

6. With both pilot air screws adjusted and both spark plug leads connected, the idle speed will probably be too fast. In such case, carefully unscrew each of the throttle stop screws equally until the idle is acceptable. Recheck the end play in the throttle cable sheaths. It should be ⅛ in. (3mm) for each cable (**Figure 62**).

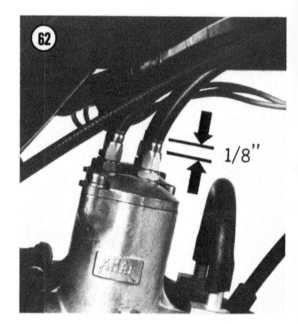

CHAPTER THREE

TROUBLESHOOTING

Diagnosing motorcycle ills is relatively simple if you use orderly procedures and keep a few basic principles in mind.

Never assume anything. Don't overlook the obvious. If you are riding along and the bike suddenly quits, check the easiest, most accessible problem spots first. Is there gasoline in the tank? Is the gas petcock in the ON or RESERVE position? Has a spark plug wire fallen off? Check the ignition switch. Sometimes the weight of keys on a key ring may turn the ignition off suddenly.

If nothing obvious turns up in a cursory check, look a little further. Learning to recognize and describe symptoms will make repairs easier for you or a mechanic at the shop. Describe problems accurately and fully. Saying that "it won't run" isn't the same as saying "it quit on the highway at high speed and wouldn't start," or that "it sat in my garage for three months and then wouldn't start."

Gather as many symptoms together as possible to aid in diagnosis. Note whether the engine lost power gradually or all at once, what color smoke (if any) came from the exhausts, and so on. Remember that the more complicated a machine is, the easier it is to troubleshoot because symptoms point to specific problems.

You don't need fancy equipment or complicated test gear to determine whether repairs can be attempted at home. A few simple checks could save a large repair bill and time lost while the bike sits in a dealer's service department. On the other hand, be realistic and don't attempt repairs beyond your abilities. Service departments tend to charge heavily for putting together a disassembled engine that may have been abused. Some won't even take on such a job—so use common sense; don't get in over your head.

OPERATING REQUIREMENTS

An engine needs three basics to run properly: correct gas/air mixture, compression, and a spark at the right time. If one or more are missing, the engine won't run. The electrical system is the weakest link of the three. More problems result from electrical breakdowns than from any other source. Keep that in mind before you begin tampering with carburetor adjustments and the like.

If a bike has been sitting for any length of time and refuses to start, check the battery for a charged condition first, and then look to the gasoline delivery system. This includes the tank, fuel petcocks, lines, and the carburetors. Rust may have formed in the tank, obstructing fuel flow. Gasoline deposits may have gummed up carburetor jets and air passages. Gasoline tends to lose its potency after standing for long periods. Condensation may contaminate it with water. Drain old gas and try starting with a fresh tankful.

Compression, or the lack of it, usually enters the picture only in the case of older machines. Worn or broken pistons, rings, and cylinder bores could prevent starting. Generally, a gradual power loss and harder and harder starting will be readily apparent in this case.

STARTING DIFFICULTIES

Check gas flow first. Remove the gas cap and look into the tank. If gas is present, pull off a fuel line at the carburetor and see if gas flows freely. If none comes out, the fuel tap may be shut off, blocked by rust or foreign matter, or the fuel line may be stopped up or kinked. If the carburetor is getting usable fuel, turn to the electrical system next.

Check that the battery is charged by turning on the lights or by beeping the horn. Refer to your owner's manual for starting procedures with a dead battery. Have the battery recharged if necessary.

Pull off a spark plug cap, remove the spark plug, and reconnect the cap. Lay the plug against the cylinder head so its base makes a good connection, and turn the engine over with the kickstarter. A fat, blue spark should jump across the electrodes. If there is no spark, or only a weak one, there is electrical system trouble. Check for a defective plug by replacing it with a known good one. Don't assume a plug is good just because it's new.

Once the plug has been cleared of guilt, but there's still no spark, start backtracking through the system. If the contact at the end of the spark plug wire can be exposed, it can be held about ⅛ inch from the head while the engine is turned over to check for a spark. Remember to hold the wire only by its insulation to avoid a nasty shock. If the plug wires are dirty, greasy, or wet, wrap a rag around them so you don't get shocked. If you do feel a shock or see sparks along the wire, clean or replace the wire and/or its connections.

If there's no spark at the plug wire, look for loose connections at the coil and battery. If all seems in order there, check next for oil or dirty contact points. Clean points with electrical contact cleaner, or a strip of paper. On battery ignition models, with the ignition switch turned on, open and close the points manually with a screwdriver.

No spark at the points with this test indicates a failure in the ignition system. Refer to Chapter Seven (*Ignition and Electrical Systems*) for checkout procedures for the entire system and individual components. Refer to the same chapter for checking and setting ignition timing.

Note that spark plugs of the incorrect heat range (too cold) may cause hard starting. Set gaps to specifications. If you have just ridden through a puddle or washed the bike and it won't start, dry off plugs and plug wires. Water may have entered the carburetor and fouled the fuel under these conditions, but wet plugs and wires are the more likely problem.

If a healthy spark occurs at the right time, and there is adequate gas flow to the carburetor, check the carburetor itself at this time. Make sure all jets and air passages are clean, check float level, and adjust if necessary. Shake the float to check for gasoline inside it, and replace or repair as indicated. Check that the carburetors are mounted snugly, and no air is leaking past the manifold. Check for a clogged air filter.

Compression may be checked in the field by turning the kickstarter by hand and noting that an adequate resistance is felt, or by removing a spark plug and placing a finger over the plug hole and feeling for pressure.

An accurate compression check gives a good idea of the condition of the basic working parts of the engine. To perform this test, you need a compression gauge. The motor should be warm.

1. Remove the plug on the cylinder to be tested and clean out any dirt or grease.

2. Insert the tip of the gauge into the hole, making sure it is seated correctly.

3. Open the throttle all the way and make sure the chokes on the carburetors are open.

4. Crank the engine several times and record the highest pressure reading on the gauge. Run the test on each of the cylinders. Refer to Chapter Four (*Engine, Primary Drive, and Clutch*) to interpret results.

POOR IDLING

Poor idling may be caused by incorrect carburetor adjustment, incorrect timing, or ignition system defects. Check the gas cap vent for an obstruction.

MISFIRING

Misfiring can be caused by a weak spark or dirty plugs. Check for fuel contamination. Run the machine at night or in a darkened garage to check for spark leaks along the plug wires and under the spark plug cap. If misfiring occurs only at certain throttle settings, refer to the carburetor chapter for the specific carburetor circuits involved. Misfiring under heavy load, as when climbing hills or accelerating, is usually caused by bad spark plugs.

FLAT SPOTS

If the engine seems to die momentarily when the throttle is opened and then recovers, check for a dirty main jet in the carburetor, water in the fuel, or an excessively lean mixture.

POWER LOSS

Poor condition of rings, pistons, or cylinders will cause a lack of power and speed. Ignition timing should be checked.

OVERHEATING

If the engine seems to run too hot all the time, be sure you are not idling it for long periods. Air-cooled engines are not designed to operate at a standstill for any length of time. Heavy stop and go traffic is hard on a motorcycle engine. Spark plugs of the wrong heat range can burn pistons. An excessively lean gas mixture may cause overheating. Check ignition timing. Don't ride in too high a gear. Broken or worn rings may permit compression gases to leak past them, heating heads and cylinders excessively. Check oil level and use the proper grade lubricants.

BACKFIRING

Check that the timing is not advanced too far. Check fuel for contamination.

ENGINE NOISES

Experience is needed to diagnose accurately in this area. Noises are hard to differentiate and harder yet to describe. Deep knocking noises usually mean main bearing failure. A slapping noise generally comes from loose pistons. A light knocking noise during acceleration may be a bad connecting rod bearing. Pinging, which sounds like marbles being shaken in a tin can, is caused by ignition advanced too far or gasoline with too low an octane rating. Pinging should be corrected immediately or damage to pistons will result. Compression leaks at the head-cylinder joint will sound like a rapid on-and-off squeal.

PISTON SEIZURE

Piston seizure is caused by incorrect piston clearances when fitted, fitting rings with improper end gap, too thin an oil being used, incorrect spark plug heat range, or incorrect ignition timing. Overheating from any cause may result in seizure.

EXCESSIVE VIBRATION

Excessive vibration may be caused by loose motor mounts, worn engine or transmission bearings, loose wheels, worn swinging arm bushings, a generally poor running engine, broken or cracked frame, or one that has been damaged in a collision. See also *Poor Handling*.

CLUTCH SLIP OR DRAG

Clutch slip may be due to worn or glazed plates, incorrect adjustment, or too much oil in the primary chaincase. A dragging clutch could result from damaged or bent plates, incorrect adjustment, uneven clutch spring pressure, or a tight primary chain.

POOR HANDLING

Poor handling may be caused by improper tire pressures, a damaged frame or swinging arm, worn shocks or front forks, weak fork springs, a bent or broken steering stem, misaligned wheels, loose or missing spokes, worn tires, bent handlebars, worn wheel bearing, or dragging brakes.

3

If shimmy, or "head shaking" occurs during deceleration, the alignment of the front tire on the rim should be checked. Misalignment, particularly for tires with a center groove, can cause the front wheel to "hunt" in a rapid oscillating motion when the weight is transferred forward during deceleration. To check for alignment, support the motorcycle so the front tire is clear of the ground. Place a short strip of masking tape on the front fender, lined up with the center groove in the tire. Now, spin the front wheel and sight along the tread, from the front of the motorcycle and visually line up the tire groove with the tape on the fender. If the alignment is correct, the groove will not waver; however, if the tire is misaligned, the groove will waver back and forth.

This condition can often be corrected by deflating the tire, soaping the seating bead, and rotating the tire 90 degrees. After the tire has been reinflated, recheck the alignment as before. Also, if the tire is rotated on the rim, the balance of the wheel should be checked and corrected if necessary.

BRAKE PROBLEMS

Sticking brakes may be caused by broken or weak return springs, improper cable or rod adjustment, or dry pivot and cam bushings. Grabbing brakes may be caused by greasy linings which must be replaced. Brake grab may also be due to out-of-round drums or linings which have broken loose from the brake shoes. Glazed linings or glazed brake pads will cause loss of stopping power.

LIGHTING PROBLEMS

Bulbs which continuously burn out may be caused by excessive vibration, loose connections that permit sudden current surges, poor battery connections, or installation of the wrong type bulb.

A dead battery or one which discharges quickly may be caused by a faulty generator or rectifier. Check for loose or corroded terminals. Shorted battery cells or broken terminals will keep a battery from charging. Low water level will decrease a battery's capacity. A battery left uncharged after installation will sulphate, rendering it useless.

A majority of light and horn or other electrical accessory problems are caused by loose or corroded ground connections. Check those first, and then substitute known good units for easier troubleshooting.

TROUBLESHOOTING GUIDE

The following "quick reference" guide summarizes the troubleshooting process. Use it to outline possible problem areas, then refer to the specific chapter or section involved.

LOSS OF POWER

Cause	Things to check	Cause	Things to check
Poor Compression	Piston rings and cylinders Head gaskets Crankcase leaks	Improper mixture	Dirty air cleaner Restricted fuel flow Gas cap vent holes
Overheated engine	Lubricating oil supply Clogged cooling fins Ignition timing Slipping clutch Carbon in combustion chamber	Miscellaneous	Dragging brakes Tight wheel bearings Defective chain Clogged exhaust system

STEERING PROBLEMS

Problem	Things to check	Cause	Things to check
Hard steering	Tire pressure Steering stem head Steering head bearings	Pulls to one side (contd.)	Defective swinging arm Defective steering head
Pulls to one side	Unbalanced shock absorbers Drive chain adjustment Front/rear wheel alignment Unbalanced tires	Shimmy	Drive chain adjustment Loose or missing spokes Deformed rims Worn wheel bearings Wheel balance

BRAKE TROUBLES

Problem	Things to check	Cause	Things to check
Poor brakes	Worn linings Brake adjustment Oil or water on brake linings Loose linkage or cables Low fluid level	Noisy brakes	Worn or scratched lining Scratched brake drums Dirt in brake housing
		Unadjustable brakes	Worn linings Worn drums Worn brake cams

GEARSHIFTING DIFFICULTIES

Cause	Things to check	Cause	Things to check
Clutch	Adjustment Friction plates Steel plates—distorted Oil quantity	Transmission	Oil quantity Oil grade Shift adjustment Shift quadrant Shift forks

3

CHAPTER FOUR

ENGINE, PRIMARY DRIVE, AND CLUTCH

The Norton twin (**Figure 1**—Commando, and **Figure 2**—Atlas and G15) is a non-unit engine; that is, the engine and transmission do not share a common case. As a result, major engine service, including a complete rebuild, can be carried out without disturbing the transmission. Upper end service (cylinder head, cylinders, pistons) can be accomplished without removing the engine from the frame. Lower end service requires that the engine be removed from the frame which in turn requires that the primary drive and clutch be removed.

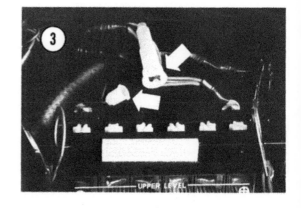

CYLINDER HEAD

Cylinder head service is presented as a complete rebuild procedure so that the relationship of the components can be better understood. Naturally, not all situations require that all parts be replaced. However, every check, inspection, and measurement called for in this section should be made each time the head is removed to ensure that all parts are within specifications. If a wear component, such as a valve guide, is satisfactory, its service procedure need not be carried out.

Removal

1. Raise or remove the seat. Disconnect the negative battery lead at the fuse holder and remove the fuse (**Figure 3**).

2. Shut off the fuel taps, disconnect the lines, and remove the fuel tank. On all original equipment Commando tanks (**Figure 4**), the front is mounted on 2 studs beneath the tank and held with self-locking nuts. The rear mount on some models uses a rubber loop which passes beneath the top frame tube, while others are held in place with a metal cross bar.

3. Straighten the lock tabs on the finned head-pipe clamps (**Figure 5**) and unscrew the clamps. Unscrew the muffler mounting nuts and the balance pipe clamp nuts (**Figures 6A and 6B**) and remove each exhaust system as a unit.

4. Refer to Chapter Six and remove carburetors.

5. Carefully pull the caps off the spark plugs and unscrew them. Remove the entire coil assembly

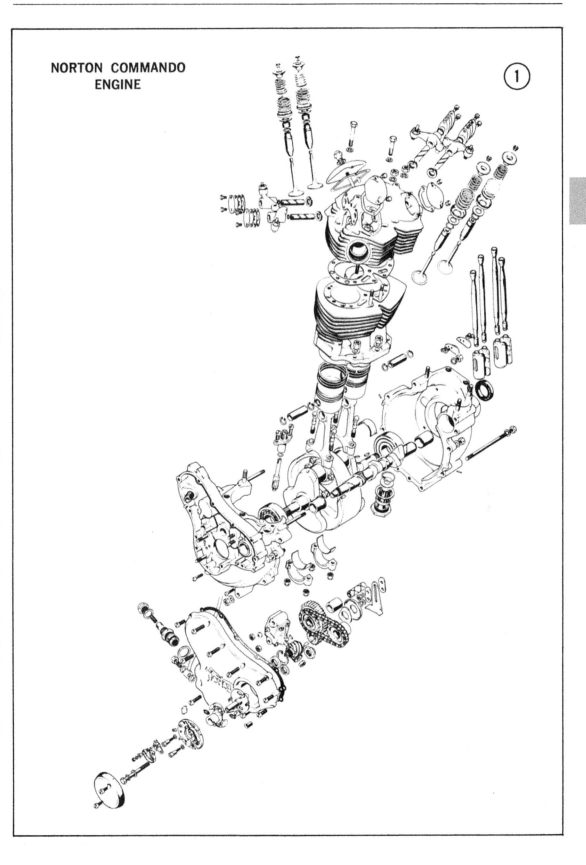

NORTON COMMANDO
ENGINE

①

4

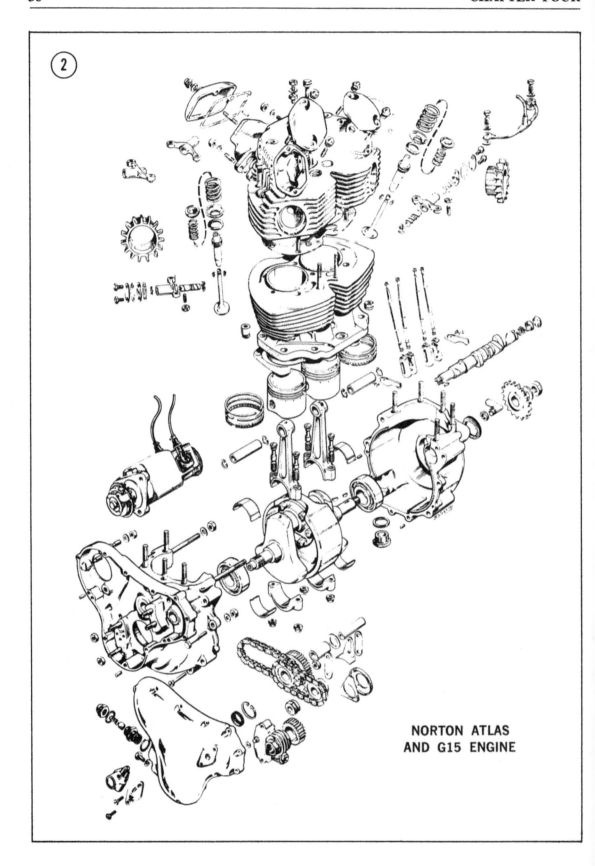

② NORTON ATLAS
AND G15 ENGINE

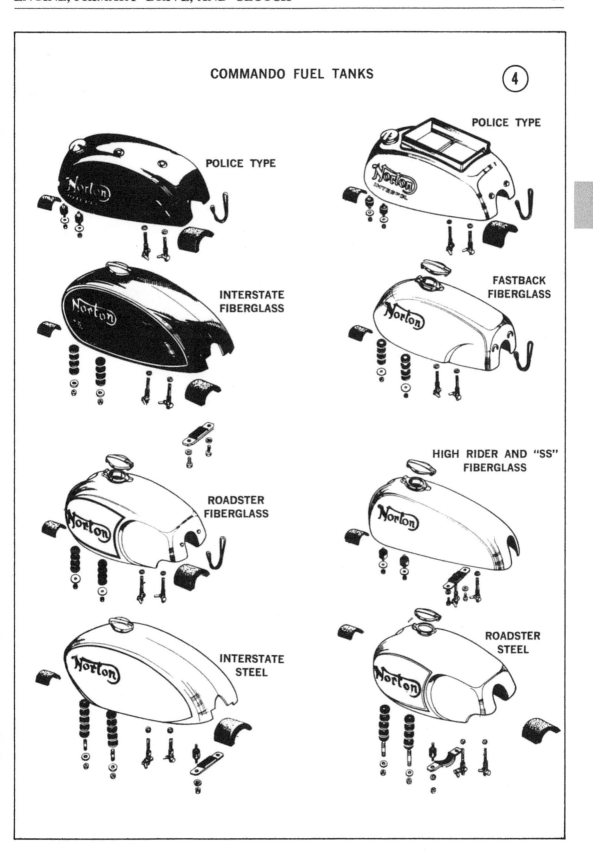

COMMANDO FUEL TANKS

POLICE TYPE

POLICE TYPE

INTERSTATE FIBERGLASS

FASTBACK FIBERGLASS

ROADSTER FIBERGLASS

HIGH RIDER AND "SS" FIBERGLASS

INTERSTATE STEEL

ROADSTER STEEL

from the frame as a unit (**Figure 7**). Instead of unplugging the leads and risking incorrect installation later on, tie the assembly to the handlebar with a piece of cord.

6. Unscrew the nuts which mount the headsteady to the frame (**Figure 8**). Unscrew the 3 Allen screws (**Figure 9**) and remove the headsteady from the head.

On 850 Mark III models, loosen the nut which holds the spring trunion to the head and disconnect the ends of the spring from the trunion. Remove the spring from the frame bracket.

7. Unscrew the bolts from the rocker feed pipe (**Figure 10**) and lift the pipe from the head. There's no need to disconnect the pipe from the engine lower end; however, it should be tied or taped out of the way.

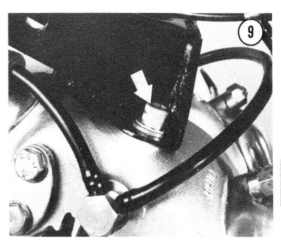

4

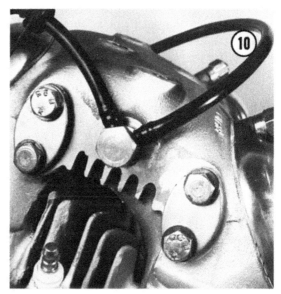

front center bolt last. Lift the head, move the pushrods as far into the head as possible (**Figure 12**), and remove it from the engine.

Disassembly

1. Remove the rocker covers and gaskets (**Figure 13**) and the spindle cover plates and gaskets (**Figure 14**).

2. Heat the cylinder head in an oven, at 300°F, for 30 minutes. Screw a slide hammer shaft (Norton tool No. 064298) into the end of the rocker spindle and withdraw the rocker spindle. If a slide hammer is not available, an inexpensive substitute extractor tool like that shown in **Figure 15** can be fabricated.

8. With the transmission in gear, rotate the rear wheel to bring the pistons to TDC. Unscrew the cylinder nuts and bolts in the pattern shown in **Figure 11**, beginning with No. 2. Remove the

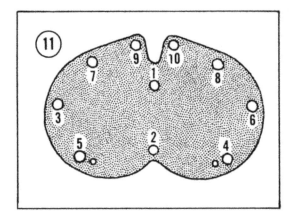

Repeat the procedure for the remaining 3 shafts. Remove the rockers and the thrust washers.

3. Double nut the inlet rocker cover stud, Tighten the nuts securely against one another, and unscrew the stud with a wrench on the bottom nut (**Figure 16**). Compress each of the valve springs in turn (**Figure 17**) and remove the keepers from the valve stem (**Figure 18**) with a

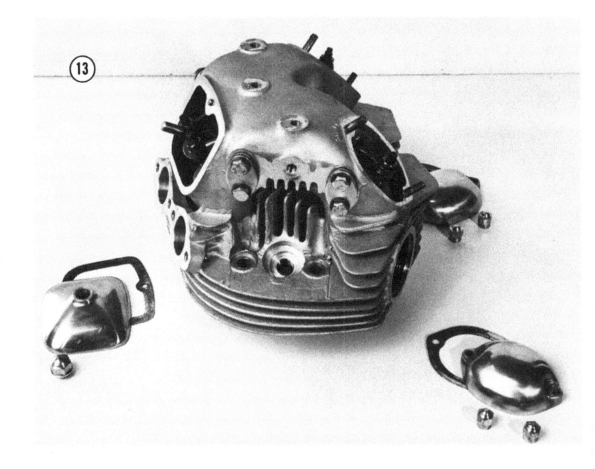

4

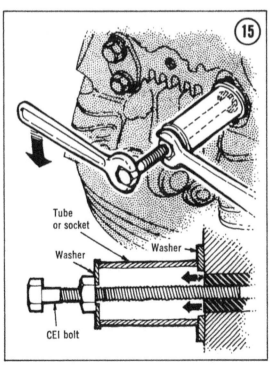

Tube
or socket

Washer

Washer

CEI bolt

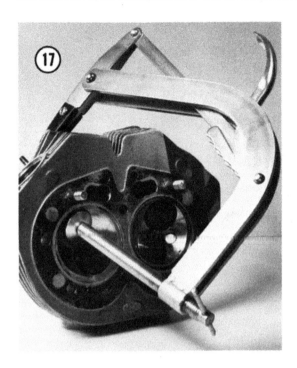

and 20. Keep each valve assembly together and mark it for location in the head.

Inspection

small magnet or screwdriver. Release the compresser and remove the valve, springs, collars, and the stem seals (inlet only). See **Figures 19**

1. Clean the carbon from the combustion chambers, exhaust ports, valves, and piston crowns

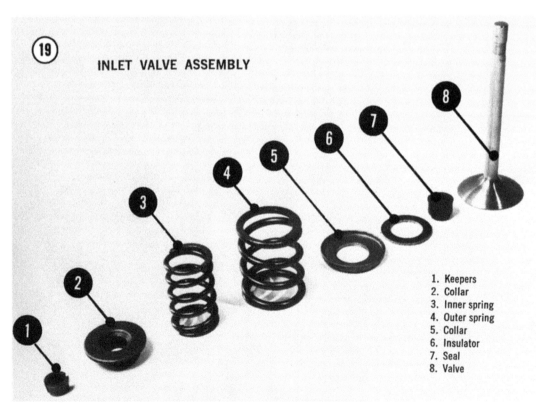

INLET VALVE ASSEMBLY

1. Keepers
2. Collar
3. Inner spring
4. Outer spring
5. Collar
6. Insulator
7. Seal
8. Valve

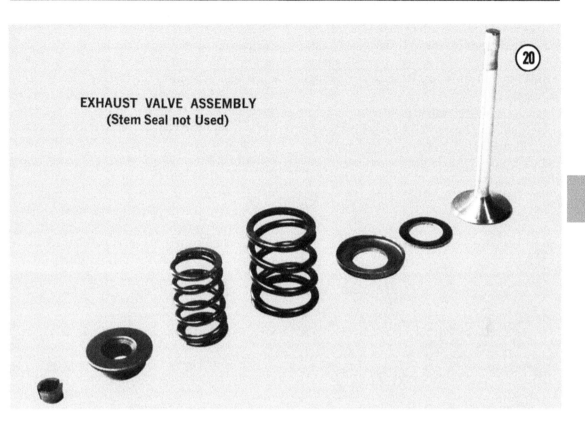

EXHAUST VALVE ASSEMBLY
(Stem Seal not Used)

4

using either a special chemical carbon remover or a scraper made of *soft* metal, such as aluminum. Then clean all of the parts in solvent and dry them.

> NOTE: *Don't remove the carbon ridges around the tops of the cylinders or from the piston above the top compression ring.*

2. Inspect the contact surface of each valve for burning (**Figure 21**). Minor roughness and pit-

ting can be removed by reseating the valve, but excessive unevenness to the contact surface is an indication the valve is not serviceable. The contact surface of the valve may be ground on a valve grinding machine, but it's best to replace a burned or damaged valve with a new one.

Inspect the valve stems for wear and roughness and measure the vertical runout of the valve face as shown in **Figure 22**. The runout should not exceed 0.001 in. (0.025mm).

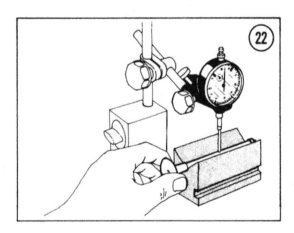

3. Measure the free length of the valve springs. The inner springs should be 1.482 in. (37.642mm) and the outer springs should be 1.618 in. (41.097mm). Replace any springs that are short.

4. Install each of the valves in its guide, in turn, and measure the clearance of the stem and the guide (**Figure 23**) in 2 axes to determine if the guide has worn oval. If the measurements are different for the 2 axes, or if the clearance is greater than 0.006 in. (0.152mm), the guide should be replaced.

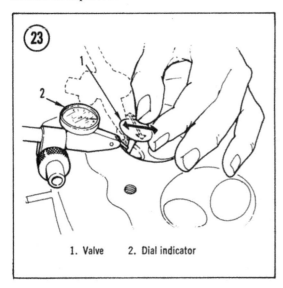

1. Valve   2. Dial indicator

5. Check the valve seats for excessive wear or pitting that can't be removed by lapping the valve. In such cases, the seat must be recut. Also, the seat will have to be recut if the valve guide is replaced.

> NOTE: *Both the replacement of a valve guide and recutting a valve seat require special tools and skills. This work should be entrusted to a Norton service shop or to a cylinder head rebuilding service.*

6. Measure the bores of the rockers and the rocker shafts. The difference (clearance) should be no greater than 0.0020 in. (0.051mm).

Assembly

1. Lap each valve in its seat using a fine grade of grinding compound and a hand lapping stick. Apply compound to the contact surface of the valve and set the valve into the head. Wet the suction cup on the end of the lapping stick and stick it to the head of the valve. Lap the valve to the seat by rotating the lapping stick in both directions (**Figure 24**). Every 5 to 10 seconds, rotate the valve head 180° in the seat, continue lapping until the contact surfaces of the valve and the seat are a uniform grey. Wash the head thoroughly with solvent, making sure no compound remains in the ports or the valve guides.

2. Oil the valve stems and install the valves in the head and reassemble the collars and springs in the order shown in Figures 19 and 20. All 1971 and earlier engines must have heat insulators installed between the bottom spring collar and the head on the inlet valves, and all models must have the insulators installed on the exhaust valves.

> NOTE: *When guide replacement is required on early engines, late model guides (**Figure 25**) should be installed to reduce oil consumption. Also, stem-to-guide seals must be installed on the inlet valve guides.*

Also, on heads equipped with the type of guide shown in **Figure 25**, the stem and guide seals must be installed on the inlet valves (Figure 19).

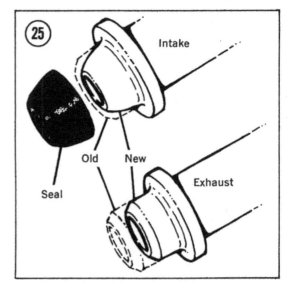

Compress each valve assembly and install the split keepers in the top collars. Make sure the keepers are firmly seated in the collar and the valve stem groove.

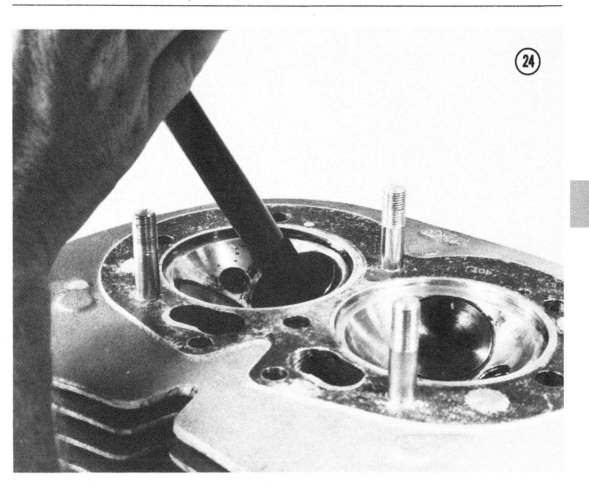

4

3. Heat the cylinder head and start one of the rocker shafts into its bore. On the exhaust rockers, the flat on the shaft (**Figure 26**) should face forward, and on the inlet rockers the flat should face to the rear. Carefully tap the shaft in until it protrudes about 1/16 in. (1.6mm) into the case (**Figure 27**). Install a plain washer on the end of the shaft and set the rocker in place. Tap the shaft in part way so that it engages the rocker and fits the spring washer between the end of the rocker and the inside wall (**Figure 28**). Tap the shaft in until it passes through the spring washer and into the opposite boss in the head. If necessary, rotate the spindle using the slit in the outer end and line up the spindle as shown (**Figure 29**). Finally, tap the spindle the rest of the way in until the spindle is flush with or slightly beneath the machined surface. Install the gaskets and plates in the order shown (**Figure 30**). Repeat the procedure for each of the other rockers.

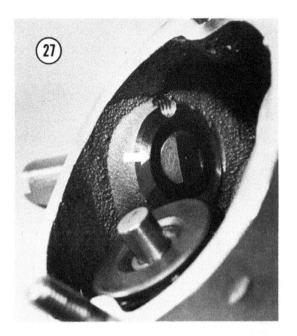

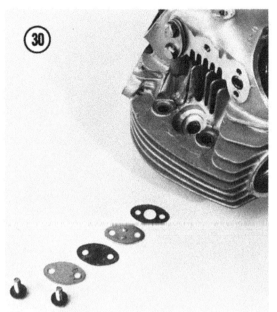

Installation

1. Make certain the mating surfaces of the head and cylinders are clean and set the head gasket on top of the cylinders. If the position of the pistons has been disturbed, bring them to TDC by rotating the rear wheel with the transmission in gear. This is necessary to provide sufficient room for setting the head in place.

> NOTE: *It is recommended that the latest head gasket with a "Flame Ring" be installed when the head is removed for service. The new gasket is an improvement over the copper and composition fiber/metal types. Before installing the gasket, clean the cylinder and head mating surfaces with a non-petroleum base solvent such as lacquer thinner and install the gasket without gasket cement. Part numbers for the "Flame Ring" gasket are: 750 — No. 063844; 850 — No. 065051.*

2. Set the long intake pushrods—cups first—in the inside bores and the shorter exhaust pushrods in the outside bores (**Figure 31**). Push the rods into the head as far as they will go. Hold the head in one hand and hold the pushrods in place with the other. Position the head over the cylinders and let the pushrods down into their bores.

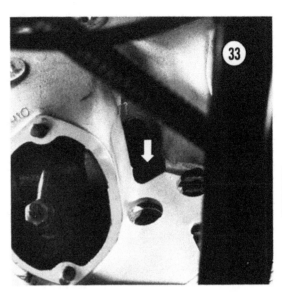

They will locate automatically on the cam followers.

3. Before the head can be set in place on the cylinder, the pushrods must be lined up with the rockers. Begin with the long rods and carefully guide them into contact with the rockers, using a small screwdriver or probe (**Figure 32**). Then do the same with the shorter pushrods.

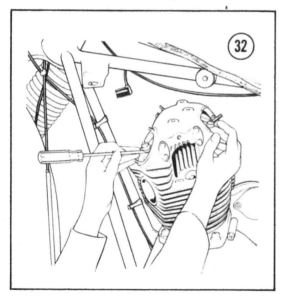

4. Screw in the short bolt located at the center of the front of the head (**Figure 33**). Make sure all the pushrod cups are engaged with the rockers and tighten this bolt to pull the head down onto the cylinders against the valve spring pressure.

5. Install the bolts and washers on each side of the spark plug holes (**Figure 34**). Screw the long nuts onto the studs beneath the exhaust ports (**Figure 35**) and conventional nuts onto the studs beneath the intake ports (**Figure 36**). None of these require washers. Next, screw the 2 nuts with washers onto the front studs (**Figure 37**). Tighten the nuts and bolts in the sequence shown in Figure 11 to 20 ft.-lb. (2.75 mkg) for the smaller (5/16 in.) bolts and 30 ft.-lb. (3.68 mkg) for the larger (⅜ in.) nuts and bolts.

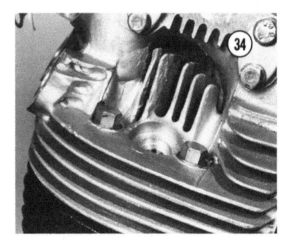

6. Adjust the valve clearances as described below and reassemble the remaining components in reverse order of their removal. The spark plugs should be installed as soon as the valve

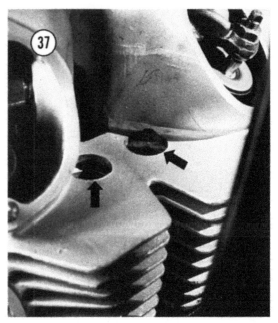

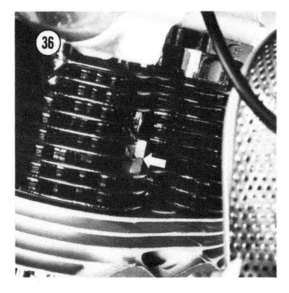

clearances have been set to prevent nuts or washers from falling into the cylinders.

Make sure the tubing to the rocker feed pipe is not crimped or in contact with the head or head-steady. Check the wiring to the coil cluster to make sure like colors are connected to like colors. Finally, don't forget to reinstall the fuse in the holder on the battery negative lead.

VALVE ADJUSTMENT

Clearance of rockers should be adjusted at 2,500-3,000-mile intervals and each time cylinder head or any part of valve train is removed

and installed. The clearances should be checked and set with the engine cold. For the standard 750 and 850 engines, the inlet clearance is 0.006 in. (0.15mm) and the exhaust clearance is 0.008 in. (0.2mm); for the 750 Combat engine, the inlet clearance is 0.008 in. (0.2mm) and the exhaust clearance is 0.010 in. (0.25mm).

1. Remove the rocker covers and unscrew the spark plugs from the head.

2. Rotate the crankshaft using the kickstarter until the left inlet valve is completely open (**Figure 38**). Check the clearance of the right inlet valve and rocker. When the clearance is correct, there will be a slight resistance on the feeler gauge when it is inserted and withdrawn.

3. To correct the clearance, back off the locknut and screw the adjuster out far enough to insert the gauge with no resistance. Screw the adjuster in until a slight resistance can be felt in the gauge. Hold the adjuster to prevent it from turning further and tighten the locknut. Recheck the clearance to make sure the adjuster didn't turn after the correct clearance was achieved.

4. Rotate the crankshaft to open the right intake valve completely and check and adjust the clearance of the left valve and rocker. Then, check and adjust the exhaust valves and rockers in the same manner.

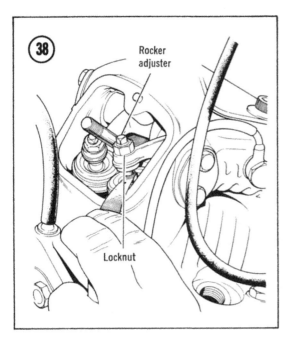

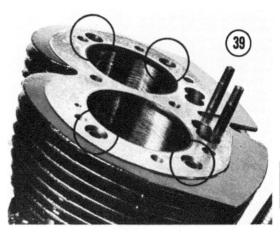

CYLINDERS AND PISTONS

Cylinder wear should be checked each time the head is removed or if such symptoms as excessive oil consumption, performance degradation, or piston slap indicate the possibility of the cylinders being worn beyond service limits.

Removal

1. Remove cylinder head as described earlier.

> NOTE: *On 750 engines, it is not necessary to remove the head from the cylinder; they can be removed together as an assembly after the spark plugs have been removed, the crankshaft is rotated to bring the pistons to* BDC, *and the cylinder base nuts are unscrewed.*

2. On 850 engines, unscrew the 4 thru-bolts (**Figure 39**) and 5 nuts (**Figures 40A and 40B**). On 750 engines, unscrew 9 nuts around the base of the cylinders. The nuts must be unscrewed progressively; it's essential to lift the cylinder assembly so that the nuts will not jam against the fin above them. Remove all washers (there are none on the front center stud) before lifting the cylinder assembly off the pistons. Prevent the rods from falling forward or backward against the crankcase as the cylinder is lifted free.

3. Stuff a clean shop rag into the top of the crankcase to prevent small parts and dirt from falling into the crankcase. A wooden block made from soft wood (like the one shown in **Figures 41 and 42**) will protect the connecting rods, crankcase opening, and piston skirts.

> NOTE: *On 850 engines, mark the pistons, "right" and "left" and "front" so that they may be installed in the locations and direction from which they were removed.*

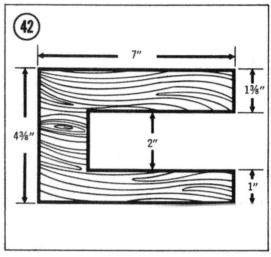

for several minutes by wrapping a rag, heated in hot water, around it.

Inspection

1. Remove the rings from the pistons by spreading the ring ends with your thumbs and lifting the rings up evenly (**Figure 44**). Carefully clean the carbon from the piston crown with a chemical remover or with a soft scraper. Try not to remove or damage the carbon ridge around the circumference of the piston above the top ring. If the pistons, rings, and cylinders are found to be dimensionally correct and can be reused, removal of the carbon ring from the tops of pistons or the carbon ridges from the tops of cylinders will promote excessive oil consumption.

4. Remove the circlips from the wrist pin bores in the pistons (**Figure 43**) and press the pins out with a drift or a small socket on an extension. The piston must be held securely so that none of the force that is required to push out the pin is applied to the connecting rod. It's advisable to have someone help you hold the piston. Under no circumstances should you hammer on the drift; if the pin can't be moved, warm the piston

CAUTION
The rail portions of the oil scraper can be very sharp. Be careful when handling them to avoid cut fingers.

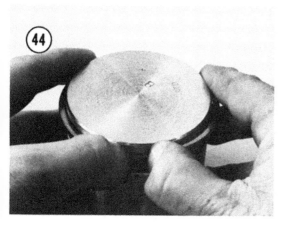

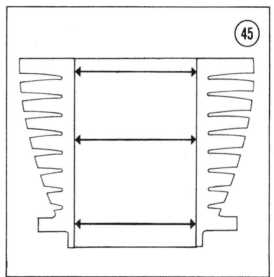

Clean the ring grooves in the pistons using either a groove scraper or a piece of old ring. Then clean the piston in solvent and dry it thoroughly.

2. Measure the cylinder bores at the points shown in **Figure 45** in 2 axes—in line with the wrist pins and at 90° to the pins. If taper or out-of-round are greater than 0.005 in. (0.127mm), the cylinders must be bored to the next oversize and new pistons installed. Bore sizes are shown in **Table 1**.

NOTE: *The new pistons should be obtained before the cylinders are bored so that pistons can be measured; slight*

manufacturing tolerances must be taken into account to determine the actual bore size and the working clearance. Piston-to-cylinder clearance should be 0.0035-0.0040 inch (0.089-0.10mm).

If a cylinder gauge or inside micrometer is not available, cylinder wear can be checked by placing a compression ring in the cylinder at the locations shown in Figure 45 and measuring the ring gap with a flat feeler gauge. The ring must be positioned squarely into the cylinder. This can be accomplished by pushing it into place with the head of a piston (**Figure 46**). This method is not as accurate as the first and if any doubt exists about the serviceability of a cylinder, the unit should be checked by a dealer.

Table 1   BORE SIZES

Engine	Piston	Bore Size
750cc	Standard:	2.875 in. (73.025mm)
	+ 0.010 in. oversize	2.885 in. (73.279mm)
	+ 0.020 in. oversize	2.895 in. (73.477mm)
	+ 0.030 in. oversize	2.905 in. (73.787mm)
	+ 0.040 in. oversize	2.915 in. (74.041mm)
850cc	Standard:	3.032 in. (77.013mm)
	+ 0.010 in. oversize	3.042 in. (77.267mm)
	+ 0.020 in. oversize	3.052 in. (77.521mm)
	+0.040 in. oversize	3.072 in. (77.970mm)
	+0.060 in. oversize	3.092 in. (78.477mm)

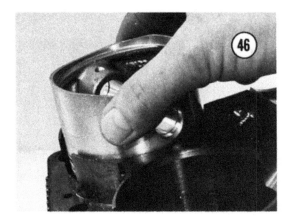

Installation

1. Make certain the threads of the crankcase studs are clean and apply a light coat of grease to them. Carefully coat the cylinder mounting flange with Loctite "Plastic Gasket" or gasket cement.

> NOTE: *Omit the base gasket (early models) to reduce oil seepage at the cylinder flange. Normal flexing of the crankcase causes the gasket to extrude after several thousand miles, promoting bothersome seepage.*

2. Install one circlip in each piston. Make sure the clip is installed with the sharp edge facing out (**Figure 47**) so the clip will lock in the groove. Heat the pistons in hot water.

3. Lightly oil the wrist pin bores in the connecting rods and pistons, set the pistons in place on the rods, and push the pins in, taking care to support the pistons so there is no side loading on the rods. Then fit the remaining circlips, again making sure the sharp edge of each clip faces outward.

> CAUTION
> *The piston stamped* RH *must be installed on the right connecting rod and the one stamped* LH *must be installed on the left. Also, the* EX *stamp on each piston must face forward, toward the exhaust ports.*

4. Rotate the crankshaft to bring the pistons into contact with the wooden block and install the rings—first the oil scraper, then the second and top compression rings (**Figure 48**). Rotate

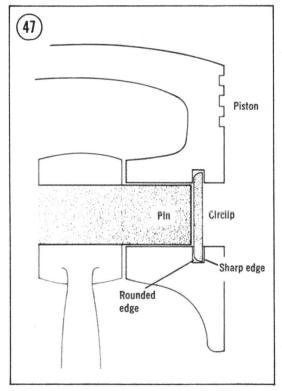

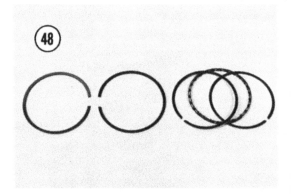

the rings in the grooves to position the end gaps as shown in **Figure 49**. When aligning the scraper ring, make sure the gaps in the rail and the expander are staggered and that the ends of the expander do not overlap (**Figure 50**).

5. Lightly oil the pistons and install the ring compressors (**Figure 51**). The compressors should be snug enough to compress the rings but not so tight that they won't slip easily down and off the rings when the cylinders are installed.

6. Set the cylinder block in place on top of the pistons and push it quickly and evenly down over

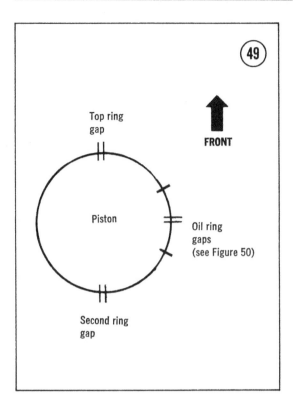

Top ring
gap

FRONT

Piston

Oil ring
gaps
(see Figure 50)

Second ring
gap

49

51

4

the pistons and rings. Remove the ring compressors and the wooden block and lower the cylinder block part way down over the studs so the washers and nuts can be installed.

7. Tighten the nuts in the sequence and to the torque values indicated in **Figure 52**.

8. Install the cylinder head as described earlier.

50

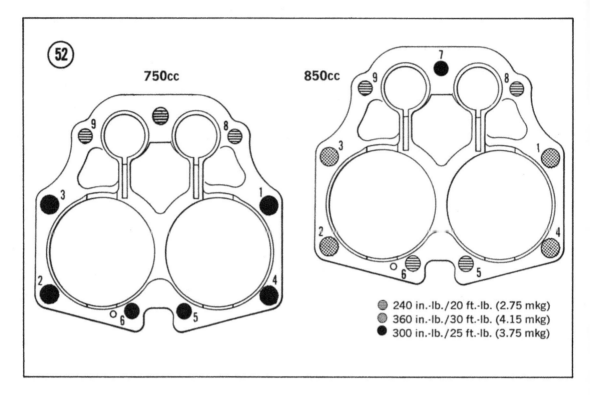

BREAK-IN

Following cylinder servicing (boring, honing, new rings, etc.) and major lower end work, the engine should be broken in just as though it were new. The performance and service life of the engine depend greatly on a careful and sensible break-in.

For the first 500 miles no more than one-third throttle should be used and speed should be varied as much as possible within the one-third throttle limit. Prolonged, steady running at one speed, no matter how modeate, is to be avoided, as is hard acceleration.

Following the 500-mile service (below) increasingly more throttle can be used but full throttle should not be used until the motorcycle has covered at least 1,000 miles and then it should be limited to short bursts until 1,500 miles have been logged.

The mono-grade oils recommended for break-in and normal use (**Table 2**) provide a superior bedding pattern for rings and cylinders than do multi-grade oils. As a result, piston ring and cylinder bore life are greatly increased. However, the factory has found that complete bedding does not occur until about 1,200 miles and

Table 2   RECOMMENDED MONO-GRADE BREAK-IN OILS

Castrol GP or HD 40
Kendall Racing 40
Shell—Aero Shell No. 80, 40W
Torco Racing 40
Valvoline Racing 40

during this period oil consumption will be higher than normal. It is therefore important to frequently check and correct the oil level. At no time, during break-in or later, should the oil level be allowed to drop below the "L" mark on the dipstick; if the oil level is low, the oil will become overheated resulting in insufficient lubrication and increased wear.

500-Mile Service

It is essential that the oil and filter be changed and the sump filter cleaned after the first 500 miles. In addition, it is a good idea to change the oil and filter and clean the sump filter at the completion of break-in (about 1,500 miles) to ensure that all of the particles produced during break-in are removed from the lubrication system. The small added expense may be considered a smart investment that will pay off in increased engine life.

PRIMARY DRIVE AND CLUTCH

The entire primary drive assembly must be removed from the motorcycle before the crankcase can be removed.

Disassembly

1. On 750 and 850 Mark II models, remove the nuts and washers which hold the left footrest and rear brake pedal assembly in place (**Figure 53**). Pull the unit off the studs and allow it to hang on the brake cable. On Mark III models it is not necessary to remove the gear selector pedal; simply remove the left footrest.

2. Place a drip pan beneath the primary chaincase. The pan should be as long as the chaincase to prevent oil from being spilled when the cover is removed. On 750 and 850 Mark II models, unscrew the nut in the center of the cover (**Figure 54**). On 850 Mark III models, unscrew the 11 screws located around the outer edge of the case. Pull the cover off the alignment dowels and away from the chaincase. Remove any oil that remains in the bottom of the chaincase.

NOTE: *If the entire primary drive assembly is to be removed, proceed in sequence with the next step and remove the alternator rotor and stator. However, if only the clutch is being serviced, there's no need to disturb the alternator. Skip Steps 3 through 7.*

3. On 750 and 850 Mark II models, reinstall the footrest/brake pedal assembly and screw the nuts on finger-tight. On all models, depress the brake pedal to lock the rear wheel and prevent the clutch from turning, unscrew the rotor nut (**Figure 55**), and remove the large washer.

4. Unplug the stator leads (**Figure 56**) and carefully pull the leads and the rubber grommet out of the primary chaincase (**Figure 57**). On 850 Mark III models, remove the shifter cross shaft (**Figure 58**) by pulling it straight out to the left.

NOTE: *For most operations the stator lead need not be removed from the case. Make sure, however, that the stator is not allowed to hang from the electrical lead.*

5. Unscrew the stator mounting nuts, remove the washers, and pull the stator off the studs (**Figure 59**). On 750 and 850 Mark II models, remove the spacers from the studs (**Figure 60**).

6. Pull the rotor off the crankshaft. If necessary, the rotor may be broken loose from the shaft by applying pressure with 2 tire irons (**Figure 61**). Remove the key, spacer, and shims from the

shaft. Set the rotor into the stator, wrap the assembly in clean newspaper and set it aside. On 850 Mark III models, straighten the lock tabs on the nuts which attach the inner frame, unscrew the nuts, and pull the frame off the studs (**Figure 62**). Remove the inner washer, bushing, starter gear, and starter intermediate drive assembly (**Figure 63**). Remove the sprag

clutch and bushing (**Figure 64**) and inner race (**Figure 65**) from the crankshaft. Unscrew the nuts from the tensioner cover plate (**Figure 66**) and remove it. Note the location of the smaller nut. Hold the top and bottom tensioner pads in place to prevent them from jumping out of the tensioner body and pull the tensioner off the studs (**Figure 67**).

7. Unscrew the clutch adjuster locknut and the adjuster screw from the center of the diaphragm (**Figure 68**).

8. Install the diaphragm compressor (Norton tool No. 060999) on the diaphragm (**Figure 69**).

The tool bolt must be threaded into the diaphragm at least 6 full turns. Hold the compressor arms to prevent them from turning and turn the compressor nut clockwise until the spring rotates freely in the clutch housing.

9. Use the tip of a small screwdriver to lift the end of the diaphragm retaining clip out of its groove (**Figure 70**) and peel it out of the groove all around the circumference. Remove the tool and the diaphragm together. If the diaphragm is in good condition and is to be reused, it's not necessary to remove the tool. If the tool must be

4

removed, however, hold the bolt with one wrench and turn the compressor nut counter-clockwise until the diaphragm is completely relaxed before unscrewing the bolt.

CAUTION
Any attempt to remove the diaphragm circlip without use of the compressor tool would be extremely hazardous and at the very least could cause severe damage to the clip and the clutch body. Installation of the diaphragm and clip without the use of the compressor is virtually impossible. If the tool cannot be purchased or borrowed, the job should be entrusted to a Norton service shop.

10. If only the clutch is being serviced, remove the plates one at a time using a small magnet to draw them out of the body and off the hub (**Figure 71**). Then refer to the inspection and assembly sections which follow. If the entire primary drive is to be removed, however, leave the plates installed and proceed with the steps which follow.

11. On 750 and 850 Mark II models, with the footrest/brake pedal assembly installed as described above, depress the brake pedal (all models) to lock the rear wheel and prevent the clutch hub from turning. Straighten the lock-washer tabs and unscrew and remove the nut, washer, and tabwasher (**Figure 72**).

12. Install the sprocket puller (Norton tool No. 064297) on the engine sprocket (**Figure 73**). Screw the puller bolts in at least 8-10 turns and then turn in the center bolt tightly against the end of the crankshaft. Rap the sprocket sharply with a steel drift and hammer to jar the sprocket

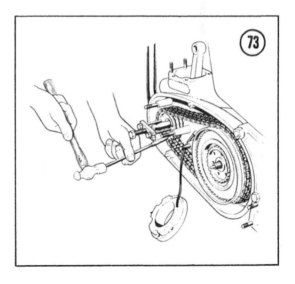

loose from the taper on the crankshaft. Remove the puller from the sprocket.

13. On 750 and 850 Mark II models, once again remove the footrest/brake pedal assembly and

pull off the engine sprocket, chain, and clutch together (**Figure 74**). Note the location and arrangement of the spacer and shims on the main transmission shaft and remove them and set them out of the way. These pieces locate the clutch sprocket in alignment with the engine sprocket and should not be substituted or eliminated during assembly.

14. On 750 and 850 Mark II models, straighten the tab washers on the chaincase mounting bolts (**Figure 75**), unscrew them, and pull the chaincase away from the engine and transmission. Screw the bottom 2 bolts back into the crankcase to prevent oil from dripping out. On 850 Mark III models, disconnect the electrical lead to the starter (**Figure 76**), unscrew the starter mounting screws (**Figure 77**), and pull the starter out of the primary case. Unscrew the nut from the center of the inside case half and pull the case off the long crankcase studs.

Inspection

1. Inspect the O-rings, rubber cover seal, and alternator cable grommet for wear or deterioration and replace any of these pieces that are defective.

2. Inspect the rollers of the primary chain and the sprocket teeth for wear and damage. Normally, these components will last for many thousands of miles; however, incorrect chain tension or extremely dirty oil can cause under-cutting of the sprocket teeth (**Figure 78**) and accelerated wear on the chain rollers.

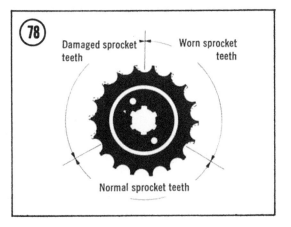

of the sprockets should be checked and corrected if necessary by adding or removing shims from the transmission main shaft, behind the clutch wheel. The outer row of teeth on each sprocket must be the same distance from the mating edge of the primary chaincase (**Figure 80**).

3. After thoroughly washing and drying the clutch plates, inspect the splines for wear (**Figure 79**) and check the contact surfaces of all the plates for signs of burning, galling, and deep grooving. All questionable plates should be replaced.

4. Check the splines on the clutch hub and in the drum for signs of wear and damage and replace as necessary.

NOTE: If one or both of the primary sprockets are replaced, the alignment

5. Inspect the chaincase and outer cover for cracks and fractures and for damage to the sealing surfaces. Repairs to either of these castings should be attempted only by a specialist experienced in the repair of non-ferrous precision castings.

The following inspection points apply only to the 850 Mark III:

6. Inspect the teeth of the starter gears for chipping and excessive wear. Minor burnishing of contact areas is acceptable but deep scarring or

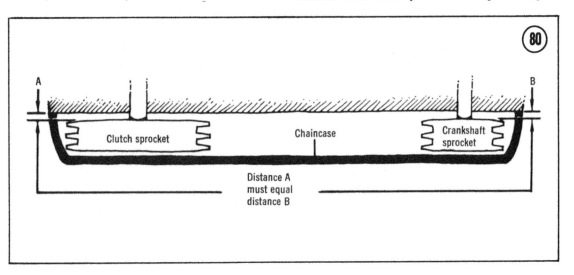

chipped teeth are reason for replacement. Inspect the rollers and cage in the crankshaft
starter gear for wear and damage and replace
the bearing assembly if necessary (**Figure 81**).

7. Check the sprag clutch (**Figure 82**) for wear
and damage. Counter-rotate the inner and outer
cages. The rollers should extend and retract
with no effort or binding.

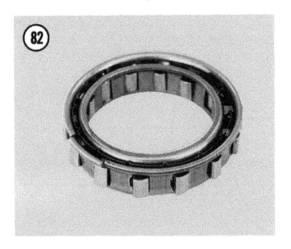

8. Inspect the gear selector cross-over assembly
for damaged teeth and splines (**Figure 83**). If
the gear teeth are damaged, resulting in shifting
difficulties, the gearset can be rotated 180° to
eliminate the damaged teeth from the mesh.

9. Check the rubber contact shoes of the chain
tensioner (**Figure 84**) for deeply worn grooves.
These pieces will wear normally and they should
be replaced before the primary chain wears
through the rubber and comes in contact with
the metal.

Assembly

1. Install the engine sprocket key in its recess
in the crankshaft (**Figure 85**). Make sure it seats
completely and that there is no roughness along
the flat which would prevent it from entering
the keyway in the sprocket.

2. Lightly coat both sides of the crankcase/chaincase gasket with Loctite "Plastic Gasket" or gasket cement and set it in place on the crankcase (**Figure 86**). Place the flat washer and shim (if one was removed) on the end of the chaincase center stud (**Figure 87**).

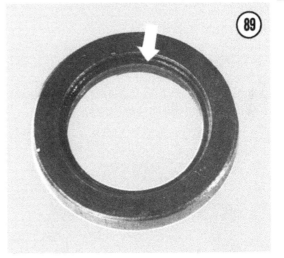

3. Set the chaincase in position, making sure the seal in the case does not snag on the transmission main shaft (**Figure 88**). On 750 and 850 Mark II models, install the 3 bolts with tab washers which hold the chaincase to the crankcase, tighten them securely, and bend the tabs over to lock them in place. On 850 Mark III models, screw in and tighten the chaincase center nut.

4. Install the clutch spacer (recess toward the transmission—**Figure 89**) on the main shaft, followed by the shim or shims (**Figure 90**). Assemble the sprocket and the chain (**Figure 91**). Set

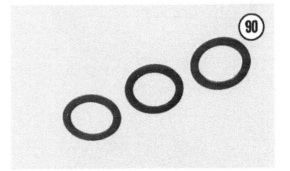

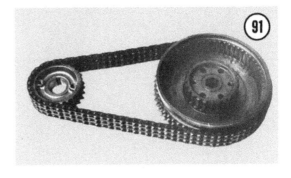

them in position on their shaft and rotate them to line up the keyway in the crankshaft sprocket with the key in the shaft and push the sprockets and chain all the way on together.

5. Line up the splines in the clutch center hub with the splines on the transmission main shaft and push it on. Install the washer and nut. If the washer is a tab type, engage 2 of the tabs with the holes in the hub. Install the footrest/brake lever assembly as before, depress the brake pedal, and tighten the center nut to 70 ft.-lb. (9.68 mkg). Bend the tab washer over to engage 2 flats on the nut.

6. Install the clutch plates in the carrier, beginning with a friction plate (splines on inner diameter — **Figure 92**), followed by a steel plate (splines on outer diameter—**Figure 93**), alternating plates until all 9 are in place. Then install the iron pressure plate (**Figure 94**).

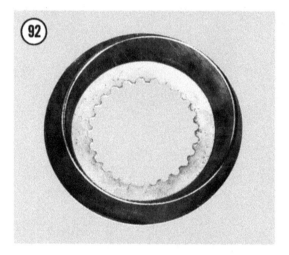

7. Reinstall the assembled compressor tool and diaphragm as before, making sure the compressor nut is backed off to permit the plate to relax until the center bolt has been threaded into the diaphragm at least 6 full turns. Then turn the compressor nut clockwise until the diaphragm is flat and set the diaphragm into the clutch as far as it will go. Start one end of the diaphragm clip into the groove in the clutch body and press it into place until it is fully seated all the way around the clutch. Turn the compressor nut counterclockwise until the diaphragm is relaxed and unscrew the tool's center bolt.

8. Remove the inspection cover from the right side of the transmission and check to see that the clutch operating lever is correctly positioned (**Figure 95**). If it is not, pull it up into place so the recess in the lever engages the roller.

9. Install the clutch adjuster screw and locknut (**Figure 96**). Turn the hand control cable adjuster all the way in (**Figure 97**) and turn the adjuster screw in until the diaphragm just begins to lift or move outward. Then turn the screw out one full turn, hold it to prevent it from turning further, and tighten the locknut securely.

10. On 850 Mark III models, install the inner race for the sprag clutch in the crankshaft sprocket with the notched side of the race facing out (**Figure 98**). Set the sprag clutch in the sprocket with the flanged side facing *in*. Tilt the clutch outward from the top and introduce the bottom edge first. Install the bushing, starter gear (with the castellated side of the hub facing out), and the washer. Refer to the disassembly instructions and install the chain tensioner and the starter intermediate drive assembly. Make certain the recess in the thrust washer covers the circlip (**Figure 99**). Move the large idler gear up into engagement with the small idler gear before attempting to fit the intermediate drive assembly.

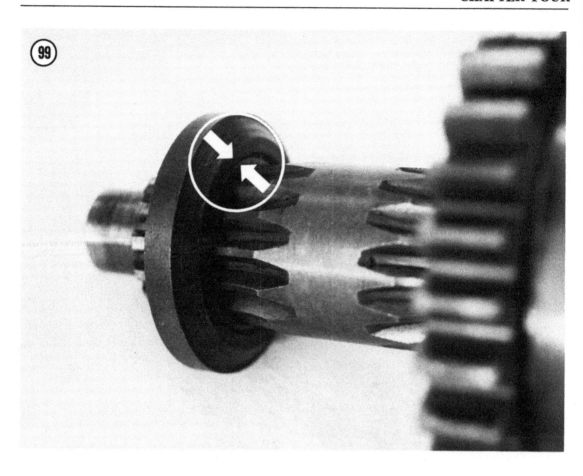

Install the inner frame on the studs, screw on and tighten the nuts, and bend the tab washers over against one flat on each nut.

11. On 750 and 850 Mark II models, install the shims and spacer (recess outward) on the crankshaft (**Figure 100**) and set the rotor key in its recess.

12. Inspect the rotor for any particles it may have attracted and clean it if necessary. Install it on the shaft, lined up with the key and with the timing marks facing out (**Figure 101**). Install the washer and nut, apply the rear brake, and tighten the nut to 70-80 ft.-lb. (9.68-11.06 mkg).

13. Thread the stator leads through the chaincase and press the grommet into the hole, making sure it is fully seated. Set the stator spacers on their studs and install the stator with the lead on the outside positioned at about 5 o'clock (**Figure 102**). Install the washers and nuts and tighten them to 15 ft.-lb. (2.07 mkg). On 750 and 850 Mark II models, check the air gap between the rotor and stator with a flat feeler gauge (**Figure 103**). Ideally, it should be even all the way around but not less than 0.008-0.010 in. (0.203-0.254mm) at any one point. If the gap is less than this, it may be possible to realign the stator by loosening the nuts, pushing the stator in the direction of the smallest gap, and retightening the nuts to the torque specified. If this does not work, one or more of the studs is probably bent. It may then be possible to correct the gap by very carefully realigning the studs. Should this fail, the studs should be replaced with new ones that are known to be straight. When the air gap is correct, reconnect the stator leads.

On 850 Mark III models, set the stator in place and push it on evenly so the stator body fits inside the machined stud bosses. Install the washers and nuts and tighten them progressively in a criss-cross pattern to draw the stator on evenly all the way around. The air gap on Mark III models is self-adjusting.

Install the gear selector cross-over shaft making sure its splines correctly engage the splines in the connector sleeve (**Figure 104**). Install the

starter with the electrical post facing up and screw in and tighten the mounting screws. Reconnect the electrical lead.

14. On 750 and 850 Mark II models, install the rubber sealing ring in the groove in the inner chaincase with the joint seal at the top (**Figure 105**). Set the outer cover in place, carefully aligning it with the dowels. Push the cover on evenly and install the mounting bolt and washer. Install the level plug, timing inspection plug, and clutch adjuster plug, with O-rings.

On 850 Mark III models, lightly coat both surfaces of the gasket with gasket cement and set the gasket in place over the hollow alignment dowels. Hold the cover up to the primary case and position the gear selector pedal in its operating position and push the cover on, moving the pedal slightly if necessary so that its gear can engage the gear on the cross-over shaft. Screw in

and tighten the case screws in a criss-cross pattern. Note the location of the long screw which passes through the case and screws into the starter motor (**Figure 106**).

15. For Mark II models, refer to Chapter Two, *Primary Chain Adjustment,* and adjust the primary chain so that there is total up-and-down movement of the chain of ⅜ in. (9.5mm) at the tightest point.

16. Pour 7 ounces (200cc) of an oil recommended in Chapter Two into the chaincase and install the fill plug with its O-ring.

17. Install the footrest/brake pedal assembly.

ENGINE LOWER END

Service to the engine lower end requires that the crankcase assembly be removed from the motorcycle. The complete primary drive assembly must first be removed and it's advisable that the head, cylinders, and pistons be removed also; they'll have to come off anyway and the task of removing them with the engine installed in the frame is considerably easier. In addition, the decrease in engine weight at the point at which the crankcase is lifted from the frame will be greatly appreciated.

Because of differences in work habits and availability or absence of certain tools such as a suitable large bench vise, or an elevated workstand, no hard-and-fast rule is set down here for the point at which the crankcase assembly should be removed from the motorcycle. It may be removed at Step 8 as described, or you may elect to wait until Step 13 has been completed.

The lower end service procedure is presented as a complete rebuild. To service and repair major components related to the lower end but not requiring its removal and disassembly, disassemble the engine up to the point at which the affected part can be reached and then reassemble from that point on beginning with the appropriate step in the *Assembly* section; just be sure that a subsequent component has not been affected by the failure or deterioration of a component located earlier in the disassembly procedure.

Removal and Disassembly

1. Refer to earlier sections in this chapter and remove the seat, fuel tank, coil assembly, exhaust system, cylinder head, cylinders, pistons, and primary drive assembly.

2. Unscrew the tachometer cable collar (**Figure 107**) and pull the cable out of the crankcase. On 750 and 850 Mark II models, unscrew the gear indicator bolt (**Figure 108**) and remove the indicator. Loosen the bolt in the gear selector pedal (**Figure 109**) and pull the pedal off the shaft. On all models, unscrew the rocker feed pipe bolt (**Figure 110**) and remove the washers from each side of the union.

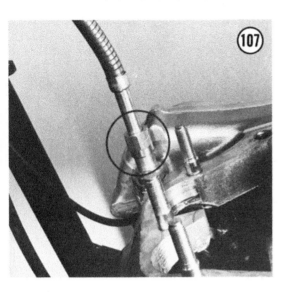

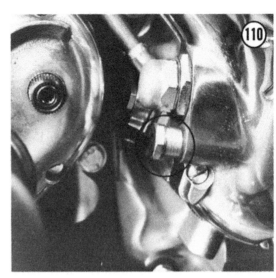

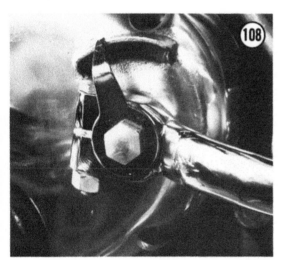

3. Place a drip pan beneath the engine. On 850 models, loosen the clamp on the crankcase breather hose (**Figure 111**) and pull the hose off the nipple and direct it so that residual oil will drain into the pan. On 750 models, remove the 2 bolts which hold the breather to the crankcase, pull the breather away and remove the gasket (**Figure 112**). Unscrew the bolt from the oil pipe junction (**Figure 113**) and remove it and the gasket.

4. Loosen the crankcase filter located beneath the engine (**Figure 114**) on 850 models, and

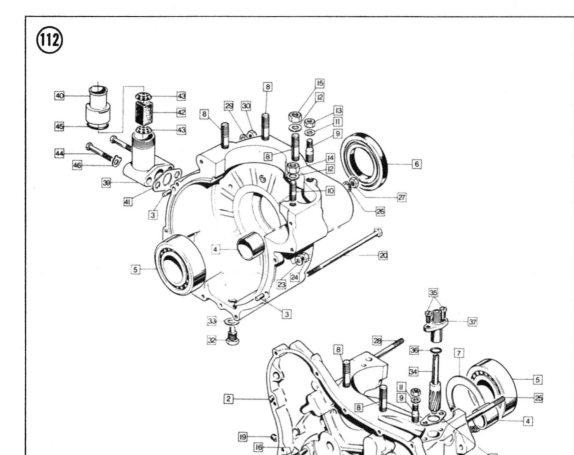

CRANKCASE — 750 ENGINE

1. Crankcase	17. Oil pump stud nut
2. Crankcase dowel	18. Grub screw-oil stop
3. Crankcase dowel	19. Junction block dowel
4. Camshaft bushing	20. Crankcase bolt, long
5. Main bearing	21. Crankcase bolt nut
6. Main bearing oil seal	22. Crankcase bolt, short
7. Main bearing shim	23. Crankcase bolt washer
8. Cylinder stud, front	24. Crankcase bolt nut
9. Cylinder stud stepped	25. Crankcase stud stop, front
10. Cylinder stud, front	26. Crankcase stud washer
11. Cylinder stud washer	27. Crankcase stud nut
12. Cylinder stud washer	28. Crankcase stud stop, rear
13. Cylinder stud nut	29. Crankcase stud washer
14. Cylinder stud nut	30. Crankcase stud nut
15. Cylinder stud nut	31. Crankcase screw, T.S. to D.S.
16. Oil pump stud	at sump

32. Magnetic sump plug
33. Sump plug washer
34. Gear tachometer
35. Screw
36. O-ring
37. Housing
38. Tachometer housing gasket
39. Breather tube
40. Sump filter body
41. Filter washer
42. Filter gauze
43. Strainer
44. Bolt
45. Filter body washer
46. Tab washer

4

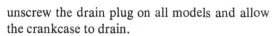

unscrew the drain plug on all models and allow the crankcase to drain.

5. Unscrew the left nut from the front mount (**Figure 115**), line up one of the flats on the right end bolt head so it will clear the timing case (**Figure 116**), and tap it about halfway out. Peel back the left cover (**Figure 117**) and remove the spacer, end cap, and shims. Unscrew the nuts from the engine mounting bolts (**Figure 118**) and remove the bolts. Remove the complete front mount (**Figure 119**).

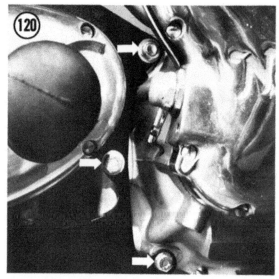

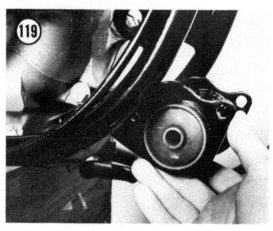

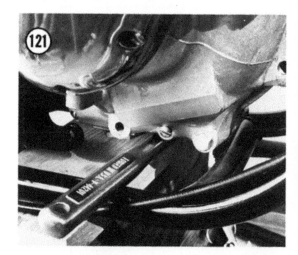

6. Remove the bottom- and center-rear engine bolts (**Figure 120**). When removing the bottom bolt it will be necessary to lift up on the crankcase so the bolt will clear the frame.

7. On models equipped with a sump filter, lift the front of the engine and place a long drift or socket extension beneath it so that it is supported on the frame rails (**Figure 121**). Remove the filter and allow the remaining oil to drain.

8. Unscrew the nut from the top rear engine bolt (Figure 120), lift up on the engine, and pull the bolt out. Lift the crankcase assembly out of the frame, taking care not to damage it on the mounts.

9. Remove the bottom front crankcase bolt (**Figure 122**). Mount the crankcase assembly in a bench vise fitted with protective jaw covers.

The timing side of the crankcase should face you and the vise should be tight enough to hold the crankcase firmly (**Figure 123**).

10. Remove the contact breaker cover, unscrew the breaker cam bolt and remove it along with the 2 washers (**Figure 124**). Screw the end of the slide hammer (Norton tool No. 064298) into the end of the cam at least 6 full turns, pull the slide sharply to free the cam from the taper on the shaft (**Figure 125**), and remove it. Unscrew the 12 screws from the timing cover (**Figure 126**). Carefully tap the cover loose with a soft mallet and remove it from the crankcase. Feed the electrical leads through the hole in the case, one at a time.

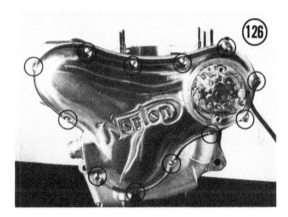

11. Unscrew the oil pump nuts (**Figure 127**) and rotate the pump drive gear bolt (**Figure 128**) clockwise with a wrench to draw the pump away from the case. Place 2 blocks of wood over the top opening in the crankcase, pass a drift through the wrist pin holes in the connecting rods, and turn the oil pump worm nut clockwise to bring the drift in contact with the blocks (**Figure 129**). Unscrew the worm nut (left-hand thread) using even pressure on the wrench. Remove the worm gear from the shaft.

12. Unscrew the camshaft nut (**Figure 130**), again using even pressure on the wrench. Loosen the nuts on the timing chain tensioner (**Figure 131**) and remove the sprockets, idle gear,

and chain together with the aid of a puller (**Figure 132**).

13. Install the extractor (Norton tool No. ET.2003) on the crankshaft pinion gear with the tool jaws lined up with the flats on the washer (**Figure 133**). Hold the tool to prevent it from rotating and turn the puller bolt clockwise to remove the pinion. Remove the key from the shaft and the washer. With a small magnet, remove the lipped seal disc (**Figure 134**). It's not necessary to remove the chain tensioner unless the contact shoe is deeply worn and must be replaced.

In such case, the tensioner should be replaced with the latest type which has a rubber facing on the slipper surface, resulting in quieter operation and reduced wear of the chain sideplates.

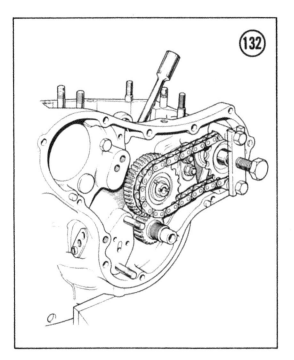

4

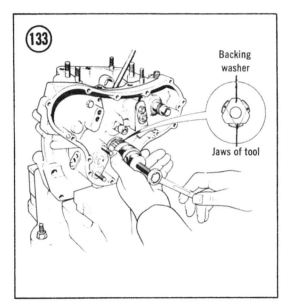

Backing
washer

Jaws of tool

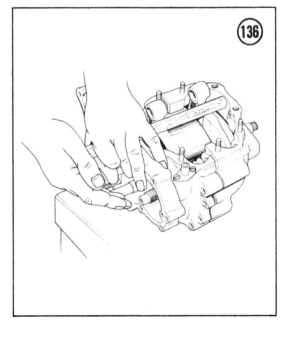

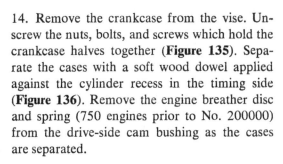

14. Remove the crankcase from the vise. Unscrew the nuts, bolts, and screws which hold the
crankcase halves together (**Figure 135**). Separate the cases with a soft wood dowel applied
against the cylinder recess in the timing side
(**Figure 136**). Remove the engine breather disc
and spring (750 engines prior to No. 200000)
from the drive-side cam bushing as the cases
are separated.

15. Remove the camshaft and thrust washer from the timing case (**Figure 137**). On engines from No. 200000 to 300000, remove the flat thrust washer.

Thrust washer

16. Fit a suitably-sized length of tubing over the end of the crankshaft (**Figure 138**). With assistance, stand the tube on end on a work bench and tap along the mating surface of the crankcase with a block of wood and a hammer to drive the case off the bearing (**Figure 139**).

> NOTE: *This is not necessary on engines with a roller bearing on the timing side of the crankshaft (after Number 200000).*

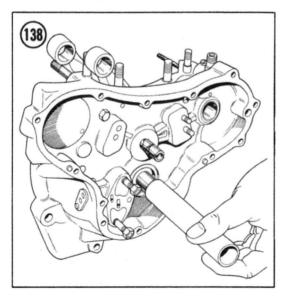

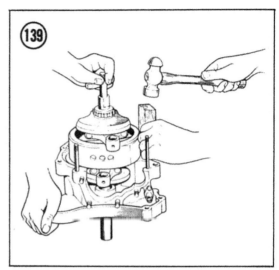

If necessary, the crankcase may be heated with a propane torch in the area of the bearing.

17. Unscrew the nuts from one of the connecting rods and remove the rod and end cap from the crank. Mark both the cap and rod for position (right or left) and direction (front). See **Figure 140**. Then remove the other rod assembly and mark it appropriately. Under no circumstances should the rods and caps be interchanged, switched from one side to the other, or reversed (along a vertical axis) when they are reinstalled.

18. Scribe a "T" on the timing side of the flywheel so that it may be installed in the same position from which it was removed. Unscrew the nuts from the timing side (**Figure 141**) and break the crank halves loose from the flywheel with a soft mallet or a hammer and soft drift. The crank should be parted over a drip pan to catch the oil remaining in the assembly.

Inspection

1. Thoroughly clean all the parts in solvent and dry them. Pay particular attention to the crank assembly which will likely have an accumulation of sludge in the crank halves and the flywheel. Blow out the oilways in the crank with air pressure. Do the same with oil feed and filter lines.

2. Inspect the crank journals for scoring and galling. Minor roughness may be removed with fine emery cloth. Measure the journals on the horizontal and vertical axes (**Figure 142**) to

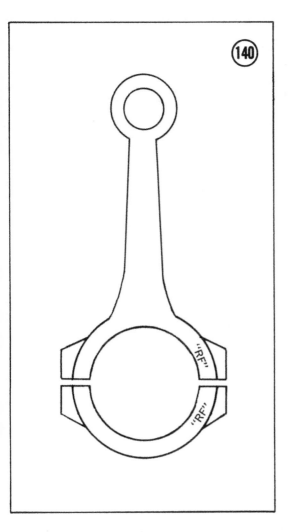

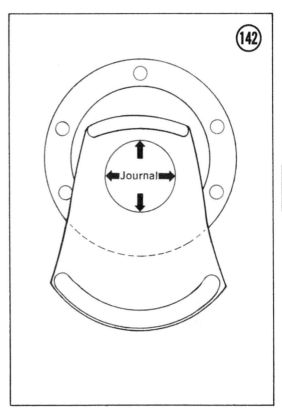

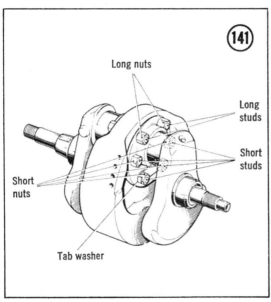

Long nuts

Long studs

Short studs

Short nuts

Tab washer

check for out-of-roundness. If the difference between the 2 axes on either journal is greater than 0.0015 in. (0.038mm), the journals must be ground to the next undersize with a 0.090 in. (2.28mm) face radius (**Figure 143** and **Table 3**) and new bearings installed.

New bearings are available in finished sizes of —0.010, —0.020, —0.030, and —0.040 to correspond to the specified regrind dimensions.

3. If the connecting rod journals do not have to be reground, inspect the connecting rod bearings for scoring and galling. Minor roughness may be removed with fine emery cloth but no attempt should be made to salvage a bearing that is deeply scored or galled. If crank journal wear is very slight, consideration may be given to replacing the bearings with a new set of the same size. However, if any doubt exists, it's preferable to have the journals reground to the next undersize and appropriate bearings installed.

4. Rotate the main bearings by hand, either in the cases or on the crankshaft, and check for roughness and radial play. The bearings should

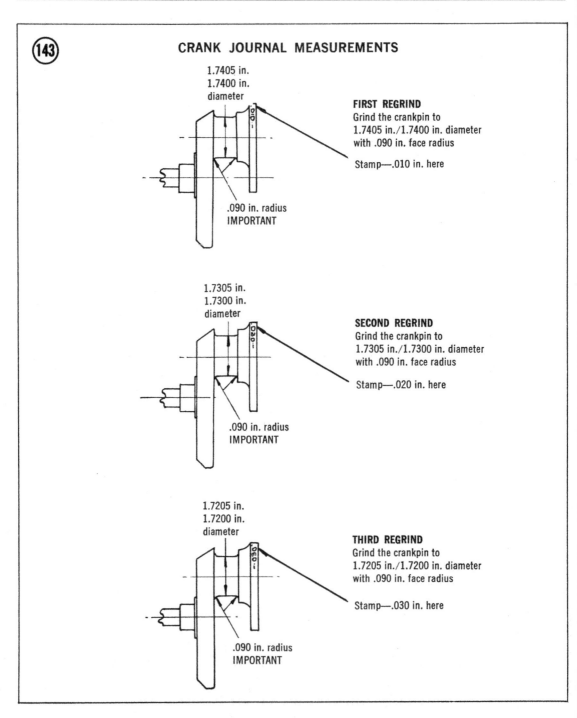

(143) CRANK JOURNAL MEASUREMENTS

1.7405 in.
1.7400 in.
diameter

FIRST REGRIND
Grind the crankpin to
1.7405 in./1.7400 in. diameter
with .090 in. face radius

Stamp—.010 in. here

.090 in. radius
IMPORTANT

1.7305 in.
1.7300 in.
diameter

SECOND REGRIND
Grind the crankpin to
1.7305 in./1.7300 in. diameter
with .090 in. face radius

Stamp—.020 in. here

.090 in. radius
IMPORTANT

1.7205 in.
1.7200 in.
diameter

THIRD REGRIND
Grind the crankpin to
1.7205 in./1.7200 in. diameter
with .090 in. face radius

Stamp—.030 in. here

.090 in. radius
IMPORTANT

Table 3   JOURNAL REGRIND SIZES

Regrind	Dimension (Crankpin dia.)	Size Stamp
First	1.7405-1.7400 in. (44.175-44.162mm)	.010
Second	1.7305-1.7300 in. (43.921-43.908mm)	.020
Third	1.7205-1.7200 in. (43.667-43.654mm)	.030

turn smoothly and exhibit no appreciable play. If a bearing is to be replaced, it's first necessary to heat the case in the area of the bearing and then dislodge the bearing either with a soft drift applied to the bearing inner race (timing side ball-type bearings) or by rapping the case sharply on a piece of soft wood.

The cast must also be heated when a bearing is being installed, and the bearing should fit completely into the bore with no effort.

CAUTION
When heating the case during removal or installation of a bearing, do not bring the flame in contact with any part of the bearing; even a brief exposure to the flame can destroy the case hardening on the bearing.

On engines equipped with roller main bearings, inner races should be removed from crankshaft with a puller (Norton tool No. 063970). See **Figure 144**.

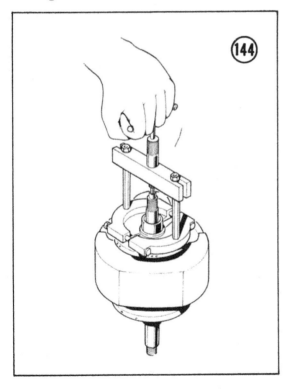

5. Check the cam bushings, journals, and lobes for wear and damage. Minor roughness on the lobes may be removed with fine emery cloth.

6. Inspect the seals in the timing cover for damage or deterioration (**Figure 145**). Oil in the contact breaker cavity is a good indication that the contact breaker seal is failing and should be replaced. The seal can be removed by prying it out with a screwdriver; be careful not to damage the case. Press a new seal evenly into the seal bore with the open side facing out (**Figure 146**), using Norton tool No. 064292.

The crankshaft seal is held in the case with a circlip. To remove the seal, first remove the circlip (**Figure 147**). Then carefully pry the seal out of the case. Install a new seal squarely in the bore, with the open side facing in (**Figure 148**), and install the circlip with its sharp edge facing out so it will lock in the groove.

7. Check the end-play of the oil pump gears. If the gears have begun to wear into the oil pump

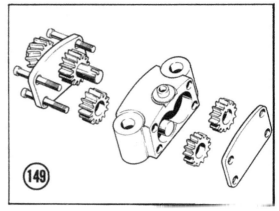

Hone the scavenging side in the same manner, checking the resistance with only the scavenging gears installed in the pump. When both sides have been corrected, carefully wash and dry all of the parts and assemble the entire pump. Squirt oil into the feed hole (**Figure 150**) and turn the pump drive gear several revolutions until the pump turns freely.

body covers, in and out movement of the drive gear will be apparent. Wearing in of the gears reduces the efficiency of the pump but the condition can be corrected by honing the sides of the pump body to compensate for the wear. Remove the 4 screws from the pump and remove the rear cover. Use a small drift to drive the shaft out of the keyed gear (**Figure 149**). Disassemble the pump and clean and dry the parts thoroughly. Beginning with the feed side of the pump (narrow gears), hone the face of the pump on a piece of fine emery cloth laid on a surface plate or a piece of glass. Periodically check your work by assembling the pump and checking its rotational movement. Sufficient material has been removed when a slight resistance to rotation can be felt.

8. On 850 Mark III models, clean and check the oil anti-return valve in the outer cover (**Figure 151**). The valve piston should move freely in its bore with a slight spring resistance.

Assembly and Installation

1. Polish the connecting rod journals with fine emery tape. Wash and dry the crank halves. Blow out the oilways with compressed air. Install 2 short studs in the inner holes in the flywheel. Set the drive side crank half over the studs on the drive side of the flywheel (opposite of side marked with "T") and install the lock plate and 2 short nuts. Set the timing side crank

half in place on the opposite ends of the studs and install the lock plate and 2 short nuts. Install the remaining 4 studs from the drive side with the long studs in the outer holes and screw on the nuts—short nuts to short studs and long nuts to long studs. Tighten the nuts, beginning with the inner ones and working diagonally, to 25 ft.-lb. (3.45 mkg). Bend the tabs over on the inner nuts and stake the others with a center punch to keep them from backing off. Screw the oilway plug into the timing side crank cheek and tighten it securely.

2. Oil the bearing shells and fit them to the connecting rods and end caps (**Figure 152**). Note that the drilled shell half is fitted to the rod and the plain one to the cap. Oil the journals and install the right side rod and cap on the right journal (timing side) with the oil hole in the rod to the outside. The front of the rod, as marked during disassembly, should be on the right side of the journal viewing the crank from the right end (**Figure 153**). Install the original bolts and new nuts and tighten them to 25 ft.-lb. (3.45 mkg). Install the left-hand rod and cap in the same manner, with the oil hole in the rod to the outside.

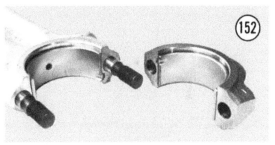

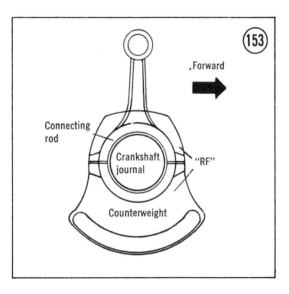

NOTE: *The self-locking nuts must be replaced each time the connecting rod assembly is parted. Also, if a new connecting rod is being installed, do not use the non-locking nuts that hold the rod assembly together in the package. Discard these and use the self-locking nuts that are also included.*

3. Oil the main bearings and the crankshaft and set the timing side of the crankcase in place over the right end of the crankshaft (**Figure 154**) and press it down into place. Make sure the connecting rods line up with the cylinder spigot cutaway in the case. Hold the rods in this position and prevent them from turning by slipping a heavy rubber band or piece of inner tube over them and pushing it down against the flywheel.

NOTE: *Main bearings should be replaced with bearing No. 064118, available through Norton parts dealers. See Sources list in Chapter Eleven.*

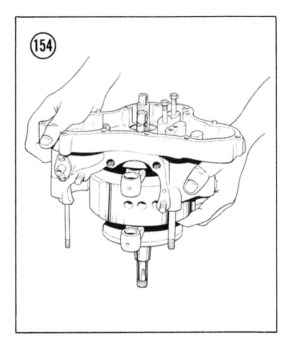

4. Dab some grease on the flat side of the camshaft thrust washer and set it in place on the bushing (**Figure 155**). On engines from No. 200000 to 300000, make sure that the flat thrust washer is located between the bushing and the chamfered washer, and that the tab on the washer engages in the hole beneath the bushing.

5. In the drive-side case half, install the rotary disc and spring on engines prior to No. 200000. Lightly oil the end of the camshaft and install it in the bushing. (On engines prior to No. 200000, make sure the camshaft engages with the dogs on the disc.)

6. Coat the mating surface of the timing-side case half with Loctite "Plastic Gasket" or gasket cement and oil the timing end of the camshaft and the drive side main bearing. Hold the camshaft in place in the drive side case and set the case in position on the crankshaft (**Figure 156**). Push the case into place, making sure the camshaft lines up with the thrust washer and the bushing and that the connecting rods are centered in the cylinder spigot recesses. If resistance is felt just before the case halves contact, check the tachometer drive to make sure it engages with the gear on the camshaft. Tap the crankcase halves together all around the mating line with a soft mallet. Check the camshaft to make sure it has some end-play.

7. Install the screws, studs, washer, nuts, and short front bolts in the crankcase. Omit the bottom bolt at this point so that the crankcase may be clamped in a vise. Tighten the nuts and screws evenly in a crosswise pattern.

8. Clamp the crankcase in a vise as for disassembly and install the oil seal disc (lip out) and the washer on the timing side of the crankshaft. Set the key in the crankshaft recess and install the crankshaft gear with the timing marks outward. Rotate the crankshaft so the timing mark is at the top and journals are at TDC (**Figure 157**).

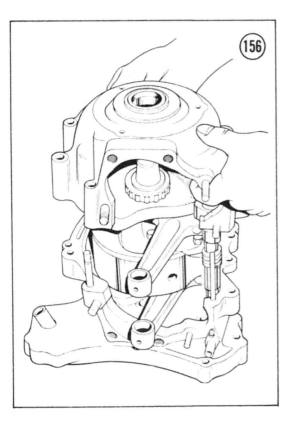

9. Assemble camshaft sprocket, gear sprocket, and chain with the marks on the sprockets located 10 rollers apart (**Figure 158**). Line up the marked space on the intermediate gear with the

marked tooth on the crankshaft gear and install the assembled sprockets and chain (**Figure 159**).

10. Install the cam chain tensioner as shown in **Figure 160**—thin plate inside with the long end down, the tensioner, and the thick plate on the outside with the long end up. Install the washers and nuts and adjust the tensioner so that the maximum up and down movement of the chain is 3/16 in. (4.8mm) at the tightest point. Rotate the crankshaft, checking the chain movement every several plates until you are certain that the adjustment is correct at the tightest position. Tighten tensioner nuts to 15 ft.-lb. (2.07 mkg).

11. Screw on the oil pump drive gear (left-hand thread) and the camshaft nut (right-hand thread). Lock the crankshaft as described in the disassembly section, this time rotating the crankshaft counterclockwise (from the right end) until the drift is stopped against the blocks. Then tighten both the drive gear and the camshaft nut.

12. Prime the oil pump by turning the drive gear and squirting oil into the pump through the feed port (**Figure 161**). Install the oil pump on the mounting studs (with a gasket if one was removed during disassembly) after making sure the studs are tight. Screw on the nuts (no washers) and tighten them to 10-12 ft.-lb. (1.38-1.66 mkg). Install a new oil seal (part No. NMT272) on the pump outlet (**Figure 162**). Omit any shims that may have been removed during disassembly, provided a gasket is used. Check the tightness of the drive gear.

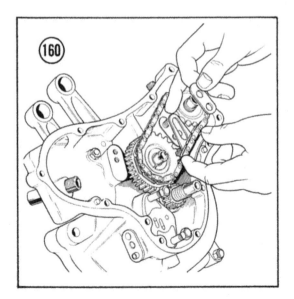

13. Install the timing cover gasket over the dowels in the crankcase. Screw a seal protector (Norton tool No. 061359) into the end of the camshaft and lightly oil it (**Figure 163**). If the

15. Insert the contact breaker electrical leads through the hole in the timing case and set the cover in place. Install the timing cover screws and tighten them in the sequence shown in **Figure 165**. Unscrew the seal protector tool if used, or remove the tape from the end of the camshaft.

16. Lightly lubricate the automatic advance mechanism with oil. Make sure the cam taper is clean and dry and set the unit in position (**Figure 166**).

tool is not available, wrap the end of the shaft with a piece of transparent mending tape and shape it into a cone. The tool or the tape are required to protect the contact breaker oil seal from damage when the timing cover is installed.

14. Remove the crankcase from the vise, tilt it onto its left side (drive side) and squirt several ounces of fresh engine oil into the flywheel through the right end of the crank (**Figure 164**). Allow a couple of minutes for the oil to run into the journals and reclamp the crankcase assembly in the vise.

17. Install the contact breaker assembly with the yellow lead to the rear (**Figure 167**). Line up the plate so that the pillar bolts are in the center of their slots and tighten them.

18. Remove the crankcase from the vise and install the bottom front bolt, the drain plug, and the oil filter on 850 models.

19. Set the crankcase in the frame, between the rear mounting plates, and install the top and center rear bolts (**Figure 168**). Lift up on the engine and install the bottom bolt from the right side. Install the washers and screw the nuts on finger-tight.

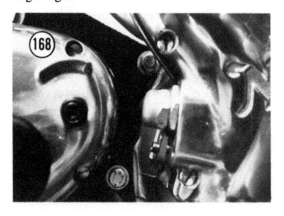

20. Set the complete front mount in place as shown in **Figure 169** and install the bottom bolt. Pivot the mount up and install the top bolt.

Refer to Chapter Ten and assemble the shims, caps, spacers, and covers. Install the right-hand set first and lift up the front of the crankcase to align the mount with the frame tab. Push the bolt into the mount, lining up one of the flats on the bolt head to clear the timing case. Before the bolt reaches the other bushing, install the left-hand set and then push the bolt the rest of the way through.

Fit the lips of the outer covers over the caps, install the washer and nut, and tighten it to 25 ft.-lb. (3.45 mkg). Tighten the nuts on the ⅜ in. studs to 25-30 ft.-lb. (3.45-4.15 mkg) and those on the 5/16 in. bolt to 15 ft.-lb. (2.07 mkg).

21. The remaining components should be assembled and installed in reverse order of disassembly. Refer to the appropriate sections and chapters to ensure that your work is correct. Following engine installation, the engine should receive a major service including fresh oil, ignition timing and adjustment, carburetor adjustment, primary and secondary chain adjustment, clutch adjustment, and brake adjustment.

4

CHAPTER FIVE

TRANSMISSION

The 4-speed Norton transmission (**Figure 1**) is not unitized with the engine and can therefore be completely dismantled and serviced without dismantling the engine. However, removal of the gearset from the transmission housing, removal of the entire transmission from the motorcycle, or removal and replacement of the countershaft sprocket requires disassembly of the primary drive.

There are few situations requiring removal of the transmission housing from the motorcycle (such as repair to a crack or fracture, renewal of the drain plug threads, etc.). Even the bearings in the main housing can be removed with an expanding type bearing puller with the housing in place.

Beginning in 1975, with the Mark III Electric Start, the gear selector mechanism was redesigned to provide left-foot shifting with a 1-down, 3-up change pattern. This change was made to accommodate international agreements for standardization of motorcycle controls.

The transmission service procedures in this chapter are arranged by subassembly (outer cover, inner cover, gasket, and selectors). It's not necessary to totally dismantle the transmission to correct many unsatisfactory conditions that may arise through normal use and wear. For instance, replacement of a worn kickstarter pawl requires no disassembly beyond that required for the inner cover. Be certain, however, that an unsatisfactory condition that's found in an early disassembly step has not affected components covered in later steps. If the possibility exists that subsequent components may have been damaged by a known failed component or assembly, continue to disassemble and inspect the transmission until you're confident that all unserviceable parts and conditions have been corrected.

OUTER COVER

The outer cover houses the clutch operating mechanism, the gear selector stop plate, and the starter return spring.

Removal

1. Place a drip pan beneath the transmission and remove the drain plug. Remove the fuse from the holder on the negative battery lead (**Figure 2**).

2. Unscrew and remove the bolt from the kickstarter lever (**Figure 3**) and pull the kickstarter lever off the shaft. It may be necessary to strike the bottom of the foot lever with a mallet to break the kickstarter loose from the shaft.

FOUR-SPEED TRANSMISSION

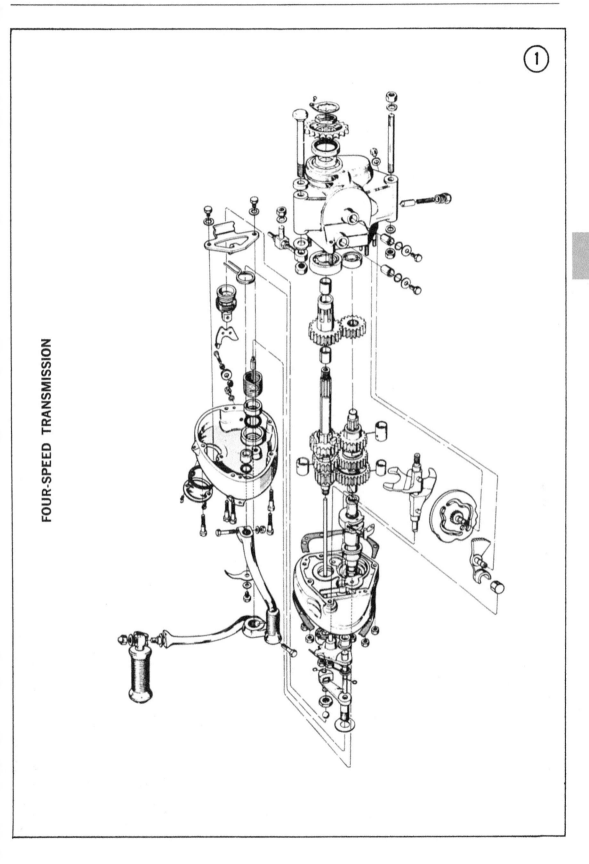

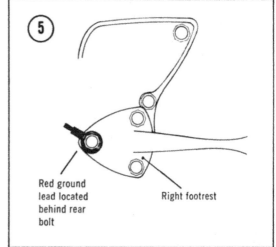

Red ground
lead located   Right footrest
behind rear
bolt

3. Unscrew the nuts and bolt from the right footrest (**Figure 4**) and remove it. Note the location of the red ground wire (**Figure 5**) so it can be installed in the same manner later.

4. On right-foot selector models, remove the setscrew and washer from the gear selector indicator (**Figure 6**) and remove the indicator but not the gear selector pedal.

5. Remove the cover inspection cap and gasket (**Figure 7**) and disconnect the end of the clutch cable from the clutch fork (**Figure 8**).

6. Remove the 5 screws which hold the cover in place (**Figure 9**). Place the drip pan beneath the outer cover. Tap gently around the edge of the outer cover with a soft mallet and pull the cover

away from the transmission with the shift lever and off the starter and gear selector shafts.

7. On right-foot selector models, unscrew and remove the bolt from the gear selector pedal and remove the pedal from the shaft. Remove the pawl assembly and the ratchet plate from the cover (**Figure 10**). On left-foot selector models, remove the pawl assembly and ratchet from the inner cover.

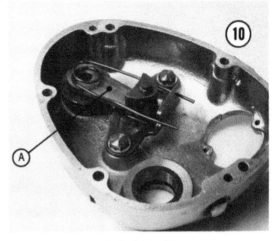

A. Pawl assembly

Inspection

1. Thoroughly clean and dry all parts. Check the O-rings and gaskets for damage and deterioration and replace them if necessary.

2. Check the teeth on the pawl (**Figure 11**) and the ratchet plate for wear. They should be even and sharp to ensure accurate engagement.

3. Install the kickstarter lever on its shaft and depress it part way to check the resiliency of the spring. The resistance of the spring should be firm and even, and when the lever is relaxed, the spring should hold it firmly against the stop.

4. Install the ratchet and pawl assembly in the outer cover (**Figure 12**) and fit the gear selector pedal onto its shaft. Alternately lift and depress the pedal and check the resiliency of the pawl spring by observing the movement of the pawl. With the pedal relaxed, the pawl should be centered with the arms of the spring not quite touching the ends of the pawl (**Figure 13**). As the pedal is lifted or depressed, the pawl should immediately begin to move toward engagement.

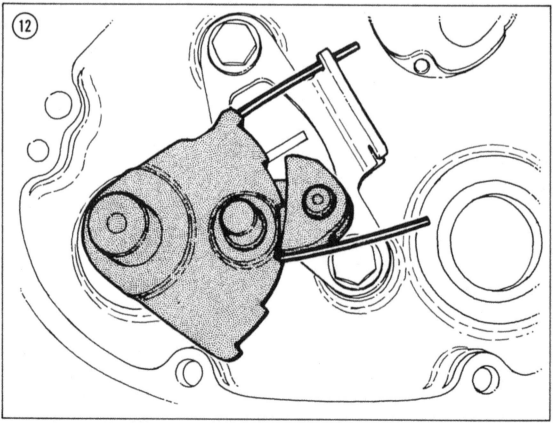

5. On 850 Mark III models, check the condition of the splines on the pawl assembly shaft for rust or damage. Light rust can be removed with a wire brush, after which the splines should be lightly coated with oil to inhibit further rust. Check the protector boot on the coupling sleeve for damage or deterioration and replace it if necessary.

6. Check the mating surfaces of the inner and outer covers for roughness and burrs. Minor damage can be removed by carefully dressing the surface with fine emery cloth. Remove any bit of gasket that may have remained attached when the gasket was removed.

Installation

1. Install new O-rings (**Figure 14**) in the starter shaft and selector shaft bores in the outer cover, and in the groove on the ratchet shaft.

2. Put the spring washer and the return spring on the selector shaft. Make sure the ends of the spring straddle the arm on the pawl assembly (**Figure 15**). Install the shaft and pawl assembly in the cover. Install the stop plate, making sure

the ends of the return spring straddle the tang on the inside of the stop plate (**Figure 16**). Then tighten the stop plate screws.

3. Install the pawl spring (**Figure 17**). The spring arm which has 2 bends must be on the bottom. Check to see that there is a very slight

clearance (0.010 in. minimum) between the spring arms and the ends of the pawl (**Figure 18**).

4. Coat the mating surfaces of the inner and outer covers with gasket cement and set the gasket in place on the inner cover.

5. Line up the outer cover with the inner cover, making sure the coil of the pawl spring goes over the ratchet shaft. Be careful not to disturb

the position of the pawl spring. Otherwise, it will rotate the pawl and disturb its alignment with the ratchet plate. Carefully push the outer cover all the way on. If there is hard resistance before the cover contacts the gasket, pull the cover away and check the alignment of the pawl spring and the pawl.

On left-foot selector models, also check the alignment of the splines on the pawl assembly shaft with the splines of the collar that connects it to the thru-shaft.

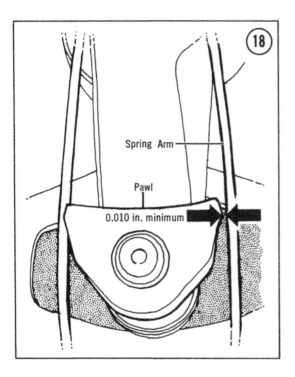

Spring Arm

Pawl

0.010 in. minimum

7. Install and tighten the transmission drain plug and fill the transmission with 0.9 pts.—U.S.; 0.75 pts. Imperial—U.K.; 420cc of oil. Connect the clutch cable to the fork and adjust the cable's free play. Install the inspection cover and gasket and check the cover mating surface and drain plug for leaks.

6. When the cover is in position, install the 5 screws and tighten them securely. Install the kickstarter lever and gear selector pedal on their shafts and check their operation. Slowly rotate the rear wheel and operate the gear selector pedal to engage each of the gears. If the operation is satisfactory, install the gear selector indicator.

INNER COVER

The inner cover functions as the right half of the transmission case in addition to housing the clutch operating mechanism and part of the gear selector mechanism.

Removal

1. Remove outer cover as described previously.

2. Unscrew the bolt which holds the clutch lever (**Figure 19**) in the clutch operating body and remove the lever.

3. Mark the location of the clutch operating body with 2 punch or scribe marks (**Figure 20**) and unscrew the lock ring from the inner cover. Remove the clutch operating body and the ball (**Figure 21**).

4. Shift the transmission into fourth gear by carefully prying the quadrant up (**Figure 22**) while slowly rotating the rear wheel. With assistance, apply the rear brake and unscrew the transmission main shaft nut (**Figure 23**). Release the brake, slowly rotate the wheel, and shift the transmission back into neutral by carefully prying the quadrant down.

5. It is not necessary to remove the kickstarter assembly from the inner cover. However, if the kickstarter pawl is suspected of being damaged or worn (see *Inspection*), carefully pry the the inboard tang of the kickstarter spring off

the post (**Figure 24**). Disengage the outboard tang from the hole in the starter shaft and remove the spring.

6. Unscrew 7 nuts which mount the inner cover to the transmission (**Figure 25**).

7. Tap gently around the edge of the inner cover with a soft mallet and pull the cover away from the transmission.

8. Withdraw the clutch operating rod from the transmission main shaft (**Figure 26**) and pull the starter shaft out of the cover (**Figure 27**).

Inspection

1. Thoroughly clean and dry all parts. Check the gasket for damage or deterioration and replace it if necessary.

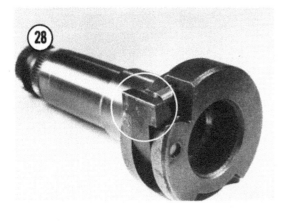

2. Check the kickstarter pawl (**Figure 28**) for wear. Also make sure it moves freely in its pivot and that the spring and plunger will hold it open.

3. Rotate the main shaft bearing (**Figure 29**) by hand and feel for roughness and radial play. Any movement other than smooth rotation should be barely perceptible. If the bearing is found to be unserviceable, warm the case and dislodge the bearing by tapping the cover on a block of soft wood. Press in a new bearing while the case is still warm, until the bearing is fully seated in the bore (**Figure 30**).

4. Check the starter shaft and layshaft bushings (**Figure 31**) for signs of wear and galling. Lightly oil the starter shaft bushing and install the shaft in the cover. Rotate the shaft and check to see that it turns smoothly and that there is only slightly perceptible radial movement. Check the layshaft bushing in the same manner, while installed in the end of the starter shaft, then while installed on the end of the layshaft (**Figure 32**). Check the bushing for the ratchet plate.

5

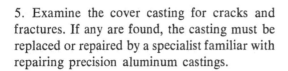

5. Examine the cover casting for cracks and fractures. If any are found, the casting must be replaced or repaired by a specialist familiar with repairing precision aluminum castings.

Installation

1. Install the kickstarter shaft and bushing in the case. Compress the pawl and rotate the shaft so that the pawl is held down by the stop (**Figure 33**). Fit the starter return spring on the shaft with the outboard tang in the hole in the shaft and the inboard tang over the post in the cover (**Figure 34**).

2. Fit the layshaft bushing into the end of the starter shaft.

3. Coat the mating surfaces of the transmission case and the inner cover with gasket cement and set the gasket in place on the transmission.

4. Check to make sure the main shaft and layshaft are completely seated in transmission case.

5. Line up the inner cover with the main shaft and the layshaft and push it into place. Screw on the 7 mounting nuts and tighten them in the pattern shown to 12 ft.-lb. (1.66 mkg). See **Figure 35**.

6. Shift the transmission into fourth gear by levering the ratchet plate carefully upward while slowly rotating the rear wheel. With assistance, apply the rear brake and tighten the transmission main shaft nut to 70 ft.-lb. (9.7 mkg). Then shift the transmission back into neutral.

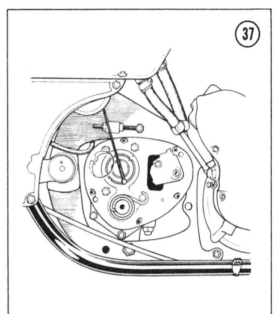

7. Assemble the clutch operating body, ball, and lock ring and install it in the cover. Line up the mark in the body with the mark in the cover (**Figure 36**) and tighten the lock ring until its notches line up with the marks. When the clutch operating body is correctly aligned, the cable and lever are in alignment so that a straight pull is applied to the lever (**Figure 37**).

8. Install the clutch lever, roller, bushing, and bolt in the end of the clutch body and screw on and tighten the locknut (**Figure 38**).

9. Install the ratchet plate in the cover (**Figure 39**), making sure the peg in the plate engages the hole in the quadrant roller. Put the O-ring on the ratchet shaft and seat it in the groove.

10. Install the outer cover referring to the procedure presented earlier.

GEARSET AND HOUSING

The main transmission housing contains the gearset, shifting forks, camplate, and quadrant. Removal of the gearset from the housing requires that the inner and outer cover assemblies be removed and the primary drive and clutch be disassembled (see Chapter Four).

Removal

1. Remove the shims, spacer, and circlip from the left end of the main shaft (**Figure 40**).

> NOTE: *Reuse of this circlip is not recommended.*

2. Remove the first gearset from the right ends of the main shaft and layshaft, along with the layshaft first gear bushing (**Figure 41**).

3. Unscrew the selector spindle (**Figure 42**) from the case and pull it out of the forks. Remove the forks from the gearset.

4. Pull the main shaft, with second and third gears (**Figure 43**), out of the case. Fourth gear will be held in the case by the countershaft sprocket and bearing. Remove the layshaft and gears. It may be necessary to warm the case in the area of the countershaft left-end bearing and remove the bearing with the shaft.

5. Unscrew the setscrew from the countershaft nut lock plate (**Figure 44**) and remove the plate. Install the left footrest and brake lever assembly

and tighten the nuts finger-tight. Apply the rear brake and unscrew the left-hand thread countershaft nut (**Figure 45**). Disconnect the rear chain and remove the sprocket from the shaft. Tap

the fourth sleeve gear (**Figure 46**) carefully into the case with a soft mallet.

6. Unscrew the detent bolt from the bottom of the case and remove the camplate detent plunger and spring (**Figure 47**).

7. Unscrew the 2 bolts which hold the camplate and quadrant in the case (**Figure 48**). Remove camplate and quadrant and the 2 O-ring seals.

8. Remove the spacer from the countershaft seal (**Figure 49**) and carefully pry the seal out of the case.

9. Warm the case in the area of the main shaft left-end bearing and remove the bearing with an expanding type bearing puller. If a puller is not available, the bearing may be removed by tapping very gently on a soft drift inserted through the left side of the case. If the bearing does not release easily, don't force it; excessive pressure

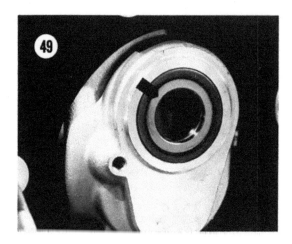

on the bearing will damage it. In such case, it's better to inspect the bearing installed in the case and remove it only if it's found to be defective.

Inspection

1. Thoroughly clean and dry all the components as well as the inside of the transmission case.

2. Rotate the bearings by hand and feel for roughness and radial and axial play between the inner and outer races. The bearings should turn smoothly and play should be barely perceptible.

3. Check both shafts for wear and damage to the splines, threads, and bearing surfaces.

4. Check each gear for chipped or worn teeth. Some bright burnishings of the contact area of the gear teeth is normal, but deep scores or chips indicate an unsatisfactory gear. It's recommended that both gears in a set (layshaft and main shaft) be replaced if one of the gears is unsatisfactory; while wear or damage on the mating gear may not be as apparent, its condition could be such that it won't permit a new gear to wear in correctly.

5. Check the engagement dogs (**Figure 50**) for excessive rounding or chipping. Check internally splined gears for wear and damage to the splines.

6. Examine the gear bushings for signs of wear or damage and check their serviceability by installing them on the shaft with their gears and attempt to rock the gear from side to side. Any appreciable rocking movement is an indication that the bushing should be replaced. Check the condition of the camplate and quadrant bushings

in a similar manner, by installing the bushings and camplate and quadrant in the case and checking for rocking movement.

7. Check each selector fork in the groove of its appropriate gear by rotating the gear and feeling for roughness or sticking (**Figure 51**). Minor roughness on the fork or in the groove can be removed with fine emery cloth but excessive wear or damage is an indication that the fork or gear or both should be replaced.

8. Check the condition of the teeth in the quadrant and on the camplate for roughness, chipping, or wear (**Figure 52**). Check the camplate tracks and the detent profile (**Figure 53**) and the detent plunger (**Figure 54**) for wear and galling.

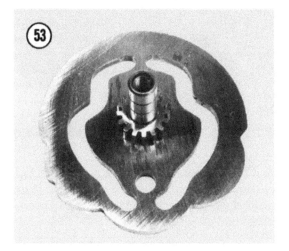

Assembly

1. Warm the transmission case in the area of the left-end bearings and press them into the case until they contact shoulders in bearing bores.

2. Allow the case to cool; then line up the countershaft seal squarely with the bore, with the lip seal facing inward, and press the seal in evenly until it is flush with the case (**Figure 55**).

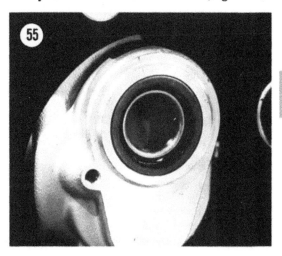

5

3. Assemble the detent plunger, spring, and acorn bolt and install them in the bottom of the case. Turn the bolt in only a couple of turns.

4. Install the quadrant in the case. Make sure the bushing is in place and that the O-ring seal is installed in the outside end of the boss (**Figure 56**) and is covered by the washer. Tighten the bolt securely.

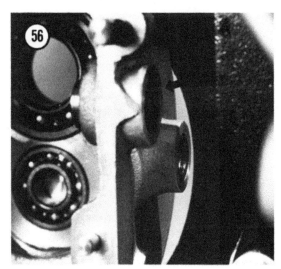

5. Line up the upper edge of the quadrant arm with the top front case stud (**Figure 57**) and install the camplate with its teeth engaged with the teeth on the quadrant with the camplate positioned as shown (**Figure 58**). The notched edge of the camplate must face the left side of the transmission and the plunger should index the bottom notch. Make sure the camplate bushing is in place in the boss in the case and install the O-ring seal, washer, and bolt. Tighten the camplate bolt securely, then tighten the acorn detent bolt in the bottom of the case.

6. Install the main shaft fourth gear in the case and install the spacer on the shaft on the seal side, inside the seal bore (**Figure 59**). Fit the countershaft sprocket to the shaft and reconnect the rear chain. Screw on the countershaft sprocket nut (left-hand thread), apply the rear

brake, and tighten the nut securely. Install the lockwasher and screw (**Figure 60**).

7. Fit main shaft into fourth gear (**Figure 61**).
Install a *new* circlip on the clutch end of the main shaft (Figure 40). Make sure it is completely seated in the groove.

8. Assemble the layshaft, third and fourth gears, and the third gear bushing, as shown in **Figure 62**. The dogs on third gear must face outward, and the flat side of fourth gear must

face third gear. Install the layshaft in the bearing in the case (**Figure 63**).

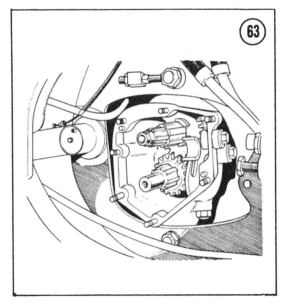

9. Rotate the camplate 2 notches counterclockwise (viewed from the rear). See **Figure 64**. Fit the main shaft shifting fork to third gear with the fork grooves facing outward and install the gear on the main shaft while engaging the inside cam track with the pin on the fork (**Figure 65**). Install the main shaft second gear and bushing on the main shaft with the shifting dogs facing inward (**Figure 66**).

10. Fit the layshaft selector fork to second gear with the fork groove facing inward and install the gear on the layshaft while engaging the outside cam track with the pin on the fork (**Figure 67**).

11. Install the fork pivot pin through the forks, threaded end first, and screw it into the case (**Figure 68**).

12. Install the bushing and first gear on the layshaft (**Figure 69**). Then install the first gear on the main shaft with the shoulder on the gear facing outward (**Figure 70**).

13. Apply a dab of grease to the quadrant roller and set it in place in the end of the quadrant with the hole in the roller positioned horizontal (**Figure 71**).

14. Refer to the earlier instructions in this chapter and install the inner and outer transmission covers. Refer to Chapter Four and install the clutch and primary drive.

CHAPTER SIX

FUEL SYSTEM

The fuel system consists of the fuel tank, air cleaners, and carburetors.

CARBURETORS

Amal concentric float bowl carburetors are used on late model Norton twins. The carburetor is shown in exploded view in **Figure 1**. On some earlier models, such as the Atlas, Amal mono-bloc carburetors are used. This carburetor is shown in **Figure 2**. Twin carburetor models are fitted with a right-hand and a left-hand carburetor. The chief difference between the right and left units is that the float "ticklers," the throttle stop screws, and air pilot screws are located on the outboard side of each carburetor. If a carburetor is to be replaced, be sure to specify either right or left when ordering.

Basic Principles

An understanding of the function of each of the carburetor components and their relationship to one another is a valuable aid for pinpointing a source of carburetor trouble.

The carburetor's purpose is to supply and atomize fuel and mix it in correct proportions with air that is drawn in through the air intake. At the primary throttle opening—at idle—a small amount of fuel is siphoned through the pilot jet by the incoming air. As the throttle is opened further, the air stream begins to siphon fuel through the main jet and needle jet. The tapered needle increases the effective flow capacity of the needle jet as it is lifted with the air slide in that it occupies decreasingly less of the area of the jet. In addition, the amount of cutaway in the leading edge of the throttle slide aids in controlling the fuel/air mixture during partial throttle openings.

At full throttle, the carburetor venturi is fully open and the needle is lifted far enough to permit the main jet to flow at full capacity.

Service

The carburetor service recommended at 10,000-mile intervals involves routine removal, disassembly, cleaning, and inspection. Alterations in jet size, throttle slide cutaway, changes in needle position, etc., should be attempted only if you're experienced in this type of "tuning" work; a bad guess could result in costly engine damage and at the very least poor performance. If after servicing the carburetors and making the adjustments described in Chapter Two, the motorcycle does not perform correctly (and assuming that other factors affecting performance are correct, such as ignition timing and condition, valve adjustment, etc.), the motorcycle should be checked by a Norton dealer or a qualified performance tuning specialist.

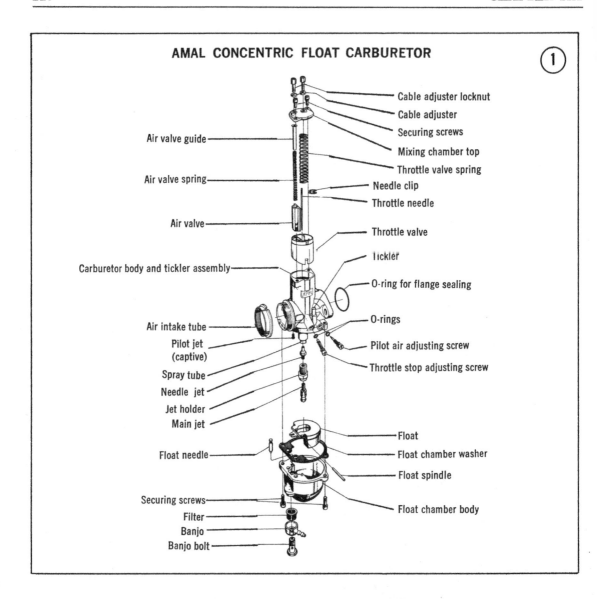

AMAL CONCENTRIC FLOAT CARBURETOR ①

Air valve guide

Air valve spring

Air valve

Carburetor body and tickler assembly

Air intake tube

Pilot jet (captive)

Spray tube

Needle jet

Jet holder

Main jet

Float needle

Securing screws

Filter

Banjo

Banjo bolt

Cable adjuster locknut

Cable adjuster

Securing screws

Mixing chamber top

Throttle valve spring

Needle clip

Throttle needle

Throttle valve

Tickler

O-ring for flange sealing

O-rings

Pilot air adjusting screw

Throttle stop adjusting screw

Float

Float chamber washer

Float spindle

Float chamber body

Removal

1. Close both fuel taps and disconnect the lines. Refer to Chapter Four and remove the fuel tank.

2. Disconnect the inlet bellows from the carburetor inlets (**Figure 3**).

3. Remove the Allen screws which hold the manifolds to the cylinder head (**Figure 4**) and remove the carburetors from the motorcycle.

4. Remove the screws from the carburetor tops and pull out the slides (**Figure 5**).

Disassembly

1. Disconnect the balance pipe from the carburetors (**Figure 6**).

AMAL MONOBLOC CARBURETOR

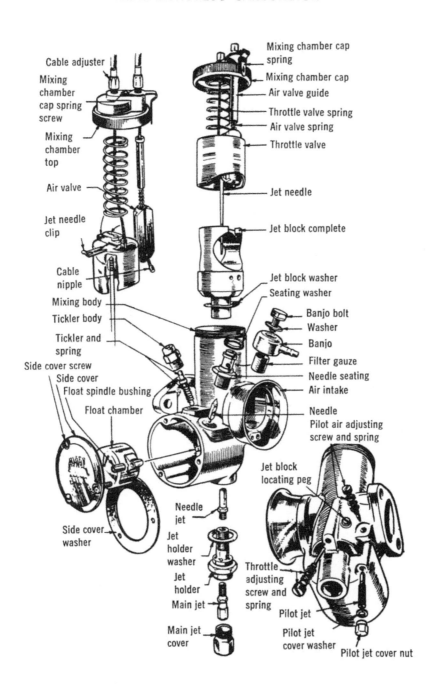

6

3. Remove the 2 screws which hold the float bowl to the carburetor and remove the bowl, float, float needle, and gasket (**Figure 8**).

4. Remove main jet and needle jet (**Figure 9**).

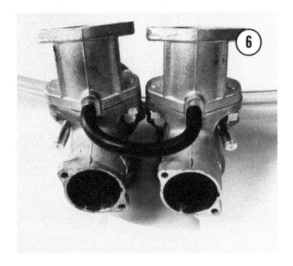

1. Needle jet   2. Jet block   3. Main jet

5. Unscrew the air pilot screw and the throttle stop screw (**Figure 10**).

6. Unscrew the drain plug from the bottom of the float bowl (**Figure 11**).

7. Disconnect the carburetor from the manifold pipe and remove the O-ring (**Figure 12**).

2. Unscrew the fuel banjo nut from the bottom of each carburetor and remove the banjo and the filter screen (**Figure 7**).

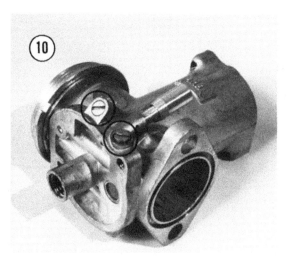

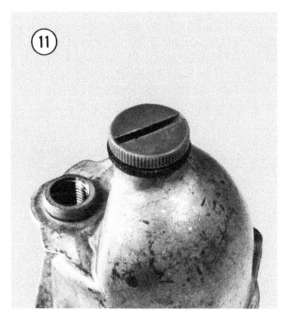

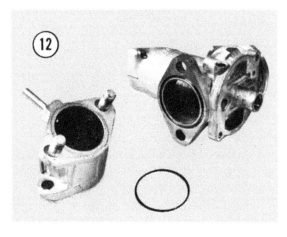

Cleaning and Inspection

1. The carburetors should be cleaned by submersing all their metal parts in a cleaner that's specially compounded for carburetors and other fuel system components. This type of cleaner is available under several brand names through automotive supply stores. In one-gallon sizes it frequently includes a handy dip basket for holding the parts while they are soaking. The parts should remain in the cleaner for at least 5 minutes. Solvent may also be used to clean the carburetors, but it is not effective for dissolving the residues that form as gasoline evaporates.

CAUTION

Don't clean the non-metal parts in a carburetor cleaning solution; it's harmful to such things as rubber O-rings, fiber gaskets, and the nylon float and needle.

2. Remove the carburetor parts from the cleaning solution and rinse them with water or solvent. Dry all of the parts thoroughly with a clean rag and blow out all the jets and the passages in the carburetor body with compressed air.

3. Check the floats for leakage by holding them up to a bright light to see if there is gasoline in them.

4. Check the tapered end of the float needle for grooves, scores, or pits which would prevent it from shutting off the fuel flowing through the needle valve.

5. Check each of the jets for blockage and any damage that might alter the size of the flow area. Don't use wire to clean blockage from the jets, but instead blow them out with compressed air.

6. Check the throttle slide for wear and scoring. Minor surface imperfections can be removed with fine emery cloth but if the slide is deeply scored or severely worn it should be replaced.

7. Check the slide bore in the carburetor body for wear or damage. Wear on the slide is often accompanied by similar wear in the bore. Also check the jet seats for damaged threads.

8. Check the throttle needle to make sure it is straight and that the taper is not worn.

9. Check the taper of the pilot air screw for wear and grooving.

Reassembly

If possible, the O-rings, fiber washers, and float bowl gasket should be replaced each time the carburetors receive major service. If any of these pieces are damaged or have deteriorated it's essential that they be discarded and replaced with new pieces.

1. Screw the pilot air screw in all the way and then back it out 1½ turns. Make certain its friction O-ring is in place.

2. Screw in the throttle stop screw until the head protrudes about 1/16 in. (**Figure 13**).

3. Screw the needle jet into the jet holder and screw the holder into the carburetor body (**Figure 14**). Screw the main jet into the holder (**Figure 15**). The jets should be tight but care must be taken not to strip the threads.

4. Invert the float bowl and set the float, float needle, and pivot pin in place (**Figure 16**). Set the gasket in place on the flange and install it on the carburetor. Screw in and tighten the 2 float bowl screws. Install a fiber washer on drain plug and screw it into the bottom of the float bowl.

5. Set the fuel strainer in the banjo fitting and install it on the bottom of the carburetor with the flange on the strainer against the carburetor body (**Figure 17**). Put a fiber washer on the banjo bolt and screw it in tight.

6. Reinstall the carburetors on the engine by reversing the steps in the removal procedure.

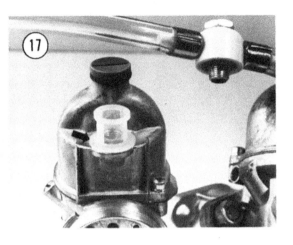

Before installing the Allen screws which mount the manifolds to the cylinder head, thoroughly clean the screw threads and apply a few drops of Loctite to them.

Assemble the needles and slides, making sure each needle clip is in the same groove from which it was removed. Install the slide assemblies in the carburetors, taking care to line up the needles with the needle jets and the vertical grooves in the slide bores with the lug at the top of the slide. Reinstall the fuel tank and reconnect the lines. Adjust the carburetors as described in Chapter Two.

AIR FILTER

Service of the air filter element is described in Chapter Two.

FUEL TANK

Removal/Installation

1. Shut off fuel taps.

2. Disconnect fuel lines.

3. Remove the fuel tank. On Commando tanks, the front of the tank mounts on 2 studs beneath the tank and is held by 2 self-locking nuts. The rear mount on some models uses a rubber loop which passes beneath the top frame tube. Others are held in place with a metal cross bar.

4. Installation is the reverse of these steps.

CHAPTER SEVEN

IGNITION AND ELECTRICAL SYSTEMS

The Norton Commando ignition system is a battery-coil type system with a centrifugally controlled automatic advance contact breaker for spark timing control. The 1972 and later models are fitted with two 6-volt coils which are isolated from the 12-volt portion of the electrical system by a ballast resistor.

The Norton Atlas and G15 CS models use a Lucas K2F magneto mounted behind the cylinders and driven by a chain connected to the camshaft drive idler in the timing side case. Routine service of this unit is covered at the end of the chapter.

The charging-lighting system in all Norton 750/850 twins uses an alternator (AC generator). A rectifier changes alternating current to direct current to charge the battery and power the lights. The charge rate to the battery is controlled by a zener diode.

IGNITION

Primary electrical power is supplied to the ignition through the battery and charging circuits. The advance mechanism and contact breaker cam are driven off the camshaft sprocket on the right of the engine. The firing point occurs at 28° BTDC (fully advanced).

CONTACT BREAKER

Two similar contact breaker assemblies are used on Norton Commandos. On both types, the contact set for each cylinder can be set independent of the other set. Different adjustment procedures are used for each type. Both types should be checked every 5,000 miles for contact condition and point gap.

The Lucas 6 CA breaker assembly is fitted to models prior to 1973. There is a single lock screw for each contact set and the fixed contact plate is moved with an eccentric adjuster. On 1973 and later models, a 10 CA breaker assembly is used which has 2 lock screws for each point set but no eccentric adjuster screws; during adjustment, the plates must be moved with a screwdriver tip.

Removal

1. Remove the ignition cover plate (**Figure 1**).
2. Unscrew the nuts from the spring anchor

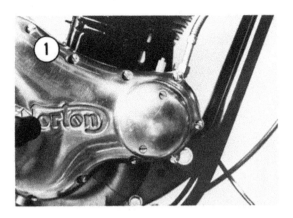

posts and remove the terminals and movable contact, along with the insulating bushing and washer (**Figure 2**).

ing washer and bushing are installed in the order shown in **Figure 4**.

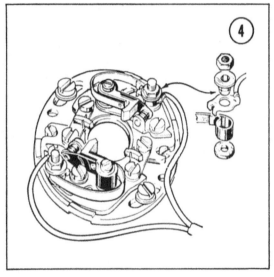

3. Unscrew the lock screws from the fixed contact plates and remove the plates (**Figure 3**).

4. Inspect the contacts for wear and pitting. If neither is excessive, the points can be dressed with a flat point file and cleaned with lacquer thinner or acetone. Don't dress the points with emery cloth; this will round the edges of the contacts and create the type of condition you may be trying to correct. If the points can't be dressed with a few strokes of the file they should be discarded and replaced with new parts.

Installation

Install the contact breaker assembly by reversing the steps above. Make sure the insulat-

1. Lightly grease the pivot posts and the breaker cam with breaker lubricant. Knead a couple of drops of oil into each of the felt lubricators.

2. Locate the tangs on the terminals over the contact spring (**Figure 5**).

Adjustment

1. Remove the spark plugs from the cylinder head and rotate the crankshaft to line up the mark on the cam with the lifter heel on one of the contacts (**Figure 6**). This is the maximum contact opening.

2. Measure contact gap with a flat feeler gauge (**Figure 7**). It should be 0.015 in. (0.38mm). If it is not, loosen the breaker lock screws (**Figure 8**) and move the plate in or out until the gap is correct. Tighten the lock screw and recheck the gap to make sure the plate did not move when the lock screw was tightened.

3. Rotate the crankshaft to line up the cam mark with the other lifter heel and adjust this contact set in the same manner as above.

AUTOMATIC ADVANCE UNIT

The automatic advance unit (**Figure 9**), located behind the contact breaker plate, advances ignition timing as the engine speed increases. The unit is spring loaded and returns to the static timing position when the engine is shut off. The unit may be serviced without removing it from the engine.

1. Mark the position of the contact breaker plate (**Figure 10**), unscrew the 2 plate-holding screws, and remove the plate with the contacts and terminals intact.

2. Lightly lubricate the pins and pivots on the advance unit.

3. Check to see that the ends of the springs are seated in their recesses in the bobweights and on the raised tangs in the fixed plate (**Figure 11**).

4. Turn the cam by hand to the fully advanced position to extend the bobweights. Then release

it. If the springs are in good condition, the bob-weights will snap back to the static position. If they do not, the springs should be replaced.

5. Reinstall the contact breaker plate, making sure the marks are aligned. Recheck the contact gap and the ignition timing.

IGNITION TIMING

Ignition timing can be accomplished in one of two ways, either statically or dynamically with the use of a timing strobe. The static method is accurate enough for most situations but strobe timing is essential for complete timing accuracy required for maximum performance.

Static Timing

If the breaker cam has been removed and reinstalled, it must first be correctly located on the shaft before the ignition can be timed. To do this, first remove the intake and right-hand rocker exhaust covers and the spark plugs. Rotate the crankshaft clockwise to bring the right piston to TDC with both valves closed (firing position). Use the wire color coding to determine which contact set fires the right cylinder. Turn the breaker cam on the shaft counterclockwise until it just begins to open the right cylinder points and tighten the bolt in the end of the shaft (**Figure 12**).

1. Check the contact gap as described earlier and adjust it if necessary.

2. Remove the inspection plug from the primary chaincase (**Figure 13**).

3. Remove the bolt and washer from the end of the breaker cam shaft and install a washer with a large enough hole to pass over the end of the shaft and contact the face of the cam (**Figure 14**). Reinstall the bolt and washer in the shaft, hold

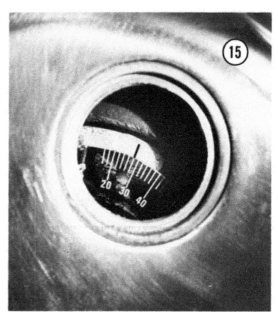

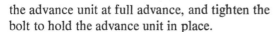

the advance unit at full advance, and tighten the bolt to hold the advance unit in place.

4. Shift the transmission into fourth gear and rotate the rear wheel to bring one of the pistons to TDC with both valves closed. Slowly rotate the wheel backward to align the timing mark with the 28° mark on the degree plate (**Figure 15**). At this position, the contact breaker for the cylinder being timed should just begin to open.

5. To determine the precise opening point, insert a strip of thin paper between the contacts. Slowly rotate the crankshaft until the paper can be withdrawn with a slight pull. This is the point at which contact opening begins.

6. Check the alignment of the timing mark with the degree plate. If the timing is correct, the mark should line up with the 28° mark.

7. If the timing is not correct, align the timing mark at 28° and loosen the 2 screws which secure the contact breaker plate (**Figure 16**). Insert

the paper between the contacts for the cylinder being timed and rotate the breaker plate clockwise or counterclockwise as required until the paper can be withdrawn with a slight pull. Then tighten the breaker plate holding screws.

8. Check the timing of the opposite cylinder in the same manner. If the timing must be adjusted for this cylinder, loosen the screws holding the secondary contact plate (**Figure 17**), move the plate as required until the timing is correct, and securely tighten the screws.

9. Double check both the timing and the contact gap or both cylinders and both sets of contacts.

Dynamic (Strobe) Timing

Before the ignition can be timed with a strobe, the contact gap and static timing must be adjusted as described above.

1. Remove the inspection plug from the primary chaincase.

2. Connect the strobe to the high-tension lead on the right cylinder.

3. Start the engine and run it at 3,000 rpm.

4. Point the strobe light at the indicator plate in the chaincase. The rotor mark should line up with the 28° mark on the degree plate. If it does not, loosen the 2 screws which secure the contact breaker plate and rotate the plate clockwise (advance) or counterclockwise (retard) until the marks line up correctly. Then tighten the screws in the contact breaker plate. Shut off the engine.

5. Connect the strobe to the high-tension lead on the left cylinder, start the engine and run it at 3,000 rpm and check the timing for the left cylinder in the same manner as for the right. If the rotor mark does not line up with the 28° mark, loosen the screws in the secondary contact plate, move the plate as required until the marks line up, and tighten the screws.

6. After both cylinders have been timed, recheck them again with the strobe to ensure that neither of the contact plates has moved.

> NOTE: *If the timing mark tends to jump around during strobe timing, check the adjustment of the camshaft chain and the tightness of the alternator rotor. Also, a loose rotor will often sound like lower end trouble.*

COILS

Both 6- and 12-volt coils are used, in pairs, on Norton Commando electrical systems. The 6-volt coils are isolated from the rest of the 12-volt electrical system by a ballast resistor. The coils should be checked to see that they are mounted securely and that their connections are clean and tight; no other service is necessary.

If a coil is suspected of being faulty it can be partially tested by checking the resistance of the primary windings. To do this, connect an ohmmeter across the primary terminals. At an ambient temperature of about 70°F the resistance in a 6-volt coil should be at least 1.7 ohms but not greater than 1.9 ohms. For a 12-volt coil, the resistance should be at least 3.3 ohms but not greater than 3.8 ohms. If the resistance of the primary windings is all right, the coil should be taken to a shop specializing in automotive and motorcycle electrical work for further testing. Even better, substitute a known good coil to check a suspected coil.

Ballast Resistor

The ballast resistor used with 6-volt coils can be checked with an ohmmeter. Resistance should be 1.8-2 ohms.

CAPACITORS

The ignition capacitors can be tested while installed on the motorcycle. First, remove the fuel tank and the contact breaker cover plate. Rotate the crankshaft to open one of the contact sets. Connect a voltmeter to the contacts and turn on the ignition switch. If no voltage is indicated the capacitor should be replaced. Rotate the crankshaft to open the other contact set and check it in the same manner.

If voltage is present but the capacitors are still suspected of not being within service limits (evidenced by hard starting or erratic running), check the contacts for signs of burning—an indication that the capacitor is failing.

Capacitors on early models can be removed and replaced by unscrewing them from the coil clips. On 1971 and later models, remove the entire coil assembly from the motorcycle and then remove the capacitor pack from the coil assembly (**Figure 18**).

7

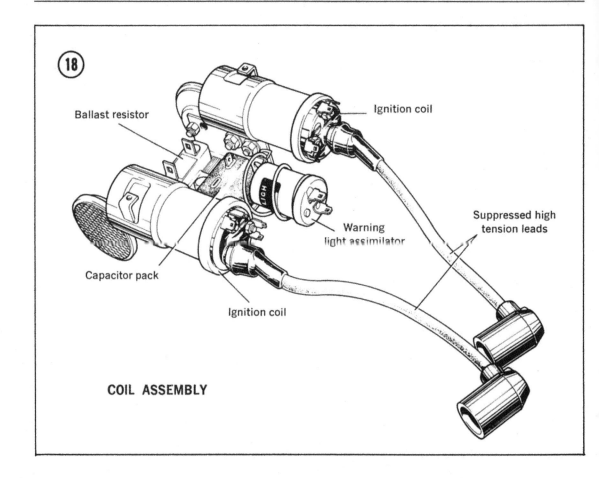

COIL ASSEMBLY

Ballast resistor · Ignition coil · Warning light assimilator · Suppressed high tension leads · Capacitor pack · Ignition coil

SPARK PLUGS

For normal use, and for all models except the 750 Combat, Champion N7Y spark plugs are recommended; however, equivalent plugs of different manufacture can be used provided their heat range corresponds to that of the N7Y. Plugs with an incorrect heat range can cause hard starting at one extreme, or preignition at the other, resulting in possible engine damage.

Champion N6Y spark plugs are recommended for use in the 750 Combat and for sustained high-speed operation.

The spark plug electrode gap should be 0.028 in. (0.71mm).

CHARGING SYSTEM

The charging system consists of the alternator, current rectifier, zener diode regulator, and heat sink. The charging system requires no service other than periodic inspection of the connections to ensure they are clean and tight.

When trouble occurs in the charging system, the components can be checked while installed on the motorcycle. The wiring diagrams at the end of this chapter will be helpful for locating and testing the components and circuits referred to in the text.

ALTERNATOR

Testing

1. Unplug the alternator leads from the wiring harness (**Figure 19**).

2. Connect a voltmeter (set to AC scale) with a one ohm resistor wired parallel to the alternator leads.

3. Start the engine and run it at a steady 3,000 rpm. The meter should indicate a minimum of 9 volts.

4. Measure the voltage between each of the leads and ground. There should be no reading.

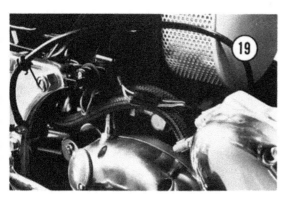

5. If there is no reading in Step 3, an open circuit is indicated in the coil circuit. In such case, the rotor should be replaced. If the reading was substantially less than 9 volts, it's again possible that the rotor is faulty. This can be verified by installing a new rotor and checking it as described above.

6. If the readings were correct, connect the black lead of an ammeter to the battery negative terminal and the red lead of the meter to the negative cable.

7. Disconnect the zener diode (**Figure 20**).

8. Start the engine and run it at a steady 3,000 rpm. With the ignition switch in the IGNITION ONLY position, the ammeter should indicate a charging rate of at least 4.5 amps. With the ignition switch in the LIGHTS AND IGNITION position, the ammeter should indicate a charging rate of at least 1.0 amp. If the ammeter readings are higher than these, the system is all right. If they are not, continue with the next steps and check the condition of the rectifier.

Removal/Installation

To remove the alternator rotor and stator, perform Steps 1-6, *Primary Drive and Clutch Disassembly*, Chapter Four. To install, perform Steps 10-15, *Primary Drive and Clutch Assembly*, Chapter Four.

RECTIFIER

Testing

1. Check to make sure the alternator leads are correctly connected to the main harness and that the connections are clean and tight.

2. Unplug the center lead from the rectifier (**Figure 21**) and connect the black (—) lead of a voltmeter (set to DC scale). Ground the voltmeter red (+) lead to the frame.

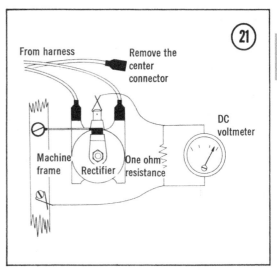

3. Start the engine and run it at a steady 3,000 rpm. The meter should indicate at least 7.5 volts. If it indicates less, the rectifier is unsatisfactory and should be replaced. If the reading is satisfactory, proceed to bench test the rectifier to verify that it permits current to flow in only one direction.

4. Using a pair of test leads, connect a headlight bulb to the battery as shown in **Figure 22**.

5. Clip the other ends of the test leads to each combination of rectifier terminals shown below first in one polarity and then in the other. Each hookup should be made for more than a few seconds at a time.

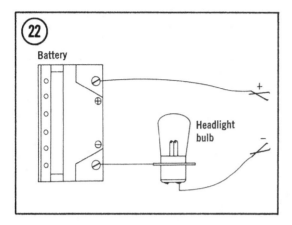

a. Terminals 1 and 2

b. Terminals 2 and 3

c. Terminal 1 and bolt

d. Terminal 3 and bolt

If the bulb lights in both directions on any one hookup, or if it lights in neither direction on any one hookup, the rectifier is defective and must be replaced.

ZENER DIODE

Testing

1. Unplug the electrical cable from the zener diode and connect the red (+) lead of a DC ammeter to the Lucar blade on the diode and the ammeter black (—) lead to the cable (Figure 20). Connect the black (—) lead of a DC voltmeter to the Lucar blade and the voltmeter red (+) lead to ground. See **Figure 23**.

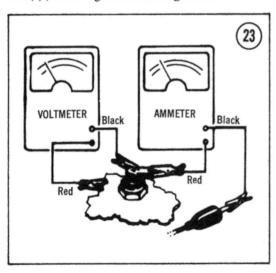

2. With all the lights switched off, start the engine. Slowly increase the engine speed and observe the 2 meters. Up to 12.75 volts, as indicated by the voltmeter, the ammeter should indicate nil current flow. From 13.5-15.5 volts, the ammeter should indicate a current flow of 2 amps. If current begins to flow before the voltmeter indicates 12.75 volts, or if the voltmeter indicates more than 15.5 volts before the ammeter reaches 2 amps, the diode is unsatisfactory and must be replaced.

Before installing the diode, lap its contact surface to the contact surface on the footrest using a fine grade of lapping compound. The footrest functions as a heat sink to help the diode dissipate heat produced by the alternator when the battery is fully charged and the lights are off. For this reason it is essential that there be a good contact between the diode and the footrest. In addition, a thin coat of silicone compound, such as G.E. Electronic Silicone Compound Transistor Z-5 No. 8101 or G.E. Silicone Compound G-640 or G-641, should be applied to the contact area before installing the diode. This compound is available through TV repair shops and electronic supply houses. When the diode has been installed, tighten the mounting nut to 24-28 in.-lb., taking care not to strip the soft copper threads.

BATTERY

Service battery as described in Chapter Two.

BATTERY CROSSBAR MODIFICATION

On many 850 Mark II models built between engine numbers 307311 and 32500, the rectifier is installed in such a manner that the battery hold-down crossbar can foul the electrical leads to the rectifier should the battery shift backward in its holder. To reduce the likelihood of this occurring, cut about one inch off each end of the battery crossbar—about midway between the notches (**Figure 24**). Dress the sawed edges with a fine file to remove burrs and sharpness and spray paint the raw metal to inhibit corrosion.

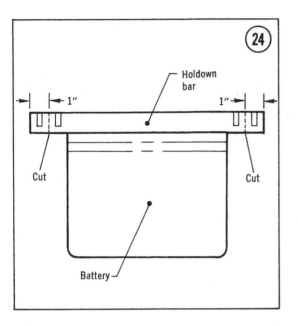

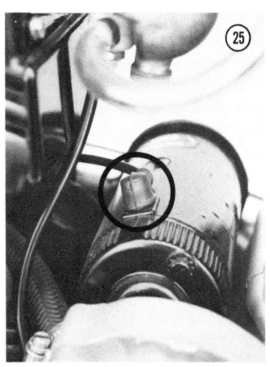

STARTER (850 MARK III)

Service to the starter is limited to replacement of the brushes. If additional work is required, the starter should be entrusted to a Norton dealer or an electrical specialist.

Removal/Installation

1. Disconnect the ground lead at the fuse holder. Disconnect the electrical lead from the starter motor (**Figure 25**).

2. Remove the 2 screws from the starter mounting boss on the primary case and the long screw from the outer chaincase cover (**Figure 26**). Pull the starter motor out of the mount.

3. Reverse the above to install the starter. Make sure the large O-ring is in good condition before installing the motor. Tighten the nut on the electrical post securely and fit the rubber cover.

Brush Replacement

1. Remove starter motor as described above.

2. Unscrew the 2 through bolts from the right end of the starter and remove the end cap (**Figure 27**). Note the position of the brush holder plate and remove it.

3. Remove the springs and brushes from the plate and install new ones. Reinstall the plate in the starter in the same position from which

it was removed, making certain the brushes are in contact with the armature.

4. Install the end cap and screw in the through bolts. Install the starter as described earlier.

Relay

If the starter relay energizes but the starter fails to turn over, check the connections of the relay-to-starter electrical lead to make sure they are tight and corrosion-free. If the starter still fails to turn over it may need new brushes.

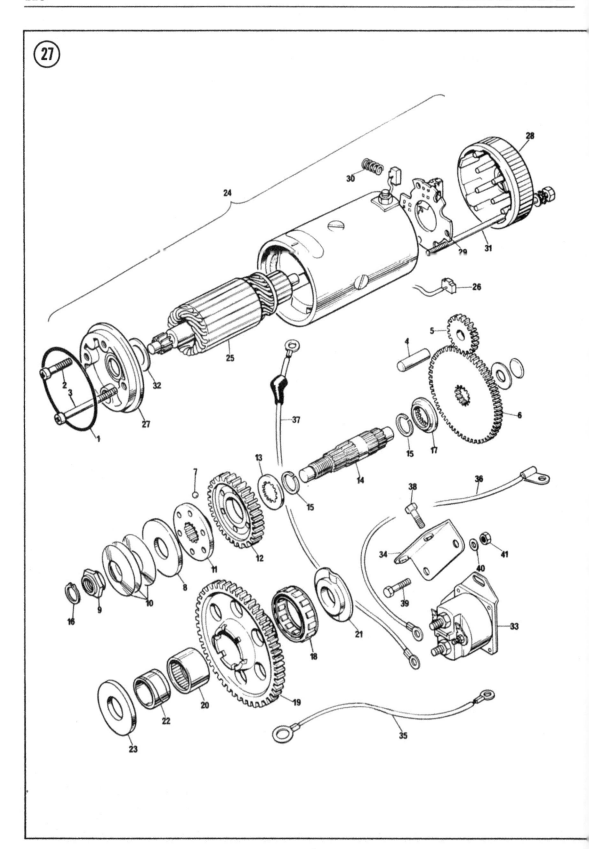

ELECTRIC STARTER
(850 MARK III)

1. O-ring
2. Screw
3. Long screw
4. Idler spindle
5. Idler gear
6. Idler gear
7. Ball
8. Thrust washer
9. Nut
10. Disc spring
11. Carrier
12. Gear
13. Thrust washer
14. Shaft
15. Circlip
16. Circlip
17. Thrust washer
18. Clutch
19. Crank gear
20. Needle roller bearing
21. Inner washer
22. Inner race
23. Outer washer
24. Starter motor
25. Armature
26. Brush
27. Head assembly (commutator end)
28. Head brush and seal (drive end)
29. Brush plate and holder
30. Spring
31. Thru-bolt
32. Thrust washer
33. Solenoid
34. Solenoid bracket
35. Starter ground lead
36. Battery to solenoid lead
37. Solenoid to starter lead
38. Solenoid bracket bolt
39. Bolt
40. Washer
41. Nut

If all of the connections are clean and tight and the relay does not energize, it is very likely faulty. In such case, remove the relay and have it checked by a Norton dealer or an electrical specialist to verify its condition.

LIGHTS AND SWITCHES

Service of the lights and switches is limited to checking connections to make sure they are clean, dry, and tight.

Fault tracing and replacement of most of the lighting and switching components is simple and straightforward. However, the adjustment of the rear brake stoplight switch must be done carefully to prevent damage to the unit, and the functional test procedure of the directional lights flasher relay should be followed to determine if the flasher unit itself is at fault when trouble occurs in this circuit.

Rear Brake Stoplight Switch

1. Loosen the bolts which mount the switch to the brake pedal (**Figure 28**).

2. Adjust the brake pedal stop (**Figure 29**) so that the pedal position is comfortable. Adjust rear brake free play, described in Chapter Two.

3. Slowly move the switch up, positioning the switch plunger in the center of the contact boss on the foot peg, until the slightest downward movement of the brake pedal causes the stoplight to light. The movement of the switch plunger, from full extension to the point at

which the pedal contacts the pedal stop, should be about 1/32 in. (0.79mm).

4. Tighten the switch mounting bolts and re-check the adjustment. The switch plunger must not be fully depressed before the pedal contacts the stop.

Directional Light Flasher

1. Check the directional lamp bulbs to make sure the filaments are not broken and check to see that all the connections are clean and tight.

2. Switch on the ignition and check the voltage of the "B" terminal on the flasher with a volt-meter. The meter should indicate battery voltage (12 volts).

3. Using a test lead, connect the "B" and "L" terminals on the flasher together and operate the flasher switch in both directions. If both lamps on one side light but do not flash, the relay is faulty and should be replaced.

MAGNETO SERVICE

Routine periodic service on the Lucas K2F magneto used on Atlas and G15 CS models, presented here, is within the ability of the hob-byist mechanic. However, major service of the unit, recommended at 10,000-mile intervals, should be referred to a Norton dealer or an electrical service specialist with experience on this particular unit.

Cleaning

1. Remove the contact breaker cover. Unscrew the long hex-head screw from the center of the breaker assembly (**Figure 30**) and pull the assembly off the shaft.

2. Inspect the contacts for burning and pitting. If their condition is reasonably good, dress them with a point file; do not use emery cloth as it will round the edges of the contacts. Clean the points thoroughly with lacquer thinner or acetone and grease the breaker cam ring with contact breaker lube. Apply a drop of oil to the end of the pivot post.

3. Reinstall the contact breaker assembly, mak-ing sure the keyway in the shaft lines up with the projection in the bore of the cam.

4. Lift the hold-down springs from the high-tension brushes and check the movement of the brushes in their holders. They should move freely in and out. If they do not, clean them with

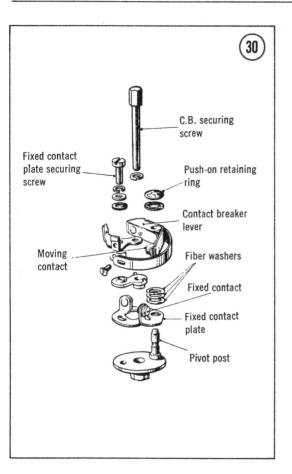

C.B. securing screw

Fixed contact plate securing screw

Push-on retaining ring

Contact breaker lever

Moving contact

Fiber washers

Fixed contact

Fixed contact plate

Pivot post

wrapped in clean cloth, in one of the brush holders and slowly rotate the engine.

6. Reinstall the brushes, making sure the hold-down springs are correctly positioned, and adjust the contacts as described below.

Adjustment

1. Remove the spark plugs from the cylinder head and rotate the engine until the contacts are fully opened. Measure the gap; it should be 0.012 in. (0.30mm) and there should be a slight resistance on the feeler gauge when it is inserted and removed.

2. If the gap is incorrect, loosen the securing screw for the fixed contact (**Figure 31**) and move it in or out until the setting is correct. Then tighten the securing screw and recheck the gap.

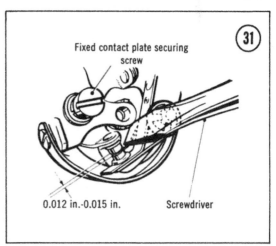

Fixed contact plate securing screw

0.012 in.-0.015 in.   Screwdriver

lacquer thinner or acetone. Check them also for wear. If they are worn down to within ⅛ in. of the shoulder, they should be replaced.

5. Inspect the contact track for dirt and corrosion. To clean, insert a small length of dowel,

WIRING DIAGRAMS

Wiring diagrams will be found on the following pages:

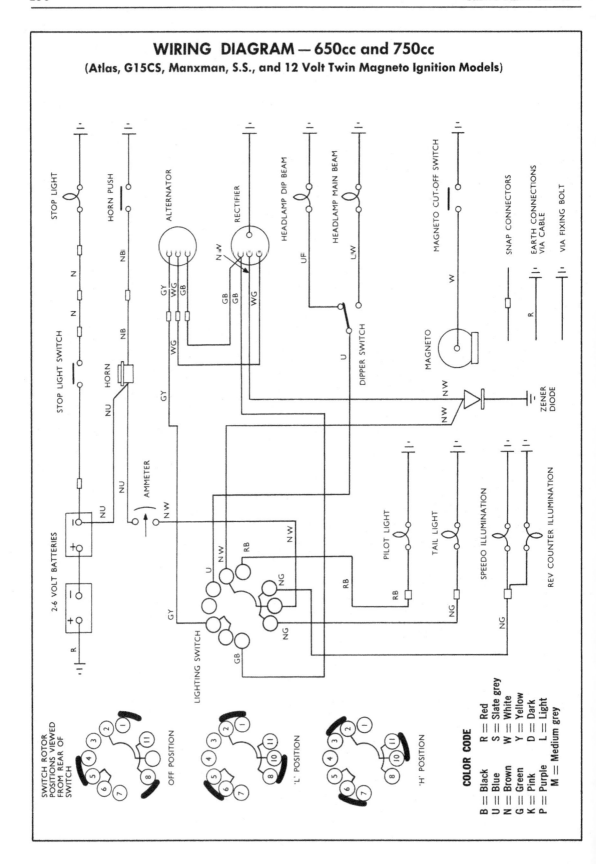

WIRING DIAGRAM — 650cc and 750cc
(Atlas, G15CS, Manxman, S.S., and 12 Volt Twin Magneto Ignition Models)

COLOR CODE

B = Black   R = Red
U = Blue   S = Slate grey
N = Brown   W = White
G = Green   Y = Yellow
K = Pink   D = Dark
P = Purple   L = Light
M = Medium grey

WIRING DIAGRAM — 750 Commando (Before 1971)

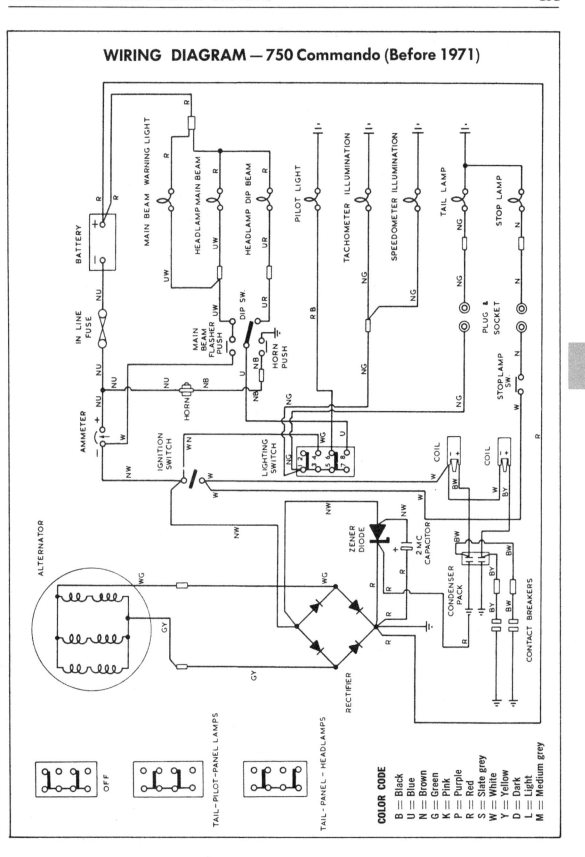

COLOR CODE

B = Black
U = Blue
N = Brown
G = Green
K = Pink
P = Purple
R = Red
S = Slate grey
W = White
Y = Yellow
D = Dark
L = Light
M = Medium grey

7

WIRING DIAGRAM

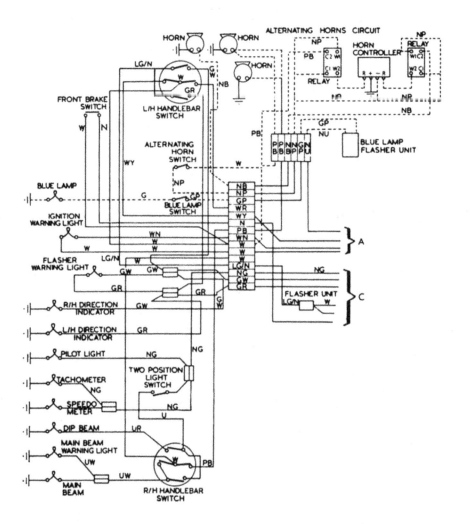

.... Denotes Interpol Circuits

1971 750 Commando

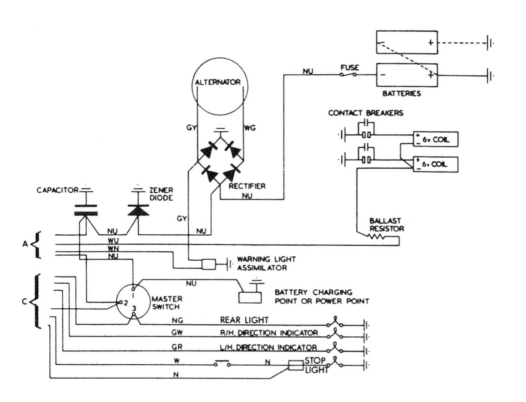

COLOR CODE

U = Blue
B = Black
N = Brown
G = Green
K = Pink
P = Purple
R = Red
S = Slate grey
W = White
Y = Yellow
D = Dark
L = Light
M = Medium grey

WIRING DIAGRAM

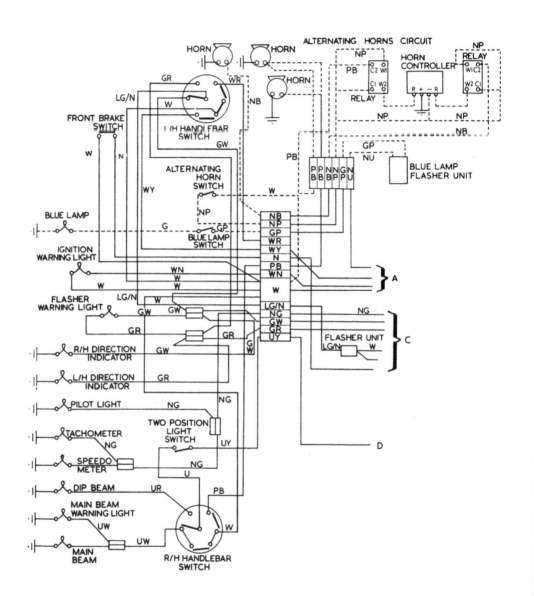

.... Denotes Interpol Circuits

1972-1974 850 Commando Mark II

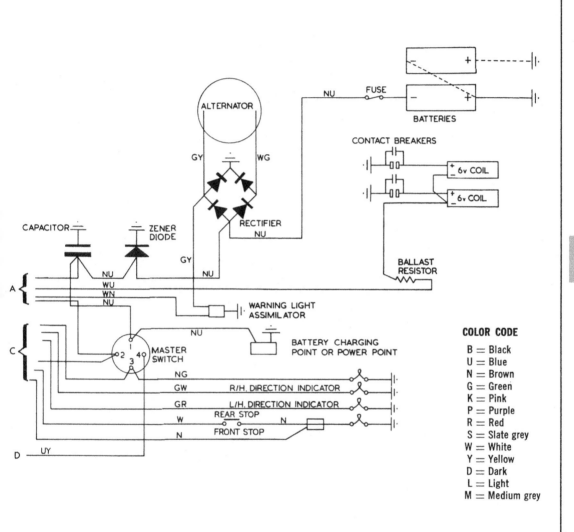

COLOR CODE

B = Black
U = Blue
N = Brown
G = Green
K = Pink
P = Purple
R = Red
S = Slate grey
W = White
Y = Yellow
D = Dark
L = Light
M = Medium grey

WIRING DIAGRAM

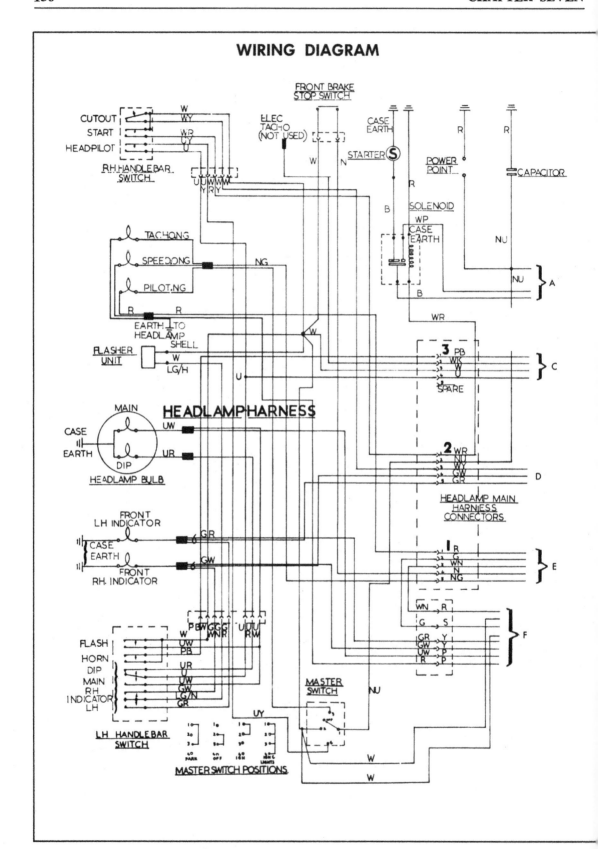

850 Commando Mark III (All Years)

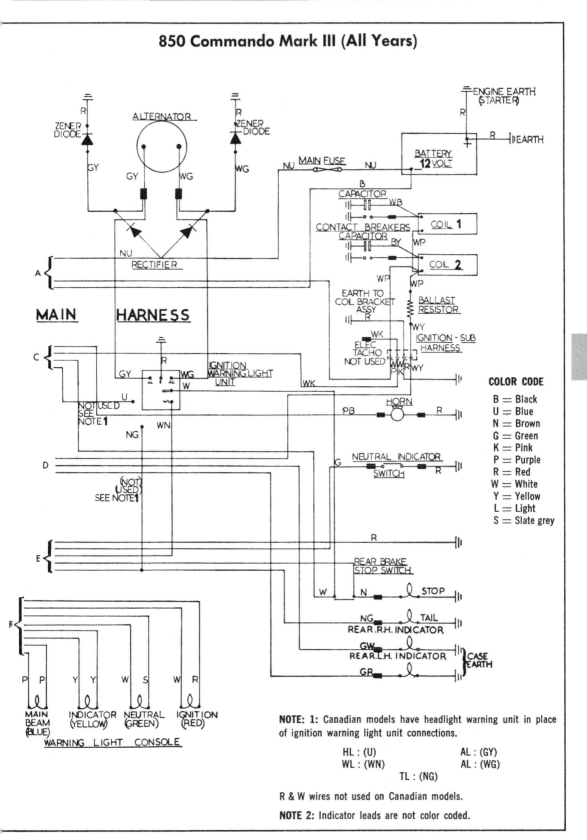

COLOR CODE

B = Black
U = Blue
N = Brown
G = Green
K = Pink
P = Purple
R = Red
W = White
Y = Yellow
L = Light
S = Slate grey

NOTE: 1: Canadian models have headlight warning unit in place of ignition warning light unit connections.

HL : (U)	AL : (GY)
WL : (WN)	AL : (WG)
TL : (NG)	

R & W wires not used on Canadian models.

NOTE 2: Indicator leads are not color coded.

CHAPTER EIGHT

FRONT WHEEL AND BRAKE

This chapter describes repair and maintenance of the front wheel, hub, drum brake, and
disc brake.

FRONT WHEEL

Removal/Installation

1. Block up the front of the motorcycle to lift
the front wheel off the ground.

2. On models with a drum type front brake, disconnect the brake cable from the lower brake
arm by removing the clevis pin (**Figure 1**). Unscrew the cable adjuster from the boss on the
brake and backing plate.

3. On all models, unscrew the axle nut (**Figure 2**). Loosen the pinch bolt at the bottom of
the left leg (**Figure 3**). Insert the shaft on a screwdriver or small bar through the hole in the left
end of the axle and pull it out of the forks and
wheel while lifting and supporting the weight
of the wheel.

4. On disc brake models, pull wheel forward to
disengage the disc from the caliper.

5. Installation is the reverse of these steps.

FRONT DRUM BRAKE

Two types of drum brakes are used on Norton
Commandos. On the standard brake, the shoes

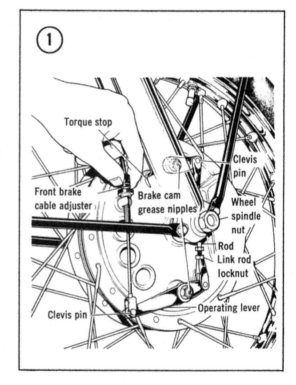

are retained on the pivots with circlips (**Figure 4**). On the modified high-performance brake,
with the expander plate, the plate and shoes are
held in position with bolts secured by tab washers (**Figure 5**). Both brakes are double-leading-
shoe type units.

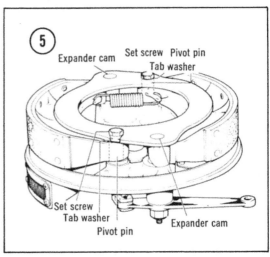

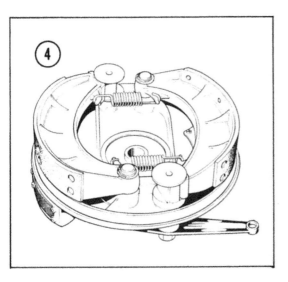

Disassembly

1. Block up the front of the motorcycle to lift the front wheel off the ground. Remove it as described in this chapter. Then remove the complete brake assembly from the drum.

2. Position the brake backing plate horizontally in a vise and clamp it by the brake anchor peg.

3a. On the standard brake, remove the circlips and washers from the shoe pivot pins.

3b. On the modified brake, straighten the tab washers on the pivot bolts and unscrew the bolts. Remove the expander plate by carefully prying the plate up with a screwdriver inserted between the plate and the shoes, to the right of each of the expander cam pivots.

4. Remove the return springs by lightly gripping the coil section with a pair of conventional pliers and extending one of the spring ends with needle-nose pliers. Then remove the brake shoes.

5. Remove the backing plate from the vise and unscrew the nuts which hold the brake arms to the cams. Remove the arms from the camshafts and pull the cams out of the backing plate.

Inspection

Thoroughly clean and dry all the parts except the linings. Check the contact surface of the drum for scoring. If there are grooves deep enough to snag a fingernail, the drum should be entrusted to a brake specialist to have it trued. In such case, the linings will have to be replaced and the new ones arced to the new drum contour.

8

Check the linings for wear as described above. If they are serviceable, inspect them for imbedded foreign material. Dirt can be removed with a stiff wire brush, but if they are soaked with oil or grease they will have to be replaced.

Inspect the cam lobes and the pivot area of the shaft for wear and galling. Minor roughness can be removed with emery cloth but if either area is deeply scored or severely worn, the cam should be replaced. In addition, check the expander cam bushings and replace them if they are worn or galled.

Assembly

Assemble and install the front brake and wheel by reversing the disassembly steps. Grease the shafts and contact areas of the expander cams, the bushings in the cam bores, and the brake shoe pivot posts with a light coat of a molybdenum disulfide grease; excess grease is likely to find its way onto the brake linings and render them unserviceable.

On a modified brake, don't enlarge the pivot or cam holes in the support plate to get it to fit more easily; to do its job, the plate must fit tightly so that it can maintain the pivots and cams in a constant relationship.

After the front wheel has been installed in the motorcycle, adjust the brake as described below. Before tightening the axle nut, spin the front wheel and apply the front brake. Hold the brake on to locate it centrally in the drum and tighten the axle nut securely.

Adjustment

1. Remove the clevis pin from the top brake arm (Figure 1).

2. With assistance, pull both of the brake arms down to bring the shoes in contact with the drum and hold them there.

3. Turn the link adjuster either in or out until the clevis can be installed with a light push fit. Tighten the locknut on the link.

4. Adjust the control cable so that the brake is applied with a slight movement of the hand lever but the shoes do not drag in the drum when the lever is relaxed. Hold the adjuster to prevent it from turning and tighten the locknut.

FRONT DISC BRAKE

The front disc brake is actuated by hydraulic fluid and is controlled by a hand lever like the drum-type brake. Unlike the drum brake, however, the disc brake does not require adjustment.

It's essential that the work area and all tools be scrupulously clean when doing any major service on the hydraulic system. Tiny particles of foreign matter and grit in the caliper assembly or the master cylinder can damage the components making the system unsafe and requiring replacement of the damaged parts. Also, sharp tools must not be used inside the calipers or on the pistons. If there is any doubt about your ability to correctly and safely carry out major service on the brake components, take the job to a Norton service shop or a brake specialist.

Master Cylinder Removal and Disassembly

1. Unplug the electrical leads from the brake light switch (**Figure 6**). Peel the rubber boot away from the switch and master cylinder to expose the brake line nut. Unscrew the nut and 4 screws which hold the master cylinder to the handlebar (**Figure 7**).

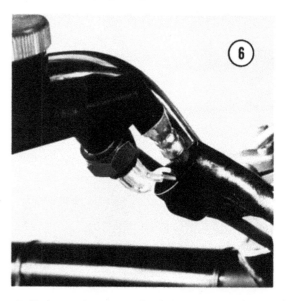

2. Remove the cap and bellows from the master cylinder and unscrew the brake light switch. Unscrew the pivot nut and bolt (**Figure 8**) and remove the lever.

3. Remove the circlip from the end of the master cylinder by carefully prying out several adjacent tangs on the clip. Remove the rubber boot and the piston.

4. Remove the cups, washer, and spreader from the master cylinder by lightly tapping the open end of the cylinder on a clean piece of wood or by applying light air pressure to the brake-line port. Do not use any hooks or sharp objects to pull the pieces out of the cylinder.

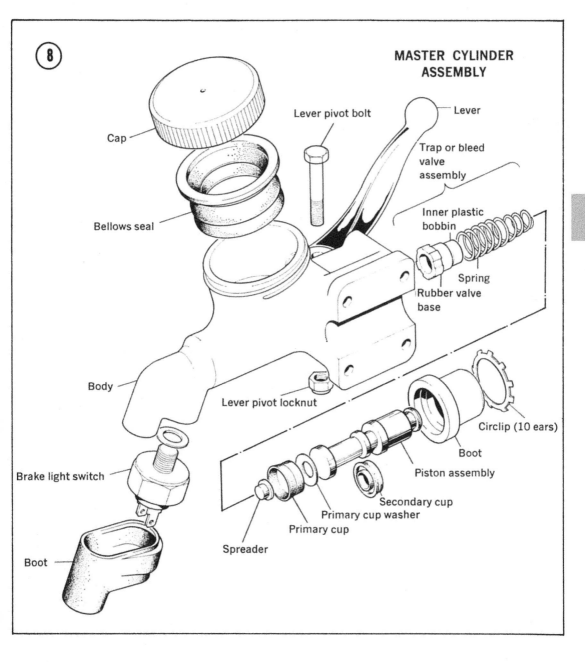

MASTER CYLINDER ASSEMBLY

Cap
Bellows seal
Body
Brake light switch
Boot
Lever pivot bolt
Lever pivot locknut
Spreader
Primary cup
Primary cup washer
Secondary cup
Piston assembly
Boot
Circlip (10 ears)
Lever
Trap or bleed valve assembly
Inner plastic bobbin
Spring
Rubber valve base

Inspection

1. Thoroughly clean the piston and master cylinder in fresh brake fluid and allow them to drain on a clean surface. Inspect the master cylinder bore and the piston contact surfaces for signs of wear and damage. If either component is less than perfect, replace it. Check the end of the piston for wear caused by the lever and check the pivot bore in the lever. Discard the bleed valve assembly, rubber boot, and cups.

2. Make sure the passages in the bottom of the brake fluid reservoir are clear. Check the reservoir cap and bellows for damage and deterioration and replace as necessary. Finally, check the condition of the threads in the bores for the brake line and the switch and examine the lever pivot lug for cracks. Any one instance of wear or damage to either the master cylinder or piston is reason for replacement of the affected component.

Assembly

1. Soak the new seal cups in fresh brake fluid for at least 15 minutes to make them pliable. Carefully work the secondary cup on the piston over the primary flange and into its groove by hand.

2. Install the boot on the piston. Make sure the outer end of the boot seats in the piston groove.

3. Assemble the trap valve, spring, and spreader. Make sure the plastic bobbin seats in the large end of the spring and the spreader seats in the small end. The hole in the bobbin must be clear.

4. Hold the cylinder upright and install the assembly, valve first. Set the primary cup into the cylinder, open end down, and set the washer on top of it with the convex surface up.

5. Coat the secondary cup with brake fluid and set the piston into the cylinder, crown first. Set the lock clip in place on the end of the boot with the clip fingers facing away from the boot.

6. Slowly turn and press in the piston assembly at the same time. Make sure that the lip of the secondary cup enters the cylinder without snagging. When the piston portion of the assembly is in the cylinder, hold it in place and fit the shoulder of the boot into the groove in the cylin-

der. Fit the boot lock clip into the groove. While still holding the piston assembly in the cylinder, engage the end of the piston assembly with the lever, line up the lever pivot holes, and install the pivot bolt and nut.

7. Reinstall the master cylinder by reversing the removal steps. Check to make sure the master cylinder is horizontal on the handlebar and fill and bleed the system.

Bleeding

1. Connect a length of tubing to the bleed valve on the brake caliper and place the other end of the tube in a clean, glass jar. The tube should be long enough so that a loop can be made higher than the bleed valve to prevent air from being drawn back into the caliper during bleeding.

2. Fill the fluid reservoir almost to the top lip. Open the bleed valve a full turn and slowly pump the hand lever. As the fluid enters the system, the level will drop in the reservoir. Maintain the level at about ½ inch from the top of the reservoir to prevent air from being drawn into the system.

3. Continue to pump the lever and fill the reservoir until there are no more air bubbles in the fluid coming from the bleed line. Then hold the lever full on and tighten the bleed valve and remove the bleed tube.

4. If necessary, add fluid to correct the level in the reservoir. It should be ½ inch from the top edge (**Figure 9**). Install the rubber bellows, closed side down, and screw on the reservoir cap.

5. Test the feel of the brake lever. It should be firm and should offer the same resistance each time it's pulled. If it feels spongy, it's likely that there is still air in the system and it must be bled again. When all air has been bled from the system, and the fluid level is correct in the reservoir, double check and tighten all the fittings and connection.

WARNING
Before riding the motorcycle, make certain that the front brake is operative by compressing the lever several times.

Brake Pad Replacement

There is no recommended mileage interval for changing the friction pads in the disc brake; service life is heavily dependent on riding conditions and habits. The disc and pads should be checked for wear every 1,000 miles and replaced when the friction material is worn down to a thickness of about ⅛ in. (3.2mm).

1. Remove front wheel as described earlier.

2. Carefully remove the pads from the caliper and discard them. Clean the pad recesses and the ends of the pistons with a soft brush. Don't use solvent, wire brush, or any hard tool which would damage the cylinders or the pistons.

3. Lightly coat the ends of the pistons and the backs of the new pads (not the friction material) with disc brake lubricant.

NOTE: *Check with your dealer to make sure the friction compound of the new pads is compatible with the disc material. Remove any roughness from the metal backs of the pads with a fine file and blow them clean with compressed air.*

4. Remove the cap and bellows from the master cylinder and slowly push the pistons into the caliper while checking the reservoir to make sure the fluid does not overflow. The pistons should move freely. If they do not and there is any evidence of them sticking in the cylinders, the caliper should be removed and serviced as described later.

5. Install the pads in the caliper and push them all the way in against the pistons.

6. Carefully remove any rust or corrosion from the disc and install the front wheel. With the front of the motorcycle still supported, spin the front wheel and activate the brake for as many times as it takes to refill the cylinders in the caliper and correctly locate the pads.

WARNING
Don't ride the motorcycle until you're sure the brake is operating correctly with full hydraulic advantage.

7. Bed the pads in gradually for the first 50 miles by using only light pressure as much as possible. Immediate hard brake applications will glaze the new friction material and greatly reduce the effectiveness of the brake.

Caliper Rebuilding

If the caliper cylinders leak, the caliper should be rebuilt. If the pistons bind in the cylinders, indicating severe wear or galling, the entire unit should be replaced. Rebuilding a leaky caliper requires special tools, a super-clean work environment, and experience. Even minor damage to the cylinders, pistons, or seals during service, or an assembly error could cause the brake to malfunction. Therefore, caliper service should be limited to removal and replacement. Take an unsatisfactory caliper to an authorized Norton dealer or to a brake specialist with experience on this particular unit for rebuilding.

Caliper Removal

1. Unscrew the pipe union at the hose bracket (**Figure 10**). Loosen or remove 2 nuts which fasten the fender bridge and hose bracket to the fork leg so the hose and pipe can be separated without bending the pipe.

2. Unscrew the caliper mounting bolts (**Figure 11**) and slide the caliper off the disc.

Caliper Installation

Install the caliper by reversing the removal steps. Fill the system with brake fluid and bleed it as described earlier. Make certain that the brake is fully operative before riding the motorcycle.

FRONT HUB

The front wheel should be removed and the hub disassembled, inspected, and the bearings cleaned and greased every 10,000 miles.

Disassembly/Assembly (Drum Type)

1. Block up the front of the motorcycle so the wheel is off the ground and remove it as described earlier.

2. Unscrew the lock ring (right-hand thread) from the left side of the hub (**Figure 12**) with Norton tool No. 063965, or by applying the end of a small drift against the side of one of the recesses and tapping the ring loose. If the ring cannot be broken loose, warm the hub with a propane torch or similar low flame.

3. Remove the seal and the spacer from the left side of the hub and install the axle from the right side. Using the axle as a drift, tap it into the hub with a soft mallet. As the right side bearing and the spacer are driven into the hub they will displace the left-side bearing. Drive it only as far as necessary for it to drop out.

4. Remove the axle and install it in the left side through the spacer tube. Drive out the right-side bearing along with the washers and seal.

5. Clean and dry the parts thoroughly. Check the bearings by hand for roughness and radial play and inspect the seals for damage or deterioration and replace any pieces that are defective. Pack the bearings with grease.

6. Line up the left-side bearing squarely with the bore in the hub and tap it into place, applying a drift only to the outer race. Install the spacer (with the flat side against the bearing), the seal, and the lock ring. Tighten the lock ring securely.

7. Install the spacer into the left-side bearing through the right side of the hub. The short end of the spacer must go into the left bearing and the spacer must be driven all the way in until its shoulder contacts the bearing inner race. Fill the area between the spacer and the inside of the hub with grease.

8. Line up the right-side bearing (double row) squarely in the right bearing bore. Using the axle as a drift, drive the bearing into the hub until it contacts the shoulder on the spacer.

9. Install the small washer, seal, and large washer. On early models it may be necessary to tap the large washer on.

10. Assemble the brake to the hub and install the wheel. Centralize the brake in the drum as described under *Front Drum Brake Assembly*.

11. Adjust the brake as described earlier.

Disassembly/Assembly (Disc Type)

1. Block up the front of the motorcycle so the wheel is off the ground and remove it as described earlier.

2. Unscrew the lock ring (left-hand thread) from the left side of the hub (**Figure 13**) using Norton tool No. 063965, or by applying the end of a small drift against the side of one of the recesses and tapping the ring loose. If the ring cannot be broken loose, warm the hub with a propane torch or similar low flame.

3. Remove the seal and spacer and heat the hub with *water* to about 212°F (100°C).

4. Insert a drift through the double-row bearing and into the spacer tube. Move the opposite end

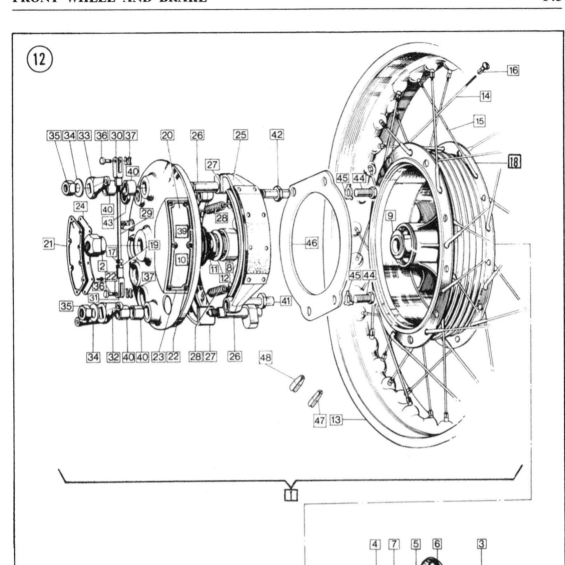

FRONT HUB
DRUM TYPE BRAKE

1. Wheel assembly	17. Nut
2. Spindle nut	18. Hub shell
3. Wheel spindle	19. Adjuster nut
4. Bearing	20. Inlet gauze
5. Bearing lock ring	21. Exit gauze
6. Dust cover	22. Air scoop screw
7. Spacer	23. Air inlet cover
8. Right-side bearing	24. Air exit cover
9. Spacer	25. Brake shoe
10. Washer	26. Expander
11. Seal	27. Slipper
12. Washer	28. Brake shoe spring
13. Rim	29. Tie rod
14. Inner spoke	30. Yoke (left-hand thread)
15. Outer spoke	31. Yoke (right-hand thread)
16. Spoke nipple	32. Long expander lever

33. Short expander lever
34. Washer
35. Expander lever nut
36. Cable pin
37. Spring clip
38. Dust cover
39. Washer
40. Expander bush ing
41. Pivot pin
42. Pivot pin/torque stop pin
43. Nut
44. Bolt
45. Tab washer
46. Support plate
47. Balance weight
48. Balance weight

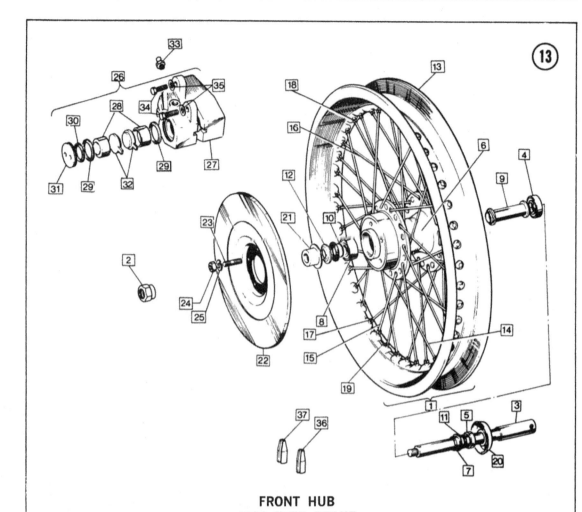

FRONT HUB
DISC TYPE BRAKE

1. Rim and hub assembly
2. Spindle nut
3. Wheel spindle
4. Left-side bearing
5. Bearing lock ring
6. Hub shell
7. Spacer
8. Right-side bearing
9. Spacer
10. Washer
11. Seal
12. Washer
13. Rim
14. Right-side inner spoke
15. Right-side outer spoke
16. Left-side inner spoke
17. Left-side outer spoke
18. Right-side spoke nipple
19. Left-side spoke nipple
20. Dust cover
21. Right-side spacer
22. Disc
23. Stud
24. Nut
25. Washer
26. Caliper assembly
27. Caliper
28. Piston
29. Seal
30. Seal
31. End plug
32. Friction pads
33. Bleed nipple
34. Screw
35. Spring washer
36. Balance weight
37. Balance weight

of the spacer to one side so the end of the drift can be applied to the bearing race. Tap the bearing out slightly and move the drift to the opposite side, moving the spacer so the drift can be applied to the bearing, and tap the bearing out, continuing the procedure—from side-to-side—until the bearing has been driven out of the hub.

5. Remove the spacer tube from the hub and drive the double-row bearing out to the right side of the hub, along with the washers and seal.

6. Clean and dry the parts thoroughly. Rotate the bearings by hand and check them for radial play and roughness. Inspect the seals for damage or deterioration and replace any pieces that are defective. Pack the bearings with grease.

7. Line up the left-side bearing (single-row) squarely with the bore in the hub and tap it into place, applying a drift only to the outer race. Install the spacer, seal, and lock ring. Tighten the lock ring securely.

8. Install the spacer tube into the left-side bearing through the right side of the hub. Fill the area between the spacer and the inside of the hub with grease.

9. Line up the right-side bearing (double-row) squarely in the right bearing bore. Using the axle as a drift, drive the bearing into the hub until it contacts the shoulder of the spacer.

10. Install the small washer, seal, large washer, and spacer in the right side of the hub and install the wheel in the motorcycle as described earlier.

CHAPTER NINE

REAR WHEEL AND BRAKE

This chapter describes repair and maintenance of the rear wheel hub, sprocket, and
drum and disc brakes.

REAR WHEEL

The rear wheel can be removed without disturbing the brake, sprocket, or drive chain.

Removal (1970 and Earlier)

1. Support the rear of the motorcycle so the
wheel clears the ground.

2. Disconnect the speedometer cable from the
drive unit (**Figure 1**).

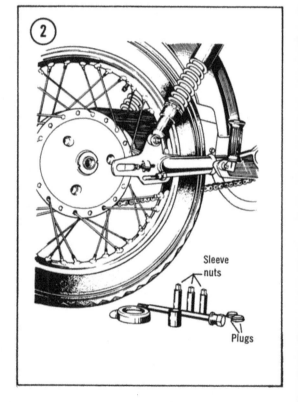

3. Carefully pry the 3 rubber plugs out of the
right side of the hub and unscrew the 3 sleeve
nuts (**Figure 2**).

4. Unscrew the axle and pull it out while lifting
up on the rear wheel. The spacer tube and the
speedometer drive should fall clear as the axle
is removed. Pull the hub away from the drum to
the right and remove wheel from motorcycle.

Removal (1971 through 1974)

1. Support the rear of the motorcycle so the rear wheel clears the ground.

2. Unscrew the axle (**Figure 3**) and pull it out while lifting up on the rear wheel. The spacer and speedometer drive should fall clear as the axle is removed.

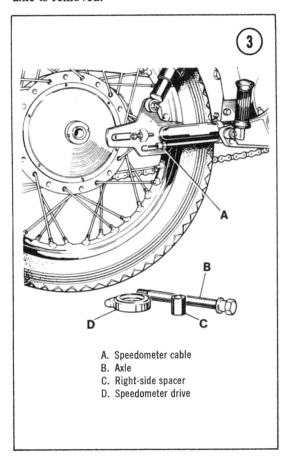

A. Speedometer cable
B. Axle
C. Right-side spacer
D. Speedometer drive

3. Pull the wheel to the right to release it from the drum and remove it to the rear.

Removal (1975 with Disc Brake)

1. Support the rear of the motorcycle so the rear wheel clears the ground.

2. Loosen the lower mounting bolt on the right rear suspension unit (**Figure 4**) and push the unit as far to the left as the circlip on the bolt will allow.

3. Unscrew the axle from the right side and pull it out while supporting the rear wheel. The right side spacer will fall clear as the axle is removed.

4. Disconnect the brake caliper mount from the shock absorber mount and hang the caliper on the hook beneath the seat rail (**Figure 5**); do not let the caliper hang on the brake line.

5. Pull the wheel to the right to release it from the sprocket and remove it to the rear.

Installation

Install the wheel by reversing the removal steps. On early models, line up one of the mounting studs in the brake drum with the swing arm so the bearing boss on the hub will pass between the other 2 studs. Then line up the studs with the hole in the hub and push the wheel to the left. On late models, line up the paddles on the drum (on the sprocket on disc brake models) with the shock absorber holes or slots in the wheel and push the wheel to the left.

On drum brake models, set the speedometer drive in place so that the tangs in the drive gear are engaged with the slots in the bearing lock ring (**Figure 6**).

Hold the spacer in place between the swinging arm and the speedometer drive and start the axle in through the arm. Push it all the way in and tighten it securely.

REAR HUB

Three hub types are used on Norton twins. On the early drum brake hub, through 1970 (**Figure 7**), the brake drum and drive sprocket bolt to the hub. On the later drum brake hub, 1971 on (**Figure 8**), 3 plastic shock absorber elements in the hub engage 3 paddles in the brake drum. In the disc brake hub, 1975 models (**Figure 9**), 5 paddles in the sprocket engage 5 rubber blocks located in the drive center in the hub.

With all types, the rear wheel can be removed without disturbing the brake, sprocket, or drive chain. For major service such as rcpacking or replacing the bearings, however, the entire assembly must be removed from the swinging arm. The recommended service interval for bearings is every 10,000-12,000 miles. The bearings and brake components should be inspected during each service and replaced as required.

The early and late drum brake wheels differ internally only in that the early wheel does not have a bearing in the brake drum. Instead, it has a double-row ball bearing assembly in the brake side of the hub and a single-row ball in the right side. The late wheel has a double-row ball assembly in the brake drum and a single-row ball assembly at each side of the hub. The procedure for removing the bearings from the hub is the same for both types.

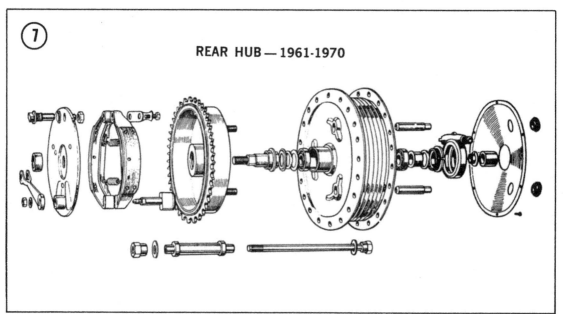

REAR HUB — 1961-1970

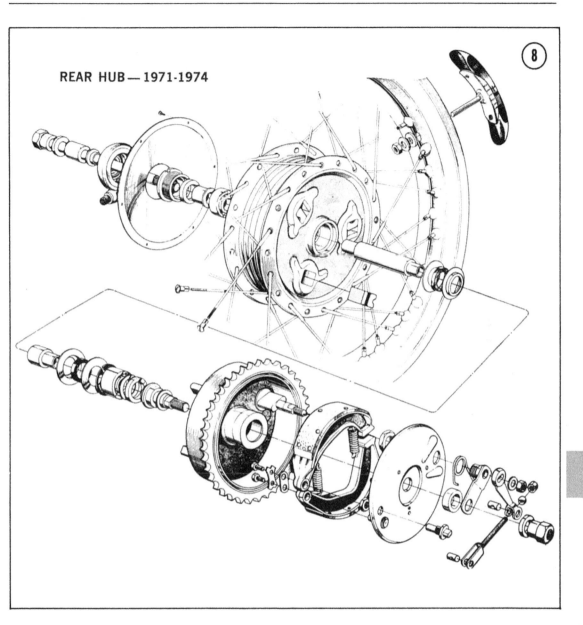

REAR HUB—1971-1974

⑧

⑨

Disassembly (Drum Brake Hubs)

1. Unscrew the lock ring from the right side of the hub (**Figure 10**). This piece has a left-hand thread. A drift or punch should not be used to break the ring loose. Instead, use either Norton tool No. 063965 or make a wrench from ⅛ in. wall tube or pipe like the one shown in **Figure 11**.

2. Remove the felt washer and spacer from the hub. Put a washer and the large right-side spacer on the rear axle. Insert the axle into the hub from the left side (**Figure 12**). Drive the axle

⑩

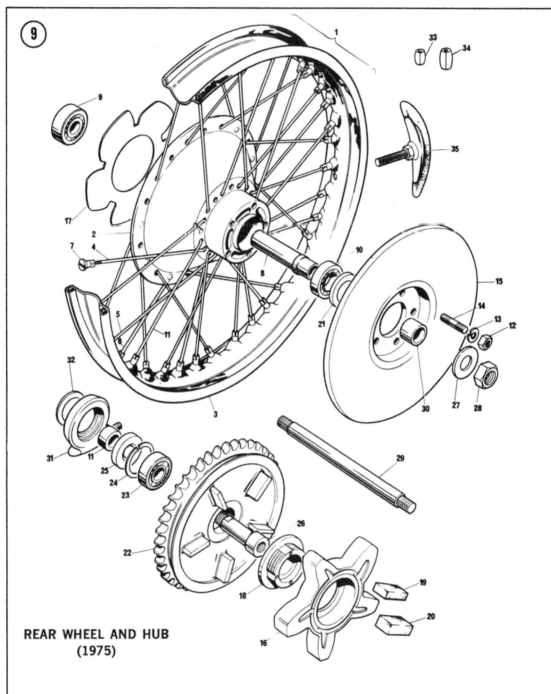

REAR WHEEL AND HUB
(1975)

1. Wheel assembly	10. Bearing	19. Drive rubber	28. Axle nut
2. Hub	11. Sprocket bearing spacer	20. Drive rubber	29. Axle assembly
3. Rim	12. Nut	21. Oil seal	30. Seal sleeve
4. Spoke, right hand	13. Washer	22. Sprocket	31. Speedometer gearbox
5. Spoke, left hand inner	14. Stud	23. Ball journal bearing	32. Washer
6. Spoke, left hand outer	15. Disc	24. Circlip	33. Balance weight
7. Spoke nipple	16. Drive center	25. Oil seal	34. Balance weight
8. Bearing sleeve	17. Plate	26. Dummy shaft	35. Security bolt
9. Bearing	18. Lock ring	27. Washer	

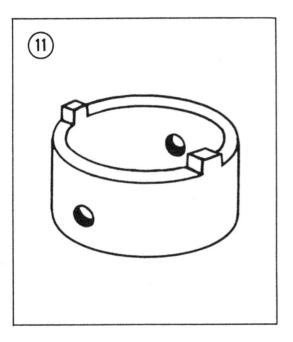

into the hub with a soft mallet until the bearing contacts the machined stop. This pushes the right-side bearing partially out of the hub. Remove the rear axle.

3. Drive the spacer out of the brake-side bearing using a shouldered drift that will pass through the bearing inner race. This drives the right-side bearing out of the hub. If a suitable drift is not available, the front axle can be used; however, make sure that the front of the frame is supported before removing the axle from the front wheel.

4. Insert the rear axle and large spacer into the brake-side bearing from the right end of the hub and drive the bearing out.

5. On late wheels, remove the drum from the brake assembly and remove the spacer, washers, and circlip from the right side of the drum (**Figure 13**). Screw the axle nut onto the left end of the short axle and drive the bearing out of the drum to the right using a soft mallet.

Disassembly (Disc Brake Hub)

1. Remove the wheel as described earlier. Unscrew the nut from the end of the dummy axle (Figure 9) and withdraw the axle from the sprocket. Collect the washer and disengage the sprocket from the chain. Pull the speedometer drive free of the sprocket.

2. Remove the bearing spacer, the seal, and the circlip from the sprocket. Insert the dummy axle into the bearing and knock it out by tapping on the dummy axle with a soft mallet.

3. Unscrew the nuts which attach the disc to the hub and remove the lockwashers, spacer, and oil seal. From the other side of the hub, unscrew the bearing lock ring and remove the drive center.

4. With the dummy axle, drive the bearing spacer into the hub from the drive side. This will knock the opposite bearing out of the hub toward the brake side. Then, invert the hub and using the dummy axle with the long axle screwed into it, knock out the drive side bearing.

Inspection

Thoroughly clean and dry all parts, including the inside of the hub and drum. Rotate bearings by hand and check for roughness and axial play. Inspect metal parts for corrosion. If there is any doubt about a bearing's condition, replace it. Replace the felt seals.

Assembly (Drum Brake Hubs)

1. Pack the bearings with a good grade of bearing grease.

2. Line up the right-side hub bearing (single-row) with the bearing bore and tap it in evenly far enough to permit the spacer, felt washer, and lock ring to be installed. Tighten the lock ring (left-hand thread) securely.

3. Tap the bearing spacer tube into the right-side bearing from the left. On 1970 and earlier

9

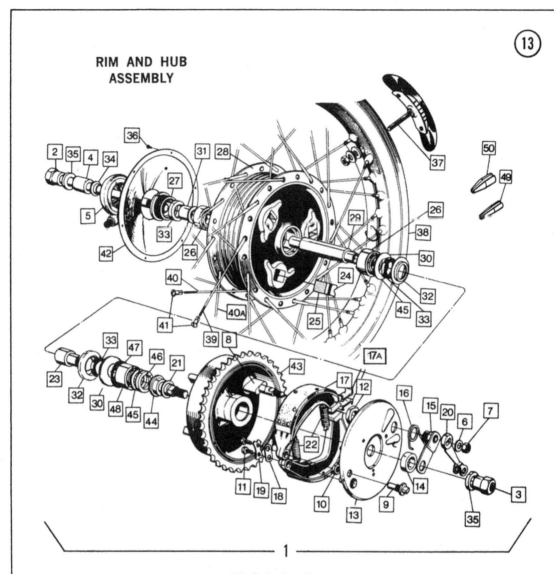

RIM AND HUB ASSEMBLY

1

1. Hub assembly
2. Wheel axle
3. Wheel axle nut
4. Axle spacer
5. Speedometer gearbox
6. Washer
7. Nut
8. Brake cam
9. Torque stop pin
10. Torque stop pin nut
11. Bolt
12. Cam bearing nut
13. Cover plate assembled
14. Cover plate outer spacer
15. Bearing cam and stay
16. Lever return spring
17. Brake shoe
17A. Brake shoe slipper
18. Plate
19. Washer
20. Lever
21. Washer
22. Spring
23. Left-side bearing spacer
24. Thick cush drive buffer
25. Thin cush drive buffer
26. Bearing
27. Lock ring
28. Hub shell
29. Inner bearing spacer
30. Felt retaining washer
31. Spacer
32. Dished washer
33. Seal
34. Gearbox spacer
35. Spindle washer
36. Disc screw
37. Security bolt
38. Rim
39. Spoke
40. Left-side inner spoke
40A. Right-side inner spoke
41. Nipple
42. Rear hub disc
43. Brake drum
44. Dummy axle
45. Felt retaining washer
46. Felt seal
47. Circlip
48. Bearing
49. Balance weight
50. Balance weight

models, the distance between the end of the spacer and the shoulder is greater on the left (brake) side than on the right to accommodate the wider double-row bearing in the left side of the hub. Make certain that the shorter end of the spacer is installed in the right-side (single-row) bearing.

4. Set the left-side bearing squarely in the left side of the hub and tap it evenly around the outer race until it is stopped by the shoulder on the spacer tube. Don't apply any pressing or tapping force to the bearing inner race.

5. Install the felt washer and seal and the outer metal washer. On early wheels, the outer washer may be tapped in with a small punch or drift.

6. On late wheels, install the washer and short axle in the drum, followed by the felt seal and the felt retainer. Line up the double-row bearing squarely with the bearing bore in the drum and tap it evenly around the outer race until its outer edge is beneath the circlip groove. Install the circlip with the outer edge facing outward and make certain it's completely seated in the groove. Install the inner felt retainer, the felt seal, the dished washer, the outer felt retainer, and the bearing spacer.

> NOTE: *An improved bearing spacer for Commando models up to engine No. 305427 can be purchased through Norton dealers as a standard replacement item. The new spacer (part No. 065290) replaces spacer No. 062070 and is interchangeable without any modifications.*

SPROCKET AND BRAKE ASSEMBLY

Removal/Installation

1. Remove the wheel as described earlier.

2. Remove the master link from the rear chain and remove the chain from the rear sprocket. Unplug the electrical leads from the brake light switch (**Figure 14**).

3. Disconnect the rear brake cable from the brake cam arm (**Figure 15**).

4. Unscrew left axle nut and remove the brake and sprocket assembly from the swinging arm.

5. Installation is the reverse of these steps.

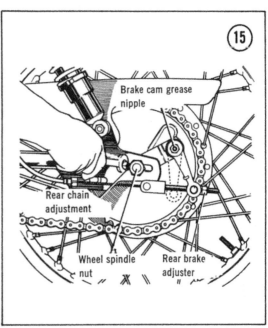

Brake cam grease nipple

Rear chain adjustment

Wheel spindle nut

Rear brake adjuster

REAR DRUM BRAKE

There is no specified interval for replacing the brake linings; the rate of wear varies greatly depending upon riding habits. The linings should be carefully examined each time the wheel bearings are serviced. If the lining is worn down to or near the rivet heads, the complete shoes and linings should be replaced. If these are not available, the existing shoes can be relined and the new linings arced to fit the drums. This is a job for a specialist and should not be attempted by a hobbyist mechanic.

Removal and Disassembly

1. Remove the rear wheel and sprocket/brake assembly as described earlier.

2. Remove the springs by lightly gripping the coil section with a conventional pair of pliers and extending one of the spring ends with needle-nose pliers (**Figure 16**).

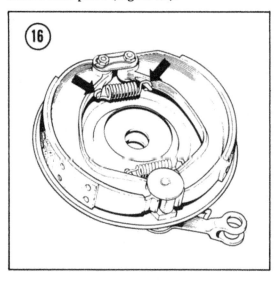

3. Straighten the tab washers and unscrew the 2 nuts from the shoe pivots. Remove the shoes from the brake backing plate.

4. Unscrew the nut from the end of the camshaft (**Figure 17**) and remove the lever. Pull the cam out of the backing plate. It is not necessary to remove the cam bearing and stay unless the bearing bore is badly worn and must be replaced.

Inspection

Thoroughly clean and dry all parts except the linings. Check the contact surface of the drum for scoring. If there are grooves deep enough to snag a fingernail, the drum should be trued by a brake specialist. In such case, the linings will have to be replaced and the new ones arced to the new drum contour.

Check the linings for wear as described above. If they are serviceable, inspect them for imbedded foreign material. Dirt can be removed with a stiff wire brush, but if they are soaked with oil or grease they will have to be replaced.

Inspect the cam lobes and the pivot area of the shaft for wear and galling. Minor roughness can be removed with emery cloth but if either area is deeply scored or severely worn, the cam should be replaced.

Assembly

1. Grease the cam bearing bore, the cam pivot area and lobes, and the pivot pins with high-temperature grease. Apply only a light coat to each of these points; excess grease is likely to find its way onto the linings and render them unserviceable.

2. Install the cam in the plate. Install the shoes on the pivot pins and against the cam. If the shoes have detachable slippers, make sure these are located between the cam and the ends of the shoes.

3. Install the tie plate, the tab washer plate, and the pivot bolts. Tighten the bolts securely and bend the tab washers over against the flats.

4. Install the return springs between the shoes. Use a short wire loop and screwdriver to stretch the springs over the ears on the shoes.

5. Reinstall the brake, sprocket, and rear wheel as described earlier. Before tightening the axle nut, connect the cable to the brake arm and apply the brake fully to centralize it in the drum. Hold the brake on and tighten the axle nut.

REAR DISC BRAKE

The rear disc brake is actuated by hydraulic fluid and is controlled by a foot lever like the drum-type brake. Unlike the drum brake, however, the disc brake does not require adjustment.

It's essential that the work area and all tools be scrupulously clean when doing any major service on the hydraulic system. Tiny particles of foreign matter and grit in the caliper assembly or the master cylinder can damage the components making the system unsafe and requiring replacement of damaged parts. Also, sharp tools must not be used inside the caliper or on the pistons. If there is any doubt about your ability to correctly and safely carry out major service on the brake components, take the job to a Norton service shop or a brake specialist.

Master Cylinder
Removal and Disassembly

1. Unscrew the bolt which attaches the brake union to the right foot peg plate (**Figure 18**). Note the position of the ground wire so it may be reinstalled in the same location.

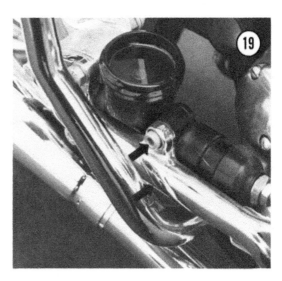

2. Slightly loosen the Allen screws which attach the master cylinder to the foot peg stalk (**Figure 19**). Depress the kickstarter and block it or tie it down so it is out of the way. As an alternative, the kickstarter pedal may be removed. Unscrew the 3 bolts which attach the foot peg stalk to the plate. Swing the stalk out far enough to gain access to the brake line union in the end of the master cylinder and unscrew it. It is a good idea to place a drip pan beneath the cylinder to catch brake fluid when the union is disconnected.

3. Loosen the forward locknut on the brake rod (**Figure 20**) and remove the cotter key from the pin which attaches the clevis to the brake pedal. Remove the pin and disconnect the clevis from

the brake pedal. Unscrew the clevis from the brake rod. Unscrew the Allen screws which attach the master cylinder to the foot peg stalk and remove the cylinder.

4. Refer to Chapter Eight, *Front Disc Brake*, for disassembly, inspection, assembly, and bleeding procedures.

Brake Pad Replacement

Brake pad replacement for the rear wheel caliper is identical to that for the front. Refer to Chapter Eight.

Caliper Rebuilding

Refer to Chapter Eight for caliper rebuilding instructions.

CHAPTER TEN

FRAME, SUSPENSION, AND STEERING

Included in this chapter are service, maintenance, and repair procedures for all chassis components such as frame, front suspension, rear suspension and swinging arm, and the Isolastic system.

FRAME

Frame service is limited to inspection for cracks in welds and tubes and restoration of the paint. If bending damage is suspected or apparent, the frame should be inspected and repaired by an authorized Norton service shop or by a specialist experienced in frame repair. **Figure 1 and Figure 2** are provided as reference for inspection and repair.

ISOLASTIC ENGINE MOUNTS

One of the Norton Commando's key features is its Isolastic system. The entire power train fastens to the rest of the motorcycle with resilient mounts, thereby preventing normal but still bothersome drive train vibrations from affecting the rider and the controls. The Isolastic mounts must be inspected and adjusted occasionally, and need to be removed and serviced only after long use.

Inspection

1. Peel back the rubber cover on the left side of the Isolastic mount.

2. Push the engine (front mount) or rear wheel (rear mount) to the right to take up the slack in the mount and measure the clearance between the plated collar and the plastic washer (**Figure 3**). Clearance should be 0.010 in. (0.25mm).

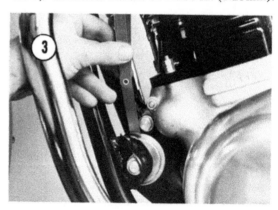

Front Mount Adjustment
(750 and 850 Mark II Models)

1. Unscrew the nut from the left side of the front mount (**Figure 4**). Line up one of the flats on the right end of the bolt so it will clear the engine case (**Figure 5**).

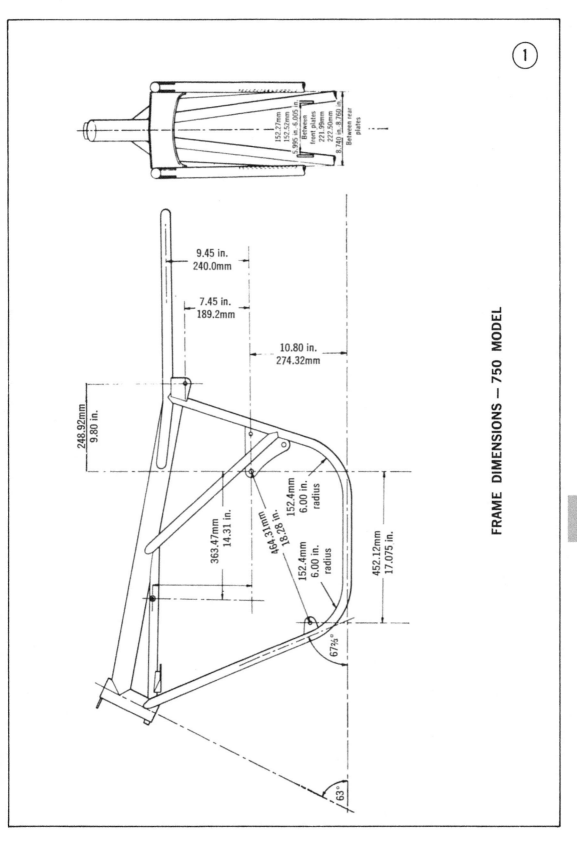

FRAME DIMENSIONS — 750 MODEL

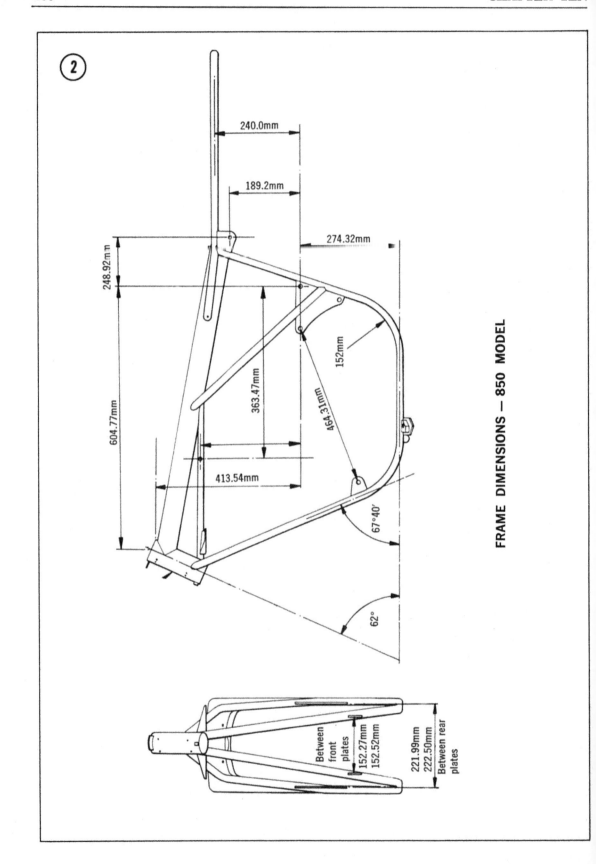

FRAME DIMENSIONS — 850 MODEL

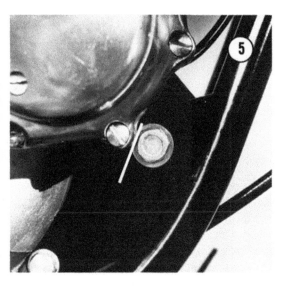

2. With a soft drift, drive the bolt into the mount from the left side far enough so that the rubber cover, spacer, and plastic washer can be removed. Remove the tube cap and the shim (**Figure 6**).

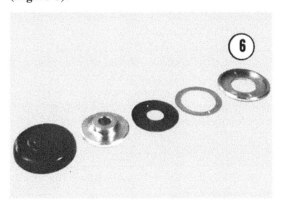

3. Install shim(s) of appropriate thickness. Shims are available in thicknesses of 0.005, 0.010, 0.020, and 0.030 in. Use as few shims as possible; for instance, use one 0.030-in. shim rather than three 0.010-in. shims.

Example:	Actual clearance	0.040 in.
	Desired clearance	0.010 in.
	Difference to be compensated	0.030 in.
	Existing shim	0.010 in.

In this instance, discard the 0.010 in. shim and substitute a 0.030 in. shim.

4. Replace the plastic washer if it shows wear.

5. Reassemble the pieces in order—shims, tube cover, plastic washer, spacer, and rubber cover. Gently tap the bolt in from the right while aligning the pieces. Screw on the nut and tighten it to 25 ft.-lb. (3.456 mkg).

Rear Mount Adjustment
(750 and 850 Mark II Models)

1. Unscrew the nut from the right end of the mounting bolt (**Figure 7**).

2. With a soft drift, drive the bolt into the mount from the right side until it protrudes about 4 inches on the left side.

3. Peel the rubber cover off the tube and push it down and back. Remove the cover, spacer, and plastic washer. Remove tube cap and shim.

4. Install shim(s) of appropriate thickness (see example above). Replace the plastic washer if it shows signs of wear. Clean the parts thoroughly and lubricate them with a light coat of silicone grease.

10

5. Install the shims and cap on the mounting tube. If more than one shim is required, the thinner shim should be outboard, against the cap. Assemble the rubber cover, spacer, and plastic washer together. Peel back the lip on the cover so that the assembly can be fitted between the mounting tube and the frame plate. It may be necessary to have someone push the rear wheel to the left to provide sufficient clearance. The rubber cover must fit completely over the end cap and the tube, and the spacer, washer, and cap should turn freely when the cap is turned.

6. With assistance, line up mounting bolt with mounting hole in the frame and carefully drive bolt in from the left side. Be sure to reinstall the spacer between the frame mounting bracket and the footrest plate. Install the flat washer and the nut. Tighten the nut to 25 ft.-lb. (3.456 mkg).

Front and Rear Mount Adjustment (850 Mark III Models)

The Isolastic mounts on 850 Mark III models are equipped with threaded adjusters and adjustment does not require the addition or removal of shims.

1. Loosen the nut on the end of the mount and peel the rubber cover away from the adjuster. Remove the spring clip (**Figure 8**).

2. Insert a 0.010 in. (0.25mm) flat feeler gauge in the mount between the thrust washer and the end cap (**Figure 9**) and turn the adjuster until there is a slight drag on the feeler gauge when it is removed. Then, without turning the adjuster further, tighten the end nut on the mount and install the spring clip and refit the rubber cover. Repeat this procedure for the remaining mount.

Front Mount Removal

In time, the Isolastic mounts require removal, cleaning, replacement of worn parts, installation, and adjustment. However, this service is not routine and is rarely required before many thousands of miles. If the Isolastic units must be rebuilt to the extent that the rubber bushings are to be replaced, a special installation tool is required (Norton tool No. 063971). If you do not have access to this tool, the bushings must be replaced by a Norton service shop after the Isolastic mounts have been removed from the motorcycle. Refer to **Figure 10** for the following procedure.

> NOTE: *Construction of the Isolastic units on 850 Mark III models is such that they require a press to remove them from the mounts. This should be entrusted to a Norton dealer. Removal of the mounts is essentially the same as for the earlier mounts. Adjustment is described above.*

1. Remove the right-side exhaust pipe and unscrew the nut from the left end of the front mounting bolt.

2. Line up one of the flats on the head of the mounting bolt so that it will clear the engine case (Figure 5) and drive the bolt out to the right side with a soft drift. The front of the engine should be supported so that the bolt threads will not be damaged by frame mounting plates as the bolt is drifted out.

3. Peel back the left-side rubber cover and remove it along with the spacer and plastic washer. Do the same to the right-side cover. Remove the tube caps and shims from both sides.

4. Unscrew 2 nuts and bolts which attach the mount to the engine (Figure 11). Remove the mount.

5. Insert the end of a drift about ½ inch into the sleeve of one of the bushings (Figure 12). Using the drift as a lever, turn the bushing in the tube (Figure 13) so it can be easily removed.

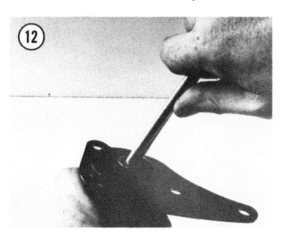

6. Remove the spacer and rubber buffers from the mounting tube. Remove the opposite bushing in the same manner as the first.

Rear Mount Removal

1. Remove the engine from the motorcycle. Refer to Chapter Four.

2. Remove the end bushings in the same manner as for the front mount. Remove the spacer tubes and buffers from each end of the mounting tube.

3. Set the end of a drift against one side of the center bushing and tap the drift with a hammer to turn the bushing in the mounting tube so it can be easily removed.

Front Mount Assembly
(750 and 850 Mark II)

1. Place the tapered bushing guide (Norton tool No. 063971) over the end of the mounting tube (Figure 14). Coat the circumference of one of the bushings with rubber lubricant (e.g., glycerine) and press it evenly into the guide.

10

2. Set the driver in place on the bushing (Figure 15) and press it into the mounting tube until the driver bottoms on the bushing guide.

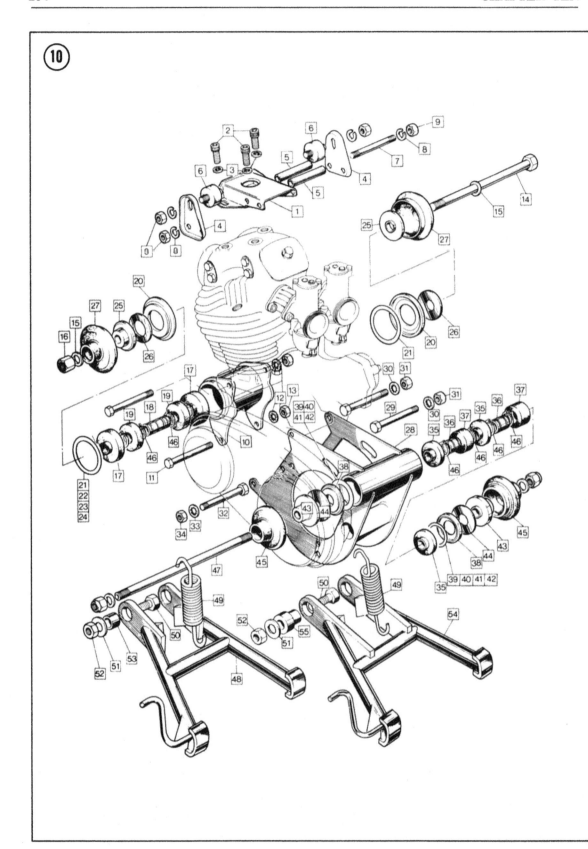

ENGINE MOUNT

1. Engine steady
2. Engine steady to head screw
3. Engine steady screw washer
4. Engine steady side-plate
5. Engine steady spacer
6. Engine steady rubber mounting
7. Engine steady stud
8. Engine steady stud washer
9. Engine steady stud nut
10. Front engine mounting
11. Front engine plate bolt
12. Front engine plate washer
13. Engine front plate nut
14. Front engine plate bolt to frame
15. Front engine plate bolt washer
16. Front engine plate bolt nut
17. Front engine mounting bushing
18. Front engine mounting spacer
19. Front engine mounting buffer
20. Front engine mounting tube cap
21. Front engine mounting shim .005 in.
22. Front engine mounting shim .010 in.
23. Front engine mounting shim .020 in.
24. Front engine mounting shim .030 in.
25. Front engine mounting collar
26. Front engine mounting washer (PTFE.)
27. Front engine mounting gaiter
28. Rear engine mounting
29. Rear engine plate bolt
30. Rear engine plate washer
31. Rear engine plate nut
32. Rear engine plate bolt bottom
33. Rear engine plate washer
34. Rear engine plate nut
35. Rear engine mounting rubber
36. Rear engine mounting spacer
37. Rear engine mounting buffer
38. Rear engine mounting tube cap
39. Rear engine mounting shim .005 in.
40. Rear engine mounting shim .010 in.
41. Rear engine mounting shim .020 in.
42. Rear engine mounting shim .030 in.
43. Rear engine mounting collar
44. Rear engine mounting washer (PTFE.)
45. Rear engine mounting gaiter
46. Circlip for isolastic buffers
47. Rear engine mounting stud
48. Center stand (750 only)
49. Center stand spring
50. Center stand bolt
51. Center stand washer
52. Center stand nut
53. Center stand spacer (750 only)
54. Center stand (850 only)
55. Center stand spacer (850 only)

3. Install the buffers and circlips on the spacer tube (**Figure 16**), turn the mounting tube over, and set the buffer assembly in place so that the tube rests on the bushing that has been installed.

4. Install the second bushing in the same manner as the first, using the guide and driver.

Rear Mount Assembly
(750 and 850 Mark II)

1. Coat the circumference of the center bushing with rubber lubricant and press it into the mounting tube by hand, using a length of tubing or pipe as a driver. The bushing must be centered in the tube, 3½ in. (82mm) from each end.

2. Install the buffer and circlips on one of the spacer tubes. Set it in place in the mounting tube.

3. Coat the circumference of one of the outer bushings with rubber lubricant and press it into the mounting tube until it is stopped against the spacer.

4. Turn the mounting tube over and install the other buffer and tube assembly and the outer bushing in the manner just described.

Front and Rear Mount Installation

Reinstall the mount and engine/transmission package by reversing the removal steps. Install new Isolastic mounts on the cylinder headsteady and check and adjust the clearance of the front and rear mounts as described earlier.

FRONT SUSPENSION

The Norton front suspension consists of a spring-controlled, hydraulically dampened telescopic fork. Before suspecting major trouble, drain the fork oil and refill with proper type and quantity. See Chapter Four. If you still have trouble, such as poor dampening, tendency to bottom out or top out, or leakage around rubber seals, then follow the service procedures in this section.

To simplify fork service and to prevent the mixing of parts, the legs should be removed, serviced, and reinstalled individually.

Removal

1. On models equipped with a front disc-brake, the caliper can be removed with the brake line attached to eliminate the need for refilling and bleeding the system. Remove the nuts which hold the brake line bracket to the right fork leg (**Figure 17**). Remove the bolts which hold the caliper to the fork leg and pull the caliper away from the disc. Insert a block, such as a piece of tubing, between the brake pads to prevent them from being ejected (**Figure 18**) by accidental movement of the brake lever. Suspend the caliper from the end of the right handlebar with a piece of cord so that it is out of the way. Be careful not to bend the brake line too sharply. As a precaution against damaging the caliper, the fuel tank, or the engine cooling fins, wrap several thicknesses of newspaper around the caliper and tape it in place.

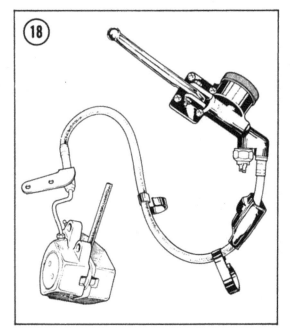

2. Block up the front of the motorcycle to lift the front wheel off the ground.

3. On models with a drum-type front brake, disconnect the brake cable from the lower brake arm by removing the clevis pin (**Figure 19**). Unscrew the cable adjuster from the boss on the brake backing plate.

4. On all models, unscrew the axle nut (**Figure 20**). Loosen the pinch bolt at the bottom of the left leg (**Figure 21**). Insert the shaft of a screwdriver or small bar through the hole in the left end of the axle and pull it out of the forks

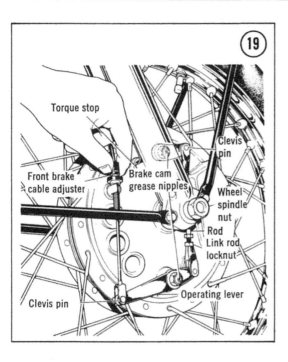

and wheel while supporting the weight of the wheel.

5. Unscrew the 4 bolts that hold the fender struts to the bottom of the fork legs, and the remaining 2 nuts that hold the center stay to the aluminum fork slider (**Figure 22**). Remove the fender and set it aside.

6. Unscrew the cable collars from the tachometer and speedometer and unplug the instrument lamp lead from the wiring harness.

7. Unscrew the caps from the tops of the forks (**Figure 23**). Hold the damper locknut with a wrench and unscrew the cap from the top of the damper rod (**Figure 24**). Remove the instruments and set them aside.

8. If the motorcycle is equipped with fork gaiters, loosen the upper clamps. Loosen the pinch bolts in the bottom fork yoke (**Figure 25**). Screw one of the caps several complete turns into the top of the fork to be removed and tap it sharply with a soft mallet to break the tube loose from the yoke (**Figure 26**). Remove the cap once again and pull the fork leg down and out of the yokes.

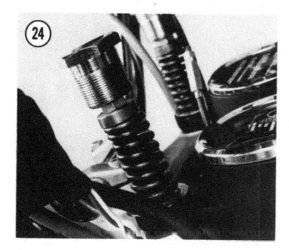

FRONT FORKS

1. Right leg assembly (drum brake)
2. Right slider (drum brake)
3. Right leg assembly (disc brake)
4. Right slider (disc brake)
5. Left leg assembly (drum brake)
6. Left slider (drum brake)
7. Left leg assembly (disc brake)
8. Left slider (disc brake)
9. Pinch bolt (drum only)
10. Gaiter
11. Collar
12. Washer
13. Damper rod seat
14. Damper rod
15. Damper rod cap nut
16. Damper rod locknut
17. Slider drain plug
18. Damper tube anchor bolt
19. Damper tube cap
20. Damper tube
21. Damper valve
22. Damper valve stop pin
23. Fork spring
24. Fork spring locating bushing
25. Damper tube anchor bolt washer
26. Damper tube anchor bolt washer
27. Main tube
28. Fork tube bushing circlip
29. Fork tube bushing
30. Oil seal
31. Oil seal washer
32. Fork tube guide bushing
33. Fork tube top bolt
34. Washer
35. Stud
36. Washer
37. Nut
38. Steering head bearing
39. Bearing spacer
40. Washer
41. Cover
42. Washer
43. Nut
44. Socket screw
45. Lower yoke (750 only)
45A. Lower yoke (850 only)
46. Plug for lower yoke
47. Upper yoke with stem (750 only)
47A. Upper yoke with stem (850 only)
48. Stem blanking plug
49. Clip
50. Screw
51. Lock/keys
52. Right headlight bracket/top cover
53. Left headlight bracket/top cover
54. O-ring
55. Bolt
56. Washer
57. Nut

Disassembly

1. Invert the fork leg over a drip pan and pour out the oil. Clamp the lower end of the fork leg in a vise with the leg positioned vertically. Wooden blocks or soft jaw pads should be used between the vise jaws and the fork leg to prevent damage.

2. Peel the dust cover away from the slider and move it up the tube (**Figure 27**). If the motorcycle is equipped with fork gaiters, loosen the bottom clamps and slide the gaiters off the tubes.

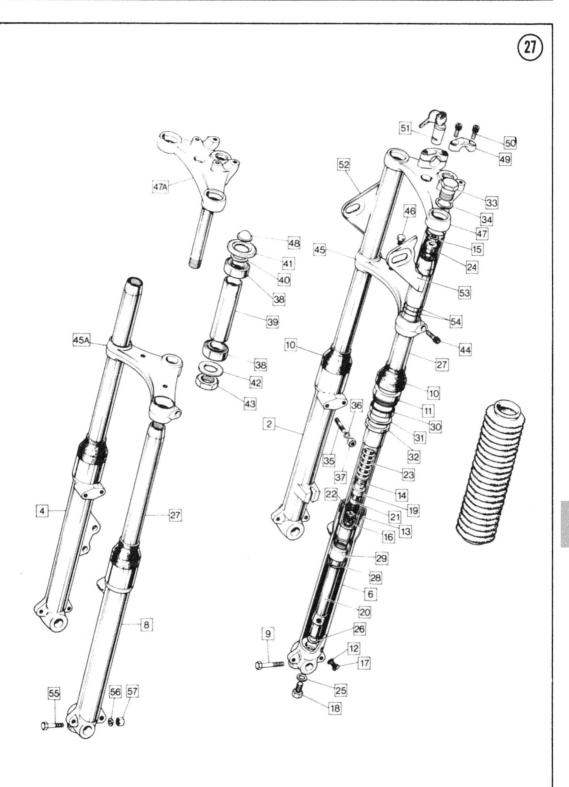

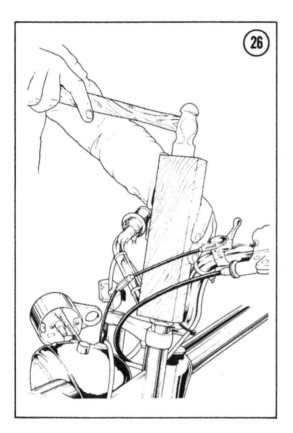

Unscrew the collar from the top of the slider and pull up on the leg to remove it from the slider.

3. Unscrew the bottom bolt from the slider and pull out the spring and damper assembly.

4. Lightly clamp the damper tube in a vise and unscrew the cap. Pull out the damper rod, clamp it in the vise and remove the locknut, washer, and valve. Remove the valve stop pin.

5. Remove the circlip and steel bushing from the end of the fork tube.

Inspection

Thoroughly clean and dry all parts. Check the bushings and the fork tube for signs of wear or galling. Check the damper rod for bending by rolling it on a flat surface. Carefully check the damper valve and the stop pin for wear or damage. Inspect the seals for scoring and nicks and loss of resiliency. If there is any doubt about their condition, pry them out and press in new seals using a large socket as a driver.

Any parts that are worn or damaged should be replaced; simply cleaning and reinstalling

unserviceable pieces will not improve performance of the front suspension.

Reassembly

Refer to Figure 27 and reassemble the fork leg components in reverse order of disassembly. When the damper assembly has been assembled and installed in the slider, oil the outside of the upper fork tube and fit it into the slider. Install the collar in the top of the slider and tighten it securely by hand. Slide the rubber dust cover down over the top of the slider.

Installation

Slide the assembled fork leg up into the lower yoke, through the headlight bracket tube, and into the top yoke. Install the instrument case and fork cap and tighten the cap to draw the fork tube up into the taper in the top yoke. When the fork tube is firmly seated in the yoke, tighten the pinch bolt in the lower yoke and unscrew the top cap.

Refer to Chapter Two. Fill the fork legs with the correct amount and grade of oil. Reconnect the top cap to the damper tube rod and securely tighten the locknut. Screw the top cap into the tube and tighten it to 40 ft.-lb. (5.83 mkg).

Reassemble the rest of the components in reverse order of disassembly. On front disc brake models, make certain that there is sufficient operating pressure by pumping the lever several times before riding the motorcycle.

STEERING

Service to the steering consists of periodically checking and correcting the tightness of the stem (1970 and earlier models) and replacing the bearings.

Adjustment (1970 and Earlier)

1. Block up the front of the motorcycle so the wheel is off the ground. Take hold of the lower fork legs and try to move them back and forth. If any movement can be felt, the steering head must be tightened.

2. Unscrew 4 Allen bolts from the handlebar retainers (**Figure 28**) and remove the handlebars.

The handlebars can be rested on the fuel tank, but the tank should first be covered with a clean cloth to prevent the handlebars from damaging the paint.

3. Loosen the cap nut on the steering stem and pinch bolts in the lower fork yoke.

4. Tighten the stem adjuster nut, turning it one flat at a time while checking the movement of the forks from lock-to-lock and as described in Step 1.

When correctly adjusted, the forks should turn freely from lock-to-lock and there should be no apparent back-and-forth movement of the fork assembly in the steering head.

WARNING

If the steering head is adjusted too tight, the front end will tend to bind under load, creating a hazardous condition.

5. When the adjustment is correct, retighten the top stem nut and the pinch bolts, and reinstall the handlebars.

Disassembly (1970 and Earlier)

1. Remove the fork legs and handlebars as described earlier. Unscrew the stem cap nut and tap the top yoke up and off the stem with a soft mallet.

2. Unscrew the stem adjuster nut and remove the dust cover. Tap the stem and bottom yoke down and out of the bearings.

3. Carefully drive the bearing assemblies out of the steering head—first the bottom bearing, then

the top. Apply the end of the drift to the inner race of each bearing.

Disassembly (1971 and Later)

1. Straighten the tab washer on the bottom stem nut and unscrew the nut (**Figure 29**).

2. Take hold of the headlight and tap the bottom yoke down and off the stem with a soft mallet. Remove the O-rings from each of the headlight bracket tubes (2 on the bottom and one on the top). To prevent damage to the headlight, wrap several thicknesses of newspaper around it and tape them in place.

3. Drive the stem and top yoke upward with a soft mallet and remove the assembly from the steering head along with the dust cover and washer.

4. Insert a drift through the inner race of the top bearing and push the interior spacer tube to the side so the end of the drift can be applied to the inner race of the bottom bearing. Carefully tap the bearing out of the steering head and remove the spacer. Remove the top bearing by driving it upward from the bottom. Again, apply the drift to the inner race.

Inspection

Thoroughly clean and dry the bearings and the inside of the steering head. Rotate the bearings by hand and check for roughness and radial play. If there is any doubt about their condition, they should be replaced.

10

Assembly (1970 and Earlier)

1. Lightly oil the outer race of one of the bearings and fit it squarely into the bottom of the steering head. Tap it all the way in with a soft mallet until it bottoms on the machined lip inside the head. Then install the top bearing in the same manner.

2. Insert the stem into the bearings from the bottom. Set dust cover over the top of stem.

3. Screw on the stem nut and tighten it until no movement can be felt in the yokes but they will still turn freely from lock-to-lock.

4. Position the headlight and brackets on the bottom yoke and set the top yoke in place. Install the washer and acorn nut on top of the yoke but do not tighten it.

5. Install the fork legs as described earlier. When the tops of the fork legs are correctly seated in the tapers in the top yoke, adjust the stem nut as described under *Adjustment (1970 and Earlier)*.

6. Reassemble the remaining components as described earlier.

Assembly (1971 and Later)

1. Lightly oil the outer race of one of the bearings and fit it squarely into the bottom of the steering head.

2. Refer to Figure 27 and set the spacer into the top of the steering head.

3. Oil the outer race of the top bearing and fit it squarely into the top of the steering head. The bearings are pressed in by the yokes as the stem nut is tightened.

4. Install the washer and dust cover on the top of the steering head. Tap the stem and top yoke down into the bearings with a soft mallet.

5. Grease 2 of the O-rings and set them in place on the underside of the top yoke. Grease the remaining 4 O-rings and set them in the recesses in the bottom yoke.

6. Hold the headlight and bracket tubes in position and line up the tops of the tubes with the top yoke. Line up the bottom yoke with the bottoms of the bracke ttubes and the stem, and push it up into place.

7. Set the tab washer in place and screw on the

bottom stem nut while checking the alignment of the bracket tubes with the yokes. Tighten the nut only enough to pull bottom yoke into place.

8. Install the fork legs as described earlier. When the tops of the fork legs are correctly seated in the tapers in the top yoke, tighten the stem nut and then the pinch bolts in the bottom yoke. Bend the tab washer over against one of the flats on the stem nut.

9. Reassemble the remaining components as described earlier.

REAR SUSPENSION

The rear suspension units are spring controlled and hydraulically damped. Spring preload can be adjusted by rotating the cam rings at the base of the springs—clockwise to increase preload, counterclockwise to decrease it. Both cams must be indexed on the same detent. The damper units are sealed and cannot be rebuilt. Service is limited to removal and replacement of either an entire unit or the damper alone after the spring has been removed.

Removal

Removal and installation of the rear suspension units is easier if they are done separately. The remaining unit will support the rear of the motorcycle and maintain the correct relationship between the top and bottom mounts.

1. Unscrew the bottom nut and bolt which attach the suspension unit to the swinging arm (**Figure 30**).

NOTE: *On 850 Mark III models, remove the circlip from the lower right bolt before removing the bolt.*

2. Remove the seat and unscrew the knob far enough to expose the top suspension nut so it can be turned with a wrench (**Figure 31**). Pull the suspension unit out of the top mount.

Disassembly

1. Clamp the lower mounting eye in a vise and rotate the preload cam to its softest position.

2. With assistance, compress the top spring and remove the 2 keeper segments (**Figure 32**). Remove the spring from the damper.

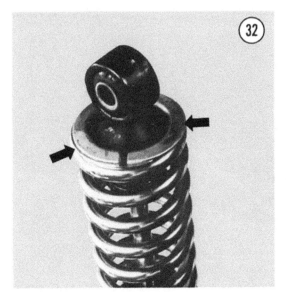

Reassembly and Installation

Reassemble and install the suspension unit by reversing the steps above.

SWINGING ARM

Under normal use and with correct periodic lubrication, the swinging arm bushings have a long service life. However, in time the bushings will wear and must be replaced. Indications of excessive bushing wear are imprecise steering, and a tendency for the motorcycle to pull to one side or the other during acceleration and braking.

Removal

1. Block up the motorcycle so the rear wheel is clear of the ground. Disconnect the rear chain and remove it from the wheel sprocket. Leave it in place on the countershaft sprocket to simplify installation later on. Refer to the section on the rear wheel and brake in this chapter and remove the wheel from the swinging arm.

2. On 750 and 850 Mark II models, place a drip pan beneath the right end of the swinging arm pivot and remove the long screw, end cap, and washers from the arm (**Figure 33**). Allow the oil to drain.

3. Remove the entire right footrest assembly, noting the location of the ground lead for the zener diode (**Figure 34**).

4. Disconnect the chain oiler on the swinging arm (**Figure 35**) and remove the bottom bolts from the rear suspension units.

5. On 750 and 850 Mark II models, remove the lock bolt from the center of the swinging arm

10

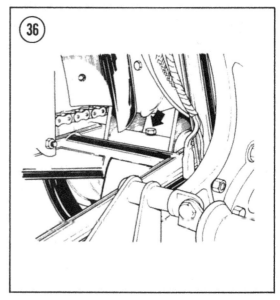

spindle (**Figure 36**). Remove the main front mounting bolt from the engine and screw the nut all the way down on the threads. Screw the bolt into the end of the pivot tube and tighten the nut to lock it in place. Slowly twist the bolt and pull the spindle out (**Figure 37**). Remove the swinging arm, left cover, O-rings, and bearings (**Figure 38**).

6. On 850 Mark III models, unscrew the nuts from the pivot lock pins, remove the rubber caps from the pin bosses, and tap the pins upward to break them loose, using a soft mallet (**Figure 39**). Remove pins. Referring to **Figure 40**, drill a hole in each of the swinging arm end caps and pry out the caps and discard them. Remove

the felt wicks from both sides and carefully tap out the pivot using a soft drift.

Inspection

Thoroughly clean and dry all parts. Slide the pivot bolt back into the swinging arm and check for radial play. If any play can be felt, the bushings and O-rings, and possibly the pivot bolt, should be replaced.

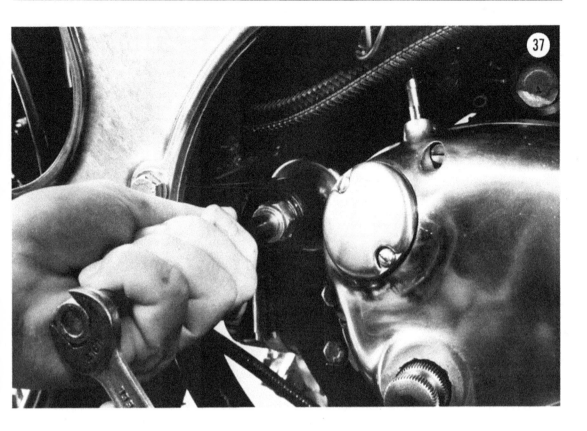

Bushing Replacement
(750 and 850 Mark II)

1. Remove the outer O-rings and press the bushings out of the arm.

2. Install the dust covers on the bushings and press them into the arm.

3. Fit the large O-rings into the dust covers and the small O-rings into the outer recesses around the bushings.

Bushing Replacement
(850 Mark III)

1. Press the bushings out of the arm.

2. Install new seals on new bushings and press them into the arm, making sure they are squarely seated in the bushing bores.

Installation (750 and 850 Mark II)

1. Oil the bushings. Lightly coat the inside face of the left-end cap with grease, and set it in place on the end of the arm pivot (**Figure 41**).

2. Carefully move the swinging arm into place, taking care not to dislodge the left cap until the

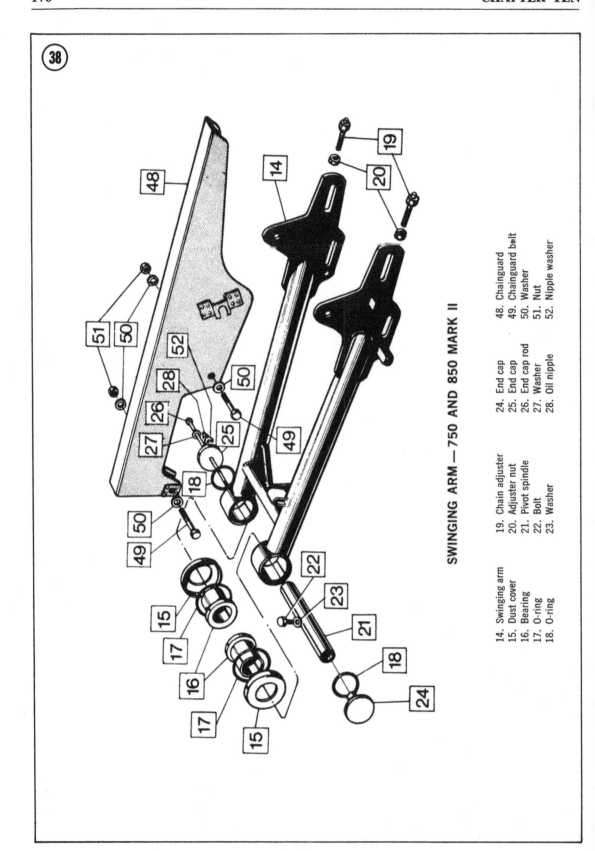

SWINGING ARM — 750 AND 850 MARK II

14. Swinging arm
15. Dust cover
16. Bearing
17. O-ring
18. O-ring
19. Chain adjuster
20. Adjuster nut
21. Pivot spindle
22. Bolt
23. Washer
24. End cap
25. End cap
26. End cap rod
27. Washer
28. Oil nipple
48. Chainguard
49. Chainguard bolt
50. Washer
51. Nut
52. Nipple washer

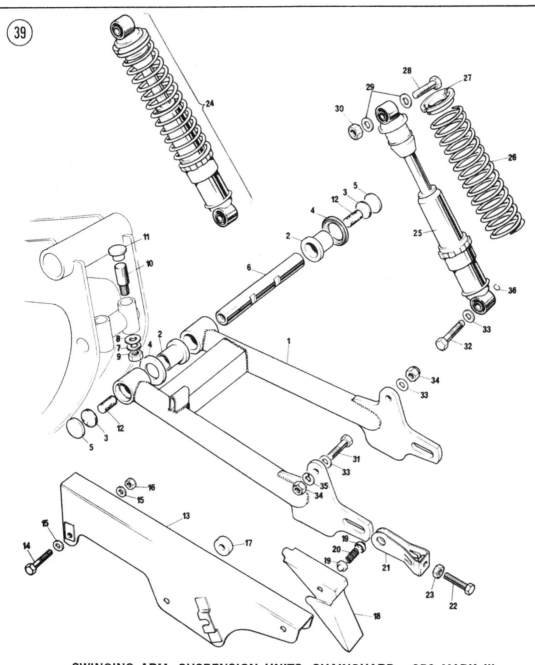

SWINGING ARM, SUSPENSION UNITS, CHAINGUARD — 850 MARK III

1. Swinging arm	10. Cotter	19. Button	28. Top suspension unit bolt
2. Bearing	11. Rubber plug	20. Spring	29. Washer
3. Disc wick	12. Oil wick	21. Chain adjuster	30. Nut
4. Seal assembly	13. Chainguard	22. Screw	31. Bolt
5. Welch plug	14. Guard to swinging arm bolt	23. Nut	32. Bolt
6. Pivot spindle	15. Washer	24. Suspension unit	33. Washer
7. Washer	16. Nut	25. Shock absorber	34. Locknut
8. Fiber washer	17. Chainguard spacer	26. Spring	35. Washer
9. Nut	18. Extension	27. Collar	36. Caliper plate to bolt circlip

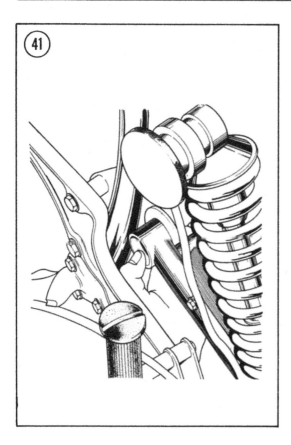

Installation (850 Mark III)

1. Lightly coat the insides of the bushings with oil and set the swinging arm in place.

2. Oil the pivot spindle and start it into the swinging arm. Line up the spindle so that the lock slots are vertical, to the rear. Tap the spindle into the arm, being careful not to rotate it, using a soft mallet and finally a soft drift. When seated, the spindle ends should be equi-distant from the ends of the swinging arm pivot tube.

3. Install the spindle lock pins with the tapered flat facing forward. Tap the pins down into place with a soft mallet to engage the notches in the spindle. Install the washers and screw on the nuts, drawing the pins down into place. Install the rubber plugs in the lock pin bores.

4. Oil the felt wicks and install them in the arm. Install new end plugs, tapping them squarely into the ends of the arm with a soft mallet.

5. Install the chain guard, bottom suspension bolts, the rear wheel, brake caliper, and the chain, reversing the disassembly steps and re-ferring to the appropriate chapters where neces-sary. Do not forget the circlip on the bottom right suspension bolt. Adjust free-play of chain.

> NOTE: *On all models, refer to Chap-ter Seven,* Zener Diode, *and install the diode as described.*

arm is in position and the cap is held by the primary chaincase.

3. Oil the pivot spindle and push it into the arm from the right side. Turn the bolt slowly as you apply pressure. Make sure that the arm is cor-rectly lined up and push the pivot spindle all the way in. Line up the pivot lock bolt hole with the hole in the mounting tube and screw in the bolt. Tighten it securely and unscrew the front engine mount bolt from the spindle. Then reinstall the bolt in the front engine mount.

4. Make sure the right-end O-ring is in place and install the cap and lock bolt. Position the lubrication fitting on the cap so it is near the 12 o'clock position (**Figure 42**) and tighten the bolt securely.

5. Reinstall the chain guard, the bottom sus-pension mounts, the rear wheel, and the chain, reversing the disassembly steps and referring to the appropriate chapters where necessary. Re-connect the rear brake and adjust it. Adjust the free-play of the chain. Fill the spindle with oil as described in Chapter Two.

APPENDIX

SPECIFICATIONS

This chapter contains specifications for the Norton Commando models covered by this book. The tables are arranged in order of increasing engine displacement. Since there are differences between various models of the same engine displacement, be sure to consult the correct table for the motorcycle in question.

750 COMMANDO

ENGINE

GENERAL

Displacement	45 cu. in. (745cc)
Bore	2.875 in. (73mm)
Stroke	3.503 in. (89mm)
Maximum torque	
Standard	48 ft.-lb. at 5,000 rpm
Combat	49 ft.-lb. at 6,000 rpm
Compression ratio	
Standard	9 : 1
Combat	10 : 1

CRANKSHAFT

Main bearing size (all)	1.18 x 2.84 x 0.75 in. (30 x 72 x 19mm)
Big end journal diameter	1.7505-1.7500 in. (44.462-44.450mm)
Permissible end float	0.005-0.015 in. (0.127-0.381mm)

CONNECTING RODS

Big end ID (less bearings)	1.895-1.8955 in. (48.133-48.145mm)
Rod side clearance	0.013-0.016 in. (0.330-0.406mm)
Rod end clearance	Less than 0.001 in. (0.0254mm)

PISTONS

Diameter (bottom of skirt) (standard)	2.8712-2.8703 in. (72.931-72.906mm)
Wrist pin diameter	0.6869-0.6867 in. (17.447-17.442mm)
Rings (type)	Two compression, one oil-control
Ring end gap	
Top compression	0.010-0.012 in. (0.254-0.305mm)
Second compression	0.008-0.012 in. (0.203-0.305mm)

CYLINDER BLOCK

Bore size	
Grade A	2.8746-2.8750 in. (72.959-72.969mm)
Grade B	2.8750-2.8754 in. (72.969-72.979mm)

CYLINDER HEAD

Valve seat angle	45°

CAMSHAFT

Bearing journal diameter	0.8735 in. (22.187mm)

CAMSHAFT BUSHINGS

Fitted ID, left	0.8750 in. (22.225mm)
Fitted ID, right	0.8750 in. (22.225mm)

ROCKER SHAFT

Diameter	0.4988-0.4985 in. (12.694-12.669mm)

(continued)

750 COMMANDO (continued)

VALVE TAPPET CLEARANCES (Engine Cold)

Inlet
Standard	0.006 in. (0.15mm)
Combat	0.008 in. (0.2mm)

Exhaust
Standard	0.008 in. (0.2mm)
Combat	0.010 in. (0.25mm)

VALVES

Inlet
Head diameter	1.490 in. (37.846mm)
Stem diameter (plated area)	0.3115-0.3105 in. (7.912-7.886mm)

Exhaust
Head diameter	1.302 in. (33.0708mm)
Stem diameter (plated area)	0.3115-0.3105 in. (7.912-7.886mm)

VALVE GUIDES

Inside diameter	0.3145-0.3135 in. (7.988-7.962mm)
Outside diameter	0.5015-0.500 in. (12.738-12.725mm)
Washer thickness	0.062 in. (1.574mm)

VALVE SPRINGS

Free length
Inner	1.482 in. (37.642mm)
Outer	1.618 in. (41.097mm)

Fitted length

Inlet inner
Standard	1.197 in. (30.40mm)
Combat	1.259 in. (31.98mm)

Inlet outer
Standard	1.259 in. (31.98mm)
Combat	1.321 in. (33.55mm)

Exhaust inner	1.222 in. (31.04mm)
Exhaust outer	1.284 in. (32.61mm)

INTERMEDIATE TIMING GEAR

Bushing diameter (finished)	0.5627-0.5620 in. (14.292-14.274mm)
Shaft diameter	0.5615-0.5610 in. (14.262-14.249mm)

CAMSHAFT CHAIN

Size	0.375 in. x 0.225 (single-row, endless)
Number of links	38
Ideal adjustment	3/16 in. (4.8mm) up and down on top run of chain

IGNITION TIMING

Fully advanced position	28° BTDC

CONTACT BREAKER

Point gap	0.014-0.016 in. (0.35-0.4mm)

(continued)

11

750 COMMANDO (continued)

SPARK PLUG	
Type	Champion N7Y or equivalent
Gap	0.023-0.028 in. (0.59-0.72mm)
CARBURETORS	
Standard	
Type	Two Amal concentric float 930 30mm
Main jet	
1970	220 with early megaphone muffler
	210 with modified megaphone
	180 with restricted megaphone
1971	220
1972	220 (210 with mute)
Needle jet	
1970	0.107
1971	0.106
1972	0.106
Needle position	
1970	Middle
1971	Middle
1972	Middle (top with mute)
Throttle valve (1970-1972)	3
Combat	
Type	Two Amal concentric float 932 32mm
Main jet	230 (220 with mute)
Needle jet	0.106
Needle position	Middle (top with mute)
Throttle valve	3
AIR FILTER	
Type	Impregnated paper

TRANSMISSION

RATIOS		
Internal		
1st		2.56 : 1
2nd		1.70 : 1
3rd		1.21 : 1
4th		1.00 : 1
Overall		
With 19T gearbox sprocket		
(U.S. 1969-1972)	1st	12.40 : 1
	2nd	8.25 : 1
	3rd	5.90 : 1
	4th	4.84 : 1
With 21T gearbox sprocket		
(U.S. and U.K. 1968)	1st	11.20 : 1
	2nd	7.45 : 1
	3rd	5.30 : 1
	4th	4.38 : 1
With 22T gearbox sprocket		
(optional)	1st	10.71 : 1
	2nd	6.84 : 1
	3rd	5.10 : 1
	4th	4.18 : 1

(continued)

750 COMMANDO (continued)

BUSHINGS

4th gear bushing OD	0.9060-0.9053 in. (23.012-22.995mm)
4th gear bushing ID	0.8145-0.8140 in. (20.688-20.675mm)
4th gear bushing fitted ID	0.8133-0.8120 in. (20.657-20.625mm)
Layshaft bushing ID	1.126-1.124 in. (28.60-28.549mm)
Main shaft second gear bushing fitted ID	0.8125-0.8115 in. (20.637-20.608mm)
Layshaft third gear bushing fitted ID	0.8125-0.81 in. (20.637-20.57mm)
Layshaft first gear bushing fitted ID	0.6885-0.6875 in. (17.488-17.462mm)
Selector spindle bushing, kickstarter case fitted ID	0.6290-0.6285 in. (15-976-15.964mm)
Camplate/quadrant bushing fitted ID	0.5005-0.4995 in. (12.713-12.687mm)
Kickstarter spindle bushing fitted ID	0.675-0.673 in. (17.145-17.094mm)

CAMPLATE PLUNGER SPRING

Free length	1.5 in. (38.1mm)
Spring rate	21 in.-lb.

SELECTOR

Spindle diameter	0.3740-0.3735 in. (9.499-9.486mm)
Fork bore	0.3755-0.3745 in. (8.537-9.525mm)

DRIVE CHAIN   92 pitches, 0.375 x 0.250 in., triple row

REAR CHAIN   99 pitches, 0.400 x 0.380 in.

CLUTCH

Type	Multi-plate, oil bath, diaphragm spring
Number of plates	5 bronze, 4 steel
Friction plate thickness	0.125 in. (0.32mm)
Pushrod length	9.813-9.803 in. (249.250-248.996mm)
Pushrod diameter	0.237-0.232 in. (5.984-5.857mm)
Operating ball diameter	½ in. (12.70mm)
Clutch adjuster length	0.904-0.894 in. (22.962-22.708mm)

CAPACITIES

FUEL TANK

Fastback	3.9 gallons (15 liters)
LR Fastback	4.8 gallons (18 liters)
Roadster	
Fiberglass	2.7 gallons (10 liters)
Steel	3.0 gallons (11 liters)
SS Hi-Rider	2.4 gallons (9 liters)
Interstate	
Fiberglass	6.3 gallons (24 liters)
Steel	6.6 gallons (25 liters)

OIL TANK   6.0 pints (2.8 liters)

GEARBOX   0.9 pints (0.42 liters)

FRONT FORKS   5 ounces (150cc) each leg

PRIMARY CHAINCASE   7 ounces (200cc)

(continued)

11

750 COMMANDO (continued)

CHASSIS

DIMENSIONS

Length	87.5 in. (222cm)
Wheelbase	56.75 in (144cm)
Width	26 in. (66cm)
Ground clearance	6 in. (15cm)
Front brake diameter	7.95 in. (20.32cm)
Rear brake diameter	7 in. (17.78cm)
Braking area, front	18.69 sq. in. (120 sq. cm)
Braking area, rear	13.60 sq. in. (88 sq. cm)
Exhaust system length (flange to muffler tip)	60 in. (154.4cm)
Front fork travel	6 in. (15.2cm)
Turning circle	17 ft. 10 in. (518.16cm)
Seat height	33-34 in. (838-863mm) depending on model

WEIGHT

Gross weight (oil and one gallon of fuel)	
Roadster	422 lb. (191.44 kg)
Interstate	436 lb. (197.75 kg)
Dry weight	
Roadster	395.4 lb. (179.3 kg)
Interstate	410 lb. (186 kg)
Distribution	
Front	45.5 percent
Rear	54.5 percent
Gross vehicle rating	859 lb. (389.6 kg)

850 COMMANDO

ENGINE

GENERAL

Displacement	50.5 cu. in. (828cc)
Bore	3.030 in. (77mm)
Stroke	3.503 in. (89mm)
Maximum torque	56 ft.-lb. at 5,000 rpm
Compression ratio	8.5 : 1

CRANKSHAFT

Main bearing size	1.18 x 2.84 x 0.75 in. (30 x 72 x 19mm)
Big end journal diameter	1.7505-1.7500 in. (44.462-44.450mm)
Permissible end float	0.005-0.015 in. (0.127-0.381mm)

CONNECTING RODS

Big end ID (less bearing)	1.895-1.8955 in. (48.133-48.145mm)
Rod side clearance	0.013-0.016 in. (0.33-0.406mm)
Rod end clearance	Less than 0.001 in. (0.0254mm)

PISTONS

Diameter (bottom of skirt—standard)	3.028-3.071 in. (76.888-76.913mm)
Wrist pin boss ID	0.6869-0.6867 in. (17.447-17.442mm)
Wrist pin diameter	0.6869-0.6867 in. (17.447-17.442mm)
Rings (type)	Two compression, one oil control
Top compression	0.010-0.012 in. (0.254-0.305mm)
Second compression	0.008-0.012 in. (0.203-0.305mm)

CYLINDER BLOCK

Bore size	
Grade A	3.0315-3.0320 in. (76.942-76.954mm)
Grade B	3.0320-3.0325 in. (76.954-76.967mm)

CYLINDER HEAD

Valve seat angle	45°

CAMSHAFT

Bearing journal diameter	0.8735 in. (22.187mm)

CAMSHAFT BUSHINGS

Fitted ID, left and right	0.8750 in. (22.225mm)

PUSHRODS

Assembled length, inlet	8.166-8.130 in. (207.416-206.466mm)
Assembled length, exhaust	7.321-7.285 in. (186.053-185.039mm)

ROCKER SHAFT

Diameter	0.4988-0.4985 in. (12.694-12.669mm)

VALVE TAPPET CLEARANCES (Engine Cold)

Inlet	0.006 in. (0.15mm)
Exhaust	0.008 in. (0.20mm)

(continued)

11

850 COMMANDO (continued)

VALVES

Inlet
    Head diameter   1.490 in. (37.846mm)
    Stem diameter (plated area)   0.3115-0.3105 in. (7.912-7.886mm)
Exhaust
    Head diameter   1.302 in. (33.0708mm)
    Stem diameter (plated area)   0.3115-0.3105 in. (7.912-7.886mm)

VALVE GUIDES

Inside diameter   0.3145-0.3135 in. (7.988-7.962mm)
Outside diameter   0.6265-0.6260 in. (15.90-15.88mm)
Washer thickness   0.062 in. (1.574mm)

VALVE SPRINGS

Free length
    Inner   1.482 in. (37.642mm)
    Outer   1.618 in. (41.097mm)
Fitted length
    Inlet inner   1.197 in. (30.40mm)
    Inlet outer   1.259 in. (31.98mm)
    Exhaust inner   1.222 in. (31.04mm)
    Exhaust outer   1.284 in. (32.61mm)

INTERMEDIATE TIMING GEAR

Bushing diameter (finished)   0.5627-0.5620 in. (14.292-14.274mm)
Shaft diameter   0.5615-0.5610 in. (14.262-14.249mm)

CAMSHAFT CHAIN

Size   0.375 in. x 0.225 in. (single-row, endless)
Number of links   38
Ideal adjustment   3/16 in. (4.8mm) up and down on top run of chain

IGNITION TIMING

Fully advanced position   28° BTDC

CONTACT BREAKER

Point gap   0.014-0.016 in. (0.35-0.4mm)
Bolt thread size   ¼ in. x 26 threads per inch

SPARK PLUG

Type   Champion N7Y or equivalent
Gap   0.023-0.028 in. (0.59-0.72mm)

CARBURETORS

Type   Two Amal concentric float 932 32mm
Main jet   260 (Mark II); 230 (Mark III)
Needle jet   0.106
Needle position   Top
Throttle valve   3½

AIR FILTER

Type   Impregnated paper (Mark II)
                                        Oil-wetted foam (Mark III)

(continued)

850 COMMANDO (continued)

TRANSMISSION

RATIOS

Internal
1st	2.56 : 1
2nd	1.70 : 1
3rd	1.21 : 1
4th	1.00 : 1

Overall

With 19T gearbox sprocket
(U.S. 1969-1972)

	1st	12.40 : 1
	2nd	8.25 : 1
	3rd	5.90 : 1
	4th	4.84 : 1

With 21T gearbox sprocket
(U.S. 1973, U.K. 1972-1973)

	1st	11.20 : 1
	2nd	7.45 : 1
	3rd	5.30 : 1
	4th	4.38 : 1

With 22T gearbox sprocket
(U.K. 1974-1975)

	1st	10.71 : 1
	2nd	6.84 : 1
	3rd	5.10 : 1
	4th	4.18 : 1

BUSHINGS

4th gear bushing OD	0.9060-0.9053 in. (23.012-22.995mm)
4th gear bushing ID	0.8145-0.8140 in. (20.688-20.675mm)
4th gear bushing fitted ID	0.8133-0.8120 in. (20.657-20.625mm)
Layshaft bushing ID	1.126-1.124 in. (28.60-28.549mm)
Main shaft second gear bushing fitted ID	0.8125-0.8115 in. (20.637-20.608mm)
Layshaft third gear bushing fitted ID	0.8125-0.81 in. (20.637-20.57mm)
Layshaft first gear bushing fitted ID	0.6885-0.6875 in. (17.488-17.462mm)
Selector spindle bushing, kickstarter case fitted ID	0.6290-0.6285 in. (15.976-15.964mm)
Camplate/quadrant bushing fitted ID	0.5005-0.4995 in. (12.713-12.687mm)
Kickstarter spindle bushing fitted ID	0.675-0.673 in. (17.145-17.094mm)

CAMPLATE PLUNGER SPRING

Free length	1.5 in. (38.1mm)
Spring rate	21 in.-lb.

SELECTOR

Spindle diameter	0.3740-0.3735 in. (9.499-9.486mm)
Fork bore	0.3755-0.3745 in. (8.537-9.525mm)

DRIVE CHAIN   92 pitches, 0.375 x 0.250 in., triple row

REAR CHAIN   99 pitches, 0.400 x 0.380 in.

CLUTCH

Type	Multi-plate, oil bath, diaphragm spring
Number of plates	5 bronze, 4 steel
Friction plate thickness	0.125 in. (0.32mm)
Pushrod length	9.813-9.803 in. (249.250-248.996mm)
Pushrod diameter	0.237-0.232 in. (5.984-5.857mm)
Operating balance diameter	½ in. (12.70mm)
Clutch adjuster length	0.904-0.894 in. (22.962-22.708mm)

(continued)

11

850 COMMANDO (continued)

CAPACITIES

FUEL TANK

Roadster
  Steel   3.0 gallons (11 liters)

  Hi-Rider   2.4 gallons (9 liters)

Interstate
  Steel   6.6 gallons (25 liters)

OIL TANK   6.0 pints (2.8 liters)

GEARBOX   0.9 pints (0.42 liters)

FRONT FORKS   5 ounces (150cc) each leg

PRIMARY CHAINCASE   7 ounces (200cc)

CHASSIS

DIMENSIONS

Length   87.5 in. (222cm)
Wheelbase   56.75 in. (144cm)
Width   26 in. (66cm)
Ground clearance   6 in. (15cm)
Front brake diameter   7.95 in. (20.32cm)
Rear brake diameter   7 in. (17.78cm)
Braking area, front   18.69 sq. in. (120 sq. cm)
Braking area, rear   13.60 sq. in. (88 sq. cm)
Exhaust system length
  (flange to muffler tip)   60 in. (154.4cm)
Front fork travel   6 in. (15.2cm)
Turning circle   17 ft. 10 in. (518.16cm)
Seat height   33-34 in. (838-863mm) depending on model

WEIGHT

Gross weight (oil and one gallon of fuel)
  Roadster   422 lb. (191.44 kg)
  Interstate   436 lb. (197.75 kg)
Dry weight
  Roadster   395.4 lb. (179.3 kg)
  Interstate   410 lb. (186 kg)
Distribution
  Front   45.5 percent
  Rear   54.5 percent
Gross vehicle rating   859 lb. (389.6 kg)

INDEX

12

12

CLYMER® *Collection Series*

BSA
500 & 650cc UNIT TWINS • 1963-1972

NORTON
750 & 850cc COMMANDOS • 1969-1975

TRIUMPH
500, 650 & 750cc TWINS • 1963-1979

CONTENTS

QUICK REFERENCE DATA

ENGINE TUNE-UP

Spark plug type	
650, 750	Champion N3
500	Champion N4
Spark plug gap	0.020-0.022 in. (0.5-0.6mm)
Breaker point gap	0.012-0.016 in. (0.3-0.4mm)
Valve clearance	
500, 650	0.002 in. (0.05mm) intake 0.004 in. (0.10mm) exhaust
750	0.008 in. (0.20mm) intake 0.006 in. (0.15mm) exhaust
Ignition timing	(See Chapter Two)

ELECTRICAL SYSTEM

Battery	Lucas PUZ 5A or MKZ9E
Fuse	35A
Replacement lamps	
Headlight	370 or 414
Parking	989
Stop/taillight	380
Instruments	281
Turn signals	382

ADJUSTMENTS

Chain play	1¾ in. (45mm)
Clutch lever free play	⅛ in. (3mm)
Front brake lever free play	⅛ in. (3mm)
Rear brake lever free play	½ in. (13mm)

TIRES

Model	Size	Pressure, psi (kg/cm²)
500		
Front	3.25 x 18	24 (1.7)
Rear	3.50 x 18	24 (1.7)
650, 750		
Front	3.25 x 19	24 (1.7)
Rear	4.00 x 18	24 (1.7)

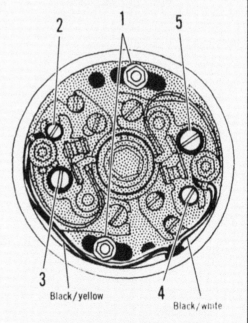

4CA BREAKER PLATE

Black/yellow   Black/white

1. Base plate nuts (to adjust ignition timing)
2. Breaker point gap adjuster (right cylinder)
3. Breaker point gap adjuster (left cylinder)

6CA BREAKER PLATE

Black/yellow   Black/white

1. Base plate nuts (to adjust ignition timing)
2. Breaker point gap adjuster (right cylinder)
3. Lock screw (right cylinder)
4. Breaker point gap adjuster (left cylinder)
5. Lock screw (right cylinder)

RECOMMENDED LUBRICANTS AND CAPACITIES

	Capacity	Type
Engine oil		
500, 750	4.8 U.S. pints (2.25 liters)	SAE 20W/50
650	7.1 U.S. pints (3 liters)	
Gearbox		
500	12.7 oz. (375cc)	EP 90 gear lube
650, 750	16.9 oz. (500cc)	
Primary chaincase		
500	5.0 oz. (150cc)	SAE 20W/50
650, 750	11.8 oz. (350cc)	
Front forks		
500	6.4 oz. (190cc)	SAE 10W/30
650, 750	6.8 oz. (200cc)	

TIGHTENING TORQUES

Item	Torque		Item	Torque	
Flywheel bolts	33 ft.-lb.	4.6 mkg	Clutch center nut		
Connecting rod bolts			750cc, 500cc	70 ft.-lb.	9.7 mkg
500cc	18 ft.-lb.	2.48 mkg	650cc	50 ft.-lb.	7.0 mkg
650, 750cc	18 ft.-lb.	2.48 mkg	Rotor fixing nut	30 ft.-lb.	4.1 mkg
Crankcase junction bolts	13 ft.-lb.	1.8 mkg	Stator fixing nuts	20 ft.-lb.	2.8 mkg
Crankcase junction studs	20 ft.-lb.	2.8 mkg	Primary cover domed nuts	10 ft.-lb.	1.4 mkg
Rocker box bolts—inner ($\frac{5}{16}$ in. dia)	10 ft.-lb.	1.38 mkg	Headlight pivot bolts	10 ft.-lb.	1.4 mkg
Cylinder block nuts	35 ft.-lb.	4.8 mkg	Head race sleeve nut pinch bolt	15 ft.-lb.	2.1 mkg
Cylinder head bolts— ($\frac{3}{8}$ in. dia.)	18 ft.-lb.	2.48 mkg	Stanchion pinch bolts	25 ft.-lb.	3.5 mkg
Cylinder head bolt ($\frac{5}{16}$ in. dia.)	16 ft.-lb.	2.17 mkg	Front wheel spindle cap bolts	25 ft.-lb.	3.5 mkg
Rocker box nuts	5 ft.-lb.	0.7 mkg	Rear brake drum to hub bolts	15 ft.-lb.	2.- mkg
Rocker box bolts	5 ft.-lb.	0.7 mkg			
Rocker spindle domed nuts	22 ft.-lb.	3.0 mkg	Brake cam spindle nuts	20 ft.-lb.	2.8 mkg
Oil pump nuts	5 ft.-lb.	0.7 mkg	Fork cap nut	80 ft.-lb.	11.2 mkg
Kickstarter ratchet pinion nut	45 ft.-lb.	6.3 mkg	Brake disc retaining bolts (750cc)	20 ft.-lb.	2.8 mkg

CHAPTER ONE

GENERAL INFORMATION

This manual provides maintenance and repair information for Triumph 750/650cc and 500cc motorcycles since 1963. The Supplement at the end of this book contains specific information for 1978 and later models. If a procedure is not mentioned in the Supplement it remains the same as on earlier models. Engines used for the various models have many similarities and share many components. The principal differences are in carburetion, ignition, and bore and stroke dimensions. Dimensions and capacities are expressed in English units, which are familiar to U.S. mechanics, as well as in metric units.

Procedures common to different models are combined to avoid duplication. Read the *Service Hints* section to make the work as easy and pleasant as possible.

MANUAL ORGANIZATION

This chapter, in addition to providing general information, also discusses equipment and tools useful both for preventive maintenance and troubleshooting.

Chapter Two explains all periodic lubrication and routine maintenance necessary to keep the motorcycle in proper running condition. Chapter Two also includes recommended tune-up procedures which eliminate the need to constantly consult chapters on the various sub-assemblies.

Chapter Three provides methods and suggestions for quick and accurate diagnosis and repair of problems. Troubleshooting procedures discuss typical symptoms and logical methods to pinpoint trouble spots.

Subsequent chapters describe specific systems such as the engine, transmission, and electrical system. Each chapter provides disassembly, repair, and assembly procedures in simple step-by-step form. If a repair is impractical for the owner-mechanic, it is so indicated. It is usually faster and less expensive to take such repairs to a dealer or competent repair shop.

Some of the procedures in this manual specify special tools. A well-equipped mechanic may find he can substitute similar tools already on hand or can fabricate his own.

The terms NOTE, CAUTION, and WARNING have specific meanings in this manual. A NOTE provides additional information to make a step or procedure easier or clearer. Disregarding a NOTE could cause inconvenience, but would not cause damage or personal injury.

A CAUTION emphasizes areas where equipment damage could result. Disregarding a CAUTION could cause permanent mechanical damage; however, personal injury is unlikely.

A WARNING emphasizes areas where personal injury or even death could result from negligence. Mechanical damage may also occur.

WARNINGS are to be taken seriously. In some cases, serious injury or death have been caused by mechanics disregarding similar warnings.

Throughout this manual, keep in mind 2 conventions. "Front" refers to the front of the bike. The front of any component such as the engine is at that end which faces toward the front of bike. The left and right side refer to a person sitting on the bike facing forward. For example, the clutch lever is on the left side. These rules are simple, but even experienced mechanics occasionally become disoriented.

SERVICE HINTS

Most of the service procedures covered are straightforward and can be performed by anyone reasonably handy with tools. It is suggested, however, that you consider your own capabilities carefully before attempting any operation involving major disassembly of the engine.

Some operations, for example, require the use of a press. It would be wiser to have these performed by a shop equipped for such work, rather than to try to do the job at home with makeshift equipment. Other procedures require precision measurements. Unless you have the skills and equipment required, have a qualified repair shop make the measurements.

Repairs go much faster and easier if the machine is clean before beginning work. There are special cleaners for washing the engine and related parts. Just brush or spray on the cleaning solution, let it stand, then rinse away with a garden hose. Clean all oily or greasy parts with cleaning solvent as they are removed.

WARNING
Never use gasoline as a cleaning agent. It presents an extreme fire hazard. Be sure to work in a well-ventilated area while using cleaning solvent. Keep a fire extinguisher, rated for gasoline and electrical fires, handy at all times.

Special tools are required for some repair procedures. These may be purchased from a dealer (or borrowed if you are on good terms with the service department personnel), or fabricated by a mechanic or machinist, often at considerable savings.

Much of the labor charge for repairs made by dealers is for the removal and disassembly of other parts to reach the defective unit. It is frequently possible to perform the preliminary operations first and then take the defective unit to a dealer for repair.

Once having decided to tackle the job at home, read the entire section in this manual which pertains to it. Make sure to identify the proper section. Study the illustrations and text to gain a good idea of what is involved in completing the job satisfactorily. If special tools are necessary, make arrangements to get them before starting. It is frustrating and time consuming to get partly into a job and then be unable to complete it.

Simple wiring checks are easily made at home, but knowledge of electronics is almost a necessity for performing tests with complicated electronic test gear.

During disassembly of parts keep a few general cautions in mind. Force is rarely needed to get things apart. If parts are a tight fit, as a magneto on a crankshaft, there is usually a tool designed to separate them. Never use a screwdriver to pry apart parts with machined surfaces such as crankcase halves and valve covers. It will mar the surfaces and cause leaks.

Make diagrams wherever similar-appearing parts are found. For instance, case cover screws are often of various lengths. Do not rely on remembering where everything came from—mistakes are costly. There is also the possibility of being sidetracked and not returning to work for days or weeks—during which interval, carefully laid out parts may become disarranged.

Tag all similar internal parts for location and mark all mating parts for position. Record number and thickness of shims as they are removed. Small parts, such as several identical bolts, can be identified by placing in plastic sandwich bags and sealing and labeling them with masking tape.

Wiring should be tagged with masking tape and marked as each wire is removed. Again, do not rely on memory alone.

Disconnect the battery ground cable before working near electrical connections and before disconnecting wires. Never run the engine with the battery disconnected; the electrical system could be seriously damaged.

Protect finished surfaces from physical damage or corrosion. Keep gasoline and cleaning solvent off painted surfaces.

Frozen or very tight bolts and screws can often be loosened by soaking with penetrating oil, then sharply striking the bolt head a few times with a hammer and punch (or screwdriver for screws). Avoid heat unless absolutely necessary, because it may melt, warp, or remove the temper from parts.

Avoid flames or sparks while working near a battery being charged or flammable liquids such as cleaning fluid, brake fluid, or gasoline.

No parts, except those assembled with a press fit, require unusual force during assembly. If a part is hard to remove or install, find out why before proceeding.

Cover all openings after removing parts to keep dirt, small tools, and other foreign matter from falling in.

While assembling 2 parts, start all fasteners and then tighten evenly to avoid warpage.

Clutch plates, wiring connections, and brake shoes and drums should be kept clean and free of grease and oil during assembly.

While assembling parts, be sure that all shims and washers are replaced exactly as they were before disassembly.

Wherever a rotating part butts against a stationary part, look for a shim or washer. Use new gaskets if there is any doubt about the condition of old ones. Generally, apply gasket cement to one mating surface only so the parts may be easily disassembled in the future. A thin coat of oil on gaskets helps them seal effectively.

Heavy grease can be used to hold small parts in place if they tend to fall out during assembly. However, keep grease and oil away from electrical components or brake shoes and drums.

High spots may be sanded off a piston dome with sandpaper, but emery cloth and oil do a much more professional job.

Carburetors are best cleaned by soaking the disassembled parts in a commercial carburetor cleaner. Never soak gaskets and rubber parts in cleaner. Never use wire to clean jets and air passages; they are easily damaged. Use compressed air to blow out the carburetor only if the float has been removed first.

A baby bottle makes a good measuring device for adding oil to forks and transmissions. Obtain one which is graduated in ounces and cubic centimeters.

Take sufficient time to do the job right. Do not forget that a newly rebuilt motorcycle engine must be broken in the same as a new one. Keep rpm within the limits given in the owner's manual.

SAFETY HINTS

Professional motorcycle mechanics can work for years and never sustain a serious injury. Observing a few rules of common sense and safety permits many safe enjoyable hours servicing your own machine. You can hurt yourself or damage the bike if you ignore the following precautions.

1. Never use gasoline as a cleaning solvent.

2. Never smoke or use a torch in the vicinity of flammable liquids such as cleaning solvent in open containers.

3. Never smoke or use a torch in an area where batteries are being charged. Highly explosive hydrogen gas is formed during the charging process.

4. If welding or brazing is required on the machine, remove the fuel tank to a safe distance (at least 50 feet away). Welding on gas tanks requires special safety procedures and must be performed only by someone skilled in the process.

5. Use the proper sized wrenches to avoid damage to nuts and personal injury.

6. While loosening a tight or stuck nut, be guided by what would happen if the wrench should slip.

7. Keep the work area clean and uncluttered.

8. Wear safety goggles during all operations involving drilling, grinding, or use of a chisel.

9. Never use worn tools.

10. Keep a fire extinguisher handy and be sure it is rated for gasoline and electrical fires.

PARTS REPLACEMENT

When ordering parts, always order by engine and frame number. Write the numbers down and carry them in your wallet. Compare new parts to old before purchasing. If they are not alike, have the parts clerk explain the difference.

TOOLS

Tool Kit

Most new bikes are equipped with fairly complete tool kits. These tools are satisfactory for most small jobs and emergency roadside repairs.

Shop Tools

For proper servicing, an assortment of ordinary hand tools is needed. As a minimum, these include the following.

a. Combination wrenches
b. Socket wrenches
c. Plastic mallet
d. Small hammer
e. Snap ring pliers
f. Pliers
g. Phillips screwdrivers
h. Slot (common) screwdrivers
i. Feeler gauges
j. Spark plug gauge
k. Spark plug wrench
l. Dial indicator

Special tools necessary are designated in the chapters covering the particular repair in which they are used.

Electrical system servicing requires a multimeter, ohmmeter, or other device for determining continuity, and a hydrometer for battery equipped machines.

Advanced tune-up and troubleshooting procedures require the additional tools listed below.

1. *Timing gauge* (**Figure 1**). Piston position may be determined by screwing this instrument into the spark plug hole. The tool shown costs approximately $20, and is available from larger dealers and mail order houses. Less expensive ones, which utilize a vernier scale instead of a dial indicator, are also available. They are satisfactory but are not quite so quick and easy to use.

2. *Hydrometer* (**Figure 2**). This instrument measures a battery's state of charge and tells much about battery condition. Such an instru-

ment is available at any auto parts store and through most larger mail order outlets. A satisfactory one costs approximately $3.

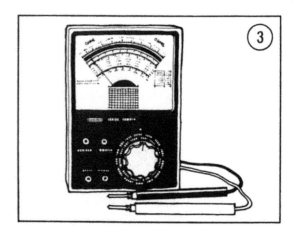

3. *Multimeter or VOM* (**Figure 3**). This instrument is invaluable for electrical system troubleshooting and service. A few of its functions may be duplicated by locally fabricated substitutes, but for the serious hobbyist, it is a must. Its uses are described in the applicable sections of this manual. Prices start at around $10 at hobbyist electronics stores and mail order outlets.

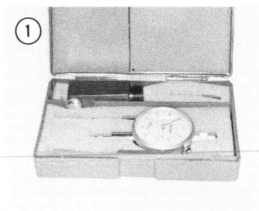

4. *Compression gauge* (**Figure 4**). An engine with low compression cannot be propery tuned and will not develop full power. A compression gauge measures engine cylinder pressure. The one shown has a flexible stem which enables it to reach cylinders where there is little clearance between the cylinder head and frame. Inexpensive ones start at approximately $3 and are available at auto accessory stores or by mail order from large catalog order firms.

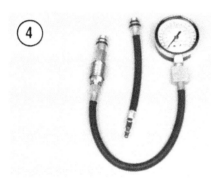

5. *Impact driver* (**Figure 5**). This tool might have been designed with the motorcyclist in mind. It makes removal of engine cover screws easy and eliminates damaged screw slots. Good ones cost approximately $12 at larger hardware stores.

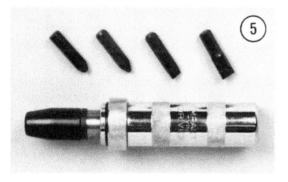

6. *Ignition gauge* (**Figure 6**). This tool measures point gap. It also has round wire gauges for measuring spark plug gap.

EXPENDABLE SUPPLIES

Required expendable supplies include grease, oil, gasket cement, wiping rags, cleaning solvent, and distilled water. Ask a Triumph dealer for the special locking compounds, silicone lubricants, and commercial chain lube products. Solvent is available at most service stations. Distilled water for the battery may be obtained at almost any supermarket.

SERIAL NUMBERS

The model serial number must be available for registration and parts ordering. These numbers can be permanently recorded by placing a sheet of paper over the imprinted area and rubbing with the side of a pencil. Some motor vehicle registration offices will accept such evidence in lieu of inspecting the bike in person.

NOTE: If you own a 1978 or later model, first check the Supplement at the back of the book for any new service information.

CHAPTER TWO

LUBRICATION AND MAINTENANCE

Regular maintenance is the best guarantee of a trouble-free motorcycle. An afternoon spent cleaning and adjusting can minimize costly mechanical problems and unexpected breakdowns on the road.

This chapter describes all required preventive maintenance and procedures for engine tune-up. Anyone with average mechanical ability can perform the procedures.

MAINTENANCE/LUBRICATION INTERVALS

Maintenance and lubrication intervals are shown in **Table 1** for bikes used on the highway and in **Table 2** for off-road use. Because of dust and dirt, off-road usage requires much more frequent service for trouble-free performance.

Figure 1 shows lubrication points and gives the correct grade or type of lubricant required for 750/650cc models. **Figure 2** gives similar information for 500cc models. Information in **Table 3** lists recommended lubricants. Selection need not be limited to these brands.

In addition to the items listed in the pertinent table (Table 1 or 2), the following parts should be carefully inspected after 24 months.

1. Front (and rear, if so equipped) brake hydraulic hose (disc brake model)

2. Brake cable
3. Brake light switches
4. Brake master cylinder(s) and primary and secondary filler caps (disc brake models)
5. Disc brake caliper piston seal
6. Carburetor rubber dust caps
7. Fuel lines

TOOLS

The basic tools needed are listed in Chapter One. In addition, equipment required for a complete tune-up include a static timing light, a strobe light, dwell tachometer, carburetor float gauge, and sets of flat and round feeler gauges.

OIL CHANGING

Engine oil should be drained while the engine is warm.

1. Remove the drain plug beneath the engine (**Figure 3** for 750/650cc models; **Figure 4** for 500cc models). Allow the oil to drain for approximately 10 minutes.

2. The oil tank must also be drained. Remove the plug from the bottom of the tank and unscrew the banjo fitting to disconnect the oil feed pipe. On older models, a side cover may have to be removed to gain access to the oil tank. On

Table 1   LUBRICATION AND MAINTENANCE   (STREET MODELS)

Service required	Months	2	4	6	8	12	16	18	20	24
	Miles	1,000	2,000	3,000	4,000	6,000	8,000	9,000	10,000	12,000
	Km	1,600	3,200	4,800	6,400	9,600	12,800	14,400	16,000	19,200
Engine										
Check level in oil tank*										
Change engine oil			X		X	X	X		X	X
Change oil filter element			X		X	X	X		X	X
Clean oil screen filter			X							
Service spark plugs			X							
Service contact breaker points			X							
Oil point cam felt wick			X¹			X²				
Adjust ignition timing			X							
Check ignition primary and secondary cables			X							
Adjust valve tappet clearances			X							
Adjust cam chain			X							
Service air cleaner			X							
Adjust carburetors**										
Check throttle valve operation					X					
Clutch										
Adjust clutch					X					
Battery										
Service battery			X							
Fuel System										
Clean fuel valve filter			X							
Check fuel tank and fuel lines					X					

(continued)

*Every 250 miles (400km)   **As required   1. Late models   2. Early models

Table 1   LUBRICATION AND MAINTENANCE (STREET MODELS)   (continued)

Months or miles, whichever comes first

Service required	2	4	6	8	12	16	18	20	24
Months / Miles / Km	1,000 / 1,600	2,000 / 3,200	3,000 / 4,800	4,000 / 6,400	6,000 / 9,600	3,000 / 12,800	9,000 / 14,400	10,000 / 16,000	12,000 / 19,200
Steering and Front Suspension									
Check steering head bearings									X
Check steering handle lock									X
Check handle bar holder				X					
Check front fork top plate					X				
Steering and Front Suspension									
Check front fork bottom case									X
Change front fork oil									X
Rear Suspension									
Grease swing arm pivot	X								
Check swing arm			X						
Check rear suspension mounting bolts					X				
Wheels and Brakes									
Check front and rear wheel spokes		X							
Check front and rear wheel rims and hubs								X	
Check front and rear wheels, bearings, and axles								X	
Check front and rear tires	X								
Check front brake caliper and pad linings					X				
Check front brake line	X								
Check brake fluid level	X								

(continued)

Table 1   LUBRICATION AND MAINTENANCE   (STREET MODELS)   (continued)

Service required	Months	2	4	6	8	12	16	18	20	24
	Miles	1,000	2,000	3,000	4,000	6,000	8,000	9,000	10,000	12,000
	Km	1,600	3,200	4,800	6,400	9,600	12,800	14,400	16,000	19,200
Wheels and Brakes (continued)										
Check and adjust brake pedal			X							
Check rear brake shoe linings							X			
Check rear brake stopper arm			X							
Lubricate control cables		X								
Chassis and Final Drive										
Check oil level in primary chaincase (chain oiler)*										
Check frame					X					
Check exhaust system					X					
Gearbox						X				
Service and adjust final drive chain		X								
Check final drive and driven sprockets		X								
Change oil in primary chaincase		X								
Lights and Accessories										
Check lights and switches			X							
Check horn			X							
Check speedometer and tachometer			X							

Months or miles, whichever comes first

*Every 250 miles (400km)

2

Table 2   LUBRICATION AND MAINTENANCE   (OFF-ROAD MODELS)

Service required	Months	2	4	6	8	12	16	18	20	24
	Miles	500	1,000	2,000	4,000	6,000	8,000	9,000	10,000	12,000
	Km	800	1,600	3,200	6,400	9,600	12,800	14,400	16,000	19,200
Engine										
Check level in oil tank*										
Change engine oil		X								
Change oil filter element		X								
Clean oil screen filter		X								
Service spark plugs		X								
Service contact breaker points			X							
Oil point cam felt wick		X				X				
Adjust ignition timing			X							
Check ignition primary and secondary cables			X							
Adjust valve tappet clearances			X							
Adjust cam chain			X							
Service air cleaner		X								
Adjust carburetors**										
Check throttle valve operation		X			X					
Clutch										
Adjust clutch					X					
Battery										
Service battery		X								

(continued)

Months or miles, whichever comes first

*Every 100 miles (160km)   **As required

Table 2   LUBRICATION AND MAINTENANCE   (OFF-ROAD MODELS)   (continued)

Months or miles, whichever comes first

Service required	Months	2	4	6	8	12	16	18	20	24
	Miles	500	1,000	2,000	4,000	6,000	8,000	9,000	10,000	12,000
	Km	800	1,600	3,200	6,400	9,600	12,800	14,400	16,000	19,200
Fuel System										
Clean fuel valve filter		X								
Check fuel tank and fuel lines		X								
Steering and Front Suspension										
Check steering head bearings		X								
Check steering handle lock										X
Check handle bar holder					X					
Check front fork top plate						X				
Check front fork bottom case										X
Change front fork oil						X				
Rear Suspension										
Grease swing arm pivot		X								
Check swing arm		X		X		X		X		X
Check rear suspension mounting bolts		X		X		X		X		X
Wheels and Brakes										
Check front and rear wheel spokes		X								
Check front and rear wheel rims and hubs		X		X		X		X		X
Check front and rear wheels, bearings, and axles		X		X		X	X	X		X
Check front and rear tires		X		X		X	X		X	X

(continued)

Table 2   LUBRICATION AND MAINTENANCE   (OFF-ROAD MODELS)   (continued)

Service required	Months	2	4	6	8	12	16	18	20	24
	Miles	500	1,000	2,000	4,000	6,000	3,000	9,000	10,000	12,000
	Km	800	1,600	3,200	6,400	9,600	12,800	14,400	16,000	19,200
Wheels and Brakes (continued)										
Check and adjust brakes		X								
Check rear brake shoe linings							X			
Check rear brake stopper arm			X							
Lubricate control cables		X								
Chassis and Final Drive										
Check oil level in primary chaincase (chain oiler)*										
Check frame			X							
Check exhaust system					X					
Gearbox						X				
Service and adjust final drive chain		X								
Check final drive and driven sprockets		X								
Change oil in primary chaincase		X								

*Every 250 miles (400km)

NOTE: Off-road motorcycles are subjected to extremes of dust, heat, water and general abuse. Ideally all critical areas should be checked after each day of riding. Table 2 is a general guide for moderate use and represents the maximum limits of service.

Months or miles, whichever comes first

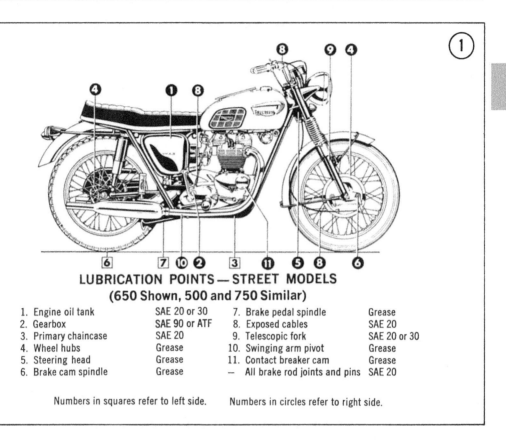

LUBRICATION POINTS — STREET MODELS
(650 Shown, 500 and 750 Similar)

1. Engine oil tank	SAE 20 or 30	7. Brake pedal spindle	Grease
2. Gearbox	SAE 90 or ATF	8. Exposed cables	SAE 20
3. Primary chaincase	SAE 20	9. Telescopic fork	SAE 20 or 30
4. Wheel hubs	Grease	10. Swinging arm pivot	Grease
5. Steering head	Grease	11. Contact breaker cam	Grease
6. Brake cam spindle	Grease	— All brake rod joints and pins	SAE 20

Numbers in squares refer to left side.   Numbers in circles refer to right side.

Table 3   RECOMMENDED LUBRICANTS

Unit	Mobil	Castrol	Exxon	Shell	Texaco
Engine and Primary Chaincase	Mobiloil Super	Castrol GTX or Castrol XL 20/50	Uniflo	Shell Super Motor Oil	Havoline Motor Oil 20W/50
Gearbox	Mobilube GX 90	Castrol Hypoy	Exxon Gear Oil GX 90/140	Shell Spirax 90 EP	Multigear Lubricant EP 90
Telescopic Fork	Mobiloil Super	Castrolite	Uniflo	Shell Super Motor Oil	Havoline Motor Oil 10W/30
Wheel Bearings, Swinging Fork and Steering Races	Mobilgrease MP or Mobilgrease Super	Castrol LM Grease	Exxon Multi-purpos Grease H	Shell Retinax A	Marfak All Purpose
Freeing Rusted Parts	Mobil Handy Oil	Castrol Penetrating Oil	Exxon Penetrating Oil	Shell Easing Oil	Graphited Penetrating Oil

NOTE: The above lubricants are recommended for all operating temperatures above 0°F (−18°C).
    Approval is given to lubricants marketed by companies other than those listed provided they meet the API Service MS Performance level.

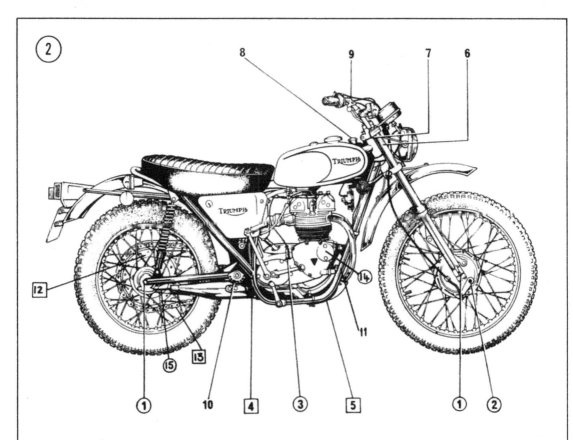

LUBRICATION POINTS — OFF-ROAD MODELS
(500 Shown; 650 and 750 Similar)

Numbers in circles refer to right side of machine

Numbers in squares refer to left side of machine

Illustration No.	Description	SAE Oil Grade
1	Wheel bearings	Grease
2	Brake cam	Grease
3	Gearbox	90
4	Brake pedal	Grease
5	Primary chaincase	As engine
6	Front forks	ATF
7	Steering head	Grease
8	Oil reservoir	20W/50
9	Cables and controls	20
10	Swing arm	Grease
11	Oil filter	—
12	Rear brake cam	Grease
13	Rear chain	Grease
14	Contact breaker	Contact breaker lube
15	Speedometer drive	Grease

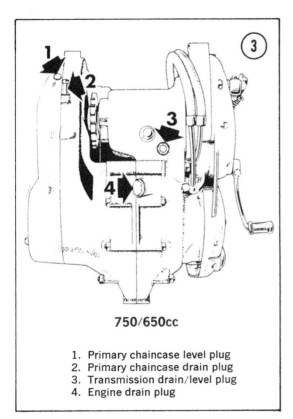

750/650cc

1. Primary chaincase level plug
2. Primary chaincase drain plug
3. Transmission drain/level plug
4. Engine drain plug

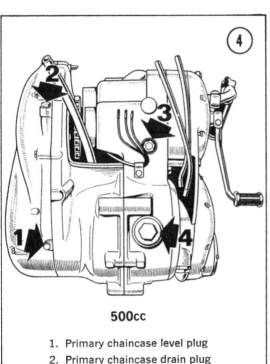

500cc

1. Primary chaincase level plug
2. Primary chaincase drain plug
3. Gearbox drain and level plug
4. Sump drain and filter plug

late models, with the oil carried in the frame, remove the oil filler cap and unscrew the drain plug from the center of the plate which is bolted to the bottom of the frame.

3. While the oil is draining, clean the filters in the sump drain plug and the oil tank. On oil-in-frame models, unscrew the 4 nuts from the reservoir plate and remove it along with the gaskets and filter. Make sure the gaskets are reinstalled with the filter. Use kerosene or solvent to clean the filters and fitting and allow to air dry.

4. Select an engine oil from the brands (or equivalent) shown in Table 3. The oil for specific models is given in Appendix I. Change oil the first time for a new bike at 600 miles (960 km) and every 2,000 miles (3,200 km) thereafter (or every 4 months, which ever occurs first). Off-road use dictates more frequent intervals based on use and severity.

5. Replace the filters and drain plugs. Be careful to reinstall the fiber washer under the oil tank filter. Do not overtighten the oil feed pipe fitting or leakage may occur.

6. Refill the oil tank to 1½ in. (38.1mm) below the filler cap. Do not overfill. On oil-in-frame models, use the dipstick to determine correct oil level.

OIL PRESSURE

Oil pressure is checked with the engine running by replacing the plug adjacent to the oil pressure release valve with an oil pressure gauge. Oil pressure should read 80 psi (5.6 kg/cm^2) with a cold engine, 65-80 psi (4.6-5.6 kg/cm^2) at a normal running speed, and 20-25 psi (1.4-1.8 kg/cm^2) at fast idle with the engine hot.

Low oil pressure may be due to any of the following: a dirty oil pressure release valve; incorrect clearance in the valve body; low oil level; oil not being returned to the oil tank; dirty sump or oil tank filter; malfunctioning oil pump; no oil getting to the pump; dirty passages in the timing cover; blocked passages from the crankcase to the oil pump and oil pipe junction block; worn oil seal in the timing cover; worn big end bearings; or long periods of running the engine at low speeds and high temperature.

OIL PRESSURE
RELIEF VALVE

The oil pressure release valve (**Figures 5 and 6**) is located at the front of the engine on the right side, adjacent to the timing cover.

This valve should be cleaned occasionally and its main spring should be measured. Replace if its free length is less than ½ in. (12.7mm).

To remove the valve:

1. Unscrew the hex nut from the crankcase.

2. Remove the cap nut to gain access to the spring and piston. Clean all parts in solvent and inspect for wear or damage. The piston should be free from scoring and the valve body filter free from damage. See Figure 5 or 6.

3. Replace both fiber washers with new ones while reassembling, and install the piston with its open end facing the spring and cap.

4. A valve with a pressure indicator was used on early models (Figure 6). The removal procedure is the same. The brass shaft nut must be unscrewed to remove the springs. Reassemble the

unit by pressing the rubber seal and retaining cup into the valve cap. Use the indicator shaft as a guide to fit the sealing rubber over the stub on the inside of the valve cap. Fit the indicator shaft correctly in place, replace the springs and the brass nut. Replace the 2 fiber washers with new ones, screw the assembly together, and reinstall in the crankcase.

CONTACT BREAKER ASSEMBLY

1. The point cam on early models is lubricated with a few drops of light oil on the felt wick situated against the cam. This operation should be performed every 6,000 miles. Do not overlubricate. See **Figures 7 and 8**.

2. Mark the contact breaker plate and housing so the plate can be replaced in the same position later. Remove by unscrewing the 2 hex pillar bolts.

3. Oil the advance mechanism lightly, replace the contact breaker plate, and reset the ignition timing if necessary.

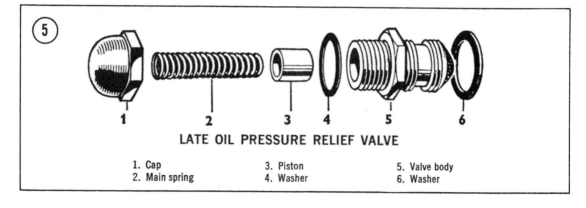

LATE OIL PRESSURE RELIEF VALVE

1. Cap	3. Piston	5. Valve body
2. Main spring	4. Washer	6. Washer

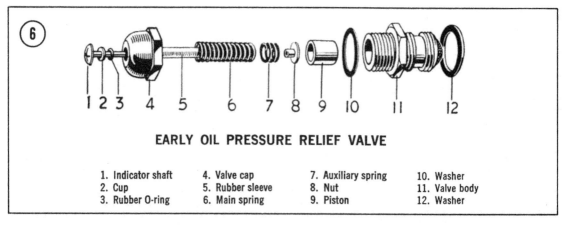

EARLY OIL PRESSURE RELIEF VALVE

1. Indicator shaft	4. Valve cap	7. Auxiliary spring	10. Washer
2. Cup	5. Rubber sleeve	8. Nut	11. Valve body
3. Rubber O-ring	6. Main spring	9. Piston	12. Washer

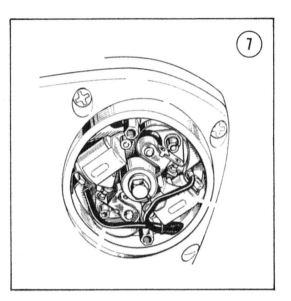

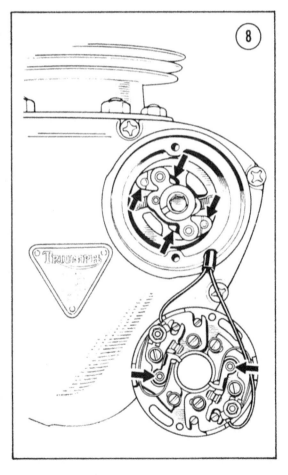

GEARBOX LUBRICATION

Gearbox lubricant should be changed every 6,000 miles (9,600 km) using a type of oil specified in Table 3. Drain old oil while the engine is warm by removing the plug on the bottom of the gearbox. Refill through the filler plug next to the kickstarter shaft. Remove the level plug from the drain plug and refill until oil drips from the level plug hole. See **Figure 9**.

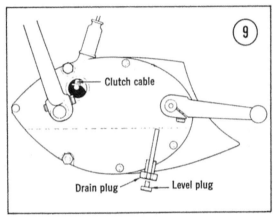

PRIMARY CHAINCASE LUBRICATION

750/650cc Models (1963-1971)

The primary chaincase (**Figure 10**) holds ⅝ pint (350cc) of oil. This oil should be changed every 1,000 miles (1,600 km). Remove the plug at the bottom of the primary case to drain the oil. On early models, the left footpeg may have to be loosened and swung downward to gain access to the plug. Refill the case through the plug next to the cylinder head or through the clutch adjustment plug.

The primary chain is lubricated by an oil feed pipe which directs dripping oil onto the front engine sprocket and the chain. It should occasionally be checked for blockage by removing the case cover and clip securing the feed pipe. Be sure the feed pipe is replaced so it is parallel to the chain's top run and is securely clipped in place.

750/650cc Models (1972 on)

The oil level in the primary chaincase (**Figure 11**) is maintained automatically by the

4. On 1972 and later models, the 2 lubricating wicks should receive about 3 drops of clean engine oil every 2,000 miles (3,000 km).

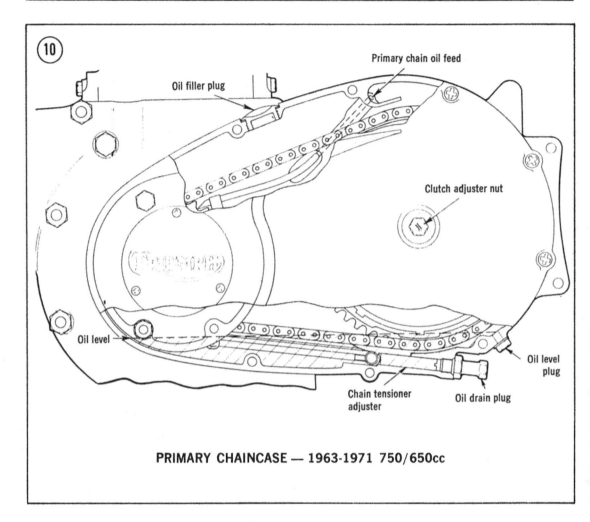

PRIMARY CHAINCASE — 1963-1971 750/650cc

crankcase breathing system. Oil enters the chaincase through the drive-side roller bearing and returns to the crankcase through 3 level holes. Oil—¾ pint (150cc)—must be added to the chaincase only when it has been completely drained for some service procedure, or when the oil has been lost during tension adjustment of the primary chain.

500cc Models

The 500cc models has no primary chain oil feed. The oil level plug is at the front of the case (**Figure 12**). The chaincase capacity is ½ pint (300cc) of SAE 20W oil.

On 1972 and later models, the primary chaincase must be "primed" with ¼ pint (150cc) of oil after it has been drained, or if oil was lost during adjustment of the primary chain tension.

REAR CHAIN LUBRICATION

1963-1971 Models

Rear chain lubrication is identical for 750/650cc and 500cc models except for Step 4.

1. Automatic chain lubricating devices are used on all Triumphs through 1971. Early models had a metering jet in the primary chaincase. Late models use a metering system in an oil junction block in the neck of the oil tank. See **Figure 13**. The screw may be tightened to reduce oil flow and loosened to increase it. Early machines require manual chain lubrication every 250 miles (400 km).

2. If the metering system is not used, a good grade of chain lubricant should be applied according to directions.

3. The chain should also be removed and cleaned in solvent occasionally. Disconnect the

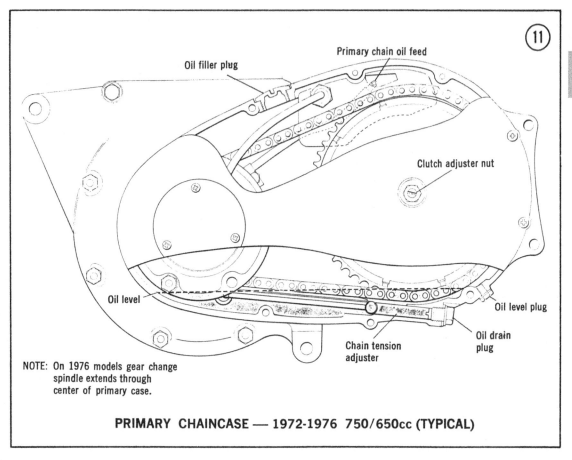

Oil filler plug

Primary chain oil feed

(11)

2

Clutch adjuster nut

Oil level

Oil level plug

Oil drain plug

Chain tension adjuster

NOTE: On 1976 models gear change spindle extends through center of primary case.

PRIMARY CHAINCASE — 1972-1976 750/650cc (TYPICAL)

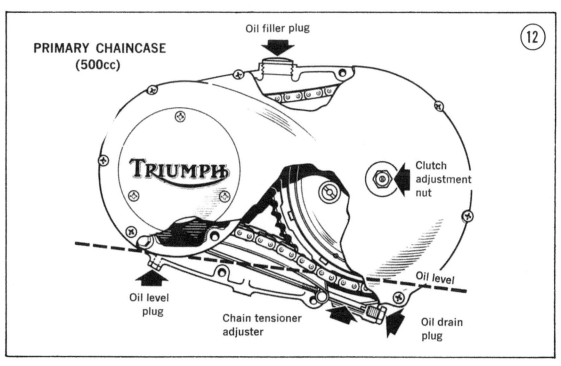

Oil filler plug

(12)

PRIMARY CHAINCASE (500cc)

TRIUMPH

Clutch adjustment nut

Oil level

Oil level plug

Chain tensioner adjuster

Oil drain plug

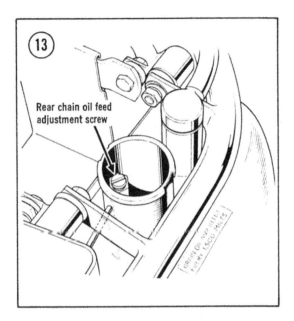

Rear chain oil feed adjustment screw

3. Lubricate the chain by immersing it in a pan of grease which has been melted by being placed over a pan of boiling water.

> **WARNING**
> *Melting grease over an open flame is hazardous and not recommended. The rising smoke ignites easily.*

4. When grease has cooled, remove the chain from the bath and wipe off excess grease with a clean cloth.

5. Reinstall chain on motorcycle with the closed end of the master link clip facing the direction of chain travel.

DRIVE CHAIN ADJUSTMENT

Typical (Except TR5T)

1. Place the motorcycle on its center stand. With the transmission in neutral, rotate the rear wheel and check the play in the chain midway between the sprockets. At its slackest point, the movement should be about 1¾ in. (44.45mm).

2. If adjustment is required, loosen the rear axle nut and the brake anchor bolt on the brake backing plate. Tighten or loosen both axle adjusters as appropriate until the free movement in the chain is correct.

3. Sight along the chain from the rear of the motorcycle to check the alignment of the rear wheel. Misalignment will be apparent with the chain angling either right or left where it ingages the sprocket at the top. Correct any misalignment by turning the adjusters in opposite directions. Then recheck the chain adjustment, correct it if necessary, and tighten adjuster locknuts, brake anchor bolt, and axle nut.

master link and remove the chain. If a piece of old chain is available, connect it to the chain on the machine before it is removed. Leave the old chain around the gearbox sprocket to make chain replacement easier.

4. The chain may be inspected for excessive wear by scribing marks on a flat surface 12.5 in. (317.5mm) apart for 750/650cc models and 12 in. (305mm) apart for 500cc models. With the chain compressed to its minimum free length, the 2 marks should be even with pivot pins and exactly 20 links apart. Extended, 20 links should not exceed the above by more than ¼ in. (6.35mm).

5. When the chain is replaced, the closed end of the master link clip should face in the direction of normal chain rotation.

1972-1976 Models

1. On late models not equipped with a chain oiler, periodically inspect the chain for dryness and lubricate with an oil soaked brush. Wipe off excess oil.

2. Every 1,000 miles (1,600km), remove the chain from the motorcycle and clean thoroughly in kerosene or solvent using a wire brush to loosen grit and grease deposits. Hang up the chain and allow it to air dry. Inspect for wear (see above) and replace if necessary.

4. Double check the chain adjustment by rotating the rear wheel until the chain is at its tightest point. Take the motorcycle off its center stand. With the motorcycle on its wheels, the free play in the chain should be ¾ in. (20mm) at the tightest point. If the free play in the chain cannot be adjusted within limits, and if inspection of the chain indicates that its condition is satisfactory, the sprockets should be checked for excessive wear and replaced if necessary.

TR5T Model

1. Support the rear of the motorcycle so the wheel is off the ground. Rotate the rear wheel to find the tightest point in the chain. The free movement of the chain, measured midway between the sprockets, should be one inch (25mm), approximately.

2. To adjust the chain, loosen the nuts on both ends of the swing arm pivot and tap the cam plates with a soft mallet to loosen them from the locating dowels. Loosen brake rod adjuster.

3. Rotate the cam plates (**Figure 14**) clockwise or counterclockwise until the free movement in the chain is correct. The adjuster cams must be located correspondingly; e.g., if on the right cam, the second hole from the top is located on the forward pin, the second hole from the top of the left cam must also be located on its forward pin.

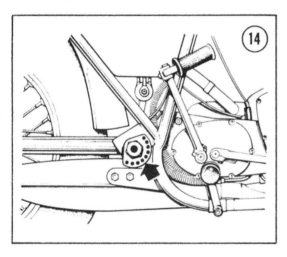

4. Tighten the swing arm pivot nuts and adjust the rear brake.

FRONT WHEEL BEARINGS

Removal/Installation (Typical, Except Conical Hub and Disc Brake)

See the following section if the front wheel has a conical hub; otherwise, follow procedures below.

1. With the front wheel removed (Chapter Eight), pull out the brake anchor plate. Unscrew the retaining ring (which has left-hand threads). See **Figure 15**.

2. Drive the right bearing out from the left side using the axle as a tool. Remove the backing ring and inner retaining disc (early models) or the combined part (late models).

3. Remove the circlip and drive out the left bearing from the right side, again using the axle as a driver. The inner and outer grease retainers will come out at the same time.

4. Clean all parts thoroughly with cleaning solvent or kerosene. Blow dry with compressed air. Inspect parts for wear or damage; replace if necessary. Repack bearings with hi-temp bearing grease and soak felt washers in motor oil.

5. Replace the left inner grease retainer, bearing, and dust cap after greasing.

6. Replace the circlip and use the shouldered end of the axle to drive the bearing and retainer up against the circlip. Pull out the axle and reinsert it with the other end in first.

7. Replace the right grease retainer disc and backing ring.

8. Drive the right bearing into position.

9. Replace the retainer ring (left-hand threads). Tap the axle from the left to bring its shoulder against the right bearing. Be sure that parts turn freely without discernible free play.

Removal/Installation (Conical Hub)

1. To remove bearings from a front wheel with a conical hub, begin by unscrewing the speedometer drive ring (left-hand thread) from the right side of the wheel and the bearing retaining ring (right-hand thread) from the left side. Use a drift the size of the one shown in **Figure 16** to knock out the inside spacer tube and one of the bearings. This can be accomplished from either side of the wheel.

2. Use the drift to remove the bearing from the spacer tube.

3. Use the tube and the drift to knock the other bearing out of the wheel.

CAUTION
If a correctly-sized drift is not used, the spacer tube is likely to be damaged and require replacement.

4. Clean all parts thoroughly with cleaning solvent or kerosene. Blow dry with compressed air.

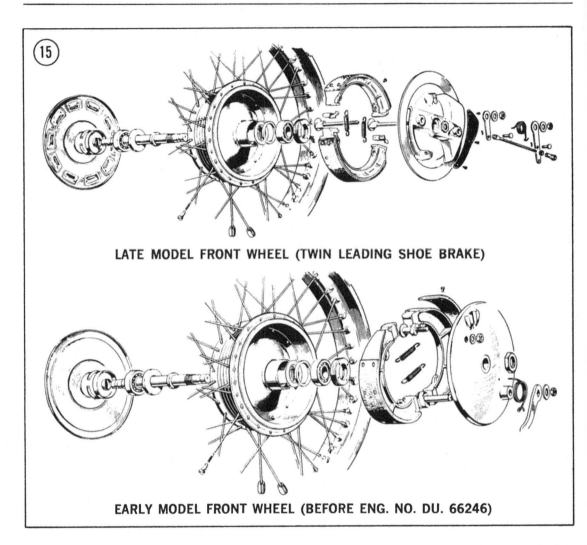

LATE MODEL FRONT WHEEL (TWIN LEADING SHOE BRAKE)

EARLY MODEL FRONT WHEEL (BEFORE ENG. NO. DU. 66246)

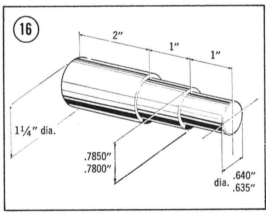

Inspect parts for wear or damage; replace if necessary. Repack bearings with grease; soak felt washers in oil.

5. Install one bearing on the spacer tube.

6. Press the assembly into the hub from the left side until the bearing contacts the grease retainer. Apply force only to outer bearing race.

7. Install the retaining ring and tighten.

8. Install the right-side bearing and the speedometer drive ring and install the remaining pieces in reverse order of removal. Be sure that parts turn freely without discernible free play.

Removal/Installation (Disc Brake)

Refer to **Figure 17** for this procedure.

1. Remove the front wheel, using the procedure given in Chapter Eight.

> NOTE: *Do not apply the front brake while the front wheel is removed. To do so could cause displacement of the pistons in the caliper cylinders.*

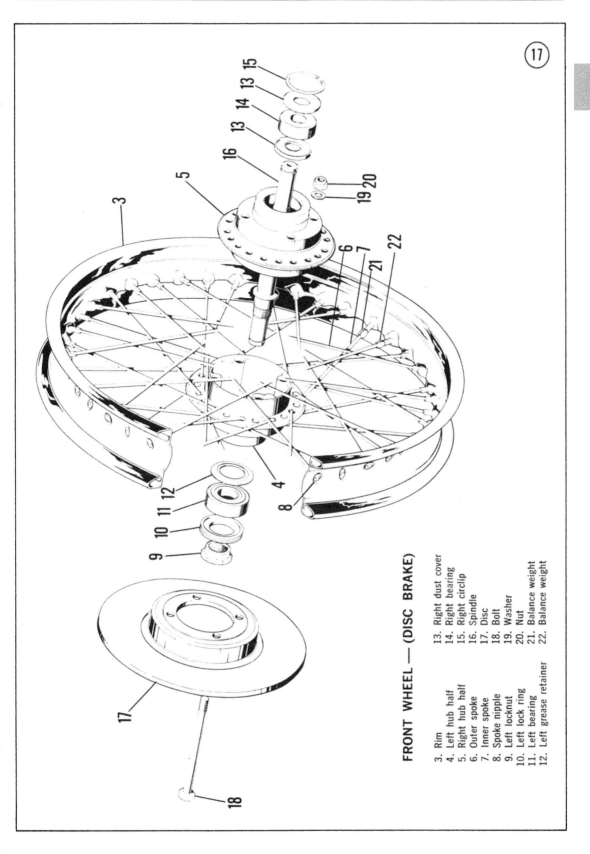

FRONT WHEEL — (DISC BRAKE)

3. Rim
4. Left hub half
5. Right hub half
6. Outer spoke
7. Inner spoke
8. Spoke nipple
9. Left locknut
10. Left lock ring
11. Left bearing
12. Left grease retainer
13. Right dust cover
14. Right bearing
15. Right circlip
16. Spindle
17. Disc
18. Bolt
19. Washer
20. Nut
21. Balance weight
22. Balance weight

2. Unscrew the locknut on the left side of the axle and remove the lock ring (right-hand threads).

3. Remove the left bearing by tapping on the right end of the axle. Remove the grease seal from the left side.

4. Remove the circlip from the right half of the hub. Remove the axle and reinsert it from the left side and drive out the right bearing and dust covers (grease seals).

5. Clean all parts thoroughly in kerosene or other solvent. Clean and dry bearings, using compressed air if available. Inspect for wear and damage, and replace if necessary. Repack the bearings with high temperature grease.

6. Reinstall the right inner grease seal, bearing, and outer seal, then reinstall the circlip. Insert the axle from the left side and tap until the bearing and seals are flush with the circlip.

7. Remove axle and reinsert from right side. Install left grease seal and then drive left bearing into place. Install and tighten lock ring (right-hand threads). Tap the axle from the right side to bring the axle shoulder flush with the left bearing and seal. Reinstall the locknut and tighten securely.

8. Reinstall wheel, using the procedure given in Chapter Eight.

REAR WHEEL BEARINGS

Different procedures are used for the early type, the quick-detach type, or the quick-detach type wheels on bikes prior to engine No. DU.13375, the conical-hub type wheel, or disc brake-equipped rear wheels. Determine which type you have and follow the appropriate directions. See **Figures 18 through 23.**

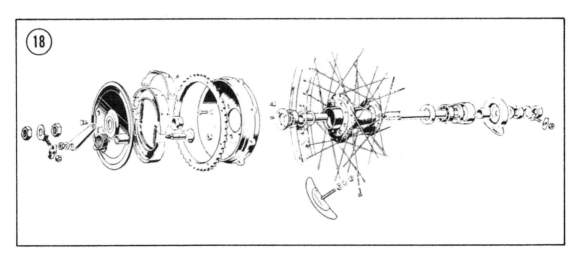

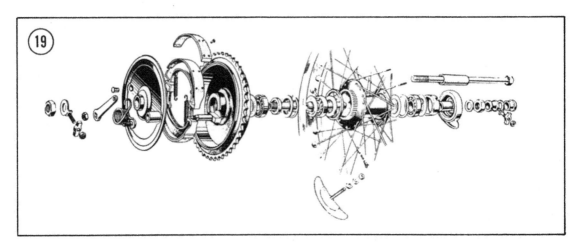

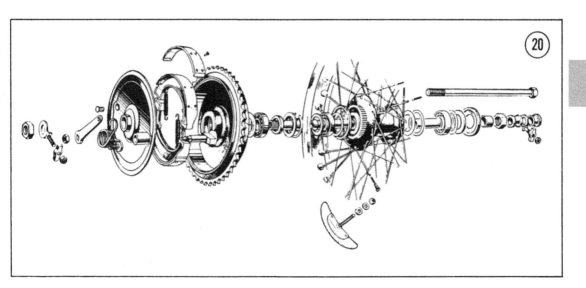

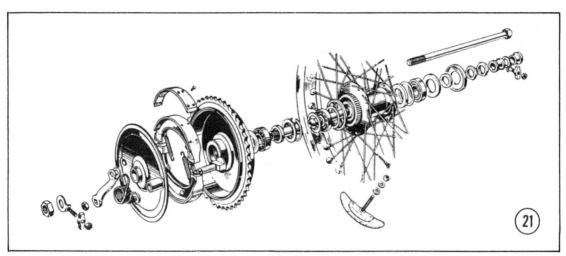

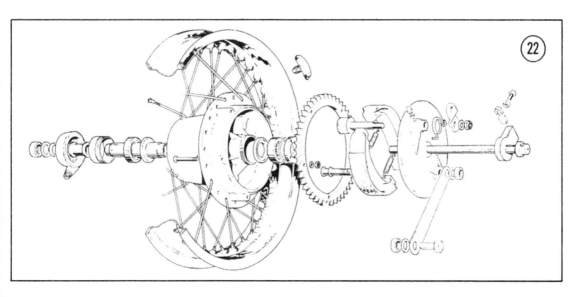

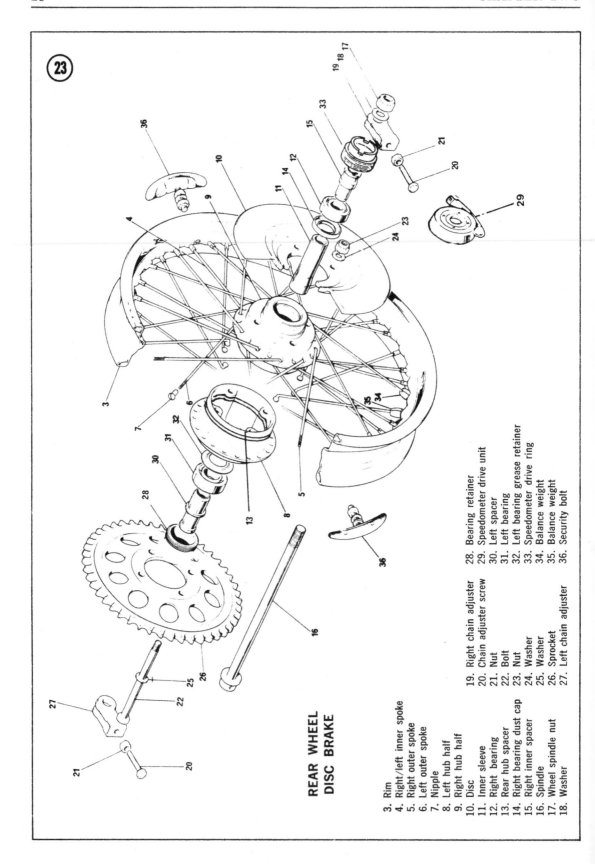

REAR WHEEL DISC BRAKE

3. Rim
4. Right/left inner spoke
5. Right outer spoke
6. Left outer spoke
7. Nipple
8. Left hub half
9. Right hub half
10. Disc
11. Inner sleeve
12. Right bearing
13. Rear hub spacer
14. Right bearing dust cap
15. Right inner spacer
16. Spindle
17. Wheel spindle nut
18. Washer
19. Right chain adjuster
20. Chain adjuster screw
21. Nut
22. Bolt
23. Nut
24. Washer
25. Washer
26. Sprocket
27. Left chain adjuster
28. Bearing retainer
29. Speedometer drive unit
30. Left spacer
31. Left bearing
32. Left bearing grease retainer
33. Speedometer drive ring
34. Balance weight
35. Balance weight
36. Security bolt

Removal/Installation (Early Type)

1. Remove the rear wheel, anchor plate retainer nut, and brake anchor plate assembly.

2. Remove the axle and unscrew the screw which locks the bearing retainer ring. Unscrew the retainer ring.

3. The spacer in the center of the hub must be moved to one side with a drift so the inner ring of the left bearing is exposed and can be reached and driven out. See **Figure 24**. The grease retainer shim will collapse when this is done, but it can be straightened and reused in most cases.

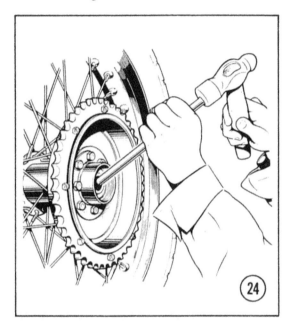

4. Remove the backing ring, grease retainer shim, and spacer.

5. The right bearing and dust cap can be driven out with a drift 1⅝ in. in diameter.

6. Clean all parts thoroughly with cleaning solvent or kerosene. Blow dry with compressed air. Inspect parts for wear or damage; replace if necessary. Repack bearings with hi-temp bearing grease and soak felt washers in oil.

7. Install by greasing the components and driving in the right inner grease retainer disc and the bearing. Press on the outer dust cap.

8. Working from the left side, insert the spacer, grease retainer shim, and backing ring.

9. Grease the bearing and press it in, lining up the spacer with the axle.

10. Replace the bearing retainer ring and its locking screw.

11. Install remaining parts in reverse order of removal. Be sure that parts turn freely without discernible free play.

**Removal/Installation
(Prior to Engine No. DU.13375)**

1. Remove the quick-detach rear wheel and the 2 locknuts on the right side of the axle sleeve. Push the sleeve out from the right.

2. Pull out the inner roller races and dust cover.

3. Carefully drive the outer races out with a soft metal drift.

4. Clean all parts thoroughly with cleaning solvent or kerosene. Blow dry with compressed air. Inspect parts for wear or damage; replace if necessary. Repack bearings with hi-temp bearing grease and soak felt washers in oil.

5. Install by pressing the left and right backing rings and grease retainers into the wheel hub.

6. Press left and right outer races into hub.

7. Grease the rollers and inner races well and put them into their proper outer races.

8. Insert the threaded end of the axle into the bearings and replace the outer dust caps.

9. Replace the right-side spacer and the 2 locknuts on the axle sleeve. Tighten the inner locknut and then loosen it ⅙ turn.

10. Tighten the outer locknut. Check that the sleeve and inner races turn freely yet have no discernible free play in the rollers.

**Removal/Installation
(After Engine No. DU.13375)**

1. Remove the quick-detach rear wheel. Secure the bearing sleeve using the slot at its tapered end and remove the nut on the right side.

2. Unscrew the locking ring (left-hand threads) and lift off the spacer, felt washer, and locating disc.

3. Remove the right bearing in the manner detailed for the left bearing in Steps 2 and 3 above for the *Early Type* wheel. The bearing must be driven out from the left side.

4. After the right bearing, backing ring, grease retainer shim, and spacer are removed, the left bearing may be driven out easily.

5. If the brake drum and sprocket must be removed, disconnect the drive chain at the master link, then the retainer and the brake rod.

6. Remove the large nut from the axle sleeve.

7. The bearing in the brake drum can be removed by pressing out the axle sleeve, and removing the circlip, retainer, and felt washer.

8. Drive the bearing out.

9. Clean all parts thoroughly with solvent or kerosene. Blow dry with compressed air. Inspect parts for wear or damage; replace if necessary. Repack bearing with hi-temp bearing grease and soak felt washers in oil.

10. Install in reverse order of removal. Be sure that the parts turn freely without discernible free play.

Removal/Installation (Conical Hub)

1. Remove the rear wheel and bearing retainers from both sides of the wheel.

2. Use a drift to knock out the inside spacer tube and one of the bearings (from either side of the wheel).

3. Use the drift to remove the bearing from the spacer tube.

4. Use the tube and drift to knock the other bearing out of the wheel. Be sure to use a correctly-sized drift to avoid spacer tube damage.

5. Clean all parts thoroughly with cleaning solvent or kerosene. Blow dry with compressed air. Inspect parts for wear or damage; replace if necessary. Repack bearings with hi-temp bearing grease and soak felt washers in oil.

6. Install one bearing on the spacer tube.

7. Press the assembly into the hub from the left side until the bearing contacts the grease retainer. Apply force only to outer bearing race.

8. Install the retaining ring and tighten.

9. Install the right-side bearing and then install the remaining pieces in reverse order of removal. Be sure that parts turn freely without discernible free play.

Removal/Installation (Rear Disc Brake Models)

1. Remove the rear wheel, then remove the left-hand bearing retainer and left-hand spacer. Refer to Figure 23.

2. Remove the speedometer drive unit and driving ring from the right side of the wheel. Then remove the right-hand inner spacer.

3. Use a proper size drift to tap out the inner sleeve and the bearing and grease seal from one side of the hub. Then reverse the sleeve and drift out the other bearing and seal.

CAUTION
Use a correct size drift to avoid damage to the inner sleeve and bearings.

NOTE: *If necessary, use drift to remove bearings from inner sleeve.*

4. Clean all parts in cleaning solvent or kerosene and blow dry with compressed air. Inspect all parts for wear or damage and replace if necessary. Repack bearings with high temperature bearing grease. Replace grease retainers if possible.

5. Install inner sleeve and grease seals in hub, then tap in bearings, using a correct size drift on the outer race. Then install the left-hand and right-hand spacers.

6. Install the bearing retainer in the left side of the hub, and the speedometer ring driving unit in the right side of the hub. Install the speedometer drive unit on the ring driving unit.

7. Reinstall the rear wheel on the bike.

FRONT FORKS

Front forks must have the same amount of oil in each fork leg to operate properly. Leakage at the junction between the top and bottom fork tubes may mean the fork seals and O-rings need replacement. Fork oil should generally be changed twice yearly. If the forks begin to malfunction, check that each has the proper level of oil.

Oil Level (750/650cc Models)

For engine Nos. DU.101-DU.5824, the forks require 1/4 pint (150cc) oil in each fork leg.

From engine No. DU.5825 up, they require ⅓ pint (200cc) in each fork leg. If a sidecar has been fitted to an earlier model and special long fork legs have been installed, they will require ⅜ pint (225cc) in each fork leg.

1. Drain fork oil by removing drain plugs in each fork leg near the front axle. Lock the front brake and pump the forks up and down to force out all the oil.

2. Replace drain plugs and refill fork tubes with 20- or 30-weight oil in the amount specified above. Oil should be pumped into the fork legs through the filling plugs located about 3 in. below the top lug, or by removing the top cap nuts. These cap nuts are 1½ in. (38.1mm) in diameter. On some models, it will be necessary to remove headlights, nacelles, or handlebars to gain access to cap nuts.

> NOTE: *For 1972 and later models, the factory recommends use of automatic transmission fluid (ATF) in the forks.*

Oil Level (500cc Models)

Before serial No. H57083, each fork leg requires ½ pint (285cc) and after H57083, ⅓ pint (200cc). After H57083, do not use oil heavier than 20-weight. Check the oil seal holders; they must be screwed down tightly on the bottom tubes.

1. To change the oil, the drain procedure is the same as for 750/650 models.

2. Replace the drain plugs and remove the cap nuts on top of the fork tubes for refilling.

It may be necessary to remove the handlebars or headlight assembly to gain access to the cap nuts. On early models, the headlight and rim assembly is removed to expose small hex filler plugs located about 3 in. from the top triple clamp. Oil is pumped into the filler holes or poured into the tubes after the cap nuts have been removed.

LUBRICATION FITTINGS

Fittings should be lubricated with a grease gun. The swing arm grease nipple on early models and the TR5T is located at the right end of the swing arm axle. See **Figure 25**. If the

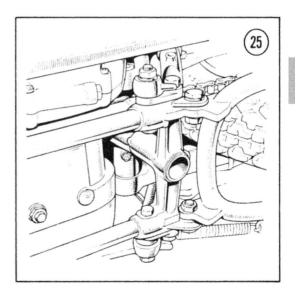

grease gun does not force grease out the pivot bearings, the swing arm will have to be disassembled as detailed in Chapter Nine, and the components cleaned and greased.

1. The brake camshaft fittings should be lubricated with a single stroke from a grease gun. Excessive lubrication will allow grease to get on the brake linings. The brake cams may have to be disassembled for proper lubrication, especially if the machine has seen hard use or is old.

2. Lubricate the fitting on the bottom of the swing arm at the front cross section. See **Figures 26 and 27**.

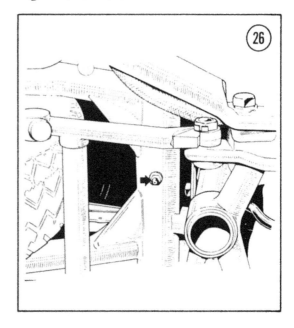

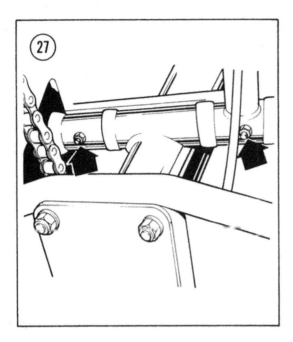

CONTROL CABLES

Control cables should be lubricated at their exposed ends every 1,000 miles (1,600 km) with light oil. Occasionally remove them and soak in oil so their entire length is properly oiled. Place in a pan filled with SAE 30 motor oil for at least 8 hours.

The front brake cable on 1972 models and the clutch cable on 1972 and later models are equipped with lubrication fittings through which oil may be forced into the cable sheath with a pump type oil gun.

SPEEDOMETER CABLE

Lubricate the speedometer cable by disconnecting it from the speedometer and removing the inner cable. Clean and grease it lightly along its entire length except for the top 6 in. Replace the inner cable, turning it to ensure its end is inserted in the drive mechanism and reconnect the outer cable at the speedometer. A jumping or vibrating speedometer needle usually means that lubrication is necessary.

BRAKE LEVER

Remove the brake lever occasionally and clean spindle and lever bore before greasing. Rust may be removed with emery cloth.

CHECKING ENGINE FOR WET SUMPING

High oil consumption, heavy smoking, and excessive oil emitting from crankcase breather may be symptomatic of *wet sumping*. To avoid this, the oil pump is designed to have a greater capacity pumping oil back to the oil tank than through the engine, so make the following checks if the above conditions exist.

1. Check for improper seating of the ball valve in the scavenge side of the oil pump.

2. Check the ball valve seat. There may be an air leak in the crankcase oil scavenge pipe, oil pump to crankcase joint, or in an oil plug at the bottom of the engine. A crankcase casting may be too porous. The above malfunctions may be checked by performing the scavenge suction test given below. Additional causes for wet sumping could be blockage in the return oil pipe (caused by a misaligned oil junction block gasket), an oil pressure release valve piston in full bypass position because of a stuck piston or broken or missing spring, or a restriction in the oil tank vent pipe.

3. Perform the scavenge suction test using a vacuum gauge calibrated in inches (or mm) of mercury. Fit a length of Triumph oil pipe to it and run the engine until it has warmed up. Remove the oil sump cap and screen and attach the vacuum gauge hose to the oil scavenge pipe. The gauge should read 18-26 in. (46-66 cm) of mercury with the engine at a fast idle. After the engine is stopped, the gauge needle should return slowly to zero.

4. If the test is not satisfactory, remove the oil pump, clean and inspect. Pay particular attention to the balls and ball seats. Reinstall the pump using a new gasket.

5. Check the crankcase scavenge tube for leaks and the crankcase itself for porosity by filling a pumper type oil can with light oil and then squirting into the pickup tube through a folded rag. Some back pressure should be felt after a few strokes of the pump.

ENGINE TUNE-UP

Engine tune-up consists of several accurate and careful adjustments to obtain maximum engine performance. Since different systems in an engine interact to affect overall performance, tune-up must be accomplished in the following order:

1. Valve clearance adjustment
2. Ignition adjustment and timing
3. Carburetor adjustment

Perform an engine tune-up every 3,000 miles or 6 months. Tune-up specifications are summarized in **Table 4**.

Table 4   ENGINE TUNE-UP

Spark plug	
Type	Appendix I
Gap	0.020-0.022″ (0.5-0.6mm)
Breaker point gap	0.012-0.016″ (0.3-0.4mm)
Valve clearance	
500cc	0.002″ (0.05mm) intake 0.004″ (0.10mm) exhaust
650cc	0.002″ (0.05mm) intake 0.004″ (0.10mm) exhaust
750cc	0.008″ (0.20mm) intake 0.006″ (0.15mm) exhaust
Ignition timing	Appendix I

Valve Clearance Adjustment

This is a series of simple mechanical adjustments which are performed while the engine is *cold*. Valve clearance for your engine must be adjusted carefully. If the clearance is too small, the valves may be burned or distorted. Large clearance results in excessive noise. In either case, engine power is reduced.

Valve Clearance (750/650cc Models)

1. Check and reset valve clearance every 3,000 miles (4,800 km).

2. Tappet clearance must be set with the engine cold. See Table 4 for proper clearance.

3. Remove the inspection caps from the rocker boxes and both spark plugs. Rotate the engine until the left exhaust valve is fully open and check the clearance on the right exhaust valve.

4. Loosen the locknut and turn the adjuster in the desired direction to bring the clearance within specification. See **Figure 28**.

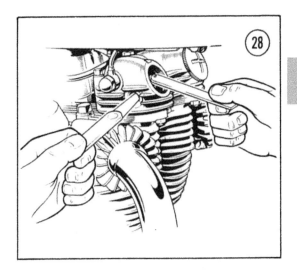

5. Tighten the locknut, check the clearance again and readjust if necessary. Rotate the engine until the right exhaust valve is fully open and adjust the left exhaust valve. Perform the operation in the same way for the intake valves, opening one and adjusting the other.

6. Another method may be used in a field situation where a feeler gauge is not available. The locknut is loosened and the adjuster is turned finger-tight so there is no up-and-down play in the rocker arm. Slightly less than ⅛ turn when slackening the adjuster will give a clearance close to 0.004 in. (0.10mm). Half of that will give a clearance close to 0.002 in. (0.05mm). These settings will usually enable the machine to be ridden until more accurate settings can be done. Set the clearance properly as soon as possible.

Valve Clearance (500cc Models)

1. Adjustment of valve clearance on late versions of 500cc models follows the same procedure given for 750/650cc models. The use of a feeler gauge on early models is difficult because of the location of the rocker arm and valve tip. Refer to Step 6 in the previous section for adjustment.

2. On T90, T100SS, T100T and later 5TA models from H40528, the inlet valve clearance should be 0.002 in. (0.05mm). This has been translated as the slightest perceptible movement of the rocker so that a click can be heard when the rocker is moved by hand. Exhaust valve

clearance on these models is 0.004 in. (0.10mm). Adjust the tappet to zero clearance by hand, then loosen the adjuster ⅛ turn (half of a flat). Tighten the locknut.

3. On 3TA and early 5TA models, the intake and exhaust valve clearance should be set to 0.010 in. (0.25mm). This is obtained by tightening the adjuster to zero clearance and then loosening the adjuster ¼ turn (one flat) and tightening the locknut.

Spark Plug Type

Spark plugs are available in various heat ranges hotter or colder than the plug originally installed at the factory.

Select plugs of a heat range designed for the loads and temperature conditions under which the engine will run. Use of incorrect heat range can cause seized pistons, scored cylinder walls, or damaged piston crowns.

In general, use a lower-numbered plug for low speeds, low loads, and low temperatures. Use a higher-numbered plug for high speeds, high engine loads, and high temperatures.

> NOTE: *Use the highest numbered plug that will not foul. In areas where seasonal temperature variations are great, the factory recommends a high-numbered plug for slower winter operation.*

The reach (length) of a plug is also important. A longer than normal plug could interfere with the piston, causing permanent and severe damage. Refer to **Figures 29 and 30**.

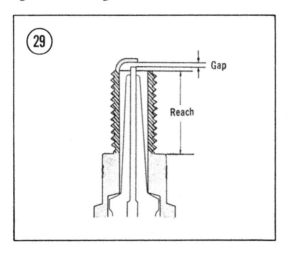

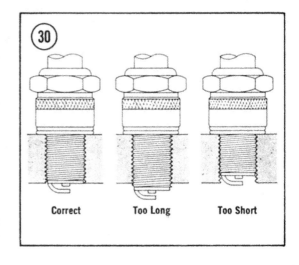

Correct   Too Long   Too Short

Spark Plug Cleaning/Replacement

1. Grasp the spark plug leads as near to the plug as possible and pull them off the plugs.

2. Blow away any dirt which has accumulated in the spark plug wells.

> CAUTION
> *Dirt could fall into the cylinders when the plugs are removed, causing serious engine damage.*

3. Remove spark plugs with spark plug wrench.

> NOTE: *If plugs are difficult to remove, apply penetrating oil around base of plugs and let it soak in about 10-20 minutes.*

4. Inspect spark plugs carefully. Look for plugs with broken center porcelain, excessively eroded electrodes, and excessive carbon or oil fouling. Replace such plugs. If deposits are light, plugs may be cleaned in solvent with a wire brush or cleaned in a special spark plug sandblast cleaner.

5. Gap plugs to 0.020-0.022 in. (0.5-0.6mm) with a *wire* feeler gauge.

6. Install plugs with a *new* gasket. First, apply a *small* drop of oil to threads. Tighten plugs finger-tight, then tighten with a spark plug wrench an additional ½ turn. If you must reuse an old gasket, tighten only an additional ⅛ turn.

> NOTE: *Do not overtighten. This will only squash the gasket and destroy its seating ability.*

Reading Spark Plugs

Much information about engine and spark plug performance can be determined by careful exmination of the spark plugs. This information is only valid after performing the following steps.

1. Ride bike a short distance at full throttle in any gear.

2. Kill engine and close throttle simultaneously.

3. Pull in clutch and coast to a stop.

4. Remove the spark plugs and examine them. Compare them to **Figure 31**.

Condenser (Capacitor)

The condenser (capacitor) is a sealed unit and requires no maintenance. Be sure the connections are clean and tight.

The only proper test is to measure the resistance of the insulation with an ohmmeter. The value should be 5,000 ohms. A make-do test is to charge the capacitor by hooking the leads, or case and lead, to a battery. After a few seconds, touch the leads together, or lead to case, and check for a spark as shown in **Figure 32**. A damaged capacitor will not store electricity.

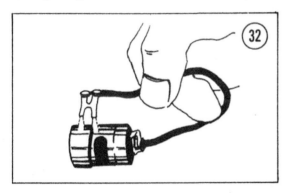

Most mechanics prefer to discard the condensers and replace them with new ones during engine tune-up.

On 500cc models after H57083, the condensers are located under the gas tank and are covered with a rubber protector.

Breaker Point Inspection and Cleaning

Check that the insulation between the breaker contacts and the contact breaker base is not defective. A short circuit will prevent the motor from running. To test for this condition, disconnect the wire or wires on the points, and with the points still blocked open, measure insulation resistance between the movable point and a good ground, using the highest range on the ohmmeter. If there is any indication at all on the ohmmeter, the points are shorted.

Through normal use, points gradually pit and burn. See **Figure 33**. If this condition is not too serious, they can be dressed with a few strokes of a clean point file. Do not use emery cloth or sandpaper, as particles remain on the points and cause arcing and burning. If a few strokes of the file do not smooth the points completely, replace them.

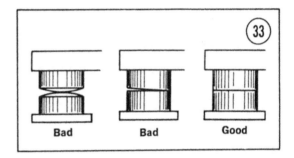

If points are still serviceable after filing, remove all residue with lacquer thinner or special contact cleaner. Close the points on a piece of clean white paper such as a business card. Continue to pull the card through the closed points until no particles or discoloration remains on the card. Finally, rotate the engine and observe the points as they open and close. If they do not meet squarely, replace them.

Adjust point gap and ignition timing as described below.

Breaker Point Replacement (750/650cc Models)

1. To remove the contact breaker unit, disconnect the battery leads or remove the fuse from the holder near the battery.

2. Remove the contact breaker cover located on the right end of the exhaust camshaft.

3. Unscrew the cam center bolt and use a puller to remove the point cam.

4. If the contact breaker unit is to be removed,

⟨31⟩ SPARK PLUG CONDITION

NORMAL
- Identified by light tan or gray deposits on the firing tip.
- Can be cleaned.

GAP BRIDGED
- Identified by deposit buildup closing gap between electrodes.
- Caused by oil or carbon fouling. If deposits are not excessive, the plug can be cleaned.

OIL FOULED
- Identified by wet black deposits on the insulator shell bore and electrodes.
- Caused by excessive oil entering combustion chamber through worn rings and pistons, excessive clearance between valve guides and stems, or worn or loose bearings. Can be cleaned. If engine is not repaired, use a hotter plug.

CARBON FOULED
- Identified by black, dry fluffy carbon deposits on insulator tips, exposed shell surfaces and electrodes.
- Caused by too cold a plug, weak ignition, dirty air cleaner, too rich a fuel mixture, or excessive idling. Can be cleaned.

LEAD FOULED
- Identified by dark gray, black, yellow, or tan deposits or a fused glazed coating on the insulator tip.
- Caused by highly leaded gasoline. Can be cleaned.

WORN
- Identified by severely eroded or worn electrodes.
- Caused by normal wear. Should be replaced.

FUSED SPOT DEPOSIT
- Identified by melted or spotty deposits resembling bubbles or blisters.
- Caused by sudden acceleration. Can be cleaned.

OVERHEATING
- Identified by a white or light gray insulator with small black or gray brown spots and with bluish-burnt appearance of electrodes.
- Caused by engine overheating, wrong type of fuel, loose spark plugs, too hot a plug, or incorrect ignition timing. Replace the plug.

PREIGNITION
- Identified by melted electrodes and possibly blistered insulator. Metallic deposits on insulator indicate engine damage.
- Caused by wrong type of fuel, incorrect ignition timing or advance, too hot a plug, burned valves, or engine overheating. Replace the plug.

disconnect the coil leads and unhook the frame clips necessary to pass the leads through the crankcase and timing cover. Make a note of the degree figure stamped on the cam unit as an aid in static timing.

5. Place a drop of light oil on the pivot pins before reassembling the contact breaker unit.

6. Align the peg on the camshaft with the slot on the point cam and press it in place, tightening the bolt.

7. For the 6CA contact breaker plate used on early models prior to DU.66246 (**Figure 34**), assemble the black/yellow leads to the rear. A few still earlier machines had the black/white leads to the rear, so it should be noted how they are routed before the unit is disassembled.

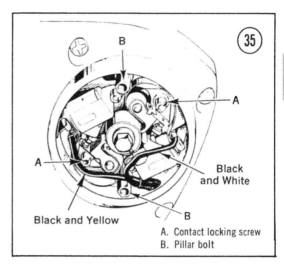

A. Contact locking screw
B. Pillar bolt

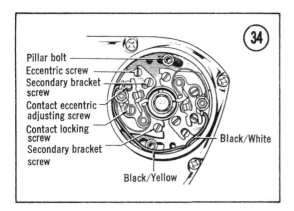

8. To adjust the point gaps, rotate the crankshaft until the scribe mark on the breaker cam is aligned with one of the nylon point lifters. Set the gap to 0.015 in. (0.38mm), using a feeler gauge, by loosening the locking screw and turning the adjusting screw. Repeat the operation for the other set of points.

9. For the 4CA contact breaker plate used on later models (**Figure 35**), assemble the black/yellow leads to the rear with the pillar bolts in the centers of their adjusting slots.

10. To adjust the point gaps, loosen the sleeve nuts and set the gaps to 0.015 in. (0.38mm).

11. On earlier models without a peg and slot on the cam unit and exhaust camshaft, the cam unit must be installed as follows. Position the base plate so the pillar bolts are in the center of their adjusting slots.

12. Remove the spark plugs and all 4 rocker caps, then turn the engine over using the rear

wheel (with the transmission in fourth gear) until the right piston is on its compression stroke.

13. Use a dial indicator and set the engine so it is 1/32 in. (0.9mm) before top dead center; turn the contact breaker cam unit until the rear set of points is just about to open.

14. Tighten the center bolt at this point. Turn the cam unit clockwise to bring it to the point where the breaker points begin to open.

Breaker Point Replacement (500cc Models)

To remove and replace the contact breaker on machines prior to H57083, follow the instructions for the 4CA type used on 750/650cc models prior to DU.66246. On machines after H57083, use the instructions given for the 6CA type used on later 750/650cc models.

Ignition Timing (750/650cc Models) After Engine No. DU.66246

1. Set the point gap to 0.015 in. (0.38mm) and remove the center bolt from the point cam assembly.

2. Place a washer large enough to fit over the cam bearing on the bolt and screw it back in. Turn the cam against the automatic advance springs to its limit and tighten the bolt to hold the assembly in full advance position.

3. The engine must be locked in position at 38° before top dead center. Use the timing plunger

shown in **Figure 36** and refer to **Figure 37** to determine which hole should be used.

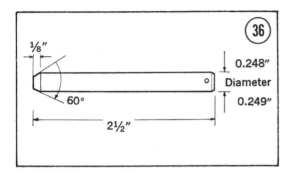

4. Remove both spark plugs and the rocker inspection plugs. Start with the transmission in fourth gear and find top dead center by turning the rear wheel until a dial gauge or marker inserted in a spark plug hole indicates the engine is at top dead center. Turn the rear wheel slowly backward until the timing plug engages in the flywheel and the engine will not turn over.

5. Look at the valves to determine which cylinder is on the compression stroke and time that set of points. The timing side cylinder is fired by the points with the black/yellow lead and the drive side is fired with the black/white lead.

6. Loosen the backing plate lock screws and turn the backing plate until the correct set of points just begin to open. Use a buzz box, a light wired to the points, or watch an ammeter; the needle will jump just as the points open (with the battery hooked up and ignition turned on).

7. Pull out the timing plunger and turn the engine over one complete revolution. Reinsert the plunger, locking the engine at 38° before top dead center with the other cylinder on its compression stroke. Adjust the other set of points at this time, using the secondary backing plate only. The pillar bolts must be left as they were adjusted to time the first cylinder.

8. If a timing light is available, more accurate results may be obtained. Engines with an inspection plate on the primary chaincase can be timed using a timing light. Some models may not have a timing pointer, but a timing plate (part No. D2014) is available and should be screwed in place. See Figure 37 and **Figures 38 and 39**. Use the "B" on 750/650cc engines.

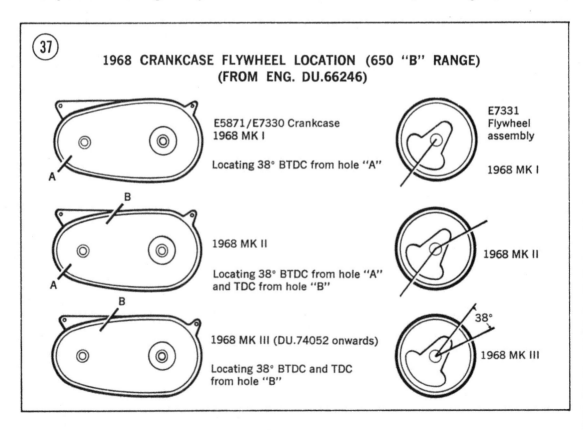

1968 CRANKCASE FLYWHEEL LOCATION (650 "B" RANGE)
(FROM ENG. DU.66246)

E5871/E7330 Crankcase
1968 MK I

Locating 38° BTDC from hole "A"

E7331
Flywheel
assembly

1968 MK I

1968 MK II

Locating 38° BTDC from hole "A"
and TDC from hole "B"

1968 MK II

1968 MK III (DU.74052 onwards)

Locating 38° BTDC and TDC
from hole "B"

38°

1968 MK III

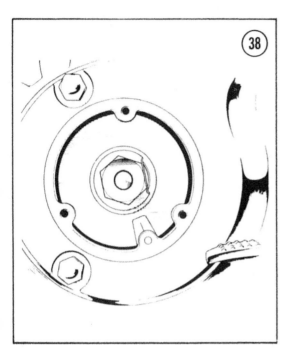

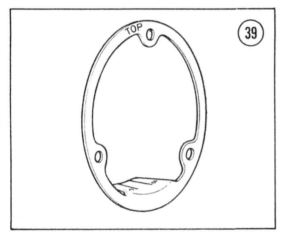

9. Connect the timing light to the right spark plug lead and start the engine. Adjust the main backing plate until the 2 marks coincide with the engine running at 2,000 rpm or more.

10. Connect the timing light to the left spark plug and adjust the auxiliary backing plate for the points with the black/white lead until the marks again coincide.

Prior to Engine No. DU.66246

1. Establish top dead center on the right cylinder by first removing the spark plugs and the rocker inspection caps. If the engine has pro-

vision for inserting a timing plug as shown in **Figure 40**, top dead center may be found by engaging fourth gear and rotating the rear wheel until the timing plug engages and locks the flywheel. Check that the valves for the right cylinder are both closed, indicating it is on the compression stroke. If there is no provision for a timing plug, find top dead center by using a dial indicator or a timing stick inserted in the spark plug hole. Rotate the rear wheel to find the point where the timing stick is at its highest position and make a mark on it, even with the top head fin. This point will indicate top dead center.

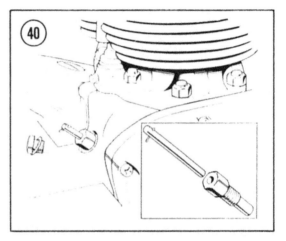

2. Fit a timing disc adapter shaft and timing disc on the camshaft auto advance unit as shown in **Figure 41**. Connect the pointer to a screw and set it to read TDC or 0° with the engine at top dead center. The indicated setting and advance range will be half that of the engine because the camshaft operates at half crankshaft speed. Some calculation will therefore be necessary. See **Table 5**.

3. Connect a timing light to the right spark plug lead and start the engine. Run the engine at 2,000 rpm or more and adjust the backing plate so the pointer and the correct setting (found in Appendix I) on the timing disc coincide. Connect the timing light to the left cylinder and adjust the point gap for the left cylinder contact points to make the correct setting and the pointer coincide. Open the contact points to advance the spark and close them to retard the spark.

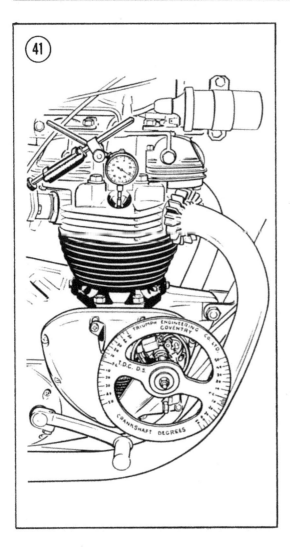

Table 5   CONVERSION CHART

Crankshaft position (BTDC)	Piston position (BTDC)	
Engine degrees	in.	mm
7	0.015	0.38
8	0.020	0.51
9	0.025	0.64
10	0.030	0.76
11	0.038	0.96
12	0.045	1.14
13	0.054	1.30
14	0.060	1.52
15	0.068	1.73
16	0.077	1.96
17	0.087	2.20
18	0.095	2.42
19	0.108	2.75
20	0.120	3.05
21	0.135	3.45

4. If a timing light is not available, obtain the correct fully advanced ignition setting from data in Appendix I. Remove the auto advance cam as detailed earlier and obtain the advance range stamped on the back of the cam. Double this figure and subtract it from the fully advanced ignition setting. This will give you the correct static ignition setting to be used when assembling the contact breaker unit, or setting the position of the contact breaker point of first opening, and for using a degree wheel or timing disc. As an example, if the figure stamped on the auto advance cam is 12°, that is doubled and subtracted from the fully advanced ignition setting found in the chart for the T120 model, i.e., 39° before top dead center.

$$(12 \times 2 = 24, \text{ and } 39 - 24 = 15)$$

The static setting is 15° before top dead center.

5. With the static setting obtained, time the engine by turning it over until the points have opened fully.

6. Set the gap on the right cylinder (black/yellow lead) to 0.015 in. (0.38mm).

7. Rotate the engine through one full revolution and set the other set of points to 0.015 in. (0.38mm).

8. With top dead center established as described earlier, rotate the engine backward until the static setting in degrees has been passed by the timing disc pointer.

9. Rotate the engine forward to exactly line up the pointer with the static setting. Passing the mark and then returning to it takes out slack in the timing gears and assures a more accurate setting.

10. Move the main point backing plate until the contact points just begin to open. This can be checked with a light hooked up with the points, a buzz box, or by watching an ammeter needle with the ignition switch turned on. The needle will jump just as the points open.

11. Rotate the engine one complete revolution, setting the pointer to the static setting again. The left cylinder points should just be opening. If not, adjust the point gap until they do.

Ignition Timing (500cc Models)

1. Prior to H57083, follow the above procedure headed *Ignition Timing Prior to Engine No. DU.66246*. After H57083, the initial procedure and timing with a timing light is the same as given above, Steps 1-10, under the heading *Ignition Timing After Engine No. DU.66246*. The line on the special tool marked "C" is used on 500cc models. Later machines have a built-in pointer and special tool does not have to be used.

2. For static ignition timing, where a timing light is not available, use the information given for 750/650cc models under the heading *Ignition Timing Prior to Engine No. DU.66246* and **Table 6**.

Table 6   IGNITION TIMING

Crankshaft position (BTDC)		Piston position (BTDC)
Degrees	in.	mm
7	.010	.25
8	.015	.38
9	.020	.51
10	.025	.64
11	.030	.76
12	.035	.89
13	.040	1.02
14	.048	1.22
15	.055	1.40
16	.060	1.52
17	.070	1.78
18	.080	2.03
19	.090	2.29
20	.100	2.54
21	.110	2.79

For machines after H57083, follow the above procedure, Steps 1-11, headed *Ignition Timing After Engine No. DU.66246*. The timing plug is inserted in the threaded hole at the rear of the cylinder. There is a hole in the flywheel for TDC and another hole for 38° BTDC. Find TDC and then turn the engine slowly backward to 38°

Carburetor Adjustment

Refer to Chapter Six for carburetor adjustment and synchronization.

DECARBONIZATION

After an engine has been run for many hours, it will probably require the removal of carbon from the piston crown and cylinder head. The best way to detect this need is if the engine has shown progressively worsening preignition or a gradual loss of power. Several new products are now being marketed to allow a simple approach to decarbonizing without the need of dismantling the engine. These products will not be as thorough, but can be used periodically. The procedure for their use is as follows.

1. Start and warm the engine to normal operating temperature.

2. Slowly pour 10 ounces (284 grams) of cleaner into the carburetor with the engine running at a fast idle of 1,500 rpm.

3. Slowly increase the speed of the engine and kill it by dumping in the remaining fluid.

4. Let the engine stand for ½ hour with the liquid still in the cylinder.

5. Start the engine and run it at full throttle, under load, for at least 5 minutes to clear out the system and remove the last traces of fluid. If this fails to completely decarbonize the cylinder then you will have to remove the head as outlined in Chapter Four.

WINTER STORAGE

Several months of inactivity can cause serious problems and a general deterioration of bike condition. This is especially true in areas of weather extremes. During the winter months it is advisable to specially prepare the bike for lay-up.

Selecting a Storage Area

Most cyclists store their bikes in their home garage. If you do not have a home garage, facilities suitable for long-term motorcycle storage are readily available for rent or lease in most areas. In selecting a building, consider the following points:

1. The storage area must be dry, free from dampness and excessive humidity. Heating is not necessary but the building should be well

insulated to minimize extreme temperature variations.

2. Buildings with large window areas should be avoided, or such windows should be masked (also a good security measure) if direct sunlight can fall on the bike.

3. Buildings in industrial areas, where factories are liable to emit corrosive fumes, are not desirable, nor are facilities near bodies of salt water.

4. The area should be selected to minimize the possibility of loss by fire, theft, or vandalism. The area should be fully insured, perhaps with a package covering fire, theft, vandalism, weather, and liability. The advice of your insurance agent should be solicited on these matters. The building should be fireproof and items such as the security of doors and windows, alarm facilities, and proximity of police should be considered.

Preparing for Storage

Careful pre-storage preparation will minimize deterioration and will ease restoring the bike to service in the spring. The following procedure is recommended.

1. Wash the bike completely making certain to remove any accumulation of road salt that may have collected during the first weeks of winter. Wax all painted and polished surfaces.

2. Run the engine for 20-30 minutes to stabilize oil temperature. Drain oil regardless of mileage since last oil change and replace with normal quantity of fresh oil.

3. Remove battery and coat cable terminals with petroleum jelly. If there is evidence of acid spillage in the battery box, neutralize with baking soda, wash clean, and repaint. Batteries should be kept in an area where they will not freeze, and where they can be recharged every 2 weeks.

4. Drain all gasoline from fuel tank, settling bowl, and carburetor float bowls. Leave fuel cock on the RESERVE position.

5. Remove spark plugs and add a small quantity of oil to each cylinder. Turn the engine a few revolutions by hand. Install spark plugs.

6. To avoid ignition point oxidation, place a paper card, lightly saturated with silicone oil, between the points.

7. Check tire pressure. Move the bike to storage area and store on center stand. If preparation is performed in an area remote from the storage facility, the bike should be trucked, not ridden, into storage.

Inspection During Storage

Try to inspect the bike weekly while it is in storage. Any deterioration should be corrected as soon as possible. For example, if corrosion of bright metal parts is observed, coat with a light film of grease or silicone spray.

Restoring to Service

A bike that has been properly prepared, and stored in a suitable area, requires only light maintenance to restore it to service. It is advisable, however, to perform a spring tune-up.

1. Before removing the bike from the storage area, re-inflate tires to the correct pressures. Air loss during the storage period may have nearly flattened the tires, and moving the bike can cause damage to tires, tubes, or rims.

2. When the bike is brought to the work area, immediately install the battery (fully charged) and fill the fuel tank. The fuel cock should be on the RESERVE position; do not move yet.

3. Check the fuel system for leaks. Remove the carburetor float bowl or open the float bowl drain cock and allow several cups of fuel to pass through the system. Move the fuel cock slowly to the CLOSE position, remove the settling bowl and empty any accumulated water.

4. Perform a normal tune-up as described earlier, adjust valve clearance, apply oil to camshaft, and while checking spark plugs add a few drops of oil to the cylinder. Be especially certain to de-grease ignition points if an oily card was used to inhibit oxidation during storage; use a non-petroleum solvent such as trichloroethylene or denatured alcohol.

5. Check safety items, i.e., lights, horn, etc., as oxidation of switch contacts and/or sockets during storage may make one or more of these critical devices inoperative.

6. Test ride and clean the motorcycle.

CHAPTER THREE

TROUBLESHOOTING

Diagnosing mechanical problems is relatively simple if orderly procedures are used and a few basic principles are kept in mind.

The troubleshooting procedures in this chapter analyze typical symptoms and show logical methods of isolating causes. These are not the only methods. There may be several ways to solve a problem, but only a systematic, methodical approach can guarantee success.

Never assume anything. Don't overlook the obvious. If while riding along, the bike suddenly quits, check the easiest, most accessible problem spots first. Is there gasoline in the tank? Is the gas petcock on? Has a spark plug wire fallen off? Check the ignition switch. Sometimes the weight of keys on a key ring may turn the ignition off suddenly.

If nothing obvious turns up in a cursory check, look a little further. Learning to recognize and describe symptoms will make repairs easier at home or by a mechanic at a shop. Saying that "it won't run" isn't the same as saying "it quit on the highway at high speed and wouldn't start,' or that "it sat in my garage for 3 months and then wouldn't start."

Gather as many symptoms together as possible to aid in diagnosis. Note whether the engine lost power gradually or all at once, what color smoke (if any) came from the exhaust, and so on. Remember that the more complicated a machine is, the easier it is to troubleshoot because symptoms point to specific problems.

After the symptoms are defined, areas which could cause the problems are tested and analyzed. Guessing at the cause of a problem may provide the solution, but it can easily lead to frustration, wasted time, and a series of expensive, unnecessary replacement of parts.

Neither fancy equipment nor complicated test gear is needed to determine whether repairs can be attempted at home. A few simple checks could save a large repair bill and time lost while the bike sits in a dealer's service department. On the other hand, be realistic and don't attempt repairs beyond your abilities. Service departments tend to charge heavily for putting together a disassembled engine that may have been abused. Some won't even take on such a job—so use common sense; do not get in too deeply.

OPERATING REQUIREMENTS

An engine needs 3 basics to run properly: correct gas/air mixture, compression, and a spark at the right time. If one or more are missing, the engine will not run. The electrical system is the weakest link of the three. More problems result from electrical breakdowns than from any other source. Keep that in mind before you begin tampering with carburetor adjustments and the like.

If a bike has been sitting for any length of time and refuses to start, check the battery for a charged condition first, and then look to the gasoline delivery system. This includes the tank, fuel petcocks, lines, and carburetors. Rust may have formed in the tank, obstructing fuel flow. Gasoline deposits may have gummed up carburetor jets and air passages. Gasoline tends to lose its potency after standing for long periods. Condensation may contaminate it with water. Drain old gas and try starting with a fresh tankful.

TROUBLESHOOTING INSTRUMENTS

Chapter One lists many of the instruments needed and detailed instructions on their use.

STARTING DIFFICULTIES

Check gas flow first. Remove the gas cup and look into the tank. If gas is present, pull off a fuel line at the carburetor and see if gas flows freely. If none comes out, the fuel tap may be shut off, blocked by rust or other foreign matter, or the fuel line may be stopped up or kinked. If the carburetor is getting usable fuel, inspect the electrical system next.

Check that the battery is charged by turning on the lights or by beeping the horn. Refer to your owner's manual for starting procedures with a dead battery. Have the battery recharged if necessary.

Pull off a spark plug cap, remove the spark plug, and reconnect the cap. Lay the plug against the cylinder head so its base makes a good connection, and turn the engine over with the kickstarter (engine rotation of at least 500 rpm). A fat, blue spark should jump across the electrodes. If there is no spark, or only a weak one, there is electrical system trouble. Check for a defective plug by replacing it with a known good one. Don't assume a plug is good just because it's new.

Once the plug has been cleared of malfunctioning but there's still no spark, start backtracking through the system. If the contact at the end of the spark plug wire can be exposed, it can be held about ⅛ inch from the head while the engine is turned over to check for a spark. Remember to hold the wire only by its insulation to avoid a nasty shock. If the plug wires are dirty, greasy, or wet, wrap a rag around them so you don't get shocked. If you do feel a shock or see sparks along the wire, clean or replace the wire and/or its connections.

If there's no spark, or only a weak spark, at the plug wire, look for loose connections at the coil and battery. If all seems in order there, check next for oily or dirty contact points. Clean points with electrical contact cleaner, or a strip of paper. On battery ignition models, with the ignition switch turned on, open and close the points manually with a screwdriver.

No spark at the points with this test indicates a failure in the ignition system. Refer to *Ignition System* section of this chapter for checkout procedures for the entire system and individual components. Refer to Chapter Two for checking and setting ignition timing.

Note that spark plugs of the incorrect heat range (too cold) may cause hard starting. Set gaps to specifications. If you have just ridden through a puddle or washed the bike and it won't start, dry off plugs and plug wires. Water may have entered the carburetor and fouled the fuel under these conditions, but wet plugs and wires are the more likely problem.

If a healthy spark occurs at the right time, and there is adequate gas flow to the carburetor, check the carburetor itself. Make sure all jets and air passages are clean. Check float level and adjust if necessary. Shake the float to check for gasoline inside it, and replace or repair as indicated. Check that the carburetors are mounted snugly and that no air is leaking past the manifold. Check for a clogged air filter.

Compression, or the lack of it, usually occurs only in the case of older machines. Worn or broken pistons, rings, and cylinder bores could prevent starting. Generally a gradual power loss and harder starting will be readily apparent in this case.

Compression may be checked in the field by turning the kickstarter by hand and noting that an adequate resistance is felt.

An accurate compression check gives a good idea of the condition of the basic working parts of the engine. To perform this test, you need a compression gauge. The motor should be warm.

1. Remove spark plug and clean out any dirt or grease.

2. Insert the tip of the gauge into the hole, making sure it is seated correctly.

3. Open the throttle all the way and make sure the choke on the carburetor is open.

4. Crank the engine several times and record the highest pressure reading on the gauge. Run the test on each of the cylinders. Normal compression value ranges for various models are given in **Table 1**. If the reading obtained is significantly lower than listed in the table, proceed to the next step.

Table 1   COMPRESSION VALUES

Model	Normal Compression	
	Psi	Kg/Cm²
T140V, TR7V	100 - 125	7.03 - 8.79
650cc	105 - 130	7.38 - 9.10
T100R, T100S, T100T	105 - 130	7.38 - 9.10
5TA	85 - 105	5.98 - 7.38
T120TT	130 - 160	9.10 - 11.25

5. Pour a tablespoon of motor oil into the cylinder and record the compression. If oil raises the compression significantly—10 psi in an old engine—the rings are worn and should be replaced.

Valve adjustments should be checked. Sticking, burned, or broken valves may hamper starting. As a last resort, check valve timing.

POOR IDLING

Poor idling may be caused by incorrect carburetor adjustment, incorrect timing, or ignition system defects. Check the gas cap vent for an obstruction.

MISFIRING

Misfiring can be caused by a weak spark or dirty plugs. Check for fuel contamination. Run the machine at night to check for spark leaks along plug wires and under the spark plug cap.

WARNING
Do not run engine in closed garage. There is considerable danger of carbon monoxide poisoning.

If misfiring occurs only at certain throttle settings, refer to the fuel system chapter for the specific carburetor problem involved. Misfiring under heavy load, as when climbing hills or accelerating, is usually caused by bad spark plugs.

FLAT SPOTS

If the engine seems to die momentarily when the throttle is opened and then recovers, check for a dirty main jet in the carburetor, water in the fuel, or an excessively lean mixture.

POWER LOSS

Poor condition of rings, pistons, or cylinders will cause a lack of power and speed. Ignition timing should be checked.

OVERHEATING

If the engine seems to run too hot all the time, be sure you are not idling it for long periods. Air-cooled engines are not designed to operate at a standstill for any length of time. Heavy stop and go traffic or slow hill-climbing is hard on a motorcycle engine. Spark plugs of the wrong heat range can burn pistons. An excessively lean gas mixture may cause overheating. Check ignition timing. Don't ride in too high a gear. Broken or worn rings may permit compression gases to leak past them, heating heads and cylinders excessively. Check oil level and use the proper grade lubricants.

ENGINE NOISES

Experience is needed to diagnose accurately in this area. Noises are hard to differentiate and harder yet to describe. Deep knocking noises usually mean main bearing failure. A slapping noise generally comes from loose pistons. A light knocking noise during acceleration may be a bad connecting rod bearing. Pinging should be corrected immediately or damage to pistons will result. Compression leaks at the head-cylinder joint will sound like a rapid on-and-off squeal.

PISTON SEIZURE

Piston seizure is caused by incorrect piston clearances when fitted, fitting rings with improper end gap, too thin an oil being used, incorrect spark plug heat range, or incorrect ignition timing. Overheating from any cause may result in seizure.

EXCESSIVE VIBRATION

Execessive vibration may be caused by loose motor mounts, worn engine or transmission bearings, loose wheels, worn swing arm bushings, a generally poor running engine, broken or cracked frame, or one that has been damaged in a collision. See also *Poor Handling*.

CLUTCH SLIP OR DRAG

Clutch slip may be due to worn plates, improper adjustment, or glazed plates. A dragging clutch could result from damaged or bent plates, improper adjustment, or uneven clutch spring pressure.

All clutch problems, except adjustments or cable replacement, require removal to identify the cause and make repairs.

1. *Slippage*—This condition is most noticeable when accelerating in high gear at relatively low speed. To check slippage, drive at a steady speed in fourth or fifth gear. Without letting up the accelerator, push in the clutch long enough to let engine speed increase (one or two seconds). Then let the clutch out rapidly. If the clutch is good, engine speed will drop quickly or the bike will jump forward. If the clutch is slipping, engine speed will drop slowly and the bike will not jump forward.

Slippage results from insufficient clutch lever free-play, worn friction plates, or weak springs. Riding the clutch can cause the disc surfaces to become glazed, resulting in slippage.

2. *Drag or failure to release*—This trouble usually causes difficult shifting and gear clash especially when downshifting. The cause may be excessive clutch lever free-play, warped or bent plates, stretched clutch cable, or broken or loose disc linings.

3. *Chatter or grabbing*—Check for worn or misaligned steel plate and clutch friction plates.

TRANSMISSION

Transmission problems are usually indicated by one or more of the following symptoms:

a. Difficulty in shifting gears

b. Gear clash when downshifting

c. Slipping out of gear

d. Excessive noise in neutral

e. Excessive noise in gear

Transmission symptoms are sometimes hard to distinguish from clutch symptoms. Be sure the clutch is not causing the trouble before working on the transmission.

POOR HANDLING

Poor handling may be caused by improper tire pressures, a damaged frame or swing arm, worn shocks or front forks, weak fork springs, a bent or broken steering stem, misaligned wheels, loose or missing spokes, worn tires, bent handlebars, worn wheel bearings, or dragging brakes.

BRAKE PROBLEMS

Sticking brakes may be caused by broken or weak return springs, improper cable or rod adjustment, or dry pivot and cam bushings. Grabbing brakes may be caused by greasy linings which must be replaced. Brake grab may also be due to out-of-round drums or linings which have broken loose from the brake shoes. Glazed linings or glazed brake pads will cause loss of stopping power. If brake fluid level drops rapidly over a short period of time, check the system for leakage.

ELECTRICAL PROBLEMS

Bulbs which continuously burn out may be caused by excessive vibration, loose connections that permit sudden current surges, poor battery connections, installation of the wrong type bulb, or a faulty voltage regulator.

A dead battery or one which discharges quickly may be caused by a faulty alternator or rectifier. Check for loose or corroded terminals. Shorted battery cells or broken terminals will keep a battery from charging. Low water level will decrease a battery's capacity. A battery left uncharged after installation will sulphate, rendering it useless.

A majority of light and horn or other electrical accessory problems are caused by loose or corroded ground connections. Check those first, and then substitute known good units for easier troubleshooting.

IGNITION SYSTEM

750/650cc Models

Troubleshooting the ignition system should be undertaken in an orderly sequence, as detailed below.

1. Clean the contact points and the spark plugs.
2. Check the low tension circuit as follows. Remove the fuel tank and disconnect the white lead connecting the sw terminals of the ignition coils. Turn the ignition switch to the IGN position and crank the engine slowly. The ammeter needle should jump from zero to the discharge side as the points open and close. The above test is made with the white lead from the wiring harness connected to the sw terminal of first the left coil and then the right coil. If the ammeter needle doesn't move as described, the low tension circuit is faulty.
3. Check the points for pitting, burning, correct setting, and general condition.
4. Use a DC voltmeter to check the entire system step-by-step. All wiring should be connected except on 12-volt systems where the white lead is disconnected from the zenor diode center terminal. Crank the engine over until both sets of points are open.
5. Check the battery ground connection by connecting the meter across the negative terminal and a bare spot on the frame. A low reading or none at all indicates a blown fuse or a poor ground connection.

6. Connect the meter between the left coil negative terminal and ground, then the right coil negative terminal and ground. No reading indicates a break between the battery and the negative terminals, or faulty switch or ammeter connections.
7. Connect the meter between both ammeter terminals in turn and a ground. No reading on the feed side indicates a bad ammeter or a poor connection between the terminals and the battery. A bad ammeter is indicated if there is a reading on the battery side only of the ammeter.
8. Connect the meter between the ignition switch input terminal and ground. The brown and white (or blue) lead should be inspected if there is no reading. Check for voltage at the lead's connections at the rectifier, ammeter, and lighting switch terminals 2 and 10 (2 and 3 on the 1972-1976 models).
9. Check the ignition switch output terminal. The switch should be replaced if no reading is found. A reading at this point and none at the coil negative terminals indicates the white lead is disconnected or broken.
10. Disconnect the black/white and black/yellow leads from the positive terminals of each coil. Connect the voltmeter across the positive terminal and ground of each coil in turn. No reading indicates a faulty coil which should be replaced.
11. Reconnect the coil leads and connect the meter across both sets of points, one set at a time. No reading indicates a broken connection or a faulty condenser. Replace the condenser with a known good one and repeat the test.
12. With 12-volt systems, reconnect the zener diode white lead and hookup the meter between the diode center terminal and ground. The meter reading should closely approximate 12 volts.

500cc Models

Procedures for troubleshooting the ignition system are same as above except, after Serial No. H57083, disconnect the brown/white lead instead of the white lead when performing Step 4, *Ignition System.*

IGNITION COILS
(ALL MODELS)

Check the coils for satisfactory operation after all the above tests, under *Ignition System*, have been made. Remove the high tension leads from both spark plugs, and with the ignition on, crank the engine until the right set of points (with the black/yellow lead) are closed. Hold the spark plug lead approximately 3/16 in. (4.8mm) from the head or block and manually open and close the ignition points. A strong spark should jump between the wire and the engine. If it does not, the coil is probably faulty. Check the left coil with the opposite set of points closed. Be sure the points are clean, gapped properly, and not grounded. Check condition of the high tension leads before replacing the coil. The coils may be removed and tested on a special coil tester. Also check the ignition suppressor caps.

EMERGENCY
STARTING CIRCUIT

1. To check emergency starting circuit on 500cc machines prior to H40528, begin by checking that the point setting and spark plug gap is correct. Insulate the points so they can't make electrical contact. Connect a DC voltmeter so its positive lead is grounded and the negative lead is connected to the moving contact spring of the front set of points.

2. Turn the ignition switch to the IGN position and check that the voltmeter reads battery voltage. Repeat the test with the voltmeter negative lead connected to the moving contact spring of the rear set of points. Disconnect the green/yellow lead from the alternator and connect the positive voltmeter lead to the green/yellow harness lead. Connect the negative lead to the frame and turn the ignition switch to the ENG position. Battery voltage should be indicated on the voltmeter. If not, check the green/yellow lead to terminal 17 on the ignition switch. The black/white lead from the ignition coil positive terminal to terminal 15 on the ignition switch should also be checked.

3. Reconnect the alternator lead, disconnect the battery, and connect the voltmeter between the frame and front moving contact spring. With the points still insulated and the switch in ENG position, turn the engine over with the kickstarter. The voltmeter should momentarily read 7-10 volts. If it does not, check the alternator as detailed later in this chapter.

CHARGING SYSTEM
750/650cc Models

1. Refer to **Figures 1, 2, and 3** for schematic diagrams of the types of charging circuits used on these models. On a machine prior to DU. 24875, check the charging rate in the 3 switch positions first. On other models, proceed in the sequence given below.

2. Check the battery input, using a fully charged or new battery. Connect a DC ammeter in series between the battery main lead (brown/blue) and the center terminal of the rectifier. Be sure the meter is not touching the frame. Run the engine at 3,000 rpm and check the meter readings for each position of the lighting switch against the data in **Tables 2 and 3**. If readings are not high enough, proceed with testing.

3. Disconnect the 3 alternator output leads at the bottom of the engine and rev the engine to 3,000 rpm. Connect an AC voltmeter with a one ohm resistor parallel with the alternator leads one at a time and check the readings. They should all be 9 volts or higher. A low reading on any group of coils indicates the leads are chafed or that some coils are shorted. Low readings on all coils indicate the green/white lead is chafed or damaged or that the rotor is demagnetized. Check that the rectifier is good and that the battery is connected properly before replacing the rotor. A zero reading for any group of coils indicates a disconnected, open, or shorted coil. If a reading is obtained between one lead and ground, a coil or connection has become grounded. Check that stator leads haven't been damaged by the chain before repairing or replacing the stator.

500cc Models

Refer to Figures 1 and 2 for schematic diagrams.

The DC input to the battery, alternator output, and rectifier is the same as detailed above for

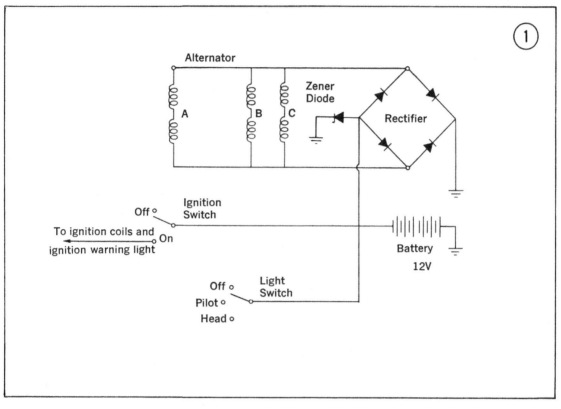

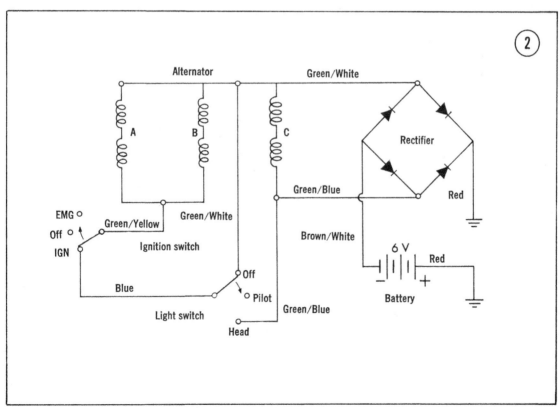

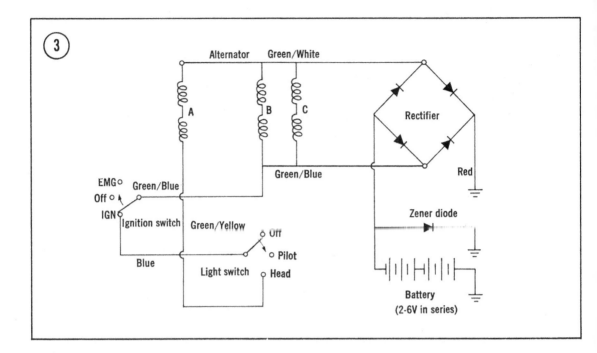

③

Alternator   Green/White

A   B   C

Rectifier

Green/Blue

Red

EMG○ Green/Blue
Off ○
IGN○
Ignition switch   Green/Yellow

Zener diode

○ Off
○ Pilot
Blue
Light switch   ○ Head

Battery
(2-6V in series)

650/750cc models. Procedures for checking the charging system follow.

1. On 6-volt matchines, check that the battery is in good condition and the alternator leads under the engine are disconnected.

2. Check the voltage at the rectifier center terminal by connecting a DC voltmeter with one ohm resistor in parallel between the rectifier center terminal and ground (positive voltmeter lead). If the reading is not equal to battery voltage, there is a bad connection in the wiring of the low tension circuit.

3. Use a jumper wire to connect the green/yellow lead from the wiring harness under the engine to the rectifier center terminal lead (brown/white). Turn the ignition switch to the IGN position and connect the voltmeter with load resistor in parallel between the green/white lead at the rectifier and the frame (ground). With the light switch OFF, the voltmeter should again read battery voltage. If it does not, check the leads to ignition switch terminals 16 and 18 and leads 4 and 5 to the light switch.

4. Use a jumper wire to connect the green/yellow lead as in Step 3 above. Turn the ignition switch to IGN and the light switch to HEAD position. Connect the DC voltmeter with resistor in

parallel between the green/black lead at the rectifier and ground. Again, the voltmeter should read battery voltage. If not, check the leads to ignition switch terminals 16 and 17 and the leads to light switch terminals 5 and 7. With the light switch turned to PILOT there should be no reading between green/black and ground or green/white and ground at the rectifier.

5. On 12-volt systems, check that the battery is charged and installed properly, with a positive ground. Check that the fuse is good. Check that battery voltage is present at the rectifier center terminal (brown/white). If not, disconnect the alternator leads at the snap connectors underneath the engine (green/black, green/white, green/yellow or green/yellow and green/white on later machines with 2 lead stators). Fit a jumper wire across the brown/white and green/yellow connections at the rectifier and measure the voltage at the snap connector. The result indicates whether or not the harness alternator lead has an open circuit. Repeat the test at the rectifier for the white/green lead.

6. If there is no voltage at the rectifier center terminal, check the voltage at the ammeter. If a reading is obtained there, the brown/white wire has an open circuit. If no reading is obtained, the ammeter has an open circuit. If there is no

Table 2 ALTERNATOR AND STATOR DETAILS — SPECIFICATIONS AND OUTPUT FIGURES
Model 650/750 Machines prior to DU.24875 and Model 500

Models	System Voltage	Ignition Type	Alternator Type	Stator No.
3TA, 5TA, T90, T100, 6T	12V			47162 (After H57083 = No. 47204)
3TA, 5TA, T90, T100, TR6SR, T120R, T120C, 6T U.S.A.	6V	Coil	RM.19	
6T, TR6, T120	6V	Coil	RM.19	47164
3TA, 5TA and 6T (Police)	6V	Coil	RM.19/20	47167
T90, T100, (E.T.), TR6SC, T120TT (Special)	6V	AC Ign.	RM.19	47188

Stator Number	System Voltage	DC Input to Battery Amps @ 3,000 rpm			Alternator Output Minimum AC Volts @ 3,000 rpm		
		Off	Pilot	Head (Main beam)	A	B	C
47162	6V	2.75	2.0	2.0	4.0	6.5	8.5
	12V	2.0[3]	2.1[3]	1.5[3]			
		4.8[4]	3.8[4]	1.8[4]			
47164	6V	2.7	0.9	1.6	4.5	7.0	9.5
47167	6V	6.6[5]	6.6[5]	13.6[5]	7.7	11.6	13.2
47188	6V	Not applicable			5.0	1.5	3.5
47204/47205	12V	Not applicable			—	—	8.5

Coil Ignition Models
A = Green/White and Green/Black.
B = Green/White and Green/Yellow.
C = Green/White and Green/Black or Green/Yellow connected.

AC Ignition Models
A = Red and Brown/Blue.
B = Black/Yellow and Black/White.
C = Black/Yellow and Brown.

Note: On machines fitted with two lead stator, only test C is
applicable as leads are colored Green/White and Green/Yellow.

1. Ignition.   2. Lights.   3. Zener in circuit.   4. Zener disconnected.   5. With boost switch in circuit.

reading at either ammeter terminal, check the brown/blue lead from the battery.

RECTIFIER

The rectifier must be kept clean and dry. See Chapter Seven for the removal/installation procedures.
1. Be sure the alternator output and charging

circuits wiring are satisfactory before testing the rectifier. Begin by disconnecting the brown/white (brown/blue on 1972-1976) lead from the rectifier center terminal. Insulate the bare lead and then connect a DC voltmeter with a one ohm load resistor parallel between the rectifier center terminal and ground. Ground the voltmeter positive lead. Disconnect the alternator green/yellow (green/black on 12V systems) lead and recon-

Table 3   ALTERNATOR AND STATOR DETAILS — SPECIFICATIONS AND OUTPUT FIGURES
Model 650/750 Machines subsequent to DU.24875

Models	System Voltage	Ignition Type	Alternator Type	Stator No.
T120, TR6, 6T, T120R, TR6R	12V	Coil	RM.19	47162 (After DU.58565 = No. 47204)
T120TT, TR6 c	6V	AC Ign.	RM.19	47188
T120R, T120RV, TR6R, TR6RV, TR6C, TR6CV, T140V, TR7V	12V	Coil	RM.21	47205

Stator Number	System Voltage	DC Input to Battery Amps @ 3,000 rpm			Alternator Output Minimum AC Volts @ 3,000 rpm		
		Off	Pilot	Head (Main beam)	A	B	C
47162	12V	1.5^3	1.0^3	1.0^3	4.0	6.5	8.5
		6.5^4	4.3^4	2.0^4			
47188	6V	Not applicable			5.0	1.5	3.5
47204/47205	12V	Not applicable			—	—	8.5

Coil Ignition Models
   A = Green/White and Green/Black.
   B = Green/White and Green/Yellow.
   C = Green/White and Green/Black or Green/Yellow connected.

   1. Ignition.   2. Lights.   3. Zener in circuit.   4. Zener disconnected.

nect it to the rectifier green/black (green/yellow on 12V systems) terminal using a jumper lead. Check that all connections will not ground themselves accidentally and start the engine.

2. Check voltmeter readings with the engine running at 3,000 rpm. The reading should be 7.5 volts or slightly more. If the readings are much more than that, check the rectifier ground connection. If the ground is good, the rectifier probably needs replacement. If the reading is zero or much lower than 7.5 volts, the rectifier or charging circuit wiring is bad. Remove the rectifier and bench test it.

3. To bench test the rectifier, disconnect the battery terminals and then disconnect and remove the rectifier. Refer to **Figure 4** and hook

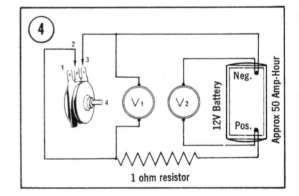

1 ohm resistor

up the rectifier, a battery, a one ohm resistor, and the voltmeter as shown. Position V1 will measure the voltage drop across the rectifier plate. V2 will measure the supply voltage and confirm it. Connect the test leads as shown in **Figure 5**. Perform Test 1 with the polarity as shown. Voltage should not be greater than 2.5 volts. Reverse the polarity and repeat measurement as shown for Test 2. Voltage should not be more than 1.5 below battery voltage. If any of the tests fail, the rectifier is faulty and should be replaced.

CHARGING CIRCUIT CONTINUITY (ALL MODELS)

1. Begin by checking that the battery is charged and correctly installed with a positive ground. Check the fuse also.

2. Connect a DC voltmeter with a one ohm load resistor in parallel between the rectifier center terminal and ground. Late model rectifiers have an end terminal, not to be confused with the center terminal. The voltmeter reading battery voltage will indicate power to the rectifier center terminal. If the meter does not read battery voltage, disconnect the alternator leads (green/black, green/white, and green/yellow) at the connectors under the engine. Place a jumper lead across the brown/white (brown/blue) and green/yellow connections at the rectifier and measure the voltage at the snap connector. This indicates presence of an open circuit at the harness alternator lead. Repeat this test at the rectifier using the white/green lead.

3. If there is no voltage at the rectifier center terminal, measure the voltage at the ammeter terminal. Voltage indicates the brown/white (brown/blue) lead has an open circuit, while no voltage at the ammeter indicates an open circuit at the ammeter. No voltage at either ammeter terminal indicates the brown/blue wire from the battery has an open circuit.

4. On machines prior to the engine number DU.24875, with 6-volt system, check voltage at the rectifier center terminal and proceed as follows. Use a jumper lead to connect the green/yellow lead from the wiring harness under the engine to the rectifier center terminal lead (brown/white). Turn the ignition switch on and connect the DC voltmeter (with load resistor) between the green/white lead and ground. With the lighting switch turned off, the voltmeter should read battery voltage. If it does not, check the leads to ignition switch terminals 16 and 18 and leads to lighting switch terminals 4 and 5 (see appropriate wiring diagram at end of book).

5. The next test can be performed on both 6- and 12-volt systems. Connect the green/yellow or green/black lead from the wiring harness under the engine to the rectifier center terminal

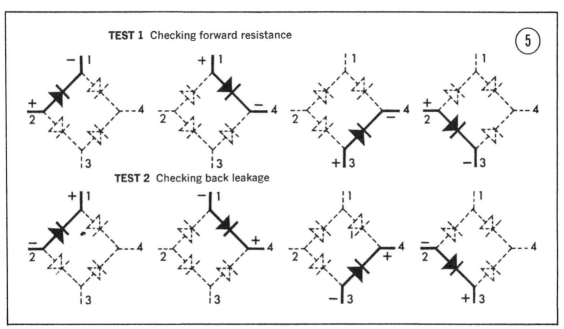

TEST 1 Checking forward resistance

TEST 2 Checking back leakage

using a jumper wire. Turn the ignition switch to IGN position and the lighting switch to HEAD position. Connect a DC voltmeter with one ohm resistor in parallel between the green/black lead (green/yellow lead on 12-volt systems) at the rectifier and ground. The reading should be the battery voltage. If it is not, check the leads to ignition terminals 16 and 17 and the lighting switch leads to terminals 5 and 7 (see wiring diagram). Switch the lighting switch to PILOT position and check the reading at the rectifier between the green/black (green/yellow on 12-volt systems), and ground and green/white and ground. The reading should be zero.

ZENER DIODE
(ALL MODELS)

The zener diode can be tested on the bike by following this procedure. Use good quality instruments; a voltmeter which reads to 18 volts and an ammeter which reads to 5 amperes. Disconnect the diode lead and connect the ammeter in series between the diode's Lucar terminal and the disconnected lead, so the ammeter's positive lead is connected to the Lucar terminal. Connect the voltmeter across the diode and the heat sink so the positive lead is attached to the heat sink and the negative lead is connected to the Lucar terminal. With all lights off, start the engine and watch both meters as engine speed is increased. The ammeter must read zero amps until the voltmeter reads 12.75 volts. Continue to increase engine speed until the ammeter reads 2 amps. At this point the voltage should read between 13.5 and 15.3 volts. Both tests must yield the above results or the diode should be replaced.

ENERGY TRANSFER IGNITION AND
LIGHTING SYSTEMS (ALL MODELS)

Refer to **Figure 6** for a schematic of the Energy Transfer system.

1. To test the system, check the spark plugs, points, and timing to see that they are set correctly and are in good condition. Disconnect both high tension spark plug leads and operate the kickstarter with one lead at a time held approximately 3/16 in. (4.8mm) from the engine head or cylinder. A fat blue spark should jump

from the lead to the engine. If not, the coil or alternator ignition supply should be checked.

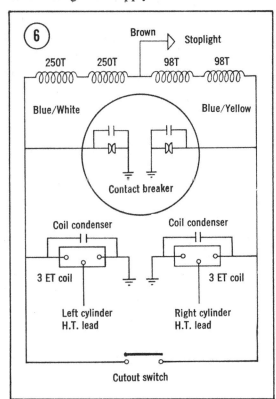

2. Check the coils using two 6-volt batteries and connecting them to the coils for the shortest time possible to avoid damage. Disconnect the 5 alternator leads beneath the engine. Unplug the black/yellow lead from the condenser at the right-hand coil underneath the gas tank. Connect the black/yellow lead to the positive terminal of a 6-volt test battery. Connect the negative battery terminal to the condenser terminal with a test lead. Unplug the black/white lead from the left-hand condenser and connect it to the positive terminal of another 6-volt test battery. Connect the second battery's negative terminal to the left condenser terminal. Remove both high tension leads from the spark plugs and open and close the contact breaker points. A fat blue spark should jump from the plug leads to ground approximately 3/16 in. (4.8mm) away. Check the connections if only a weak spark or none is produced, and repeat the test. If no spark can be induced, replace first the condensers and then the coils until a good spark is obtained.

ALTERNATOR OUTPUT

1. To check alternator output, the preferred method is to bolt a pair of MA6 coils together and connect the positive leads from the machine (black/white, black/yellow) to the positive terminals on the coils. Link the coil negative terminals together and connect them to the negative terminal of a test battery. Connect the positive battery terminal to the coil cases. Connect a wire from the positive battery terminal to a ground on the motorcycle frame. Disconnect the 5 alternator leads at the connectors under the engine, start the engine, and run it at 3,000 rpm. Connect an AC voltmeter with a one ohm resistor in parallel between the pairs of alternator leads given in Tables 2 and 3.

2. If the readings are the same as, or higher than, those given in the tables, alternator output is adequate. If there is a low reading on any group of coils, the leads may be damaged or the coil wiring could be shorted. If low readings are found at all coils, the rotor is probably demagnetized and must be replaced. A zero reading for any group of coils indicates an open circuit. The stator should be replaced. Any reading obtained between a stator lead and ground indicates coil turnings which have become grounded to the engine. Clean the coils with solvent or kerosene and test again. If a reading is still found, the stator will have to be replaced. In any case, check stator leads for chain damage before replacing the stator. See Tables 2 and 3.

Direct Lighting System

The lighting system is powered by the red, brown, and brown/blue leads from the alternator. Before the alternator is suspected in case of a loss or weakening of any lights, check the wiring carefully for loose or intermittent connections. Check the ground connections; in particular, check the red ground lead from the headlight and alternator. The way the system is wired, a short in the brown stoplight lead will cause a failure of the ignition system. The stoplight switch connections must therefore always be kept clean and dry. See Chapter Seven for headlight adjustment.

Horn

If the horn fails to operate, check the mounting bolts and wiring. A partially discharged battery may also cause a malfunction of the horn. If all connections are good, try adjusting the adjustment screw shown in **Figure 7**. Turn the screw out until the horn just stops sounding and then turn the screw back ¼ -½ turn.

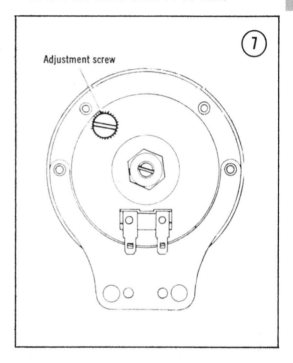

Adjustment screw

Fuses

A fuse is fitted to the brown/blue lead from the battery after DU.66246. Before this number, the fuse is fitted to the red ground lead from the battery.

Any coil-ignition Triumph can, and should, be fitted with a fuse between the battery positive terminal and ground. The original British fuse is rated "35 amps" but this means it will blow at 35 amps. The closest American fuse replacement is a 14 amp fuse.

TROUBLESHOOTING GUIDE

The following quick reference guide (**Table 4**) summarizes part of the troubleshooting process. Use it to outline possible problem areas, then refer to the specific chapter or section that is involved.

Table 4   TROUBLESHOOTING GUIDE

LOSS OF POWER

Cause	Things to Check	Cause	Things to check
Poor compression	Piston rings and cylinders Head gaskets Crankcase leaks	Improper mixture	Dirty air cleaner Restricted fuel flow Gas cap vent holes
Overheated engine	Lubricating oil supply Clogged cooling fins Ignition timing Slipping clutch Carbon in combustion chamber	Miscellaneous	Dragging brakes Tight wheel bearings Defective chain Clogged exhaust system

GEARSHIFTING DIFFICULTIES

Cause	Things to check	Cause	Things to check
Clutch	Adjustment Springs Friction plates Steel plates	Transmission	Oil quantity Oil grade Return spring or pin Change lever or spring Drum position plate Change drum Change forks

STEERING PROBLEMS

Problem	Things to check	Problem	Things to check
Hard steering	Tire pressure Steering damper adjustment Steering stem head Steering head bearings Steering oil damper	Pulls to one side (cont.)	Defective swinging arm Defective steering head Defective steering oil damper
Pulls to one side	Unbalanced shock absorbers Drive chain adjustment Front/rear wheel alignment Unbalanced tires	Shimmy	Drive chain adjustment Loose or missing spokes Deformed rims Worn wheel bearings Wheel balance

BRAKE TROUBLES

Problem	Things to check	Problem	Things to check
Poor brakes	Worn linings or pads Brake adjustment Oil or water on brake linings Loose linkage or cables Loss of disc brake fluid	Noisy brakes	Worn or scratched lining Scratched brake drums Dirt in brake housing Disc distortion
		Unadjustable brakes	Worn linings Worn drums Worn brake cams

NOTE: If you own a 1978 or later model, first check the Supplement at the back of the book for any new service information.

CHAPTER FOUR

4

ENGINE

Service of the Triumph 750/650cc and 500cc engines is basically similar. Engine service procedures for all 3 models are given in this chapter. Refer to Figures 1 and 2, Chapter Two, for illustrations of street and off-road models.

ENGINE
REMOVAL/INSTALLATION

750/650cc Models

1. Shut off the gas at the fuel tap and disconnect the fuel lines.

2. Remove the fuel tank mounting bolts and remove the tank. Earlier models with a headlight nacelle will require removal of the 2 rear nacelle screws to permit gas tank removal. On models with centerpole mount fuel tanks, remove the rubber plugs from the top of the tank, unscrew the retaining nut, and lift off the tank.

> NOTE: On 750cc models, remove the crossbrace located under the bottom front of the tank before lifting it off.

3. Disconnect the fuse at its holder or disconnect battery leads at their terminals, removing the ground wire first. Disconnect the connectors at the coils, remove the top and bottom coil mounts, and remove the coils. Handle the coils carefully.

4. Unscrew the 4 nuts holding the torque stays to the cylinder head. Remove the front and rear torque stay mounting bolts, the spacers, and the stays.

5. Disconnect the speedometer cable from the speedometer, removing the headlight if necessary to gain access to the bottom of the speedometer. If a tachometer is fitted, disconnect the tachometer cable from the right angle drive gearbox on the crankcase. See **Figure 1**.

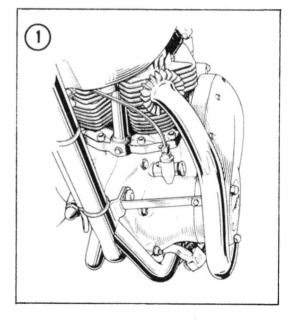

6. Early models were fitted with rear enclosure panels. Remove by unscrewing all mounting bolts and nuts. Set aside all spacers.

7. Remove the air cleaner, then remove the carburetor mounting nuts and take off the carburetors. Unhook any cable clips necessary to swing the carburetors out of the way.

8. Drain the engine oil, oil tank, primary case, and gearbox by removing the drain plugs. Early models may not have a drain plug on the oil tank; in that case remove the oil feed pipe from the tank.

9. Remove the rocker oil feed line by unscrewing the 2 cap nuts from the spindles. Move it carefully out of the way, bending the line as little as possible.

10. Disconnect the clutch cable from the operating arm by slackening the handlebar clutch adjuster as far as possible. Pull off the rubber cover at the point where the clutch cable enters the gearbox. Unscrew the abutment. Remove the slotted plug on the case cover and disconnect the cable end from the clutch operation arm. Earlier models do not have an inspection hole in the cover; the cable may be removed from the arm after the abutment is unscrewed. See **Figure 2**.

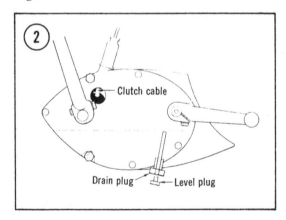

11. Disconnect the engine breather tubes at the rear of the crankcase.

12. Loosen the finned clips and muffler bolts and take off the front exhaust pipe nuts and bolts. Tap the exhaust pipes free with a plastic mallet. If you have a model with "siamesed" pipes, loosen the junction clip before attempting removal. It may also be necessary to remove the

footpegs on some models. Remove the brake pedal and the 2 bolts holding the footpegs, then give the pegs a tap with a mallet to break loose the tapered fit.

13. Remove the rear chainguard and disconnect the rear chain at its master link. Pull the chain off the rear sprocket.

14. Disconnect the generator leads underneath the engine.

15. Remove bolts and nut holding each of the left and right rear engine mounting plates. Remove the plates. Remove a nut and washer from each of the front and the bottom engine mounting studs.

16. Remove the 2 right side rocker box-to-torque stay bolts, the 2 right side screws holding the front and rear rocker cap retainer springs, and the left side bolts holding the rear frame and front frame.

17. The engine weighs about 135 lb., so use a hoist or have someone help with the next steps. Slide out the front and lower engine mounting studs and remove the engine from the left side. If necessary, the rocker boxes can be removed if the engine cannot be removed from the frame using the above method.

> NOTE: *Check the location and sizes of the spacers for reference during reassembly. The larger spacers should be reinstalled on the right side.*

18. Install the engine in the reverse order of removal, remembering to fit the bolts in Step 16 above while the engine is still loose in the frame. Replace the coils so their connector end terminal is toward the rear of the engine. Replace all spacers and mounting plates. Refill the engine, transmission, and primary case with oil before starting the engine.

500cc Model

1. Take off the gas tank and disconnect the spark plug leads.

2. Disconnect the battery leads and the Lucar connectors from the ignition coils. Remove the coil mounting bolts and spacers. Take the coils away, disconnecting the connectors between the points and the condensers. Be careful not to damage the outer casing of the coils (**Figure 3**).

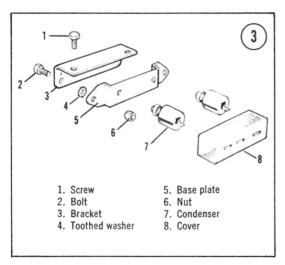

1. Screw
2. Bolt
3. Bracket
4. Toothed washer
5. Base plate
6. Nut
7. Condenser
8. Cover

3. Unscrew the 2 nuts holding the torque stays to the cylinder head. Remove the stay mounting bolt and the torque stays. If a tachometer is fitted, disconnect the tachometer drive cable at the tachometer gearbox, shown in **Figure 4**, by unscrewing the union nut and pulling the cable free.

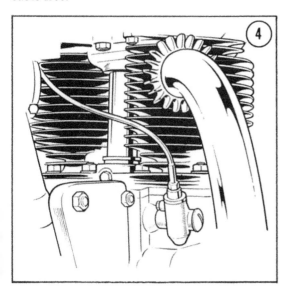

4. Remove the carburetor mounting nuts and move the carburetor out of the way, removing the cable from its clips as necessary. Remove the domed nuts and the oil feed bolts at the rocker boxes. Remove the oil feed pipe, bending it as little as possible.

5. Drain all oil at this point by removing the oil tank, gearbox, primary case drain plugs, and the filter drain plug underneath the engine. Disconnect the oil feed and return lines from the oil tank.

6. Loosen the clutch adjuster at the handlebar and pull the rubber cover away from the clutch abutment at the lower end of the cable. Unscrew the abutment and remove the clutch cable.

7. Loosen the finned clip bolts, the muffler clip bolts, and the nuts securing the exhaust pipes under the engine. If a "siamesed" pipe is fitted, loosen the junction clip. Free the exhaust pipes with a rubber mallet. See **Figures 5, 6, and 7**.

> NOTE: *On the TR5T, it is necessary to remove the skid plate before removing the exhaust system.*

8. Disconnect the drive chain at the master link and pull it off the engine sprocket.

9. Disconnect the generator leads at their snap connectors beneath the engine. On the TR5T, disconnect the rear breather tube from the primary chaincase.

10. Remove the 4 bolts holding the front engine plates to the frame and engine, then remove the plates. Remove the stud securing the underside of the engine to the frame. Remove the bolt which holds the rear engine plates to the transmission case.

11. Remove the 2 nuts holding the right rear engine plate to the frame and remove the plate. Remove the left front stud which holds the torque stay to the engine. Remove the right footrest (left for TR5T).

12. Remove the engine from the right side (left side for TR5T). It weighs approximately 105 lb. A hoist or an assistant will probably be necessary.

13. Replacing the engine in the frame is the reverse of the disassembly instructions, but remember the following points. Insert the bottom frame bolt into place first, then position the right rear engine plate and tighten the nuts finger-tight. The front engine plate goes in next and the bolts should be tightened after all are in place. Remember that the connector end of the coils goes toward the rear of the machine. Check the appropriate wiring diagram at the end of Chapter Seven when connecting the wiring. Pay particular attention to the spacers on the coil

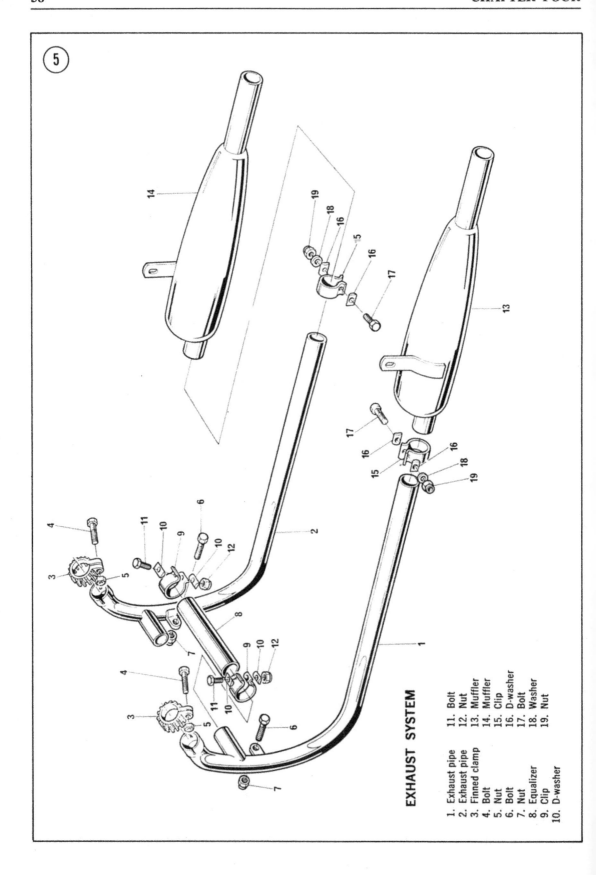

⑤

EXHAUST SYSTEM

1. Exhaust pipe
2. Exhaust pipe
3. Finned clamp
4. Bolt
5. Nut
6. Bolt
7. Nut
8. Equalizer
9. Clip
10. D-washer
11. Bolt
12. Nut
13. Muffler
14. Muffler
15. Clip
16. D-washer
17. Bolt
18. Washer
19. Nut

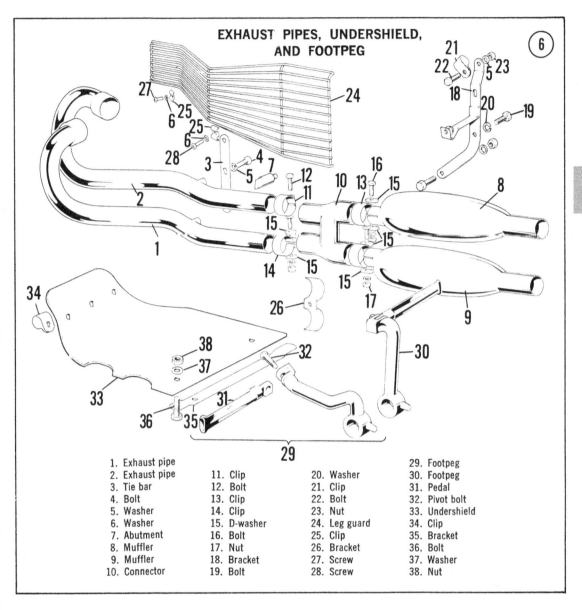

EXHAUST PIPES, UNDERSHIELD, AND FOOTPEG

6

1. Exhaust pipe
2. Exhaust pipe
3. Tie bar
4. Bolt
5. Washer
6. Washer
7. Abutment
8. Muffler
9. Muffler
10. Connector
11. Clip
12. Bolt
13. Clip
14. Clip
15. D-washer
16. Bolt
17. Nut
18. Bracket
19. Bolt
20. Washer
21. Clip
22. Bolt
23. Nut
24. Leg guard
25. Clip
26. Bracket
27. Screw
28. Screw
29. Footpeg
30. Footpeg
31. Pedal
32. Pivot bolt
33. Undershield
34. Clip
35. Bracket
36. Bolt
37. Washer
38. Nut

mounting bolts, the torque stay mounting bolts, and the bottom engine mounting stud. Fill the crankcase with oil prior to starting the engine for the first time.

ROCKER BOXES

750/650cc Models

1. To remove the rocker boxes (**Figures 8, 9, and 10**) with the engine still in the frame, follow Steps 1, 2, 3, 4, and 9 given above for engine removal (750/650cc models). Remove the rocker inspection covers.

2. Unscrew the 3 nuts from the rocker box studs and remove the outer exhaust box securing bolts and the center cylinder head bolts to free the rocker box. The inlet rocker box can be removed in the same manner. However, the 2 outer securing nuts may have to be unscrewed at the same time as the rocker box is lifted free to provide sufficient clearance for removal. Note that there is a washer under each of the underside securing nuts which sometimes sticks to the cylinder head flange.

3. Remove the pushrods and mark them so they can be replaced in their original locations.

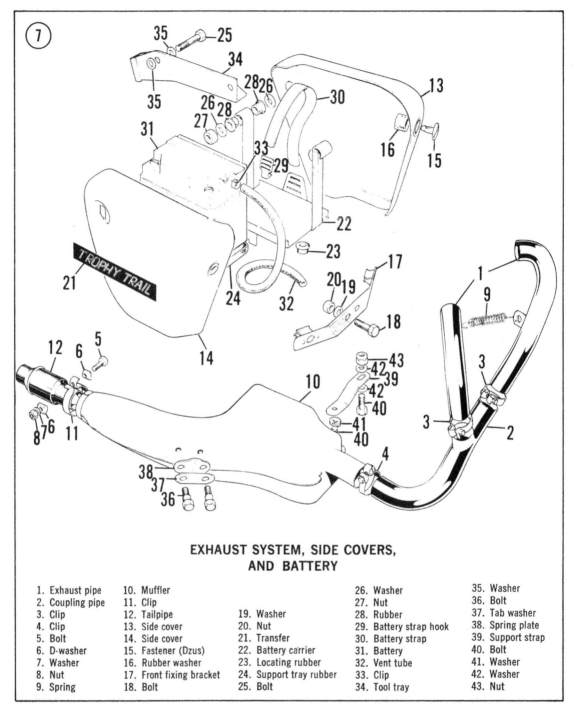

EXHAUST SYSTEM, SIDE COVERS,
AND BATTERY

1. Exhaust pipe	10. Muffler		26. Washer	35. Washer
2. Coupling pipe	11. Clip		27. Nut	36. Bolt
3. Clip	12. Tailpipe	19. Washer	28. Rubber	37. Tab washer
4. Clip	13. Side cover	20. Nut	29. Battery strap hook	38. Spring plate
5. Bolt	14. Side cover	21. Transfer	30. Battery strap	39. Support strap
6. D-washer	15. Fastener (Dzus)	22. Battery carrier	31. Battery	40. Bolt
7. Washer	16. Rubber washer	23. Locating rubber	32. Vent tube	41. Washer
8. Nut	17. Front fixing bracket	24. Support tray rubber	33. Clip	42. Washer
9. Spring	18. Bolt	25. Bolt	34. Tool tray	43. Nut

4. Clean the mating surfaces of the rocker boxes and cylinder head to remove any old gasket material or cement. Use new gaskets and cement when reassembling.

5. During reassembly, fit the pushrods in place by greasing their bottom cups and then lowering them onto the tappet balls. Test them by attempting to lift them off the tappet balls. The grease will offer a resistance to lifting if they are properly placed.

6. Remove the spark plugs and turn the engine over by hand until the intake pushrods are level

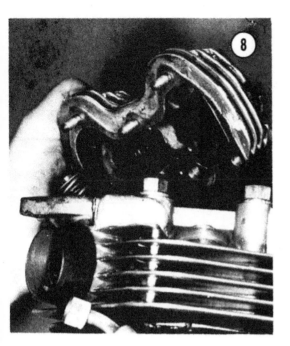

with each other and at the bottom of the stroke. Install the intake rocker box and then repeat the procedure for the exhaust rocker box. Tighten the 4 central bolts first, following the tightening sequence given in this chapter for head removal and replacement. Refer to the tables in Appendix I for torque specifications. Check that the valves are operating by slowly turning the engine over. Finally, tighten the underside nuts.

7. Replace all spacers on the torque stay mounting bolts and on the coil mounting bolts.

8. Use new copper washers when replacing the rocker oil feed pipe if possible. Old washers may sometimes be reused if they are heated to a cherry red and quenched in cold water. Brush off any scale which forms on the washers.

9. Pushrods should be inspected for worn or damaged ends and for straightness. Replace if necessary.

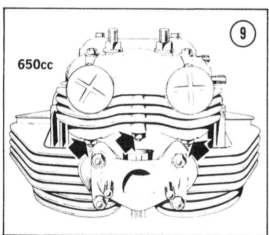

650cc

500cc Models

Remove and replace the rocker boxes as detailed previously, with the following minor differences. There are 2 nuts securing the torque stay on the exhaust rocker box. After the nuts are removed from the underside of each rocker box, unscrew the 2 Phillips head screws on top of each rocker box. Loosen all head bolts, unscrewing the central head bolts completely. When replacing the rocker boxes, tighten the central head bolts first, following the sequence given in **Figure 11**. Tighten the underside nuts and Phillips head screws after all head bolts are torqued down. Torque specifications are given in Appendix I. See **Figure 12**.

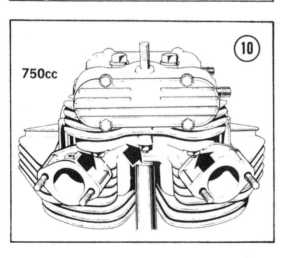

750cc

Disassembly/Assembly (All Models)

1. Drive the rocker spindles out of the boxes using a soft drift of the correct size. Remove the rocker arms and the various washers, noting their respective locations (**Figure 13**). Clean all parts in solvent. Make sure oil passages in the spindles are clear.

2. If oil leaks at rocker box-to-cylinder head and mating surface cannot be cured by replacing the gasket, the mating surface may be restored. Remove the rocker box studs and dress

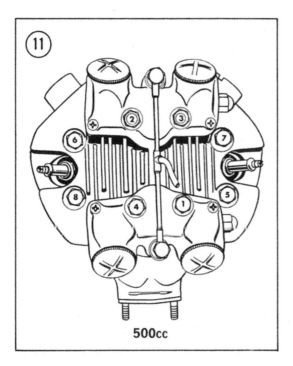

500cc

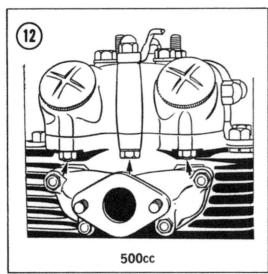

500cc

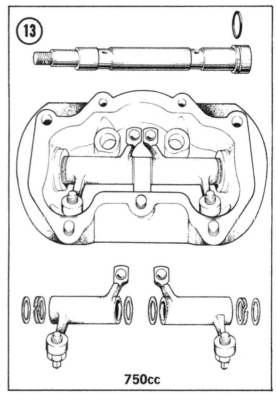

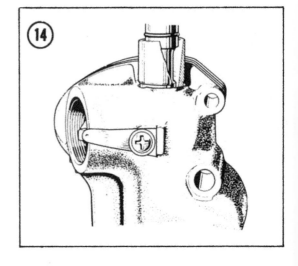

750cc

3. Remove and replace the spindle oil seals with new ones.

4. Rocker ball pins may be replaced if excessively worn by driving them out and pressing new ones in. The drilled flat faces the rocker spindle.

5. Compress the spindle oil seal so that it slides easily into place. A special Triumph tool is available to simplify this task. See **Figure 14**.

the mating surface on a piece of emery cloth laid on a surface plate or a piece of plate glass.

> NOTE: *To remove the rocker box studs, double-nut each stud in turn and unscrew it with a box-end wrench applied to the nut nearest the rocker box. To install the studs, apply the wrench to the nut farthest from the rocker box. Do not attempt to remove studs with pliers or a Vise Grip. Damage to the threads will result.*

6. A guide to align the spindle end as it is slid in place can be made by grinding a taper on the end of a 7/16 x 6 in. bolt.

7. Lay out the washers and rockers on a workbench in correct order prior to assembly. Starting at the left side (the oil seal end) of the spindle, there is a plain washer and a spring washer, then the left rocker, then a plain washer for each side of the spindle bearing or mount, the right rocker, a spring washer, and finally a washer with a smaller inside diameter than the others.

> NOTE: *On 1972 and later (750 and 650cc) models, both outer washers have smaller inside diameter than the inner washers.*

8. Coat 2 plain washers with grease so they will cling to the center bearing. Refer to Figure 14. Place the left rocker arm in position, holding it with the tapered bolt tool, and carefully insert a plain washer and spring washer against the rocker. Slide the spindle into the rocker box far enough to hold these in place. Insert the right rocker, a spring washer, and the plain washer with the small inside diameter. Push the lubricated spindle as far as possible. The spindle will have to be tapped into position with a soft metal drift and a hammer. Repeat the operation for the other rocker box.

CARBURETOR/AIR CLEANER

Carburetor and air cleaner service is given in Chapter Six. If it is necessary to remove the carburetor intake manifold, refer to **Figure 15** for 750/650cc models or **Figure 16** for 500cc models.

CYLINDER HEAD

Removal/Installation
(750/650cc Models)

Refer to **Figures 17 and 18** for this procedure.

1. Remove the rocker boxes and pushrods as described earlier in this chapter.

2. On 1971 and earlier models, remove the exhaust pipe by loosening the finned clip bolts and the muffler clip bolts. Remove the exhaust pipe bracket nuts and bolts. Tap the right ex-

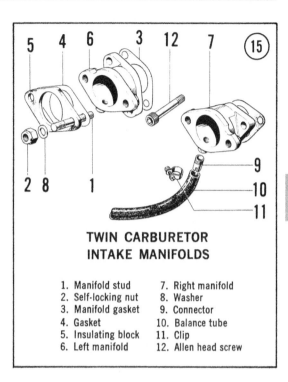

**TWIN CARBURETOR
INTAKE MANIFOLDS**

1. Manifold stud
2. Self-locking nut
3. Manifold gasket
4. Gasket
5. Insulating block
6. Left manifold
7. Right manifold
8. Washer
9. Connector
10. Balance tube
11. Clip
12. Allen head screw

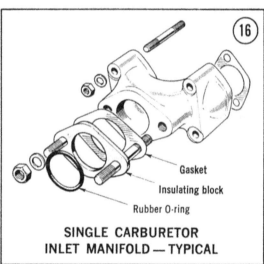

Gasket
Insulating block
Rubber O-ring

**SINGLE CARBURETOR
INLET MANIFOLD — TYPICAL**

haust pipe free with a rubber mallet, then the left exhaust pipe. If "siamesed" exhausts are used, slacken the joining clip before attempting to tap the pipes free.

3. On 1972 and later models, loosen the clamp bolts on the balance pipe and the finned clamps at the head. Loosen the head pipe-to-muffler clamps and disconnect the head pipes from the support brackets. Tap both head pipes away from engine with a soft mallet.

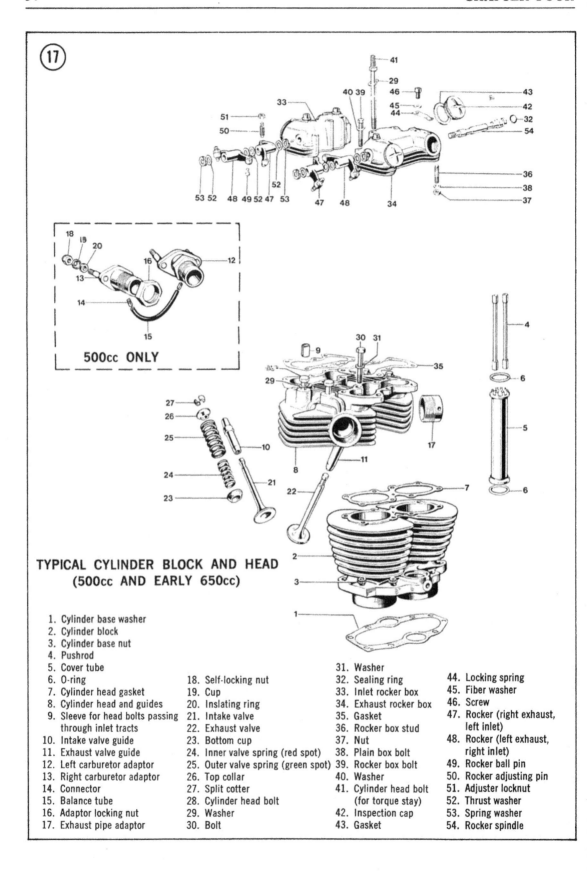

⑰

500cc ONLY

TYPICAL CYLINDER BLOCK AND HEAD
(500cc AND EARLY 650cc)

1. Cylinder base washer
2. Cylinder block
3. Cylinder base nut
4. Pushrod
5. Cover tube
6. O-ring
7. Cylinder head gasket
8. Cylinder head and guides
9. Sleeve for head bolts passing
   through inlet tracts
10. Intake valve guide
11. Exhaust valve guide
12. Left carburetor adaptor
13. Right carburetor adaptor
14. Connector
15. Balance tube
16. Adaptor locking nut
17. Exhaust pipe adaptor

18. Self-locking nut
19. Cup
20. Inslating ring
21. Intake valve
22. Exhaust valve
23. Bottom cup
24. Inner valve spring (red spot)
25. Outer valve spring (green spot)
26. Top collar
27. Split cotter
28. Cylinder head bolt
29. Washer
30. Bolt
31. Washer
32. Sealing ring
33. Inlet rocker box
34. Exhaust rocker box
35. Gasket
36. Rocker box stud
37. Nut
38. Plain box bolt
39. Rocker box bolt
40. Washer
41. Cylinder head bolt
    (for torque stay)
42. Inspection cap
43. Gasket

44. Locking spring
45. Fiber washer
46. Screw
47. Rocker (right exhaust,
    left inlet)
48. Rocker (left exhaust,
    right inlet)
49. Rocker ball pin
50. Rocker adjusting pin
51. Adjuster locknut
52. Thrust washer
53. Spring washer
54. Rocker spindle

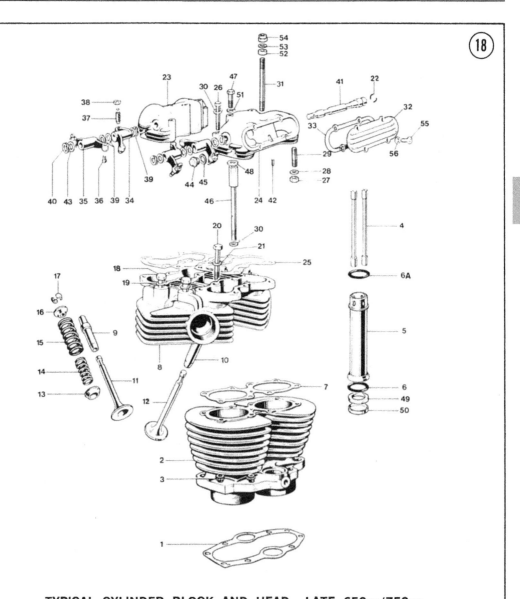

TYPICAL CYLINDER BLOCK AND HEAD—LATE 650cc/750cc

1. Cylinder base washer
2. Cylinder block
3. Cylinder base nut
4. Pushrod
5. Cover tube
6. Pushrod cover tube O-ring (bottom)
6a. Pushrod cover tube O-ring (top)
7. Cylinder head gasket
8. Cylinder head
9. Intake valve guide
10. Exhaust valve guide
11. Intake valve
12. Exhaust valve
13. Bottom cup
14. Inner valve spring (red spot)
15. Outer valve spring (green spot)
16. Top collar
17. Cotter pin
18. Cylinder head bolt
19. Washer
20. Bolt
21. Washer
22. Sealing ring
23. Intake rocker box
24. Exhaust rocker box
25. Gasket
26. Rocker box bolt
27. Rocker box nut
28. Washer
29. Rocker box stud
30. Washer
31. Cylinder head stud
32. Inspection cover
33. Gasket
34. Rocker (right exhaust, left intake)
35. Rocker (left exhaust, right intake)
36. Rocker ball pin
37. Rocker adjusting pin
38. Adjuster locknut
39. Thrust washer
40. Spring washer
41. Rocker spindle
42. Rocker box dowel
43. Thrust washer
44. Access plug
45. Washer
46. Cylinder head bolt
47. Rocker box bolt
48. Spacer
49. Sealing ring
50. Sleeve
51. Washer
52. Spacer
53. Washer
54. Nut
55. Screw
56. Washer

4. Remove the carburetors as described in *Carburetor Overhaul,* Chapter Six. On single carburetor models, the intake manifold must be removed. Take care not to lose the washers which are under the manifold mounting nuts.

5. Unscrew the remaining cylinder head bolts. Loosen them in sequence to prevent warping the head. See **Figures 19 and 20**. Lift off the head.

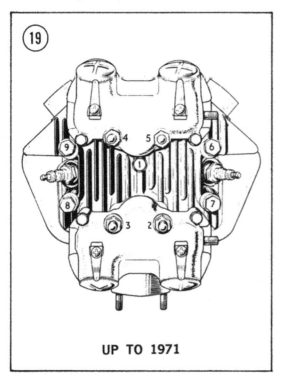

UP TO 1971

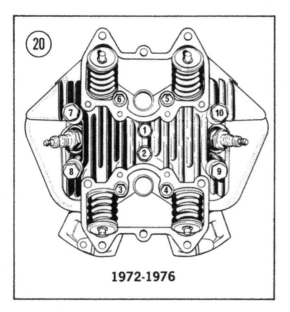

1972-1976

6. Remove the pushrod cover tubes. Replace the O-rings when reassembling. Check the tappet guide blocks for sharp edges which could damage the O-rings; smooth with emery cloth if necessary.

7. Check the copper head gasket for serviceability. If possible, replace with a new one; if not, anneal by heating it until cherry red and quenching in water. Use emery cloth to remove any scale that may have formed.

8. Clean the cylinder-to-head mating surfaces thoroughly and grease the gasket prior to assembly. Place the gasket on the cylinder and check that all 9 bolt holes are aligned. Position pushrod cover tubes with their top and bottom oil seals and cups in place.

Check the pushrod bores in the head for roughness and smooth with emery cloth if necessary.

9. Lower cylinder head in place and tighten the 4 outer and the center head bolts finger-tight. Align the pushrod tubes as shown in **Figure 21**.

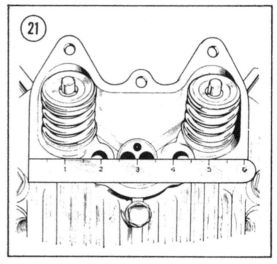

10. On 1971 and earlier engines, slowly rotate the crankshaft until the intake pushrods are at the bottom of their stroke. Fit the intake rocker box in place and insert the 2 center head bolts and tighten finger-tight. Reinstall the 2 outer rocker box bolts and 3 underside retaining nuts.

11. On 1972 and later engines, tighten the nuts and bolts in the order shown (Figure 20) to the torque specified in Appendix I. Install the rocker boxes as described earlier.

12. Repeat the above procedure for the exhaust rocker box.

13. On engines prior to 1972, tighten the head bolts in the sequence shown in Figure 19 and to the torque specified in Appendix I.

14. Complete assembly in the reverse order of removal. Set the tappet clearance as described earlier in this chapter.

15. An improved oil seal arrangement can be fitted to engines prior to engine No. DU.24875. See your Triumph dealer for details. Refer to **Figure 22**.

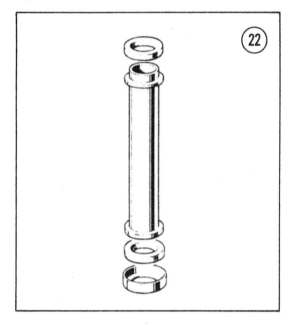

Removal/Installation (500cc Model)

1. Refer to **Figure 23** for removal of the cylinder head. Remove the rocker boxes as described earlier.

2. Remove the exhaust pipes as described in the *Engine Removal* section.

3. Remove the manifold securing nuts or socket screws. Leave the manifold in place until the head is removed.

4. Loosen the remaining cylinder head bolts one turn at a time until the head is loose. Lift the head and slide it forward to release the manifold.

5. Remove the pushrod cover tubes and replace the rubber seals. The copper cylinder head gasket should be replaced, or heated cherry red and then quenched in water to anneal it. Clean off

any scale which results from the annealing process with fine emery cloth.

6. Carefully clean the head and cylinder mating surfaces. Grease the gasket and position it so all 8 bolt holes are lined up. Place the pushrod cover tubes (**Figure 24**) into position on the tappet guide block.

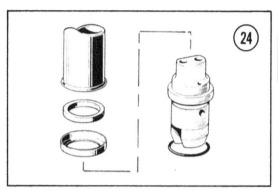

7. Lower the head into place and tighten the 4 outer head bolts finger-tight.

8. Replace the pushrods.

9. Turn the engine over (using the rear wheel with the transmission in fourth gear) until the intake pushrods are at the bottom of their stroke.

10. Lower the intake rocker box into position with the pushrods engaged and screw in the 2 central head bolts finger-tight.

11. Replace the Phillips head screws, underside nuts, and washers. Do not tighten fully at this time. Do the same for the exhaust rocker box, noting that the 2 rocker box head bolts are double ended and are also used to secure the torque arms.

12. Replace the oil feed bolts and rocker feed pipe.

13. Tighten the head bolts in the sequence shown in Figure 11, 19, or 20. Tighten the rocker box nuts and Phillips head screws. Refer to Appendix I for torque values.

14. Continue assembly in the reverse order of disassembly.

VALVES

Removal/Installation

1. Remove the cylinder head as described in the previous section.

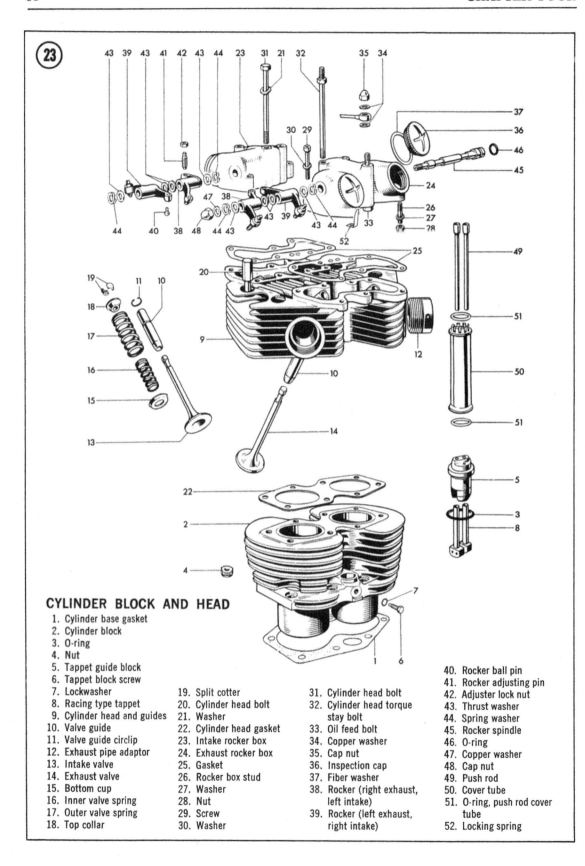

CYLINDER BLOCK AND HEAD

1. Cylinder base gasket
2. Cylinder block
3. O-ring
4. Nut
5. Tappet guide block
6. Tappet block screw
7. Lockwasher
8. Racing type tappet
9. Cylinder head and guides
10. Valve guide
11. Valve guide circlip
12. Exhaust pipe adaptor
13. Intake valve
14. Exhaust valve
15. Bottom cup
16. Inner valve spring
17. Outer valve spring
18. Top collar
19. Split cotter
20. Cylinder head bolt
21. Washer
22. Cylinder head gasket
23. Intake rocker box
24. Exhaust rocker box
25. Gasket
26. Rocker box stud
27. Washer
28. Nut
29. Screw
30. Washer
31. Cylinder head bolt
32. Cylinder head torque
    stay bolt
33. Oil feed bolt
34. Copper washer
35. Cap nut
36. Inspection cap
37. Fiber washer
38. Rocker (right exhaust,
    left intake)
39. Rocker (left exhaust,
    right intake)
40. Rocker ball pin
41. Rocker adjusting pin
42. Adjuster lock nut
43. Thrust washer
44. Spring washer
45. Rocker spindle
46. O-ring
47. Copper washer
48. Cap nut
49. Push rod
50. Cover tube
51. O-ring, push rod cover
    tube
52. Locking spring

2. Remove the valves using a C-type valve spring compressor. Compress the springs and remove the split keepers to free the valves. Keep track of the valves because they *must* be replaced in their original positions. See **Figure 25** for 750/650cc models or **Figure 26** for 500cc model valve components.

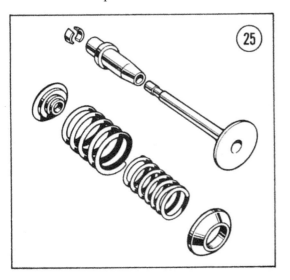

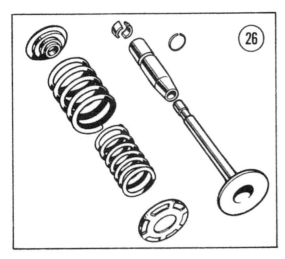

3. Check the valve springs against the specifications given in Appendix I.

4. Check the valve guides and stems for wear and replace if necessary. Guides may be pressed or driven out; the first method is preferred. A drift for driving the guides out can be made from a length of ½ in. mild steel bar by machining one end to 5/16 in. diameter for one inch. Also, Triumph makes a special tool.

5. If new guides are fitted, the valve seats will have to be recut and ground in as described below.

6. Note that if bronze valve guides are used, the 2 short ones are fitted in the intake positions. If cast iron valve guides are used, the long ones are fitted in the intake positions.

7. Service tools are available for cutting new valve seats. A cutter is used to cut the seats at a 45° angle, then a blending cutter is used to obtain an even seat 3/32 in. (2.4mm) wide.

8. Grind the valves by smearing lapping compound on the valve seat and rotating the valve with a suction tool held between the palms. Apply fresh lapping compound from time to time and continue lapping, occasionally raising the valve from its seat and turning it 90° to ensure a good seal.

Continue until the seat is a uniform gray color and all pits have disappeared from both the valve and seat. The surface can be checked by applying mechanic's blueing to the valve seat and rotating the valve one complete turn. A uniform ring of blue indicates a good seal and lapping is complete. Remove all traces of lapping compound from the valve and seat using a clean rag and solvent, prior to assembly.

9. Assemble in reverse order of disassembly.

CYLINDER BLOCK AND TAPPETS

Removal/Installation

1. To remove the cylinder or barrel, wedge a small rubber tube or block between the intake and exhaust tappets to prevent their falling into the crankcase as the barrel is lifted off.

2. Turn the engine over until the pistons are at top dead center and remove the 8 cylinder base nuts and washers. Lift the barrels off to clear the pistons. As soon as sufficient room is available, have an assistant hold the connecting rods to keep them from banging against the mouth of crankcase. Short pieces of garden hose should be fitted to the 4 central studs to protect the rods against damage. Stuff a clean rag into the crankcase opening to keep foreign matter out. Check that the 2 locating dowels are in place on the crankcase. See **Figure 27**.

4

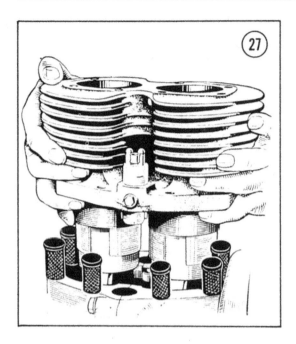

3. Remove the tappets from the cylinder block. Store them so they can be replaced in their original positions.

4. Clean all parts thoroughly in solvent and service the cylinder bores as detailed under *Pistons, Rings and Cylinders*, Step 7, this chapter.

5. Make sure the cylinder base gasket is correctly aligned on the dowels. It should not block the tappet lubrication drillways. See **Figure 28**.

6. Lubricate the tappets and replace in their original positions, wedging them in place as for removal. Use piston ring compressors to enable the barrels to be fitted over the piston rings.

7. Replace and tighten the cylinder base nuts and washers. Note that the 4 smaller nuts go on the center studs.

8. If an older machine is not equipped with the new type 12-point base nuts, these may be replaced along with the studs. Ask your Triumph parts dealer for the correct parts.

Lubrication

1. The lubrication supply to the exhaust tappets is critical. Ensure that they are correctly installed before replacing the cylinder block. Also check that the cylinder base gasket is not installed so as to block the oil passages drilled in the crankcase or cylinder block. See Figure 27.

2. Refer to **Figure 29**. Ensure that the tappets are installed with the cutaways away from each other so oil can get to them through the holes drilled through the tappet block. Intake exhaust tappets and guide blocks cannot be interchanged.

3. Various systems have been used on Triumph engines in an effort to obtain the best possible lubrication for the exhaust cam followers. Refer to **Tables 1 and 2** to determine which system is used on your engine. If a new system is to be

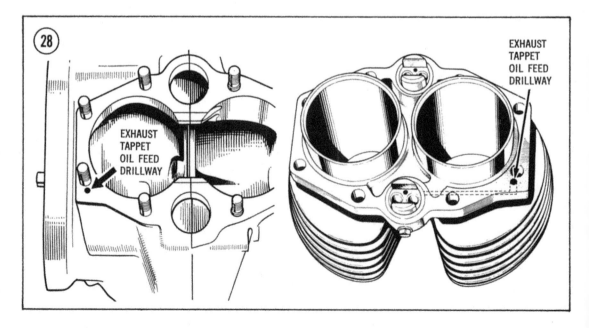

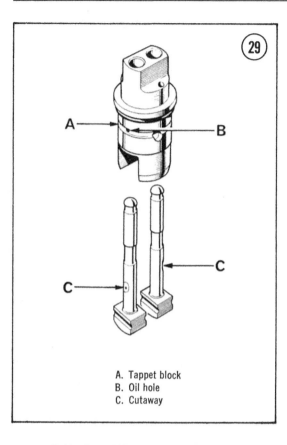

A. Tappet block
B. Oil hole
C. Cutaway

Table 1   EXHAUST CAM FOLLOWERS

E6329 6T/TR6	use	E8895 6T/TR6
E6490 T120	use	E8801 T120

installed on an older engine, new cam followers and metering plug must be installed. To install the newest type cam followers on engines from DU.24875 to DU.63043, remove the metering jet by screwing in a timing cover screw and pulling out the jet. Replace the E6348 metering jet with part No. T989. For exhaust cam followers, use part No. E8895 for the 6T and TR6 models, and part No. E8801 for the T120.

Inspection

1. Tappets should be inspected visually for an indentation on the center of their tips. If the indentation is more than 3/32 in. (2.4mm) wide, replace the tappet.

2. The guide blocks are checked without removing them from the cylinder block. Grasp the tappet while it is in position and try to rock it back and forth. It should move only up and down. Exact clearances are given previously in this chapter.

Guide Block Replacement

1. Guide blocks can be driven out with service tool Z23 (or 61-6008) after the locking screw is removed. On newer machines, O-rings are fitted between the guide block and cylinder block. Replace the O-ring if the guide blocks are removed. The 500cc model was not fitted with an O-ring before engine No. H50783.

Table 2   TAPPET SYSTEMS

	Type	Timing cover dowel	Cam follower exhaust	Cam follower inlet	Tappet block
To Engine No. DU.24875	Non lubricated	Blind dowel None	E3059	E3059	E1477
From DU.24875 To DU.44394	Lubricated	Dowel E6348	E6329 6T/TR6 E6490 T120	E3059 6T/TR6 E3059R T120	E5861
From DU.44394 To DU.63043	Lubricated	Dowel and pin E6803/6800	E6490 T120/TR6	E3059R	E5861
From DU.63043 To DU.66246	Lubricated	Dowel T989	E8801	E3059R	E5861
1972 (650cc)	Lubricated	Dowel T989	E8801	E3059R	E9353 (Exh.) E9352 (Inl.)
1973 on (750cc)	Lubricated	Dowel 57-0989	70-8801	70-3059	70-9352

2. Support the cylinder securely on the base flange when replacing the guide block. Grease the guide block and drive it into place after aligning the holes in the block and cylinder base. See **Figure 30**. Do not line up the guide block using the ears on the block or they will crack.

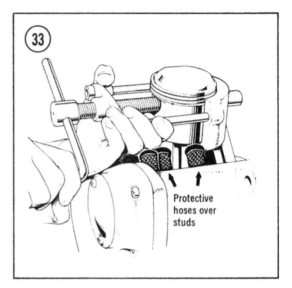

Protective hoses over studs

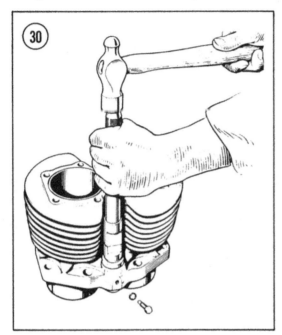

PISTONS, RINGS, AND CYLINDERS

Removal/Installation

Refer to **Figure 31** (650/750cc) or **Figure 32** (500cc) for these procedures.

1. Before removing the pistons, cover the 4 center studs on the crankcase with pieces of garden hose or other protective device. See **Figure 33**. Use a piston pin remover after taking out both circlips from the ends of the piston pin (**Figure 34**). If the pin must be driven out with a drift instead of a pin remover, use care to avoid damaging the connecting rod. Heat piston to 212°F (100°C) to make pin removal easier.

2. Mark the pistons inside to identify their location (right or left) for assembly.

3. When replacing the pins, insert one circlip to act as a stop and again heat the piston unless a pin inserting tool is used.

4. Always use new pin clips and check that they are securely engaged in their grooves.

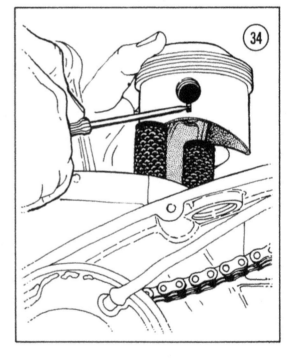

5. Remove piston rings by lifting out one end of the ring and inserting a thin strip of metal between the ring and piston. Slide the metal strip around the piston, lifting the ring slightly to keep it away from its groove. A piston ring expander may be used to make the job easier.

6. If new rings are to be fitted, the cylinder bores must first be honed to remove the glaze which forms after prolonged running. Check the rings for proper end gap by inserting them one at a

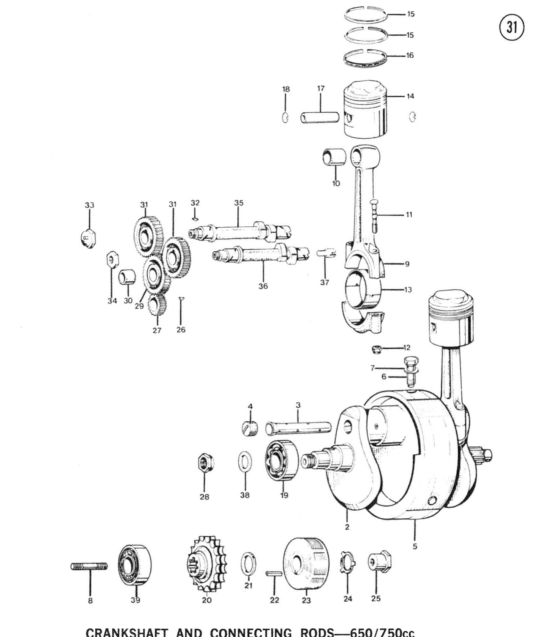

CRANKSHAFT AND CONNECTING RODS—650/750cc

1. Crankshaft and flywheel
2. Crankshaft
3. Oil tube
4. Screwed plug
5. Flywheel
6. Flywheel bolt
7. Washer
8. Crankshaft stud
9. Connecting rod
10. Small end bushing

11. Connecting rod bolt
12. Self-locking nut
13. Big end bearing
14. Piston
15. Tapered compression ring
16. Oil control ring
17. Piston pin (short)
18. Circlip
19. Right main bearing
20. Main sprocket

21. Spacer
22. Rotor key
23. Rotor
24. Tab washer
25. Nut
26. Key
27. Timing pinion
28. Nut
29. Intermediate wheel
30. Intermediate wheel bushing

31. Camshaft pinion
32. Camshaft key
33. Intake camshaft nut
34. Exhaust camshaft nut
35. Intake camshaft
36. Exhaust camshaft
37. Tachometer drive plug
38. Clamping washer
39. Left main bearing

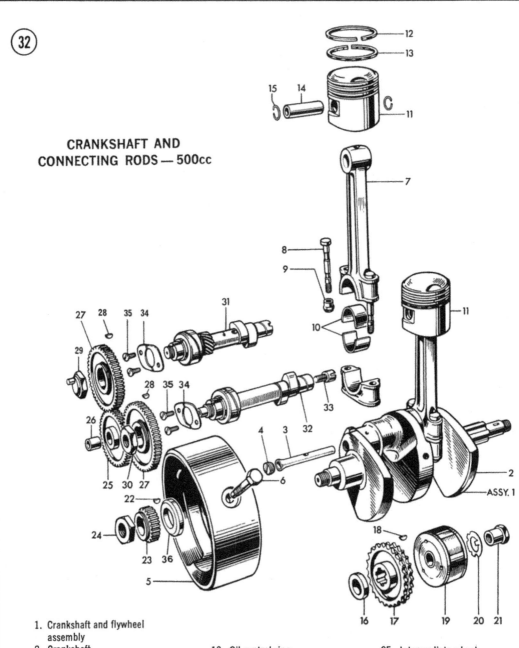

CRANKSHAFT AND
CONNECTING RODS — 500cc

1. Crankshaft and flywheel
   assembly
2. Crankshaft
3. Oil tube
4. Threaded plug
5. Flywheel
6. Flywheel bolt
7. Connecting rod
8. Connecting rod bolt
9. Self-locking nut
10. Big end bearing
11. Piston assembly
12. Tapered compression
    ring
13. Oil control ring
14. Piston pin
15. Circlip
16. Spacer
17. Engine sprocket
18. Rotor key
19. Rotor
20. Tab washer
21. Nut
22. Key
23. Timing pinion
24. Nut
25. Intermediate wheel
26. Bushing
27. Camshaft wheel
28. Camshaft key
29. Intake camshaft nut
30. Exhaust camshaft nut
31. Intake camshaft
32. Exhaust camshaft
33. Tachometer drive plug
34. Retaining plate
35. Screw
36. Seal

time into the cylinder. Invert the piston and use it to push the ring squarely to the bottom of the bore. Check end gap with a feeler gauge. New rings should have a gap of 0.010-0.014 in (0.25-0.35mm). The oil scraper ring fits in the bottom groove. The 2 compression rings fit in the top grooves. The compression rings are marked TOP; this side should be installed up.

7. Check pistons and cylinder bores for wear. See **Table 3**.

Inspect the cylinder bore for scratches or other damage and check the piston skirt for evidence of seizure. Note that the top of the piston has a lot of clearance and the top ring will be visible when viewed from above. Measure the cylinder bore at right angles to the piston pin at several points up and down the bore. A difference of more than 0.005 in. (0.13mm) in these measurements indicates need for a rebore. Use **Table 3** to determine the correct size pistons needed after boring to the specifications given. See a Triumph dealer to order oversize rings.

8. Replace small end bushings in 650cc engines using a bolt and a piece of tubing 1 1/4 in. (30 mm) long with an inside diameter of 7/8 in. (22mm). Place a washer on the bolt; then the new bushing. Insert the bolt through the old bushing still in the connecting rod (**Figure 35**). Place the piece of tubing over the bolt end, ad a washer and nut, and tighten the nut so the old bushing is pushed out by the new one. Align the oilway in the new bushing with that of the connecting rod as it is inserted. Ream the new bushing to the size given in Appendix I.

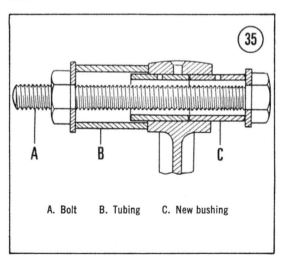

(35)

A. Bolt   B. Tubing   C. New bushing

Table 3   SUITABLE REBORE SIZES

Piston marking in. (mm)	Suitable bore sizes in.	mm
650cc **Standard**	2.7948	70.993
	2.7953	71.006
Oversizes		
+ 0.010 (0.254)	2.8048	71.247
	2.8053	71.260
+ 0.020 (0.508)	2.8148	71.501
	2.8153	71.514
+ 0.040 (1.016)	2.8348	72.009
	2.8353	72.022
750cc **Standard**	2.9911*	75.973*
	2.9921	76.000
Oversizes		
+ 0.010 (0.254)	3.0010	76.2254
	3.0021	76.2533
+ 0.020 (0.508)	3.0110	76.4794
	3.0121	76.5073
+ 0.030 (0.726)	3.0210	76.7334
	3.0221	76.7613
+ 0.040 (1.016)	3.0310	76.9514
	3.0321	76.9793
500cc **Standard**	2.716	69.00
Oversizes		
+ 0.010 (0.254)	2.726	69.254
+ 0.020 (0.508)	2.736	69.508
+ 0.040 (1.016)	2.756	70.00

*Dimensions are for smallest of Grade L and largest of Grade H of three standard piston and bore size grades. These grades are important only insofar as standard bore size is concerned.

OIL PUMP

Removal/Installation

1. Remove the oil pump and the scavenge and feed plungers (**Figure 36**). Unscrew the 2 square caps from the end of the oil pump to free the springs and balls.

2. Clean all parts in solvent and inspect the plungers and springs against the specifications in Appendix I. Replace as necessary.

3. Use oil liberally when reassembling pump components. Check the pump before reinstallation by placing approximately one cubic centimeter of oil in both ports. Press the plungers

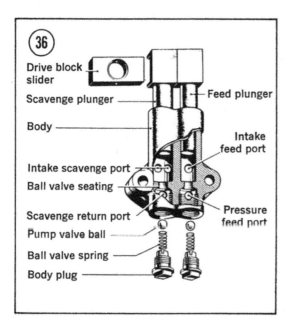

Drive block slider
Scavenge plunger
Body
Feed plunger
Intake feed port
Intake scavenge port
Ball valve seating
Scavenge return port
Pressure feed port
Pump valve ball
Ball valve spring
Body plug

until oil appears at the outlet ports and covers the intake ports. Withdraw the plungers slightly and check that the oil in the outlet ports remains at the same level. If the level falls, the ball valves are not seating properly and must be cleaned again or replaced. If the pump has a brass body, the balls may be tapped lightly to seat them properly. This must not be done to a pump with a cast iron body or the body will have to be replaced.

4. Check the aluminum drive block slider for wear on the bore and in the plunger cross-head.

5. Use a new gasket when replacing the oil pump body. Note that the cones of the conical nut and washers fit inside the countersunk holes in the pump body.

6. Clean the mating surfaces of the timing cover and crankcase. Use gasket cement when replacing the cover.

OIL SYSTEM

1. To remove the oil pipe junction block, drain the gearbox oil by removing the drain plug under the gearbox. Remove the transmission outer cover as detailed in Chapter Five. Remove the right exhaust pipe and right footpeg if necessary on older models.

2. Drain the oil tank by removing the drain plug or the oil pipe junction block.

3. Disconnect the oil pipes from the junction block and clean the block thoroughly in solvent. Check the oil pipes and block for damage. Replace the oil pipes if they show signs of wear or deterioration.

4. Use care when reinstalling the oil pipes to avoid cutting small chunks of rubber loose from the insides of the hoses which could block the oil passages. Silicone spray lubricant will ease this operation. Replace all clamps and use a new gasket when replacing the junction block.

ROCKER OIL FEED PIPE

1. Remove the rocker oil feed pipe by disconnecting from the oil tank and the 2 ends of the rocker spindles. On early models, it may be necessary to remove the right body panel.

2. The pipe may be secured to the frame by clips. Remove it carefully, bending as little as possible.

3. Flush the pipe with solvent and check for blockage by sealing first one banjo connection and then the other and blowing in the opposite end. Use new copper washers if possible when refitting banjo connections.

VALVE TIMING GEARS

Removal/Installation
(750/650cc Models)

1. Remove the rocker boxes, contact breaker assembly, timing cover, and oil pump as detailed earlier. The rocker boxes are removed to give sufficient slack in the valve mechanism. It is possible to loosen all the valve clearance adjusters, but the risk of unseating a pushrod exists.

2. Put the transmission in fourth gear and apply the rear brake to ease unscrewing gear retainer nuts. The camshaft gear nuts have left-hand threads; the crankshaft gear nut has a right-hand thread. Remove the intermediate gear.

3. Refer to **Figure 37**. Use a crankshaft gear puller to remove the gear. A Triumph special service tool aids in replacing the crankshaft gear. Use a gear puller to remove the camshaft gears.

4. Replace the camshaft gears. Make sure the keys in the camshafts are in place.

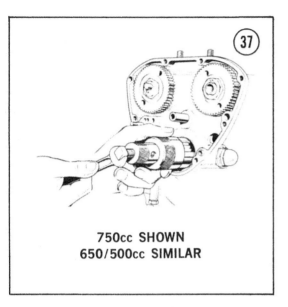

750cc SHOWN
650/500cc SIMILAR

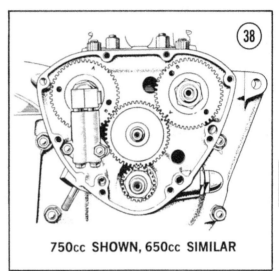

750cc SHOWN, 650cc SIMILAR

5. To set valve timing, rotate the camshaft and crankshaft gears until the timing marks are as shown in **Figure 38**. The exhaust camshaft gear mark should line up with the intermediate wheel mark and the crankshaft gear mark should line up between the 2 marks on the intermediate wheel. The intake camshaft gear mark should line up with the long mark (T120, TR6 models) or the short mark (6T models) on the intermediate wheel. With the engine in fourth gear, step on the rear brake and tighten the gear nuts to the torques specified in Appendix I.

> NOTE: *If the engine is turned over after the intermediate wheel has been positioned, the marks will appear to be out of alignment. Because of the number of teeth on the intermediate gear, the marks will only line up again after exactly 94 engine revolutions.*

Removal/Installation (500cc Models)

1. Remove the valve timing gears by following the procedures given above in Steps 1 and 2.

2. Remove the crankshaft gear. Save the crankshaft key after removal. See Figure 37.

3. To replace the gear, smear its bore with grease, align the key and keyway, and drive the gear into place using a hollow drift.

4. Remove the camshaft gears (the securing nuts have left-hand threads).

5. To replace the intermediate wheel on gear, refer to Figure 38. Turn the camshafts and crankshaft until the timing marks on the gears are aligned as shown, and fit the gear in place. With the transmission in gear, apply the rear brake and tighten the gear retaining nuts to the torques specified in Appendix I. The camshaft gears are located with keyway directly opposite appropriate timing mark.

CAMSHAFTS

Removal/Installation

Camshafts can be removed and replaced without splitting the cases.

1. Remove the rocker boxes as detailed earlier. Remove the timing cover and the oil pump. Holes in the crankcase face can be blocked so the engine will not drain out, but do not forget to clear the holes before the pump is replaced.

2. Remove the camshaft timing gears as detailed in the previous section. Remove the camshaft retaining plates. The retaining plate screws are centerpunched on assembly to prevent their backing out, so the indentations may have to be drilled out before the screws can be removed. See **Figure 39**.

3. Lean the machine over on its left side to keep the cam followers from falling into the engine. Pull the camshafts out, but do not drop the breather disc and spring behind the inlet camshaft into the engine.

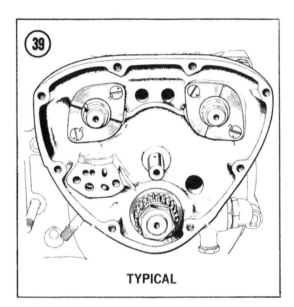

TYPICAL

4. Replace camshafts by putting the breather valve and spring into the camshaft bushing. Be sure the inlet camshaft and breather valve are engaged. New screws should be used if possible on the camshaft retaining plates. Centerpunch the screws and plates to lock the screws in place.

CRANKCASES

Disassembly/Assembly (750/650cc Models)

Refer to **Figure 40**.

1. The easiest procedure for splitting the cases is to partially disassemble the engine while it is still mounted in the frame. Follow the steps detailed earlier for removing the engine from the frame, but leave the rear chain connected and do not remove the front and bottom engine mounting bolts.

2. Remove the outer primary cover as detailed in this chapter.

3. Disconnect the generator leads (2 or 3 snap connectors) but do not withdraw the leads at this time.

4. Remove the 3 stator mounting nuts and the stator.

5. Refer to the *Clutch* section in Chapter Five and remove the pressure plate and clutch plates. Engage fourth gear and have a helper apply the rear brake while the clutch hub nut is unscrewed.

Remove the clutch hub as detailed in the *Clutch* section, Chapter Five. When the primary chain is free, unscrew the sleeve nut and withdraw the stator leads.

6. Refer to the *Transmission* section in Chapter Five, remove the gearbox cover and dismantle the gearbox.

7. Remove the rocker boxes, cylinder head, barrels, and pistons as detailed earlier in this chapter.

8. Remove the contact breaker unit, the timing cover, and oil pump as detailed earlier.

9. Remove the crankshaft gear. If the camshafts or camshaft bushings need inspection or service, remove the gears.

10. Remove the engine mounting studs and the rear chain. Lift the crankcase from the frame.

11. Remove the crankcase filter and the plug on the bottom of the crankcase. Drain any oil remaining in the cases.

12. Use the bottom mounting lug to mount the cases in a vise. Unscrew the 3 bolts and 2 screws shown in **Figure 41**. Remove the remaining 4 studs and the 2 nuts adjacent to the gearbox.

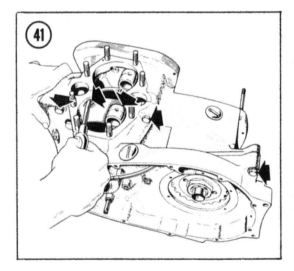

13. Split the cases as shown in **Figure 42**.

14. Carefully remove the crankshaft assembly.

15. Take out the rotary breather valve from the inlet camshaft bushing in the left crankcase. See **Figure 43**.

16. Clean the cases thoroughly. Make sure all oil passages are clear.

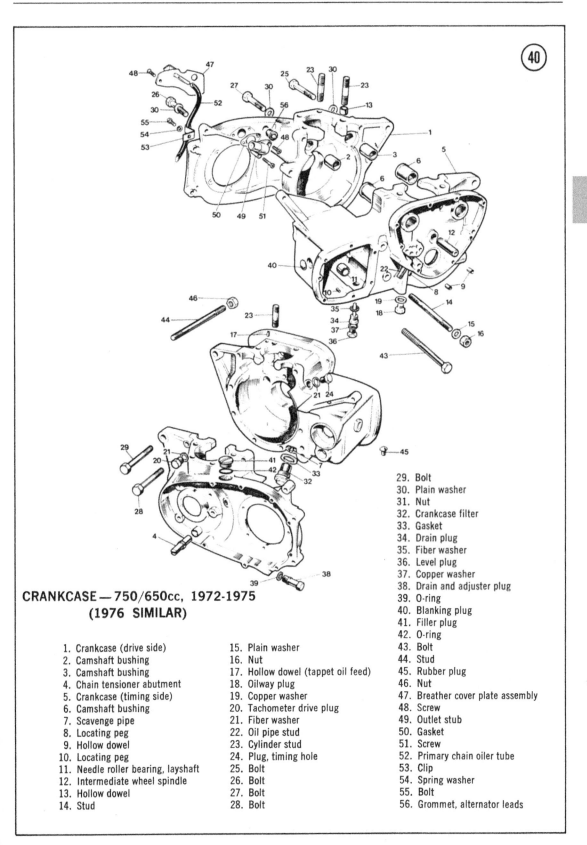

CRANKCASE — 750/650cc, 1972-1975
(1976 SIMILAR)

1. Crankcase (drive side)
2. Camshaft bushing
3. Camshaft bushing
4. Chain tensioner abutment
5. Crankcase (timing side)
6. Camshaft bushing
7. Scavenge pipe
8. Locating peg
9. Hollow dowel
10. Locating peg
11. Needle roller bearing, layshaft
12. Intermediate wheel spindle
13. Hollow dowel
14. Stud
15. Plain washer
16. Nut
17. Hollow dowel (tappet oil feed)
18. Oilway plug
19. Copper washer
20. Tachometer drive plug
21. Fiber washer
22. Oil pipe stud
23. Cylinder stud
24. Plug, timing hole
25. Bolt
26. Bolt
27. Bolt
28. Bolt
29. Bolt
30. Plain washer
31. Nut
32. Crankcase filter
33. Gasket
34. Drain plug
35. Fiber washer
36. Level plug
37. Copper washer
38. Drain and adjuster plug
39. O-ring
40. Blanking plug
41. Filler plug
42. O-ring
43. Bolt
44. Stud
45. Rubber plug
46. Nut
47. Breather cover plate assembly
48. Screw
49. Outlet stub
50. Gasket
51. Screw
52. Primary chain oiler tube
53. Clip
54. Spring washer
55. Bolt
56. Grommet, alternator leads

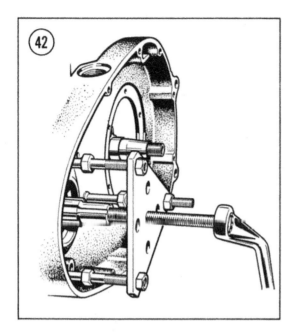

17. Reassemble after cleaning the crankcase mating surfaces thoroughly. Make sure the locating pegs and corresponding holes are free of old gasket cement or sludge. Replace the plug removed from the right crankcase half along with the crankcase filter.

18. Place the left crankcase half on its side on blocks or on a bench with a hole in it so the crankshaft will be free to slide in place.

19. Oil main bearings and camshaft bushings.

20. Replace the rotary breather valve and spring into the camshaft bushing and insert both camshafts. Be sure the intake camshaft engages the rotary disc valve.

21. Insert the crankshaft and seat it by tapping with a plastic mallet. Check that all crankcase positioning dowels are in place.

22. Apply an even coating of gasket cement to the mating surface of the left crankcase half and position the connecting rods centrally.

23. Lower the right crankcase half into place and check the crankshaft and camshafts for free rotation.

24. Replace crankcase bolts and studs finger-tight and check that the 2 crankcase halves are lined up. They may be tapped with a plastic mallet at the front or rear to get a level surface at the joint. Check the crankshaft and camshafts as before. Tighten the bolts (Figure 51) to the torque figures given in Appendix I.

25. Reassemble the remaining components in the reverse order of disassembly. Pour ⅙ pint (79cc) of clean engine oil into the crankcase before replacing the cylinder block.

Disassembly/Assembly (500cc Models)

Split the cases (**Figures 44 and 45**) as detailed above. Disregard Step 11 during disassembly and reassembly. On the 500cc engine, there are 2 screws and one bolt to unscrew just before the cases are taken apart instead of 3 bolts and 2 screws as on the 650cc. One stud and the 2 nuts adjacent to the gearbox remains at this point instead of the 4 studs mentioned in the 650cc section.

CRANKSHAFT

Removal/Installation

1. Mark the connecting rods so the end caps and rods can be replaced in their original positions relative to the crankshaft. Remove the cap nuts evenly, a little at a time, and separate the caps and connecting rods. See *Rods and Main Bearings* in this chapter for inspection and service.

2. Remove the oil tube retaining plug from the right end of the crankshaft journal. Use an im-

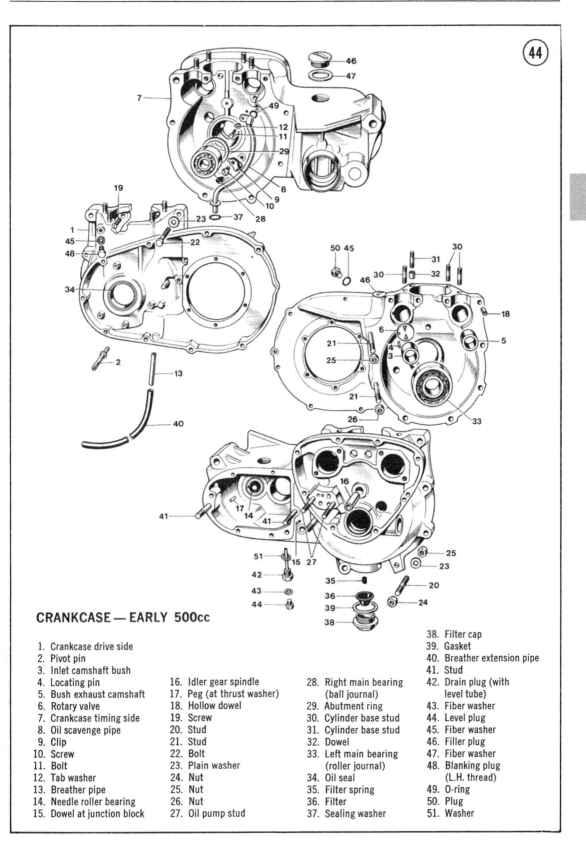

CRANKCASE — EARLY 500cc

1. Crankcase drive side
2. Pivot pin
3. Inlet camshaft bush
4. Locating pin
5. Bush exhaust camshaft
6. Rotary valve
7. Crankcase timing side
8. Oil scavenge pipe
9. Clip
10. Screw
11. Bolt
12. Tab washer
13. Breather pipe
14. Needle roller bearing
15. Dowel at junction block

16. Idler gear spindle
17. Peg (at thrust washer)
18. Hollow dowel
19. Screw
20. Stud
21. Stud
22. Bolt
23. Plain washer
24. Nut
25. Nut
26. Nut
27. Oil pump stud

28. Right main bearing
    (ball journal)
29. Abutment ring
30. Cylinder base stud
31. Cylinder base stud
32. Dowel
33. Left main bearing
    (roller journal)
34. Oil seal
35. Filter spring
36. Filter
37. Sealing washer

38. Filter cap
39. Gasket
40. Breather extension pipe
41. Stud
42. Drain plug (with
    level tube)
43. Fiber washer
44. Level plug
45. Fiber washer
46. Filler plug
47. Fiber washer
48. Blanking plug
    (L.H. thread)
49. O-ring
50. Plug
51. Washer

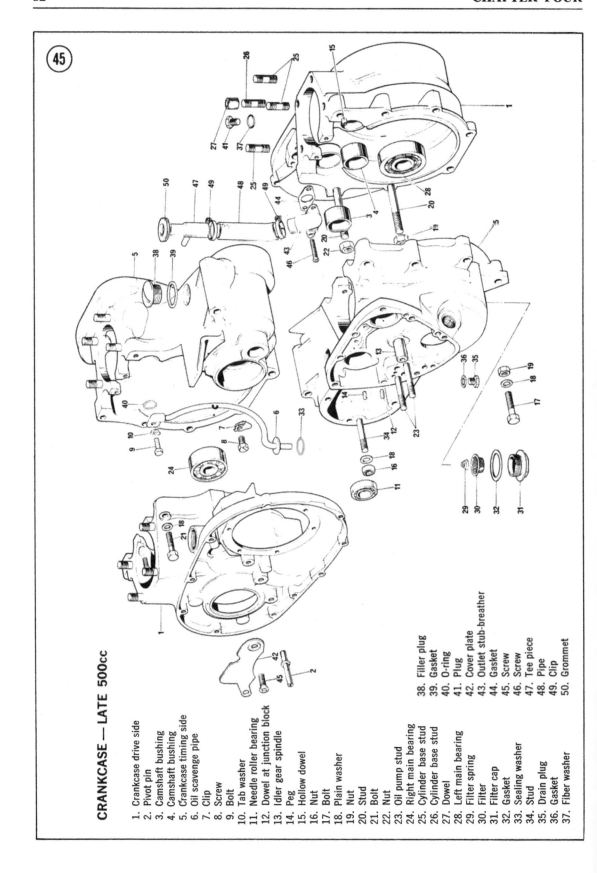

CRANKCASE — LATE 500cc

1. Crankcase drive side
2. Pivot pin
3. Camshaft bushing
4. Camshaft bushing
5. Crankcase timing side
6. Oil scavenge pipe
7. Clip
8. Screw
9. Bolt
10. Tab washer
11. Needle roller bearing
12. Dowel at junction block
13. Idler gear spindle
14. Peg
15. Hollow dowel
16. Nut
17. Bolt
18. Plain washer
19. Nut
20. Stud
21. Bolt
22. Nut
23. Oil pump stud
24. Right main bearing
25. Cylinder base stud
26. Cylinder base stud
27. Dowel
28. Left main bearing
29. Filter spring
30. Filter
31. Filter cap
32. Gasket
33. Sealing washer
34. Stud
35. Drain plug
36. Gasket
37. Fiber washer
38. Filler plug
39. Gasket
40. O-ring
41. Plug
42. Cover plate
43. Outlet stub-breather
44. Gasket
45. Screw
46. Screw
47. Tee piece
48. Pipe
49. Clip
50. Grommet

pact screwdriver if necessary. A ⅛ in. (3.2mm) diameter hole may be drilled to remove the punched indentation which holds the plug in place. Unscrew the flywheel bolt and fish out the oil tube. Refer to **Figure 46**. Clean all components in solvent or kerosene and blow out the oil passages with compressed air.

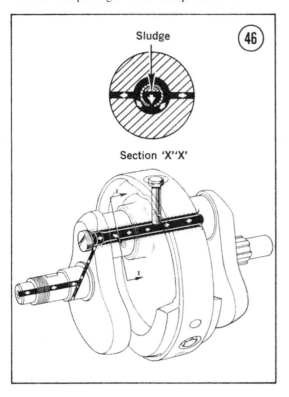

3. If the crankshaft needs further dismantling, mark the flywheel so it can be replaced in its original position. Remove the crankshaft after the remaining flywheel bolts have been unscrewed.

4. To assemble the crankshaft, replace the oil tube, using a flywheel bolt to hold it in place. Replace the oil tube plug and lock it securely by centerpunching a depression where the plug and crankshaft are adjacent to each other. Heat the flywheel to 212°F (100°C) and assemble onto the crankshaft. The flywheel will have to be turned 180 degrees from its final position during assembly to clear the crankshaft web. Align the bolt holes and replace the flywheel bolts, tightening them to the specifications given at the end of this chapter. Use a sealant such as Loctite on the bolt threads.

5. If a new crankshaft or flywheel has been installed, the assembly will have to be balanced. Refer to **Figure 47** and support the crankshaft on horizontal knife edges. Mark the point at which it comes to rest and drill a ⅜ in. (9.5mm) hole ½ in. (12.7mm) deep in the flywheel. Triumph balance weights (24.3 oz. each, 689 grams) must be used. Repeat the balancing procedure until the assembly comes to rest at random. Crankshaft balancing should be undertaken only by those familiar with the process. If any doubt exists, have a dealer do it.

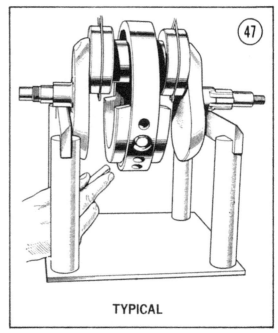

TYPICAL

6. A new, lighter flywheel has been used since engine No. DU.24875 and can be installed on earlier models if the newer type flywheel bolts are also used.

7. Check all oil passages before reinstalling the crankshaft assembly.

RODS AND MAIN BEARINGS

Removal/Installation
(750/650cc Models)

1. Remove the timing side main bearing by heating the case to 212°F (100°C) and using a drift (made from a 1¼ in. steel bar, turning down ½ in. of its end to a diameter of 1⅛ in.) to drive out the bearing.

2. Earlier models require the same method to remove the drive side bearing. The outer spool of the drive side bearing on all models must be driven out while the crankcase is heated.

3. The left crankcase oil seal can be removed after the bearing has been taken out and should be replaced whenever removed, regardless of the apparent lack of wear.

4. Clean the bearings. Measure the crankshaft using the data given in Appendix I. Main bearings should be a press fit in the crankcases and a press fit on the crankshaft. Check for smoothness of operation and look for damaged balls or bearing races.

5. Check the big end bearings for scoring and measure them against the specifications in Appendix I. If the bearing journals are scored they may have to be reground and undersize bearing shells fitted. Refer to **Tables 4 and 5**. Replace with new bearings if necessary.

Table 4   750/650cc BEARING SHELL SIZES

Shell bearing marking	Suitable crankshaft size in.	mm
Standard:	1.6235	41.237
	1.6240	41.250
Undersize:	1.6135	40.983
— 0.010	1.6140	40.996
	1.6035	40.729
— 0.020	1.6040	40.742

Table 5   JOURNAL DIAMETER

	in.	mm
Standard	1.4375	36.512
	1.4380	36.525
— 0.010″	1.4275	36.258
	1.4280	36.271
— 0.020″	1.4175	36.004
	1.4180	36.017
— 0.030″	1.4075	35.750
	1.4080	35.763

6. Be sure the journals, shells, connecting rod, and cap are all clean. Fit the shells in place with their tabs in the slots. Oil the bearing surfaces with engine or reassembly oil and assemble the

caps to their respective rods. Refer to **Figure 48** and note that the bearing shell tabs are adjacent to each other. Tighten the cap nuts to the specifications in Appendix I. With both rods in place, force oil through the oil passage in the right end of the crankshaft until it comes out around the big end bearings.

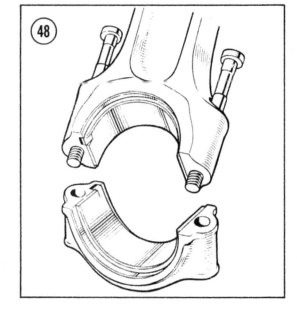

7. The drive side bearing (roller type) can be installed on earlier models. Prior to engine No. DU.13375, it is a direct replacement. Between engine Nos. DU.13375 and DU.24874, the crankshaft gear and clamp washer must be changed. See dealer for necessary parts.

8. Replace the bearings, again heating the crankcases. A drift 2¾ in. in diameter should be used. Pressing is the best method for replacing main bearings.

Removal/Installation (500cc Models)

1. Check the timing side plain bearing for wear by trying to move the right end of the crankshaft or check with a dial indicator. Any score marks on the crankshaft will necessitate regrinding.

2. Heat the cases to 212°F (100°C) and drive out the left side main bearing.

NOTE: *On engines after H65573, heat the crankcase to 212°F (100°C), install an extractor and remove the bearing* (**Figure 49**).

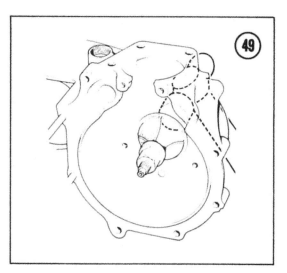

3. Drive out the right side main bearing with a suitable drift after the cases have been heated as above.

4. Remove the oil seal on the left crankcase half. Replace the seal regardless of apparent wear.

5. Connecting rod service is the same as previously detailed for 750/650cc models.

6. Be sure the bearing housings are clean. Use a drift 2¾ in. in diameter and approximately 6 in. long to press or drive the left bearing into place after the cases have been heated as above.

7. Press the right bearing in. Ream it to size after the cases have cooled (on engines prior to H65573).

8. Press the oil seal, with the open face outward (**Figure 50**), into the left case half.

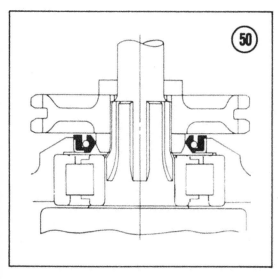

CAMSHAFT AND BUSHINGS

Removal/Installation (Typical)

1. Clean the camshafts and examine for signs of scoring.

2. Check for bushing wear by inserting the camshaft and moving it up and down.

3. Measure the camshaft and bushings against the data in Appendix I.

4. Check the cam lobes for grooving. If there is extensive wear, the cam and possibly the tappet followers will have to be replaced. More than 0.010 in. (0.25mm) wear on the cam face necessitates replacement.

5. Remove cam bushings from the right crankcase by heating the case to 212°F (100°C). The right camshaft ends have no bushings on 500cc models. Use a drift to remove the bushing from the outside. Replace with a new bushing while the cases are still hot, aligning the oil feed holes. Make a drift by machining a 1⅛ in. steel bar so one end is turned down to ⅞ in. for a distance of one inch from the end.

6. Remove cam bushings from the left crankcase by cutting a thread in the old bushing, using a suitable tap. Insert a bolt in the threaded bushing and heat the crankcase as described before. Secure the bolt in a vise and tap the crankcases with a mallet to drive the bushing free. Do not attempt to pry the bolt and bushing out, as damage to the cases will result. There is a breather valve porting disc behind the inlet camshaft bushing which must be located correctly on its peg when the bushing is replaced.

7. Refer to Appendix I for bushing clearance specifications. If the bushings are reamed, clean all metal particles from the cases using solvent and compressed air.

TACHOMETER DRIVE

Removal/Installation

The tachometer drive gearbox (**Figure 51**) can be removed without splitting the crankcase.

1. Remove the slotted end cap (**Figure 52**) and turn the engine over sharply to eject the drive gear, or pull it out with long nosed pliers.

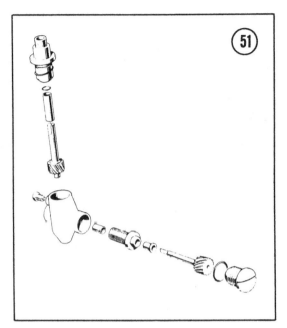

2. Remove the bolt (left-hand thread) securing the gearbox to the crankcase with a 3/16 in. 4.8mm) thin box end Whitworth wrench.

3. Remove the driven gear housing which is secured by a pin. Remove tachometer drive plug located in the end of the exhaust camshaft.

4. Use a punch to drive the damaged thimble back into the camshaft center at least one inch.

5. Use a magnet to remove the broken-off ears of the old thimble.

6. Screw a new thimble through the tachometer drive hole and into the camshaft. Drive the plug flush with the end of the camshaft with a drift made as shown in **Figure 53**.

7. Continue installation in the reverse order of removal.

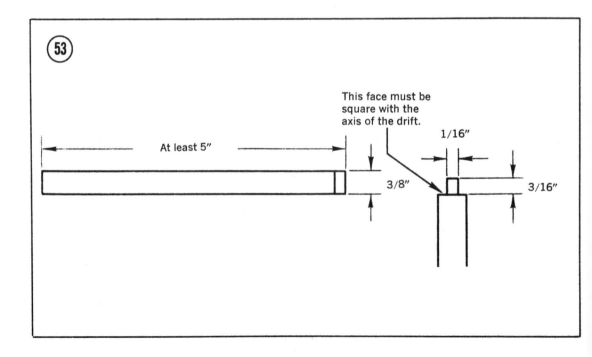

This face must be square with the axis of the drift.

At least 5"

1/16"

3/8"

3/16"

CHAPTER FIVE

CLUTCH AND TRANSMISSION

PRIMARY CHAIN

Adjustment (Typical)

1. Check primary chain tension by removing the top inspection plug next to the cylinder. With the engine stopped, feel the chain tension (**Figure 1**). On 1971 and earlier models, the free play should be ½ in. (12mm). On 1972 and later models, it should be ⅜ in. (9.5mm).

2. Adjust free play by removing the hex head plug used to drain the chaincase. Catch the oil in a pan. Certain earlier models require that the left footpeg be loosend and moved downward.

3. Use the hex head screwdriver (blade type for 750cc and 500cc models) supplied with the motorcycle tool kit and adjust chain tension. Replace the plug and refill the primary chaincase with oil. See Chapter Two.

PRIMARY COVER

Removal/Installation (Typical)

1. Loosen the left finned clip bolt, left muffler clip bolt, and remove the nut and bolt holding the front left exhaust pipe bracket. On 1972 and later models, loosen the left clamp on the balance pipe. Tap the exhaust pipe free with a plastic mallet to separate.

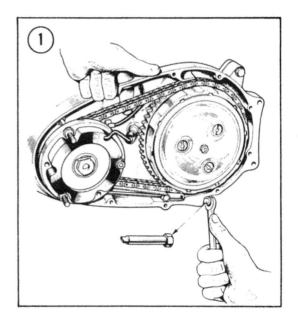

2. Loosen the rear brake adjuster so the brake pedal clears the primary cover.

3. Remove the left footpeg. Remove the drain plug at the bottom of the cover and drain the oil. On 1976 and later models, loosen the bolt (**Figure 2**) and remove the gearshift pedal. Remove the Phillips head screws, the 2 nuts and their washers, and pull the cover free. On 500cc models, remove the tension adjuster before

taking the cover off. Never use a screwdriver to pry the cover loose; the mating surface will be damaged.

4. Use a new gasket when replacing the cover. Use only a thin coating of grease to help seal the gasket. If gasket cement is used, coat only the crankcase side of the gasket to make later disassembly easy. Replace the screws and nuts, tightening them evenly. Refill chaincase with the correct amount and grade of oil.

CLUTCH

Refer to **Figures 3 and 4** for details of the clutch and clutch operating mechanism.

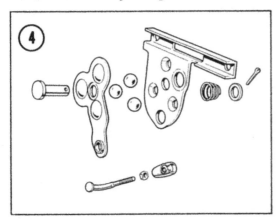

Removal/Installation (Typical)

1. Remove primary cover as detailed earlier.

2. Refer to Figure 3 and remove the 3 adjuster nuts on the clutch pressure plate, sticking a

screwdriver between the nuts and their cups to keep the small locking projections from engaging the nuts (**Figure 5**). A notched screwdriver similar to the one shown can be made from a conventional screwdriver by notching the blade with a file or grinder.

3. Take out the clutch springs, cups, and the pressure plate. The clutch plates can be removed easily using a pair of wire hooks. See **Figure 6**.

4. Clean all parts in solvent.

5. Check the friction plates (driving plates) for excessive wear or glazing. Check the driven plates for flatness on a surface plate or sheet of

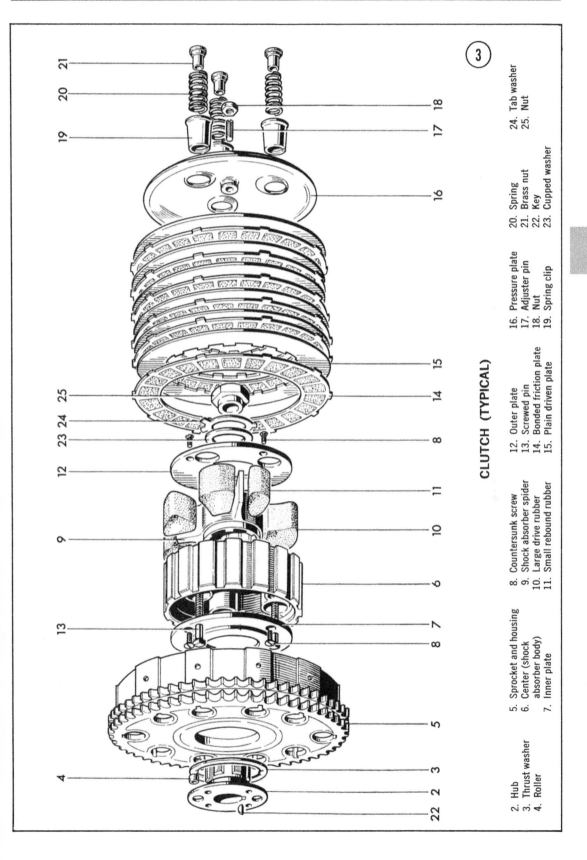

CLUTCH (TYPICAL)

2. Hub
3. Thrust washer
4. Roller
5. Sprocket and housing
6. Center (shock absorber body)
7. Inner plate
8. Countersunk screw
9. Shock absorber spider
10. Large drive rubber
11. Small rebound rubber
12. Outer plate
13. Screwed pin
14. Bonded friction plate
15. Plain driven plate
16. Pressure plate
17. Adjuster pin
18. Nut
19. Spring clip
20. Spring
21. Brass nut
22. Key
23. Cupped washer
24. Tab washer
25. Nut

plate glass. Friction discs on the driving plates should not be worn more than 0.030 in. (0.76mm) from the specifications given in Appendix I.

6. Check for wear of the driving tags in the clutch housing. Check radial clearance of the plates on the shock absorber housing.

7. Compare the clutch springs with the data in Appendix I. Springs 0.1 in. (2.5mm) or more below standard indicate the entire set should be replaced.

8. Replace the clutch plates, springs, cups, and nuts. Tighten the nuts so the clutch pins are flush with the heads of the nuts.

9. Pull in the clutch lever and turn the engine over with the kickstarter. Watch the pressure plate as it spins. If it wobbles, the springs need adjustment. Tighten the appropriate nuts until the wobble disappears. Reassemble the primary cover as detailed previously.

Adjustment (Typical)

1. Remove the inspection cap from the primary cover to expose the adjuster mechanism.

2. Turn the hand lever adjuster in until the cable is slack.

3. Loosen the locknut at the end of the rod and turn the adjusting screw clockwise until the pressure plate just begins to move forward. Turn the adjuster one full turn (½ turn for 500cc models) counterclockwise and lock it in place with the locking nut.

4. Adjust the hand lever adjuster to give ⅛ in. (3mm) free play in the cable. If the clutch continues to malfunction, slip, or grab, adjust the pressure plate as described previously.

5. A loud click coming from the clutch mechanism indicates a need for careful adjustment. On early models, before engine No. DU.66246, check the cable end and the connector for wear.

PRIMARY DRIVE SHOCK ABSORBER

Removal/Installation

1. The primary drive shock absorber is located behind the clutch. Remove the clutch as previously described.

2. Remove the 3 screws holding the shock absorber cover.

3. Pry out the shock absorber rubbers, starting with the smaller ones. Inspect for deterioration and replace if necessary (**Figure 7**).

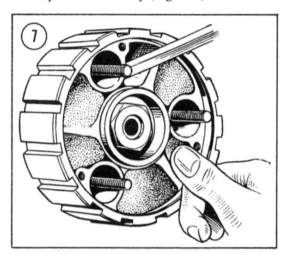

4. Install the large rubbers first and use prying bars to insert the smaller ones.

5. Secure the cover, using Loctite or a similar thread-locking compound on the cover screws.

ROTOR AND STATOR

Removal/Installation

1. Remove the primary cover as previously described.

2. Disconnect the 2 or 3 stator leads at their snap connectors under the engine (top rear of the primary case on 1972 and later models).

3. Unscrew the 3 stator nuts and pull stator free.

4. Pull the lead wire from the sleeve nut. See **Figure 8** (through 1975) **or 9** (1976 and later).

5. Flatten the locking tab washer and remove the rotor nut.

6. Use a mallet and wrench or put the transmission in gear and have a helper depress the rear brake pedal while the nut is loosened.

7. Check the rotor for cracks.

8. Replace the rotor with the key in place and tighten nut to torque specified in Appendix I.

9. Replace the stator so its leads are to the outside and tighten the 3 mounting nuts to the torque specified in Appendix I.

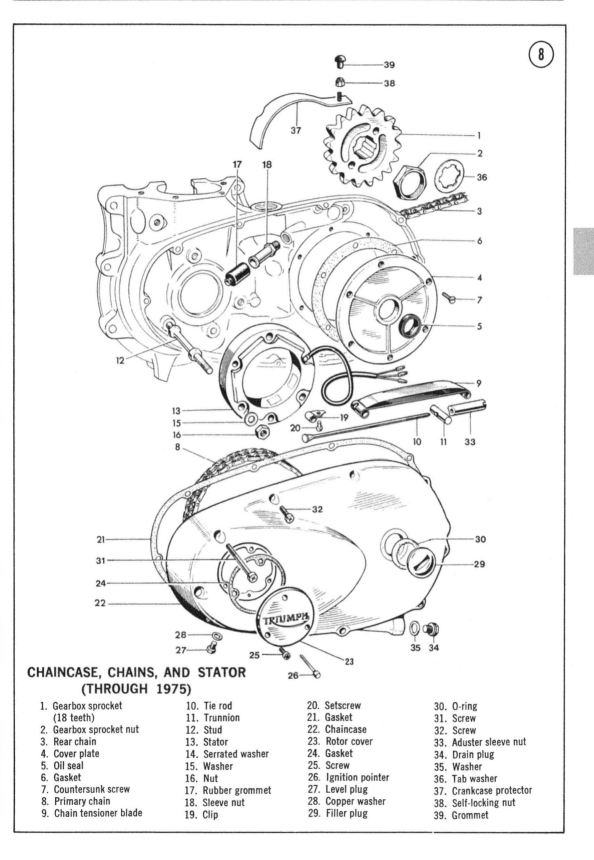

CHAINCASE, CHAINS, AND STATOR
(THROUGH 1975)

1. Gearbox sprocket
   (18 teeth)
2. Gearbox sprocket nut
3. Rear chain
4. Cover plate
5. Oil seal
6. Gasket
7. Countersunk screw
8. Primary chain
9. Chain tensioner blade

10. Tie rod
11. Trunnion
12. Stud
13. Stator
14. Serrated washer
15. Washer
16. Nut
17. Rubber grommet
18. Sleeve nut
19. Clip

20. Setscrew
21. Gasket
22. Chaincase
23. Rotor cover
24. Gasket
25. Screw
26. Ignition pointer
27. Level plug
28. Copper washer
29. Filler plug

30. O-ring
31. Screw
32. Screw
33. Aduster sleeve nut
34. Drain plug
35. Washer
36. Tab washer
37. Crankcase protector
38. Self-locking nut
39. Grommet

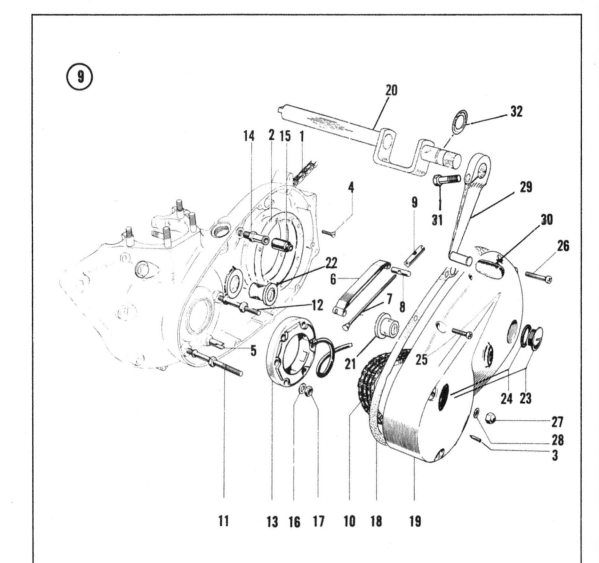

CHAINCASE AND CHAINS — 1976

1. Rear chain
2. Gasket
3. Timing pointer
4. Countersunk screw
5. Chain tensioner abutment
6. Chain tensioner blade
7. Tie rod
8. Trunnion
9. Adjuster sleeve nut
10. Primary chain
11. Stator stud
12. Stator stud
13. Stator
14. Sleeve nut
15. Rubber grommet
16. Plain washer
17. Self-locking nut
18. Gasket
19. Chaincase
20. Foot change spindle
21. Bushing
22. Bushing
23. Filler plug
24. O-ring
25. Screw
26. Screw
27. Domed nut
28. Copper washer
29. Gearchange pedal
30. Pedal rubber
31. Bolt
32. O-ring shift spindle

10. Thread the lead into the sleeve nut and re-connect the disconnected lead wires on the outside of the engine.

11. Make sure the stator lead is well clear of the primary chain. Turn the engine over and check that the rotor and stator do not touch. A 0.008 in. (0.2mm) feeler gauge should fit between the stator pole pieces and the rotor.

CLUTCH AND ENGINE SPROCKETS (ALL MODELS)

1. Remove the primary case cover, clutch plates, rotor, and stator as described previously.

2. Remove the rotor spacer and loosen the primary chain adjuster.

3. Remove the clutch hub nut and washer. Two special pullers must be used for this operation (**Figure 10**). A clutch driving and driven plate can be welded or bolted together to use as a clutch locking tool. Early machines may have a tab washer and a cupped washer instead of a locknut to secure the clutch hub.

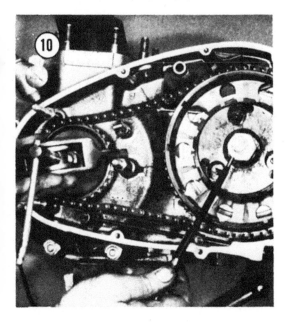

4. Screw the clutch puller into the clutch hub all the way.

5. Tighten the center bolt on the puller until the clutch hub comes free.

6. Insert the engine sprocket puller and loosen the engine sprocket.

7. Remove the sprocket simultaneously along with the one-piece chain.

8. Disassemble the clutch hub shock absorber and the sprocket, washer, rollers, and pins.

9. Remove the key from the gearbox main shaft and check the oil seal in the primary chain inner cover by removing the circular cover. See **Figure 11**. Use a new gasket and check that the oil seal is installed correctly when replacing the cover.

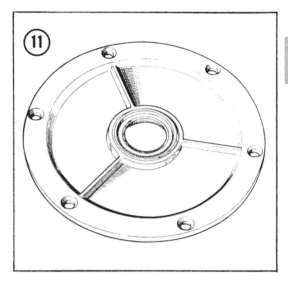

10. Clean all parts in solvent and check for wear or damage.

11. Check the primary chain for excessive wear by marking 2 points 12 in. (30.5 cm) apart on a level surface. Clean the chain and place it next to these marks. With the chain compressed to its minimum length the 2 marks should line up with pivot pin centers 32 links apart. Extend the chain and check to see that the extension is not more than ¼ in. (6.35mm).

12. Check all sprocket teeth for abnormal wear patterns or damage.

13. The shock absorber spider should be a push fit on the clutch hub with no radial movement. The engine sprocket should have no radial movement on the crankshaft.

14. Check the clutch hub bearing diameter, roller diameter, and the clutch sprocket bearing for excessive wear. Check the clutch hub and sprockets against the specifications given in Appendix I.

15. If the rollers are replaced, be sure they are the correct length, as given in Appendix I.

16. Check the fit of the shock absorber spider in the inner and outer retaining plates; make sure that it has not scored the inside faces of the plates.

17. Check the clutch operating rod for straightness and correct length, as given in Appendix I.

18. Grease the clutch hub and assemble the thrust washer with 20 rollers.

19. Position the sprocket and fit the shock absorber with its 3 threaded pins. Use Loctite or a similar compound on the threads.

20. Replace the gearbox main shaft key and tap the clutch hub carefully into place.

21. With the primary chain on the clutch sprocket, position the engine sprocket with its tapered boss toward the main bearing and engage it in the primary chain, and drive the engine sprocket into place.

22. Insert the clutch locking tool as used in disassembly.

23. Replace the cup washer with its cup side out, the tab washer with its tab in the hole in the shock absorber spider, and the clutch securing nut. Tighten the securing nut as specified in Appendix I.

24. When replacing the engine and clutch sprockets, replace the tapered spacer behind the engine sprocket with the taper toward the crankshaft main bearing and oil seal.

25. Install the rotor locating key or the sprocket dowel peg, whichever is used.

26. Reassemble the remaining components in the reverse order of disassembly. Be sure to replace the spacer between the engine sprocket and rotor, and the key in the crankshaft which locates the rotor.

27. Fill the primary case with oil.

TRANSMISSION OUTER COVER

Removal/Installation
(750/650cc Models)

1. Remove the right exhaust pipe and right footpeg (**Figure 12 or 13**).

2. Remove the right panel, if so fitted.

3. Disconnect the clutch cable by loosening the cable adjuster at the handlebar and removing the cable end from the lever.

4. Unscrew the abutment at the gearbox.

5. Remove the large slotted plug from the transmission cover and free the lower cable end from the clutch operating arm. On earlier models without the slotted plug, the cable end can be disconnected after the abutment has been unscrewed.

6. Drain the transmission oil as described in Chapter Two. Remove transmission filler plug.

7. Place the transmission in top gear.

8. Remove the hex nuts and screws securing the transmission cover. Push the kickstarter lever down slightly, and tap the cover with a plastic mallet until it comes free. Do not pry cover off.

9. Carefully move the gearshift pedal to release the plungers and springs from the gearshift quadrant. On 1976 and later models, use an improvised tool in the slotted end of the quadrant.

10. Clean the mating surfaces of the cover and apply gasket cement. Move the kickstarter until it is approximately halfway down and put the cover in position.

11. Tighten all screws and nuts.

12. Replace the drain plug and fill the transmission, referring to data in Appendix I for the proper amount and grade of oil.

13. Reassemble the remaining components in the reverse order of disassembly.

Removal/Installation (500cc Models)

1. Remove the right exhaust pipe and the right footpeg (**Figure 14**).

2. Loosen the clutch cable adjuster at the handlebar and remove the slotted adapter at the abutment on the gearbox. Unscrew the abutment and disconnect the cable at its lower end.

3. Drain the gearbox oil. See Chapter Two.

4. Put the transmission into fourth gear and apply the rear brake.

5. Remove the 2 nuts and 4 screws from the gearbox cover and take off the kickstarter lever.

6. Pull the gearshift lever with one hand and free cover with a plastic mallet. See **Figure 15**.

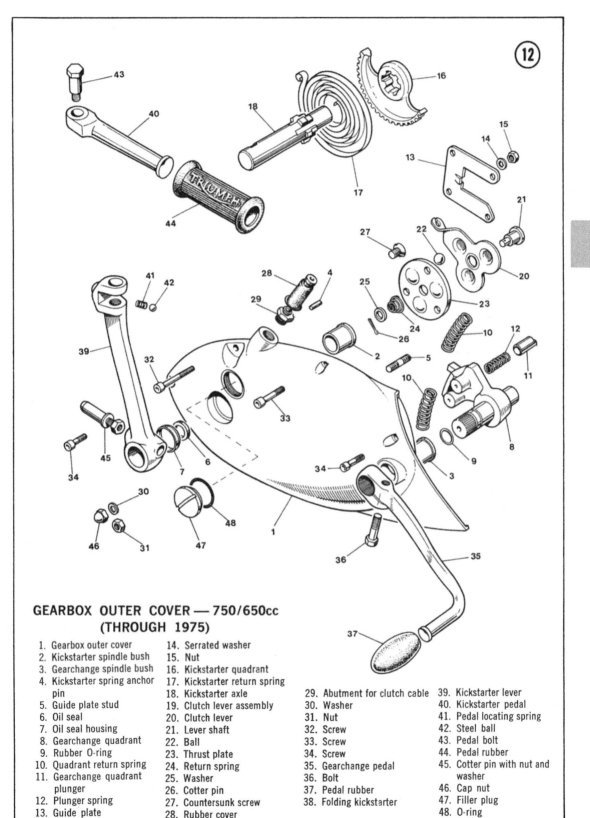

GEARBOX OUTER COVER — 750/650cc (THROUGH 1975)

1. Gearbox outer cover
2. Kickstarter spindle bush
3. Gearchange spindle bush
4. Kickstarter spring anchor pin
5. Guide plate stud
6. Oil seal
7. Oil seal housing
8. Gearchange quadrant
9. Rubber O-ring
10. Quadrant return spring
11. Gearchange quadrant plunger
12. Plunger spring
13. Guide plate
14. Serrated washer
15. Nut
16. Kickstarter quadrant
17. Kickstarter return spring
18. Kickstarter axle
19. Clutch lever assembly
20. Clutch lever
21. Lever shaft
22. Ball
23. Thrust plate
24. Return spring
25. Washer
26. Cotter pin
27. Countersunk screw
28. Rubber cover
29. Abutment for clutch cable
30. Washer
31. Nut
32. Screw
33. Screw
34. Screw
35. Gearchange pedal
36. Bolt
37. Pedal rubber
38. Folding kickstarter
39. Kickstarter lever
40. Kickstarter pedal
41. Pedal locating spring
42. Steel ball
43. Pedal bolt
44. Pedal rubber
45. Cotter pin with nut and washer
46. Cap nut
47. Filler plug
48. O-ring

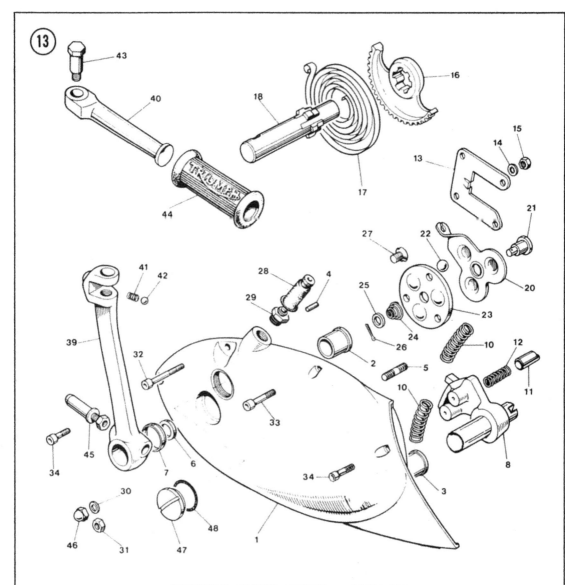

GEARBOX OUTER COVER — 1976 ON

1. Gearbox outer cover
2. Kickstarter spindle bushing
3. Gearchange spindle bushing
4. Kickstarter spring anchor pin
5. Guide plate stud
6. Oil seal
7. Oil seal housing
8. Gearchange quadrant
10. Quadrant return spring
11. Gearchange quadrant plunger
12. Plunger spring
13. Guide plate
14. Serrated washer
15. Nut
16. Kickstarter quadrant
17. Kickstarter return spring

18. Kickstarter axle
19. Clutch lever assembly
    (items 20-26)
20. Clutch lever
21. Lever shaft
22. Ball
23. Thrust plate
24. Return spring
25. Plain washer
26. Split pin
27. Countersunk screw
28. Rubber cover
29. Abutment for clutch cable
30. Plain washer

31. Nut
32. Screw
33. Screw
34. Screw
38. Folding kickstarter
    (items 39-43)
39. Kickstarter lever
40. Kickstarter pedal
41. Pedal locating spring
42. Steel ball
43. Pedal bolt
44. Pedal rubber
45. Cotter pin with nut and washer
46. Domed nut
47. Filler plug
48. O-ring

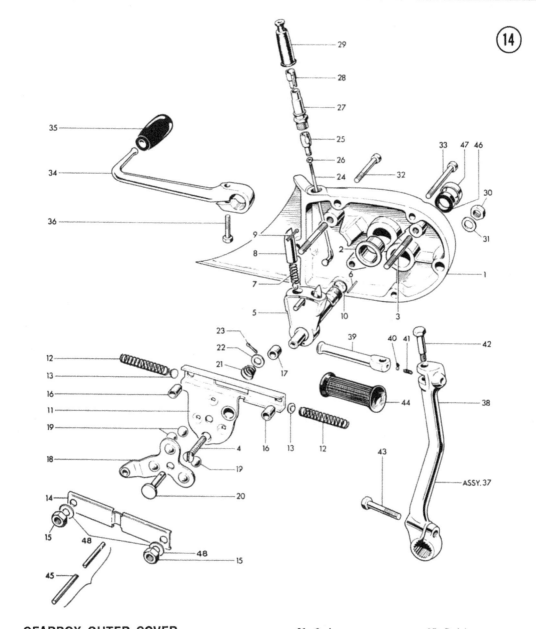

GEARBOX OUTER COVER
AND LEVERS — 500cc

1. Gearbox outer cover
2. Gearchange spindle bush
3. Stud
4. Screw
5. Gearchange quadrant
6. Spring retaining pin
7. Plunger spring
8. Selector plunger
9. Cotter pin
10. Sealing ring
11. Return spring housing
12. Return spring
13. Thrust button
14. Cover
15. Nut
16. Spacer
17. Spacer
18. Clutch lever
19. Ball
20. Shaft
21. Spring
22. Washer
23. Split pin
24. Spoke
25. Connector
26. Nut
27. Abutment
28. Adaptors
29. Rubber cover
30. Nut
31. Washer
32. Screw
33. Screw
34. Gearchange lever
35. Pedal
36. Bolt
37. Kickstarter lever
38. Kickstarter crank
39. Kickstarter pedal
40. Steel ball
41. Index spring
42. Pivot bolt
43. Clamp bolt
44. Pedal rubber
45. Clutch operating rod
46. Oil seal
47. Oil seal cover
48. Serrated washer

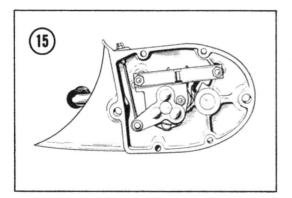

7. Make sure the mating surfaces are clean and apply a coat of gasket cement. Check that the locating pegs are in place and that the kickstarter return spring and the spring plate are correctly seated. Replace the cover and kickstarter lever.

8. Replace transmission oil as described in Chapter Two.

KICKSTARTER

Removal/Installation (Typical)

1. Loosen the kickstarter crank cotter pin nut a few turns. Use a hammer and drift to knock it free from its taper and pull the kickstarter lever off its shaft.

2. Pull out the gear and spring assembly.

3. Put the transmission in gear, apply the rear brake, and flatten the locking tabs on the end of the main shaft.

4. Remove the kickstarter ratchet pinion nut.

5. Remove the pinion, ratchet, spring, and sleeve. See **Figure 16**.

6. Clean all parts thoroughly.

7. Inspect the quadrant for broken or excessively worn teeth and for a good fit on the spindle.

8. Check the return spring for wear or cracks at the point where it engages on the spindle.

9. Check the amount of play in the spindle bushing.

10. Check the ratchet mechanism for rounded or chipped teeth.

11. Check the condition of the bushing and spring in the kickstart pinion.

12. Check that the kickstarter stop peg is not damaged and is a good fit inside the inner cover.

13. If the kickstarter spindle bushing needs to be replaced, remove all parts from the outer cover and heat the cover to 212°F (100°C). Press out the old bushing and fit the new one while the cover is still hot. This job is best left for a shop unless a hydraulic press is available.

14. If the kickstarter quadrant is to be replaced, fit it as shown in **Figure 17** so it will be in the proper relation to the flat on the shaft.

15. Replace the sleeve, spring, pinion, and ratchet on the main shaft.

16. Fit the tab washer and tighten the nut to specification, Appendix I. Do not overtighten or the sleeve may collapse. See **Figure 18** for 500cc model kickstarter.

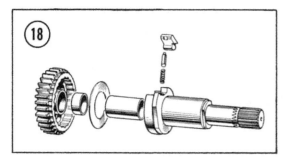

17. Replace the return spring on the kickstarter quadrant and put the spindle into the kickstarter bushing.

18. Hook the return spring onto the anchor peg on the inside cover.

19. Replace the oil seal on the shaft from the outside and fit the kickstarter lever, locking it with the tapered pin.

20. Fill the transmission with the amount and grade of oil specified in Chapter Two.

CLUTCH AND GEARSHIFT OPERATING MECHANISMS

Removal/Installation (750/650cc Models)

1. On 1975 and earlier models, loosen the bolt holding the gearshift lever to its shaft and remove the lever.

2. Remove the 4 nuts and lockwashers holding the guide plate.

3. Remove the guide plate, plunger quadrant, and return springs.

4. Wash all parts thoroughly in solvent.

5. Check the gearshift plungers for excessive wear and see that they are a proper fit in the quadrant. Compare to tables in Appendix I.

6. Measure the plunger springs and compare with specifications given in Appendix I.

7. Check the guide plate for wear or grooving and replace if necessary. Corroded pedal return springs should be replaced.

8. Insert the gearshift quadrant into the bushing and check for excessive play. Inspect the tips of the plungers and the camplate operating quadrant teeth for chipping and wear. If the camplate quadrant needs replacing, remove the gearbox inner cover as detailed below.

9. Pull out the 2 split pins and the spindle.

10. If the gearshift spindle bushings must be replaced, proceed as detailed in Step 13 of *Kickstarter* section above for the outer cover bushing.

11. If the inner cover bushing is not usable, remove the inner cover and disconnect the camplate operating quadrant as detailed below.

12. Cut threads in the bushing with a suitably sized tap. Replace the tap with a bolt and heat the inner cover to 212°F (100°C). Hold the bolt in a vise and tap the cover with a plastic mallet to free the bushing. Drive in the new bushing while the cover is still hot.

13. Place a new O-ring on the spindle and insert the spindle into the cover bushing.

14. Fit the quadrant return springs in place.

15. On 1975 and earlier models, assembly may be easier if the shifter pedal is installed at this point, so the quadrant can be easily turned to get the springs in place. On 1976 and later models, a tool can be devised to fit into the mating slot on the end of the quadrant. Use this tool to turn the quadrant to get the springs into place.

16. Replace the retainer plate with its 4 washers and nuts.

17. Replace the gearshift quadrant springs and plungers.

18. Refer to **Figure 19** for assembly and disassembly details of clutch operating mechanism.

Removal/Installation (500cc Models)

1. Refer to **Figure 20**. Remove the 2 nuts inside the gearbox outer cover.

2. Remove the cover plate and gearshift return springs, 2 thrust buttons, and the spacers.

3. Remove the countersunk screw holding the clutch operating mechanism and separate the assembly from the cover.

4. Remove the gearshift lever from the shaft and pull out the shaft.

5. Remove the cotter pin from the clutch operating shaft to free the operating balls.

6. Remove the 2 cotter pins from the gearshift quadrant.

7. Clean all parts and examine for wear. See **Figure 21**.

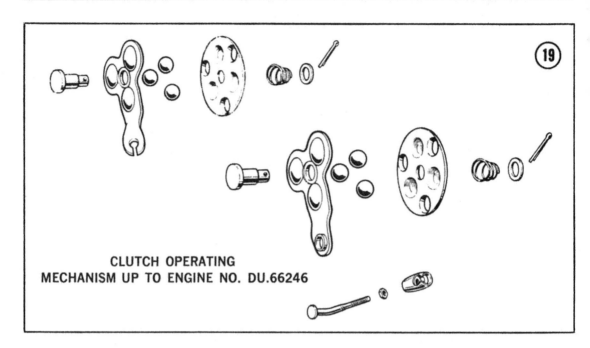

CLUTCH OPERATING
MECHANISM UP TO ENGINE NO. DU.66246

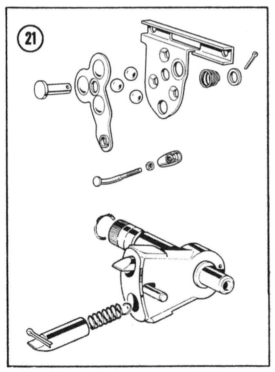

10. Check the plunger tips and the camplate cutaways for wear.

11. Reassemble the gearshift mechanism and replace it in the outer cover bushing, using a new O-ring and lubrication.

12. If the bushing needs to be replaced, heat the outer cover (with all components removed) to 212°F (100°C) and drive out the bushing with a drift made from a 3 in. diameter bar turned down to ⅝ in. for a distance of ¾ in. at one end. Press in the new bushing while cover is still hot.

13. Reassemble the clutch operating mechanism and replace the spacer on the end of the gearshift quadrant shaft.

14. Fit the clutch operating mechanism to the outer cover and secure with a countersunk screw.

15. Replace spacers on the studs and assemble the gearshift return springs and thrust buttons.

16. Replace the return spring cover plate and tighten the retaining nuts.

8. Check the gearshift plungers and springs. proper spring length is given in Appendix I. Check the gearshift lever return springs for fatigue or corrosion.

9. Insert the quadrant into the quadrant bushing and check for excessive play.

TRANSMISSION

**Removal/Installation
(750/650cc Models)**

1. Take off the gearbox outer cover, as detailed previously, leaving the transmission in top gear.

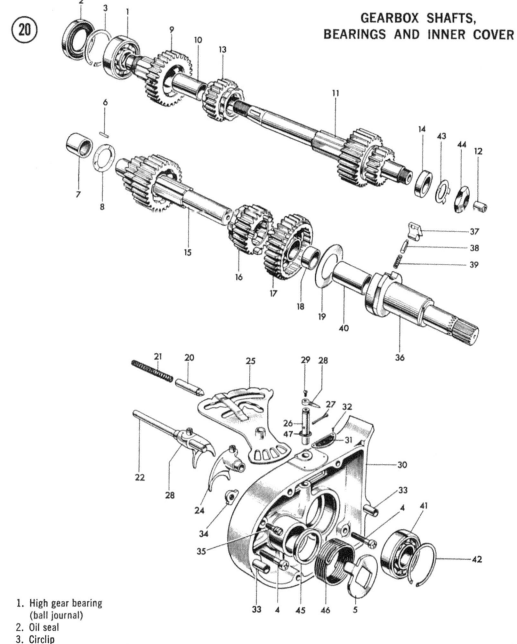

GEARBOX SHAFTS, BEARINGS AND INNER COVER

1. High gear bearing (ball journal)
2. Oil seal
3. Circlip
4. Screw
5. Return spring plate
6. Peg
7. Layshaft bearing
8. Thrust washer
9. Main shaft high gear
10. High gear bushing
11. Main shaft with third and low gears
12. Clutch rod bushing
13. Main shaft second gear
14. Main shaft spacer
15. Layshaft with second and high gears
16. Layshaft third gear
17. Layshaft low gear
18. Low gear bushing
19. Pawl retaining disc
20. Camplate index plunger
21. Index plunger spring
22. Selector fork spindle
23. Main shaft selector fork
24. Layshaft selector fork
25. Gear selector camplate
26. Camplate spindle
27. Cotter pin
28. Gear indicator pointer
29. Screw
30. Inner cover
31. Gear indicator plate
32. Indicator plate rivet
33. Hollow dowel
34. Kickstarter stop plate
35. Stop plate and anchor screw
36. Kickstarter spindle
37. Kickstarter pawl
38. Plunger
39. Plunger spring
40. Needle roller bearing
41. Main shaft bearing (ball journal)
42. Circlip
43. Tab washer
44. Main shaft nut
45. Spacer
46. Kickstarter return spring
47. O-ring

2. Remove the necessary nuts and bolts to take off the right rear engine mounting plate.

3. Straighten the tabs on the lockwasher and remove the kickstarter pinion ratchet nut from the main shaft. See Figure 16.

4. Remove the speedometer cable union nut and pull out the cable from the speedometer drive shaft on engines prior to DU.24875.

5. Remove the primary cover and dismantle the clutch; remove the hub and key from the transmission main shaft. On 1976 models, withdraw the foot change spindle (Figure 13).

6. Remove the acorn nuts under the transmission and pull out the camplate indexing plunger and spring.

7. Remove the Allen head screw, Phillips head screw, and bolt shown in **Figure 22**.

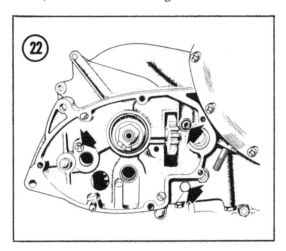

8. Tap the transmission inner cover loose with a plastic mallet; never pry the cover off.

9. Remove the selector fork spindle.

10. Pull out the main shaft. The countershaft and other gears can then be removed.

11. Remove the camplate and spindle assembly and the 2 brass thrust washers over the needle roller bearings. See **Figure 23**.

12. If the main shaft high gear or the oil seal behind it requires replacement, remove the cover from the gearbox sprocket and straighten the tabs on the locking washer.

13. Remove the nut as shown in **Figure 24**.

14. Drive the gear and spindle out through the transmission side.

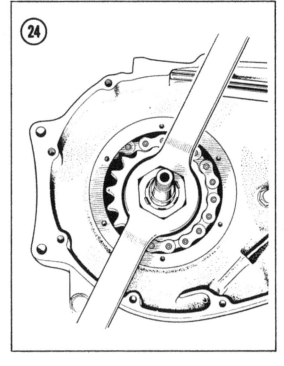

15. Check the gear case and cover for cracks or damage to mating surfaces. There are 4 locating dowels in the gear case and inner cover.

16. Check the main shaft and countershaft for damage or wear. Measure the bearing diameters and compare with the specifications in Appendix I. Insert the countershaft in the bearing and check for excessive wear.

17. Check the main shaft bearing races for pitting or wear. There should be no side play in the center track of the bearing. In 5-speed transmissions, the top gear bearing is fitted into the press fit roller bearing in the right-side case. If wear on the top gear bearing is excessive, replace the bearing and gear as a unit. The main shaft and the gear should be a hand press fit in the cover bearing.

18. Check the gears for broken or worn teeth and all splines for binding. Refer to Appendix I and check the main shaft high gear and the countershaft low gear bushings for wear.

19. The selector fork rod must be a snug fit in the case and cover. Check that the fork rod does not have grooves. Check the selector fork running faces for wear. The camplate spindle should fit snugly in its housing. If the camplate gear wheel is worn, changing gears will be difficult and the gears may become damaged.

20. Check the camplate plunger for smooth operation and measure its spring using the specifications given in Appendix I. Clean any corrosion from moving parts.

21. Check the main shaft high gear bushing using specifications given in Appendix I.

22. Remove the large circlip and heat the cases to 212°F (100°C) and drive out the right main shaft bearing. Replace the new bearing while the cases are still hot, using a drift of 2.5 in. (6.35mm) outside diameter to tap it in place. Replace the circlip.

23. Pry out the oil seal and remove the circlip from the high gear bearing. Heat the case as before and drive out the bearing. Fit the new one while the case is still hot.

24. Press out the main shaft high gear bushing from the teeth end of the gear with a drift made for the job. Machine a ⅞ in. diameter steel bar so it has a 13/16 in. pilot ¾ in. long at one end. Press the new bushing in with its oil groove at the teeth end of the gear. Ream the bushing to the size given in Appendix I.

In 5-speed transmission, the caged needle bearings located in the ends of the top gear can be pressed out with a drift as shown in **Figure 25**.

25. Heat the case cover to 212°F (100°C) and drive out the right needle roller bearing. See

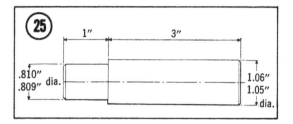

illustration in **Figure 26**. On earlier models, remove the speedometer driving shaft by removing its locating screw and driving out the shaft and bushing. Replace the bearing so that the bearing protrudes the specified amount as given in Appendix I.

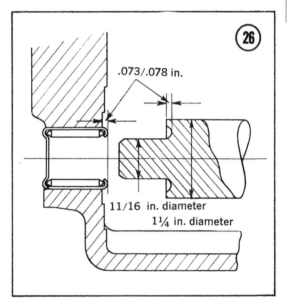

26. Heat the case as before and drive the left needle roller bearing from the sprocket cover plate aperture into the gearbox. Replace the bearing while the case is still hot, pressing until 0.073-0.078 in. (1.85-1.98mm) of the bearing protrudes above the case surface. Use care to avoid damaging the bearing. Seal the outer portion of the bearing with gasket cement where it protrudes from the case.

27. Drive a new oil seal into the main bearing. Fit the lip and spring against the bearing. Fit high gear into the bearing. Oil the sprocket boss and fit it into the high gear, tightening the securing nut by hand. Replace the rear drive chain and have a helper apply the rear brake so the securing nut can be tightened.

28. Oil the nose of the high gear (bronze bushing on main shaft high gear on earlier models) and replace the circular cover plate using a new gasket. Oil the camplate spindle and insert it. Place the camplate plunger mechanism into the gearbox lower case with the fiber washer under the plunger retaining nut. Set the camplate so the plunger is in the notch between second and third gear positions (neutral position for 5-speed transmission) as shown in **Figure 27**.

> NOTE: *Steps 29 through 32 apply to 4-speed transmissions only.*

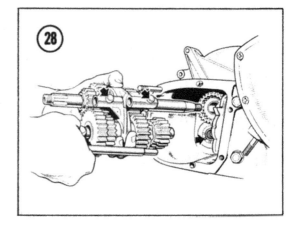

29. Set the countershaft thrust washer over the inner roller bearing, holding it in place with chilled grease. The grooved surface of the washer should face the countershaft.

30. Oil and assemble the gear clusters as shown in **Figure 28**. Use grease to hold the camplate rollers in place on the selector forks and hold the selector forks in the grooves between the gears as shown. The selector fork for the main shaft has a smaller radius than the countershaft fork.

31. As the gear clusters and shafts are moved into position inside the gearbox, fit the main

shaft and countershaft into the bearings. At the same time, slide and align the gears so the selector fork rollers fit into their tracks on the camplate. Oil the selector fork spindle, align the selector forks and insert the spindle, shoulder end first, until it seats in the case. Check that the main shaft selector fork is on the far side of countershaft fork.

32. Check that the camplate quadrant is free to move; fit the inner cover. Use grease to hold the countershaft thrust washer in place over the

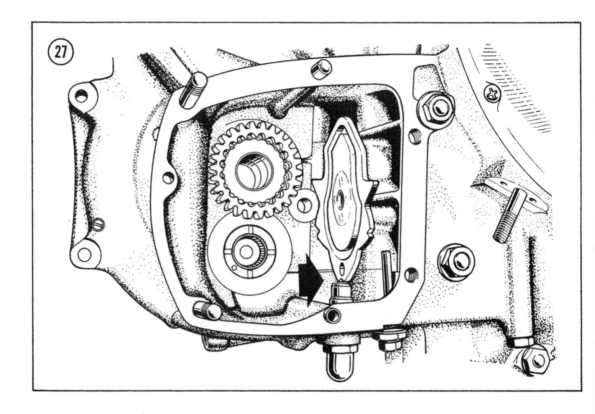

bearing in the inner cover. Continue installation in reverse order of removal.

> NOTE: *Steps 33 through 38 apply to 5-speed transmissions.*

33. Grease the smooth surface of the bronze thrust washer and place it over the roller bearing in the left case. The grooved face of the washer must face inward (**Figure 29**).

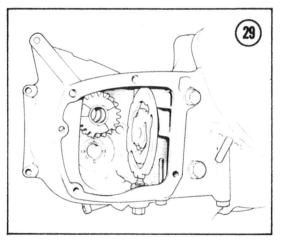

34. Lubricate top gear bearings and countershaft bearing with transmission oil. Place the main shaft top gear and its selector fork (this fork has no cutaway in the housing) on the main shaft. Assemble the main shaft in high gear. Grease the arc and knob of the selector fork to hold it in the gear track and the camplate groove.

35. Install the countershaft with fourth and fifth gears in position in the transmission. Engage the gears with fourth and fifth gears on the main shaft. The camplate should still be in the neutral position and none of the sliding dogs should be engaged. Install the countershaft third gear and its selector fork (this fork has a large pin and a cutaway on the housing). Install the main shaft third gear and engage it with the countershaft third gear. Lubricate the bushing in the countershaft second gear and install it on the shaft. Install the combined first/second gear on the main shaft. Install the countershaft first gear, along with its selector fork (small knob, cutaway in housing matches cutaway in third gear selector). Install the selector rod. Install the circlip on the end of the countershaft and slip the engaging dog onto the shaft against the circlip. Rotate the

camplate clockwise (as viewed from the rear) one notch to engage first gear. The engaging dogs should be meshed with the dogs on the countershaft first gear.

36. Make sure the camplate operating quadrant moves freely. Grease the smooth surface of the bronze thrust washer and place it, with the grooved face inward, over the roller bearing in the outer cover.

37. Lubricate all moving parts in the transmission and apply gasket compound to the transmission sealing surface. Check that the alignment dowels are in place and fit the inner cover onto the transmission until it is approximately ¼ in. from making contact with the sealing surface. Turn the camplate quadrant to the middle point of its travel and hold it there while the cover is installed. To position the camplate quadrant, line up the top edge of the second tooth with the centerline of the spindle bore for the gear selector pedal. Replace the inner cover screws and bolt, then check that gear changing is correct by replacing the outer cover and gearshift lever. Turn the rear wheel and operate the lever. If the gear changing sequence is incorrect, remove the outer and inner covers. Check that the camplate is set so the plunger is in the notch between second and third gears as detailed earlier. Replace the inner cover so that the appropriate tooth on the camplate quadrant lines up on the main shaft centerline. Refer to **Figure 30**.

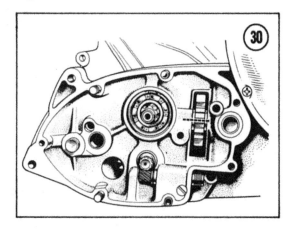

38. Reassemble the kickstarter and then refit the transmission outer cover. Reinstall the foot change spindle. Make sure the tongue on the

end of the spindle meshes with the slot in the end of the gear change quadrant. Turn the spindle slightly, if required, to ensure proper meshing. Reassemble the clutch and drive mechanism as detailed above. Fill primary and transmission cases with quantities and grades of oil given in Chapter Two.

Removal/Installation (500cc Models)

1. Remove the outer cover as detailed before (transmission in fourth gear). Remove the primary drive components as detailed previously.

2. Straighten the lockwasher tabs and remove the main shaft nut. Have a helper apply the rear brake while this is done. Remove the 2 screws and tap the clutch end of the main shaft with a plastic mallet to free the cover.

3. Remove inner cover as detailed before. Remove the cotter pin and take out the camplate spindle. Pry off the kickstarter return spring and the spacer so the kickstarter spindle can be removed from the inner cover.

4. Remove the camplate index plunger and spring from the gearbox. Remove the selector fork spindle and disengage the selector forks from the camplate. Pull the countershaft out of the inner cover and then drive the main shaft out of the bearing with the aid of a plastic mallet.

The kickstarter can be disassembled as shown in Figure 18.

5. Inspection of transmission components is the same as detailed earlier. Be sure to check the kickstarter pawl and countershaft low gear dogs for chips or cracks.

6. Reassemble the transmission as follows. Hold the countershaft thrust washer in place over the needle roller cage with grease. Oil all gears, assemble the main shaft (see following section on *Main Shaft and Countershaft Bearings*), and fit it into the inner cover. Replace the plunger spring, plunger, and kickstarter pawl on the kickstarter spindle, then insert the spindle into the inner cover. Slide the countershaft assembly into the kickstarter bearing. There is a spacer between the main shaft assembly and the bearing in the inner cover.

7. Refer to **Figure 31** and replace the shifter forks on their shafts. Hold them in place by inserting the selector fork spindle. Assemble the camplate into the outer cover and fit the selector fork rollers in the camplate grooves. Replace the camplate spindle and secure with cotter pin.

8. Refer to **Figure 32** and fit the camplate index plunger and spring. The assembly can then be inserted into the gearbox case as shown. Before the outer cover is installed, operate the camplate

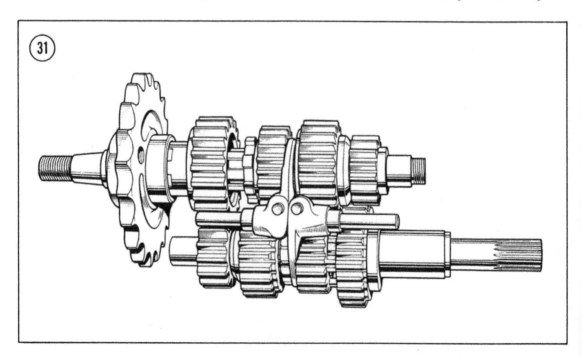

(31)

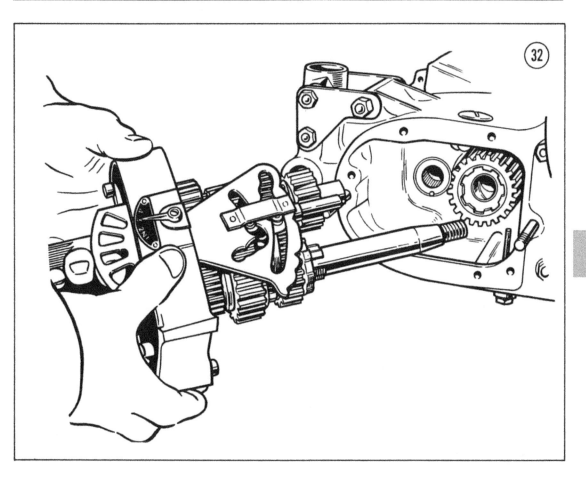

by hand and check that the selector rollers move all the way in the camplate grooves in both directions. If either of the rollers will not go all the way, the selector forks will have to be removed and their positions exchanged.

9. Replace the spacer on the kickstarter shaft and replace the screw which holds one end of the kickstarter spring. Tension the spring with a screwdriver and replace the return spring plate as shown in **Figure 33**.

10. Before replacing the cover, lubricate all moving parts inside with motor oil.

11. Use gasket cement on the cover mating surfaces. Replace the camplate index plunger and spring in its housing. Position the camplate so the selector fork rollers are midway along the camplate grooves and position the cover against the gearbox. Tap the cover into place and tighten the retaining screws.

12. Replace the spacer, tab washer, and locking nut on the main shaft. Place the transmission in

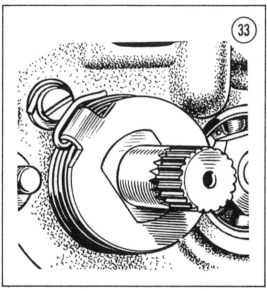

fourth gear and with the rear brake applied, tighten the main shaft nut. Bend over the tab washer.

13. Continue installation in reverse order of removal.

MAIN SHAFT AND COUNTERSHAFT BEARINGS

Replacement (500cc)

1. Replacement is the same as previously described for 750/650cc models except for removal of the right needle roller bearing and the countershaft left side bushings.

2. Heat kickstarter spindle to 212°F (100°C) and strike the spindle on a block of wood to remove the right needle roller bearing. On 3TA and 5TA models, a floating bronze bushing is used. The left countershaft bushings on 3TA and 5TA models can be driven out after the filler plug is driven out by heating the cases and using a suitable drift. Drive new bushings in; use a sealant and a new filler plug.

OTHER TRANSMISSION COMPONENTS (500cc)

Inspection, disassembly, assembly, and servicing of other 500cc model transmission components are the same as for the 750cc models given previously.

GEARBOX SPROCKET

Removal/Installation

The gearbox sprocket can be changed without disassembling the transmission.

1. Disassemble the clutch as detailed previously, remove the circular cover over the sprocket and follow the instructions in the previous section for removing the sprocket.

2. On earlier models with the speedometer drive taken from the countershaft, the speedometer drive gears will have to be changed if the number of gearbox sprocket teeth are changed for any reason.

3. After engine No. DU.13374, a different spline arrangement was used on the main shaft high gear and gearbox sprocket. If either the gear or the sprocket is replaced, the correct item must be used or they must be replaced as a set. The later models also require a gearbox sprocket cover plate with a large center hole and oil seals.

NOTE: If you own a 1978 or later model, first check the Supplement at the back of the book for any new service information.

CHAPTER SIX

FUEL SYSTEM

The fuel system includes the carburetor, air cleaner, and fuel tank.

CARBURETION

For proper operation, a gasoline engine must be supplied with fuel and air, mixed in the proper proportions by weight. A mixture in which there is an excess of fuel is said to be rich. A lean mixture is one which contains insufficient fuel. It is the function of the carburetor to supply the proper mixture to the engine under all operating conditions.

Triumph motorcycles are equipped with Amal carburetors of either monobloc (**Figure 1**) or concentric (**Figure 2**) type.

CARBURETOR OPERATION

Figure 3 is an exploded view of a typical Amal carburetor. The essential functional parts are a float and float valve mechanism for maintaining a constant fuel level in the float bowl, a pilot system for supplying fuel at low speeds, a main fuel system which supplies the engine at medium and high speeds, a tickler system, which supplies the very rich mixture needed to start a cold engine. The operation of each system is discussed in the following paragraphs.

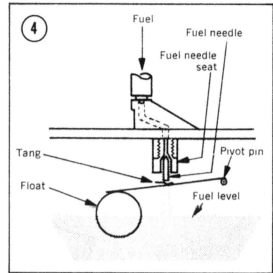

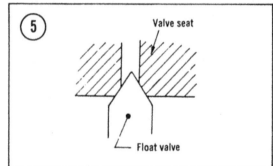

Float Mechanism

Figure 4 illustrates a typical float mechanism. Proper operation of the carburetor is dependent on maintaining a constant fuel level in the carburetor bowl. As fuel is drawn from the float bowl, the float drops. When the float drops, the float moves away from its seat and allows fuel to flow past the valve and seat into the float bowl. As this occurs, the float is then raised, pressing the valve against its seat, thereby shutting off the flow of fuel. It can be seen from this discussion that a small piece of dirt can be trapped between the valve and seat (**Figures 5 and 6**), preventing the valve from closing and allowing fuel to rise beyond the normal level, resulting in flooding.

Pilot System

Under idle or low speed conditions, at less than ⅛ throttle, the engine doesn't require much

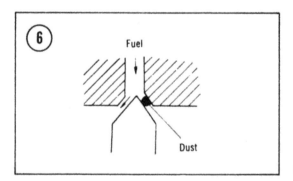

fuel or air and the throttle valve is almost closed. A separate pilot system is required for operation under such conditions. **Figure 7** illustrates the operation of the pilot system. Air is drawn through the pilot air inlet and controlled by the pilot air screw. The air is then mixed with fuel drawn through the pilot jet. The air/fuel mixture then travels from the pilot outlet into the main air passage, where it is further mixed

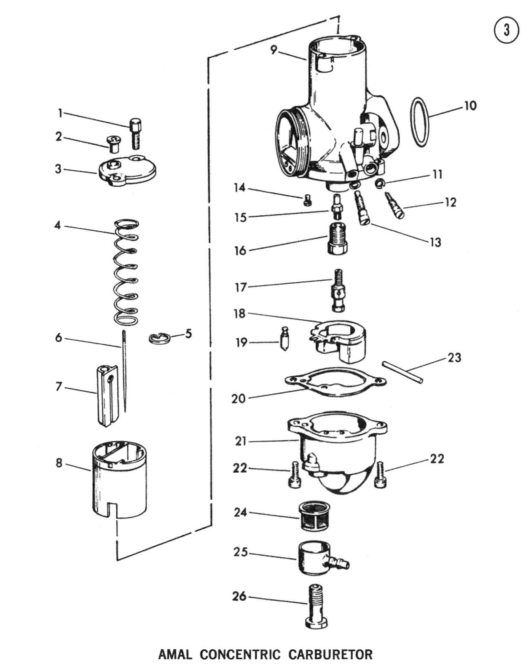

AMAL CONCENTRIC CARBURETOR

1. Screw
2. Ferrule
3. Mixing chamber cap
4. Spring
5. Jet needle cilp
6. Jet needle
7. Choke valve
8. Throttle valve
9. Mixing chamber body

10. O-ring
11. O-ring
12. Pilot air screw
13. Throttle stop screw
14. Pilot jet
15. Needle jet
16. Jet holder
17. Main jet
18. Float

19. Float needle
20. Gasket
21. Float bowl
22. Screw
23. Float pivot
24. Filter screen
25. Banjo
26. Banjo bolt

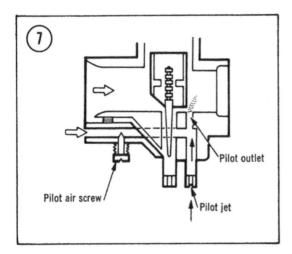

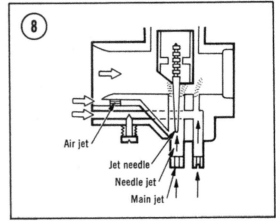

with air prior to being drawn into the engine. The pilot air screw controls the idle mixture.

If proper idle and low speed mixture cannot be obtained within the normal adjustment range of the idle mixture screw, refer to **Table 1** for possible causes.

Table 1   IDLE MIXTURE

Too rich:	Clogged pilot air intake Clogged air passage Clogged air bleed opening Pilot jet loose
Too lean:	Obstructed pilot jet Obstructed jet outlet Worn throttle valve Carburetor mounting loose

Main Fuel System

As the throttle is opened (up to approximately ¼ open), the pilot circuit begins to supply less of the mixture to the engine as the main fuel system total, illustrated in **Figure 8**, begins to function. The main jet, the needle jet, the needle, and the air jet make up the main fuel circuit.

As the throttle valve opens more than approximately ⅛ of its travel, air is drawn through the main port, and passes under the throttle valve in the main bore. The velocity of the air stream results in reduced pressure around the jet needle. Fuel then passes through the main jet, past the needle jet and needle, and into the air stream where it is atomized and sent to the cylinder. As the throttle valve opens,

more air flows through the carburetor. The needle, which is attached to the throttle slide, rises to permit more fuel to flow.

A portion of the air bled past the jet needle is mixed with the main air stream and atomized.

Air flow at small throttle openings is controlled primarily by the cutaway on the throttle slide.

As the throttle is opened still more, up to approximately ¾ open, the circuit draws air from 2 sources, as shown in **Figure 9**. The first source of air is through the venturi; the second source is through the air jet. Air passing through the venturi draws fuel through the needle jet. The needle is tapered, and therefore allows more fuel to pass. Air passing through the air jet passes to the needle jet to aid atomization of the fuel there.

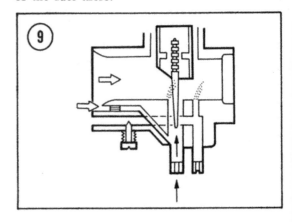

Figure 10 illustrates the circuit at high speeds. The needle is withdrawn almost completely from the needle jet. Fuel flow is then controlled

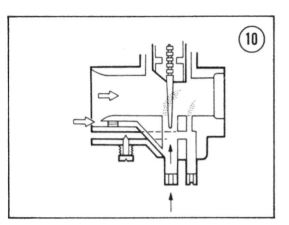

by the main jet. Air passing through the air jet continues to aid atomization of the fuel as described in the foregoing paragraphs.

Any dirt which collects in the main jet or in the needle jet obstructs fuel flow and causes a lean mixture. Any clogged air passage, such as the air bleed opening or air jet, may result in an overrich mixture. Other causes for a rich mixture are a worn or loose needle jet, or loose main jet. If the jet needle is worn, it should be replaced. However, it may be possible to effect a temporary repair by placing the needle clip in a higher groove.

Tickler System

A cold engine requires a mixture which is far richer than normal. A tickler system provides it. When the rider presses the tickler button, the float is forced downward, causing the float needle valve to open, and thereby allowing extra fuel to flow into the float chamber.

CARBURETOR OVERHAUL

There is no set rule regarding frequency of carburetor overhaul. A carburetor used primarily for street riding may go 5,000 miles (8,000 km) without attention. If the bike is used off-road, the carburetor might need an overhaul in less than 1,000 miles (1,600 km). Poor engine performance, hesitation, and little response to idle mixture adjustment are all symptoms of possible carburetor malfunctions. As a general rule, it is good practice to overhaul the carburetor each time you perform a routine decarbonization of the engine.

Amal Concentric Dissassembly/Assembly

1. Remove the mixing chamber cap (**Figure 11**).

2. Withdraw the throttle valve (**Figure 12**). Do not lose the spring or spring plate. Note the position of each part as it is removed.

3. Remove the float bowl (**Figure 13**).

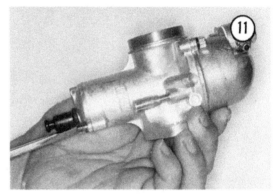

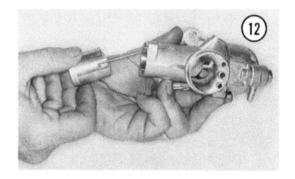

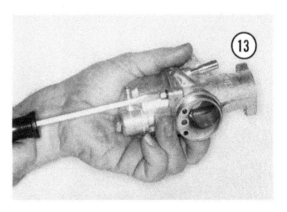

4. Remove the float and float needle together (**Figure 14**).

5. Remove the banjo bolt (**Figure 15**) and the fuel inlet banjo fitting (**Figure 16**).

6. Carefully remove filter screen (**Figure 17**).

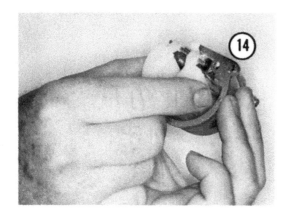

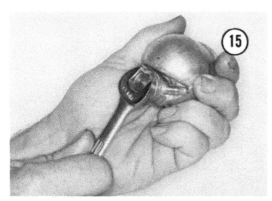

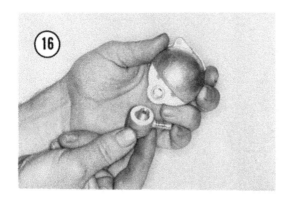

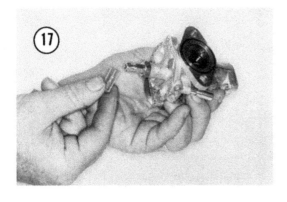

7. Remove the main jet (**Figure 18**).

8. Remove the jet holder (**Figure 19**). Unscrew the needle jet if necessary.

9. Remove the pilot jet (**Figure 20**).

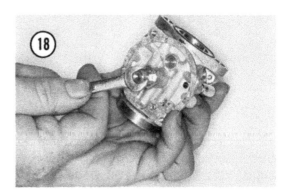

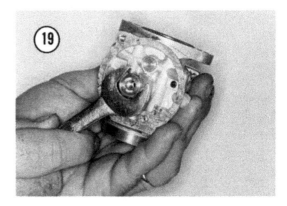

10. Remove the mounting flange O-ring.

11. Remove the pilot air screw, throttle stop screw, and tickler assembly.

12. Reverse the disassembly procedure to reassemble the carburetor. Always use new gaskets upon reassembly.

Amal Monobloc Disassembly/Assembly

Figure 21 is an exploded view of a typical Monobloc carburetor. Refer to this illustration during carburetor disassembly.

1. Remove the top ring nut (**Figure 22**).

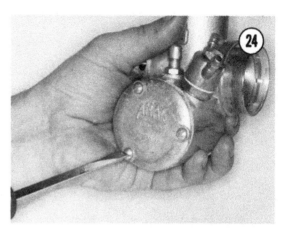

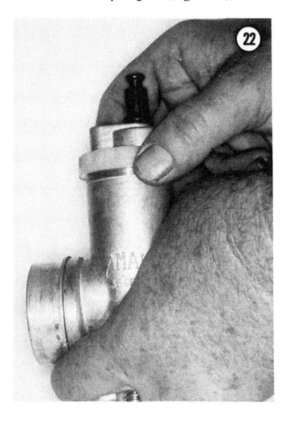

2. Remove the throttle slide assembly. See **Figure 23**.

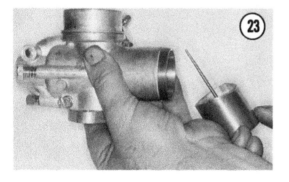

3. Note which needle groove the clip is in, then remove the needle clip and pull the needle from the throttle valve.

4. Remove the float chamber cover (**Figure 24**).

5. Remove the spacer from the float pivot shaft (**Figure 25**). Note the manner in which the float is installed, then pull it from the pivot shaft (**Figure 26**).

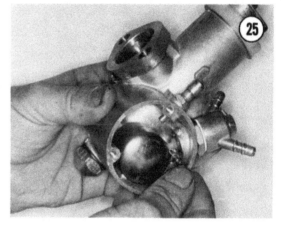

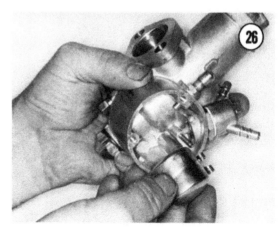

6. Remove the float needle (**Figure 27**), then the float needle seat.

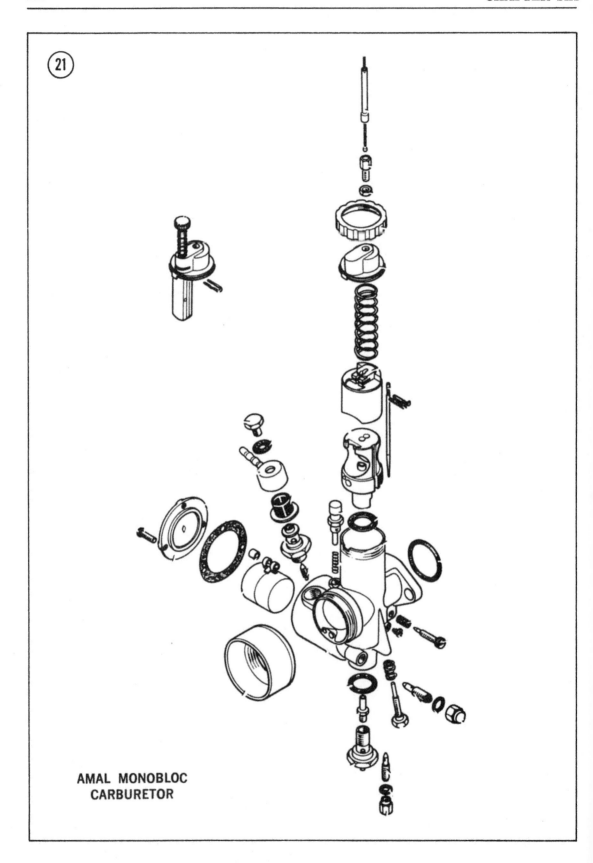

AMAL MONOBLOC
CARBURETOR

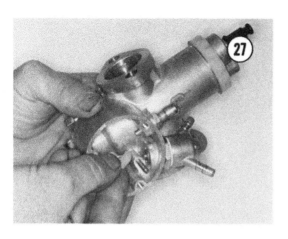

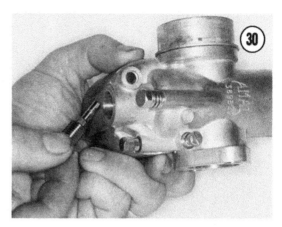

7. Remove the main jet cover (**Figure 28**), then remove the main jet together with its gasket (**Figure 29**).

8. Remove the jet holder (**Figure 30**).

9. Separate the needle jet from the jet holder (**Figure 31**).

10. Remove the pilot jet cover (**Figure 32**), then the pilot jet (**Figure 33**).

6

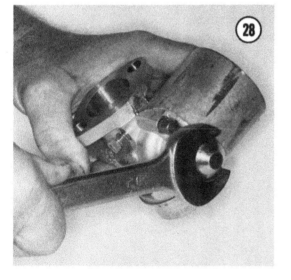

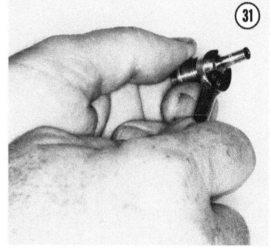

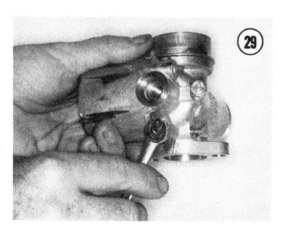

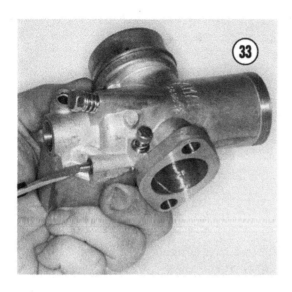

11. Remove the banjo bolt, then the fuel inlet fitting (**Figure 34**).

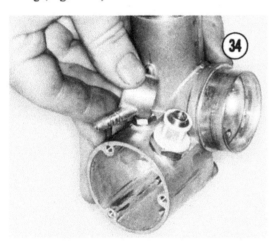

12. Remove the fuel strainer (**Figure 35**), then the float needle seat.

13. Remove the pilot air screw, tickler button, and throttle stop screw.

14. Reverse the disassembly procedure to reassemble the carburetor. Always use new gaskets upon reassembly.

CARBURETOR ADJUSTMENT

Carburetor adjustment is not normally required except for occasional adjustment of idling speed, or at time of carburetor overhaul. See the next section for synchronization of twin carburetors.

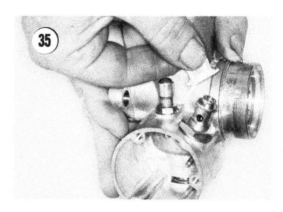

Float Level

The machine was delivered with the float level adjusted correctly. Floats on Amal carburetors do not have provision for adjustment.

Float Inspection

Shake the float to check for gasoline inside (**Figure 36**). If fuel leaks into the float, the float chamber fuel level will rise, resulting in an over-rich mixture. Replace the float if it is deformed or leaking.

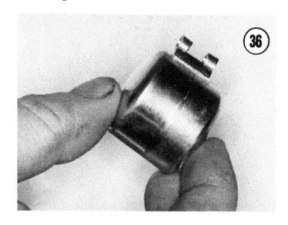

Replace the float valve if its seating end is scratched or worn. Press the float valve gently with your finger and make sure that the valve seats properly. If the float valve does not seat properly, fuel will overflow, causing a rich mixture and flooding the float chamber whenever the fuel petcock is open.

Component Parts

Clean all parts in carburetor cleaning solvent. Dry the parts with compressed air. Clean the jets and other delicate parts with compressed air after the float bowl has been removed. Never attempt to clean jets or passages by running a wire through them. To do so will cause damage and destroy their calibration. Do not use compressed air to clean an assembled carburetor, since the float and float valve can be damaged.

Speed Range Adjustments

The carburetor was designed to provide the proper mixture under all operating conditions. Little or no benefit will result from experimenting. However, unusual operating conditions such as sustained operation at high altitudes or unusually high or low temperatures may make modifications to the standard specifications desirable. The adjustments described in the following paragraphs should only be undertaken if the rider has definite reason to believe they are required. Make the tests and adjustments in the order specified.

Figure 37 illustrates typical carburetor components which may be changed to meet individual operating conditions. Shown left to right are the main jet, needle jet, needle and clip, and throttle valve.

Make a road test at full throttle for final determination of main jet size. To make such a test, operate the motorcycle at full throttle for at least 2 minutes, then shut the engine off, release the clutch, and stop the bike.

If at full throttle, the engine runs "heavily," the main jet is too large. If the engine runs better by closing the throttle slightly, the main jet is too small. The engine will run evenly at full throttle if the main jet is of the correct size.

After each such test, remove and examine the spark plug. The insulator should have a light tan color. If the insulator has black sooty deposits, the mixture is too rich. If there are signs of intense heat, such as a blistered white appearance, the mixture is too lean.

As a general rule, main jet size should be reduced approximately 5% for each 3,000 feet (914 meters) above sea level.

Table 2 lists symptoms caused by rich and lean mixtures.

Table 2   CARBURETOR MIXTURE

Condition	Symptom
Rich Mixture	Rough idle Black exhaust smoke Hard starting, especially when hot "Blubbering" under acceleration Black deposits in exhaust pipe Gas-fouled spark plug Poor gas mileage Engine performs worse as it warms up
Lean mixture	Backfiring Rough idle Overheating Hesitation upon acceleration Engine speed varies at fixed throttle Loss of power White color on spark plug insulator Poor acceleration

Adjust the pilot screw as follows.

1. Turn the pilot air screw in until it seats lightly, then back it out about 1½ turns.

2. Start the engine and warm it to normal operating temperature.

3. Turn the idle speed screw until the engine runs slower and begins to falter.

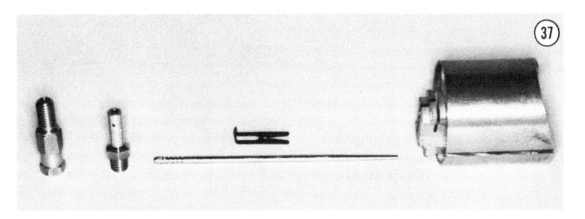

4. Adjust the pilot screw as required to make the engine run smoothly.

5. Repeat Steps 3 and 4 to achieve the lowest stable idle speed.

Determine the proper throttle valve cutaway size. With the engine running at idle, open the throttle. If the engine does not accelerate smoothly from idle, turn the pilot air screw in (clockwise) slightly to enrich the mixture. If the condition still exists, return the air screw to its original position and replace the throttle valve with one which has a smaller cutaway. If engine operation is reduced by turning the air screw, replace the throttle valve with one having a larger cutaway.

6. For operation at ¼-¾ throttle opening, adjustment is made with the jet needle. Operate the engine at ½ throttle in a manner similar to that for full throttle tests described earlier. To enrich the mixture, place the jet needle clip in a lower groove. Conversely, placing the clip in a higher groove leans the mixture.

7. A summary of carburetor adjustments is given in **Table 3**.

Table 3   CARBURETOR ADJUSTMENTS

Throttle Opening	Adjustment	If too Rich	If too Lean
0 - ⅛	Air screw	Turn out	Turn in
⅛ - ¼	Throttle valve cutaway	Use larger cutaway	Use smaller cutaway
¼ - ¾	Jet needle	Raise clip	Lower clip
¾ - full	Main jet	Use smaller number	Use larger number

Carburetor Synchronization

1. First adjust throttle cable play to a minimum at each carburetor so that both throttle slides begin to move at the same time when the throttle is opened.

2. Start the engine and remove the left spark plug lead so the motor is running on the right cylinder only. Use gloves to avoid a shock.

3. Adjust the right carburetor, using the pilot air screw and the throttle stop screws so the motor is running at as smooth an idle as possible.

4. Replace the left spark plug lead and perform the same adjustments on the left carburetor.

5. Replace the right plug lead and unscrew both throttle stop screws the same amount until the engine idles just fast enough to keep it running with the throttle closed.

AIR FILTER

Many Triumph motorcycles come equipped with paper filters. These cannot be cleaned, but excess dirt should be shaken out at frequent intervals.

Accessory houses have wet foam filters, made to fit any bike, which are more efficient and can be cleaned in kerosene. Allow the filter to air dry, dip in lightweight oil, squeeze out excess and insert in the stock housing.

Dirt riding will make it necessary to check the filter after every ride.

CAUTION
Even minute particles of dust can cause severe wear, so never run without a filter.

If a felt element is used, clean in kerosene and dry thoroughly. Replace any excessively dirty or damaged felt element.

GAS TANK

Removal/Installation

1. Shut off petcocks and disconnect fuel lines.

2. Early model gas tanks are removed by unscrewing the front mounting bolts or nuts. See **Figure 38**. Some early models used lock wire to secure the bolts. Models between Serial Nos. DU.66245 and DU.77670 use reflectors mounted on the front tank bolts. The reflectors may be removed so an open-end wrench will reach the bolts. Pry off the chrome rim, the reflector lens, and then the retainer. Late models use studs and self-locking nuts.

Tanks on the latest models have centerpole mounts which can be unscrewed after the rubber plug has been removed from the top of the tank. The 750cc models have an additional crosspiece beneath the tank.

3. Installation is the reverse of removal.

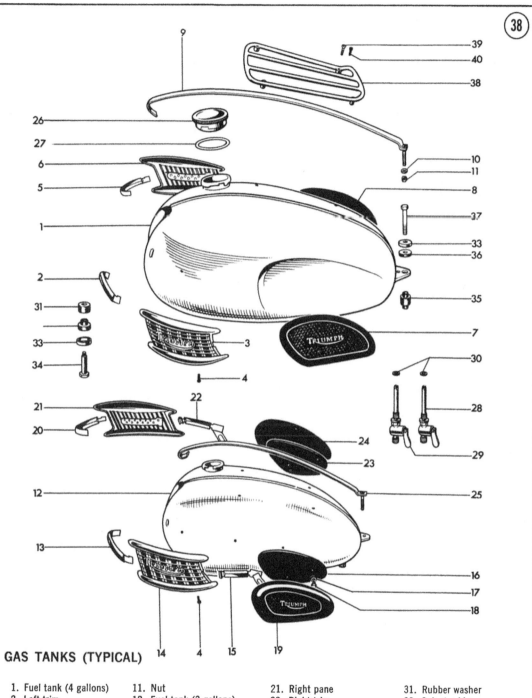

GAS TANKS (TYPICAL)

1. Fuel tank (4 gallons)	11. Nut	21. Right pane	31. Rubber washer
2. Left trim	12. Fuel tank (3 gallons)	22. Right trim	32. Spigot rubber
3. Left panel	13. Left trim	23. Right knee grip	33. Cup
4. Screw	14. Left panel	24. Right knee grip	34. Bolt
5. Right trim	15. Left trim	25. Center trim	35. Double spigot rubber
6. Right panel	16. Left knee grip	26. Filler cap	36. Rubber washer
7. Left knee grip	17. Washer	27. Gasket	37. Bolt
8. Right knee grip	18. Bolt	28. Fuel cap	38. Parcel tray
9. Center trim	19. Left knee grip	29. Fuel tap	39. Screw
10. Serrated washer	20. Right front trim	30. Fiber washer	40. Rubber blanking plug

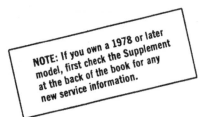
NOTE: If you own a 1978 or later model, first check the Supplement at the back of the book for any new service information.

CHAPTER SEVEN

ELECTRICAL SYSTEM

The electrical system includes the battery, ignition system, charging system, lighting, and horn. Specifications for service and replacement parts are given in Appendix I. Wiring diagrams will be found at the end of the book.

Any part of the electrical system may be repaired with a minimum of special tools by following the procedures described.

BATTERY

Battery specifications for all models are given in Chapter Nine. If two 6-volt batteries are connected in series for a 12-volt system, they must be connected as shown in **Figure 1**.

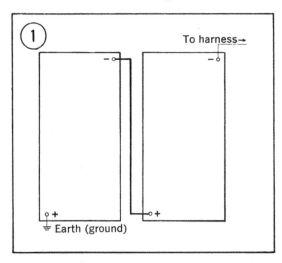

Battery construction is shown in **Figure 2**.

Be sure to check battery electrolyte level, especially during hot weather.

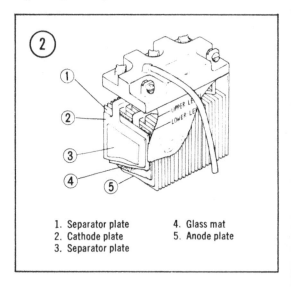

1. Separator plate
2. Cathode plate
3. Separator plate
4. Glass mat
5. Anode plate

Removal

1. Lift the seat and remove the battery retaining strap.
2. Disconnect the ground, or negative (—) cable first, then the positive (+) cable.
3. Lift the battery from the carrier. Note that the battery vent tube is routed under the motorcycle and behind the swing arm lug.

Safety Precautions

When working with batteries, use extreme care to avoid spilling or splashing the electrolyte. Electrolyte contains sulphuric acid which can destroy clothing and cause serious chemical burns. If any electrolyte is spilled or splashed on clothing, body, or other surfaces, neutralize it *immediately* with a solution of baking soda and water, then flush with plenty of clean water.

> WARNING
> *Electrolyte splashed into the eyes is extremely dangerous. Safety glasses should always be worn when working with batteries. If electrolyte is splashed into the eye, call a physician immediately, force the eye open, and flood with cool, clean water for about 5 minutes.*

While batteries are being charged, highly explosive hydrogen gas forms in each cell. Some of this gas escapes through the filler openings and may form an explosive atmosphere around the battery. This explosive atmosphere may exist for several hours. Sparks, open flame, or even a lighted cigarette can ignite this gas, causing an internal explosion and possible serious personal injury. The following precautions should be taken to prevent an explosion.

1. Do not smoke or permit open flame near any battery being charged or which has been recently charged.

2. Do not disconnect live circuits at battery terminals because a spark usually occurs when a live circuit is broken. Care must always be taken while connecting or disconnecting any battery charger; be sure its power switch is off before making or breaking connections. Poor connections are a common cause of electrical arcs which cause explosions.

Battery Inspection and Service

1. Measure the specific gravity of the battery electrolyte with a hydrometer. The specific gravity is calibrated on the hydrometer float stem. The reading is taken at the fluid surface level with the float buoyant in the fluid (see **Figure 3**).

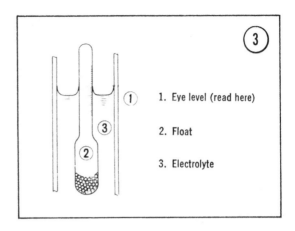

1. Eye level (read here)

2. Float

3. Electrolyte

2. If the reading is less than 1.20 with the temperature corrected to 68°F, recharge the battery. See **Figure 4** for a graph of specific gravity vs. residual capacity.

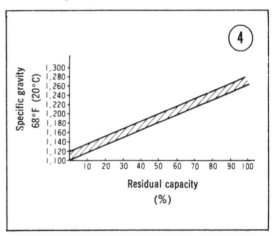

3. If any cell's electrolyte level is below the lower mark on the battery case, fill with distilled water to the upper mark.

4. Replace the battery if the case is cracked or damaged. Corrosion on the battery terminals causes leakage of current. Clean with a wire brush or with a solution of baking soda and water.

5. Check the battery terminal connections. If corrosion is present, the connection is poor. Clean the terminal and connector and coat with Vaseline and reinstall.

6. Vibration causes the corrosion of the battery plates to flake off forming a paste on the bottom (**Figure 5**). Replace the battery when the paste builds up considerably.

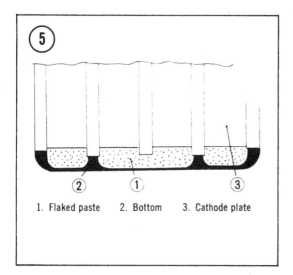

1. Flaked paste   2. Bottom   3. Cathode plate

Table 1   BATTERY CHARGE

Charging current rate	Maximum of 10% of ampere-hour rating.
Checking for full charge	(1) Specific gravity: 1.260-1.280 at 20°C (68°F) maintained constant for one hour. (2) 7.5V-8.3V maintained constant at terminals, checked with voltmeter.
Charging duration	By this method, a battery with specific gravity of electrolyte below 1.220 at 20°C (68°F) will be fully charged in approximately 10-13 hours.

Battery Charging

Batteries are not designed for high charge or discharge rates. For this reason, it is recommended that a battery be charged at a rate not exceeding 10% of its ampere-hour capacity. That is, do not exceed 0.5 ampere charging rate for a 5 ampere-hour battery, or 1.5 amperes for a 15 ampere-hour battery. This charge rate should continue for 10-13 hours if the battery is completely discharged or until specific gravity of each cell is up to 1.260-1.280, corrected for temperature. If after prolonged charging, specific gravity of one or more cells does not come up to at least 1.230, the battery will not perform as well as it should, but it may continue to provide satisfactory service for a time.

Some temperature rise is normal as a battery is being charged. Do not allow the electrolyte temperature to exceed 110°F (43.3°C). Should temperature reach that figure, discontinue charging until the battery cools, then resume charging at a lower rate.

If possible, always slow-charge a battery (see **Table 1**). Quick-charging will shorten the battery service life. Use a quick-charge only if *absolutely* necessary.

After removing the battery, use the following procedure for charging.

1. Hook the battery to a charger by connecting the positive lead to the positive terminal on the battery and the negative lead to the negative terminal. To do otherwise could cause severe damage to the battery and result in injury if the battery explodes.

2. The electrolyte will begin bubbling, signifying that explosive hydrogen gas is being released. Make sure the area is adequately ventilated and that there are no open flames.

3. It will normally take at least 8 hours to bring the battery to a full charge. Test the electrolyte periodically with a hydrometer to see if the specific gravity is within the standard range of 1.26 to 1.28. If the reading remains constant for more than an hour, the battery is charged. See Table 1.

Installation

1. Wash the battery with water to remove spilled electrolyte. Coat the terminals with Vaseline or light grease before installing.

2. When replacing the battery, be careful to route the vent tube so that it is not crimped. Connect the positive terminal first, then the negative one. Don't overtighten the clamps.

3. Remeasure the specific gravity of the electrolyte with a bulb hydrometer, reading it as shown.

RECTIFIER (ALL MODELS)

The rectifier must be clean and dry at all times. When removing it for service, be sure that you hold the nut shown in **Figure 6** to avoid twisting the plates.

CAUTION
Never remove or loosen the nut holding the plates together as this breaks internal connections and destroys the rectifier.

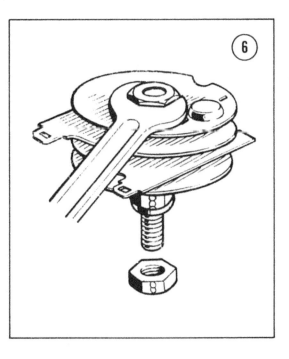

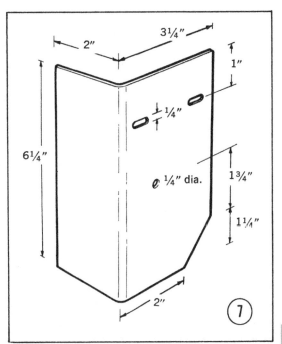

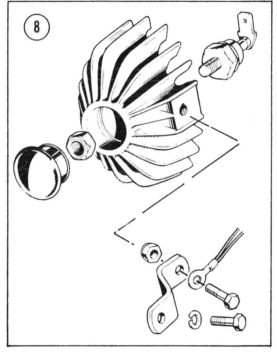

ZENER DIODE
(ALL MODELS)

Various mountings used for the zener diode are shown in **Figures 7, 8, and 9**. Machines fitted with the early aluminum-sheet heat sink can be converted to a new type. A Triumph dealer can supply the necessary parts for the conversion.

1. The diode can be removed without taking off the heat sink. Refer to Figure 8. Disconnect the Lucar connector and remove the black plastic plug from the nose of the heat sink. This exposes the diode mounting nut. To remove the heat sink, unscrew the bolt from the retaining bracket which also mounts the double ground wire. When reconnecting the unit, do not mount the ground wire between the heat sink and the diode.

2. On 1972 and later models, the diode is mounted in the right half of the air filter box, which acts as a heat sink (Figure 9). To remove the diode, remove the outer cover and the right side filter and intake hose. Unplug the terminal and unscrew the diode from the filter housing.

When reinstalling the diode, make sure the contact surfaces of the diode and the housing are clean. Foreign matter will create an air gap which will reduce heat conductivity and cause the diode to overheat. Do not tighten the diode to more than 28 in.-lb.

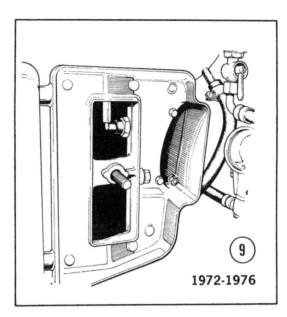

1972-1976

Table 2 IGNITION TIMING

Dowel Location	Ignition Timing Full-Advanced	Dowel Remarks
S	37° BTDC	Standard
R	41° BTDC	Racing
M	39° BTDC	Mid position

the switch in either the headlight bracket or the switch panel. Disconnect the 3 connectors and the switch. The lock assembly can be removed from the switch, after the switch has been removed from the machine, by inserting a paper clip in the small hole in the side of the switch body and pulling the lock assembly out.

3. Earlier models have the diode mounted in the left switch panel. The batteries must be removed before the panel can be taken off. Note how the ground strap is connected to the diode. It must be replaced so it is between the heat sink and the diode securing nut; not between the diode body and the heat sink.

4. Prior to engine No. DU.24875, the diode was mounted on the front gas tank mounting bolts. Removal and replacement are the same as detailed above.

ENERGY TRANSFER IGNITION AND LIGHTING SYSTEMS (ALL MODELS)

1. The rotor is located on the crankshaft and can be installed in 3 positions. Refer to **Table 2** to determine which position suits your needs. It is very important with this type ignition system that the coil, condensers, and high-tension leads are kept clean and dry at all times. The spark plugs and ignition points must be kept clean and properly gapped.

IGNITION SWITCH

Before removing the ignition switch, disconnect the battery ground lead to avoid short circuits. Remove the large retaining nut holding

KILL SWITCH

The handlebar-mounted kill switch may be one of 2 types depending upon the ignition system used on your motorcycle. The coil ignition and AC magneto types look the same externally but are connected differently internally. The correct switch *must* be used as a replacement.

SPARK PLUGS

See *Spark Plug* section of Chapter Two for removal, cleaning, and installation of plugs.

CAPACITOR IGNITION SYSTEM MODIFICATION

The Model 2MC capacitor ignition system is used mainly for competition models. The capacitor kit is available under part No. C.P.210.

1. The capacitor should be fitted into its spring as shown in **Figure 10**. Place the capacitor through the large open end of the spring and then mount it with the small coil of the spring engaged in the groove at the terminal end of the capacitor.

2. The capacitor negative terminal and the zener diode must be connected to the rectifier center terminal (brown/white). The positive terminal is connected to the center bolt grounding terminal. Attach the mounting spring to any point under the seat.

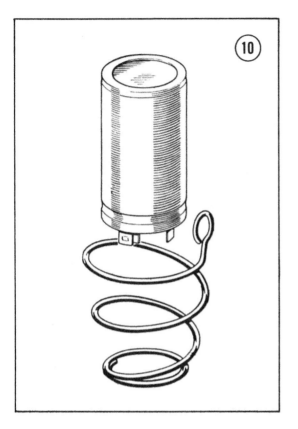

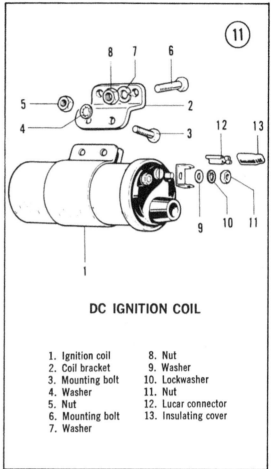

DC IGNITION COIL

1. Ignition coil
2. Coil bracket
3. Mounting bolt
4. Washer
5. Nut
6. Mounting bolt
7. Washer
8. Nut
9. Washer
10. Lockwasher
11. Nut
12. Lucar connector
13. Insulating cover

3. The zener diode heat sink must be of the late model type, larger than the one previously fitted. The diode and heat sink must be mounted so they receive a stream of cool, clean air.

4. Reconnect the alternator so the external green/black and green/yellow leads are joined. Use a double snap connector. This gives full output in all lighting switch positions.

5. Before the machine is run, if the battery is left in place, make sure that the negative battery lead will not short to the frame and possibly ruin the capacitor. Take out the fuse from its holder and replace it with a piece of ¼ in. dowel.

6. Occasionally check that the capacitor is working properly by disconnecting the battery (if used); see that machine continues to run.

IGNITION COIL

Removal/Installation

Refer to **Figure 11** for an illustration of a typical coil.

1. Remove the gas tank. See Chapter Six, *Gas Tank Removal/Installation.*

2. Disconnect battery ground lead.

3. Disconnect leads to coil.

4. Detach coil from mounting bracket and remove.

5. Install in reverse order of removal.

BREAKER POINTS

See *Breaker Point Service* section of Chapter Two for removal and installation of breaker points.

ROTOR AND STATOR

See *Rotor and Stator* section of Chapter Five for removal and installation.

Headlight Adjustment

Proper headlight adjustment is essential to safe night riding. If the lights are set too low, the road will not be visible. If set too high, they will blind oncoming vehicles. Adjustment is very simple; proceed as follows.

1. Place the machine approximately 16 ft. from a white or light-colored wall. Refer to **Figure 12**.
2. Be sure bike is on level ground and pointing directly ahead.

3. Measurements should be made with one rider sitting on the bike and both wheels on the ground.
4. Draw a cross on the wall equal in height to the center of the headlight.
5. Put on the high beam. The cross should be centered in the concentrated beam of light.
6. If the light does not correspond to the mark, loosen the bolts and adjust. Tighten the bolts and recheck positioning.

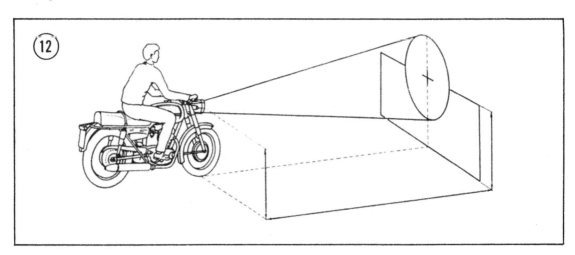

NOTE: If you own a 1978 or later model, first check the Supplement at the back of the book for any new service information.

CHAPTER EIGHT

FRONT BRAKE, WHEEL, AND SUSPENSION

FRONT DRUM BRAKES

Removal/Installation (DU.66246 Up)

1. Support the motorcycle so the front wheel is 6 inches from the ground. Disconnect the front brake cable by loosening the cable adjuster and removing the cable at the brake arm. Remove the front axle cap bolts on each fork leg and remove the wheel.

2. Unscrew the center nut which holds the brake plate in position (**Figures 1, 2, and 3**).

3. Lift up one brake shoe until it is off the pivot and cam; disconnect the return springs. The other shoe will then come off. Remove the split pin from the pivot pin at each end of the adjustment rod and move the pivot pins out of the way.

4. Remove the brake cam nuts and washers and take off the return spring from the front cam. Pull off each lever and the brake cams will come out of the anchor plate. See **Figure 4**.

5. Check the anchor plate for cracks and replace if necessary.

6. Check the brake return springs for fatigue or damage.

7. Clean and lightly grease the brake cam and shoe contacts. Do not overgrease or the linings will become contaminated and will then have to be replaced.

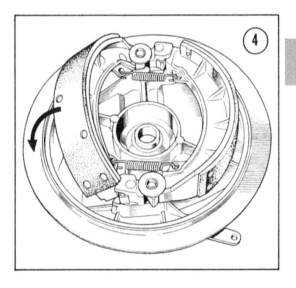

8. Check the brake drums for scoring or an out-of-round condition. Repair or replace them if necessary.

9. If the brake linings have worn to a point that they are level with the rivets, they must be replaced.

10. Lightly grease the spindles and place the cams with the wedges toward the outside.

11. Replace the outside return spring to the front cam and replace the levers at the same angle, securing them with the washers and nuts.

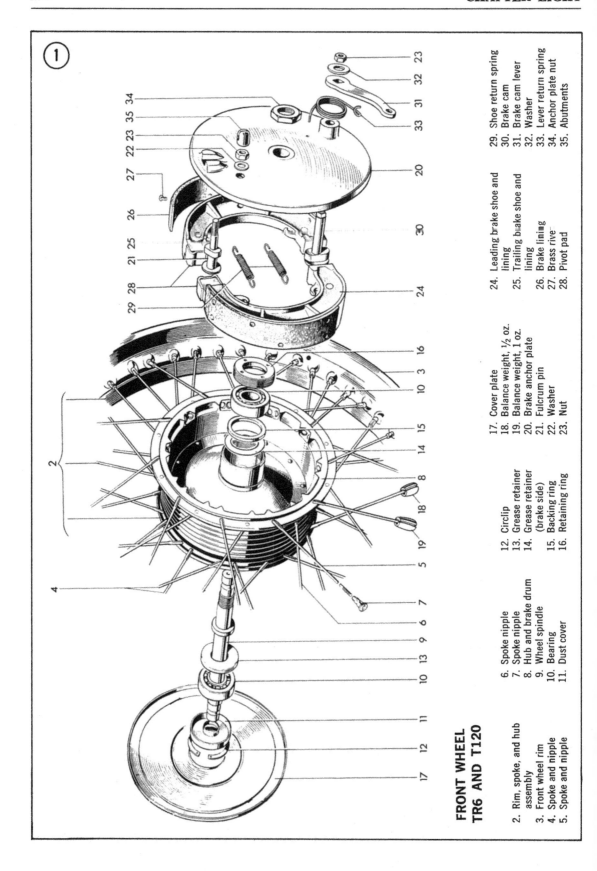

**FRONT WHEEL
TR6 AND T120**

2. Rim, spoke, and hub assembly
3. Front wheel rim
4. Spoke and nipple
5. Spoke and nipple
6. Spoke nipple
7. Spoke nipple
8. Hub and brake drum
9. Wheel spindle
10. Bearing
11. Dust cover
12. Circlip
13. Grease retainer
14. Grease retainer (brake side)
15. Backing ring
16. Retaining ring
17. Cover plate
18. Balance weight, 1/2 oz.
19. Balance weight, 1 oz.
20. Brake anchor plate
21. Fulcrum pin
22. Washer
23. Nut
24. Leading brake shoe and lining
25. Trailing brake shoe and lining
26. Brake lining
27. Brass rivet
28. Pivot pad
29. Shoe return spring
30. Brake cam
31. Brake cam lever
32. Washer
33. Lever return spring
34. Anchor plate nut
35. Abutments

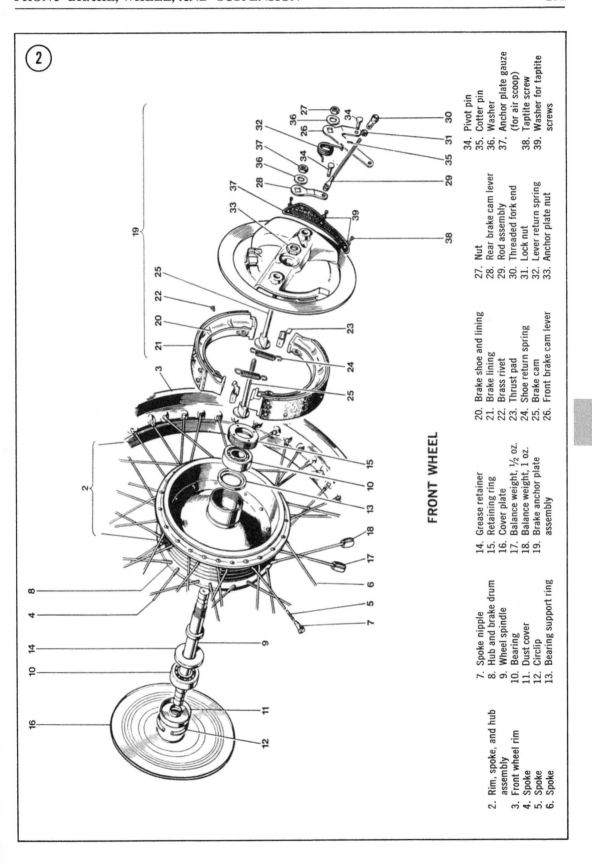

FRONT WHEEL

2. Rim, spoke, and hub
   assembly
3. Front wheel rim
4. Spoke
5. Spoke
6. Spoke

7. Spoke nipple
8. Hub and brake drum
9. Wheel spindle
10. Bearing
11. Dust cover
12. Circlip
13. Bearing support ring

14. Grease retainer
15. Retaining ring
16. Cover plate
17. Balance weight, ½ oz.
18. Balance weight, 1 oz.
19. Brake anchor plate
    assembly

20. Brake shoe and lining
21. Brake lining
22. Brass rivet
23. Thrust pad
24. Shoe return spring
25. Brake cam
26. Front brake cam lever

27. Nut
28. Rear brake cam lever
29. Rod assembly
30. Threaded fork end
31. Lock nut
32. Lever return spring
33. Anchor plate nut

34. Pivot pin
35. Cotter pin
36. Washer
37. Anchor plate gauze
    (for air scoop)
38. Taptite screw
39. Washer for taptite
    screws

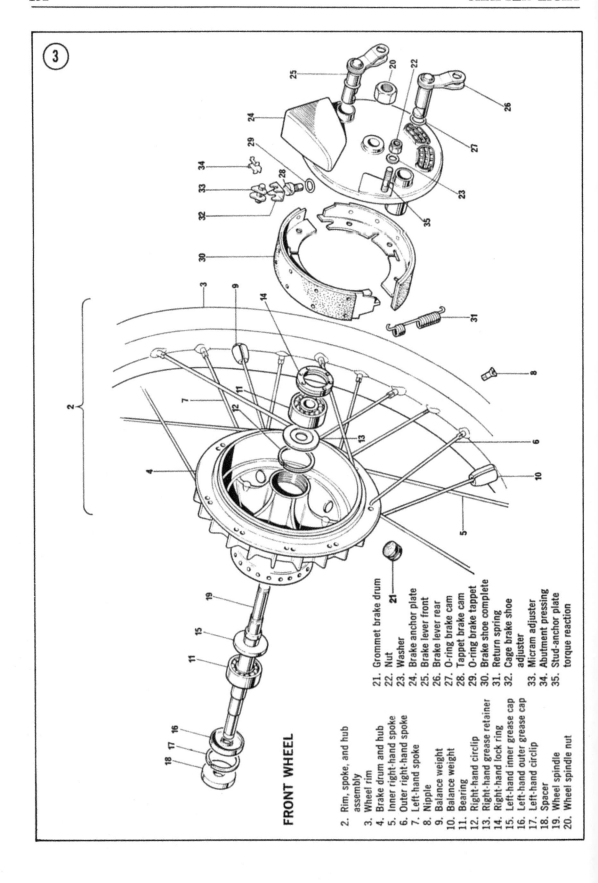

FRONT WHEEL

2. Rim, spoke, and hub assembly
3. Wheel rim
4. Brake drum and hub
5. Inner right-hand spoke
6. Outer right-hand spoke
7. Left-hand spoke
8. Nipple
9. Balance weight
10. Balance weight
11. Bearing
12. Right-hand circlip
13. Right-hand grease retainer
14. Right-hand lock ring
15. Left-hand inner grease cap
16. Left-hand outer grease cap
17. Left-hand circlip
18. Spacer
19. Wheel spindle
20. Wheel spindle nut
21. Grommet brake drum
22. Nut
23. Washer
24. Brake anchor plate
25. Brake lever front
26. Brake lever rear
27. O-ring brake cam
28. Tappet brake cam
29. O-ring brake tappet
30. Brake shoe complete
31. Return spring
32. Cage brake shoe adjuster
33. Micram adjuster
34. Abutment pressing
35. Stud-anchor plate torque reaction

12. Replace the abutment plates to the anchor plate with the tag side next to the anchor plate.

13. Hook the brake shoes together with their return springs, place one shoe in position and pull the other shoe into place.

14. Reposition the anchor plate and secure it with the nut.

Removal/Installation
(Prior to DU.66246)

1. Follow Steps 1 and 2 above. Turn the brake arm to release pressure on the brake drum and pull the brake plate assembly from the spindle.

2. Release the brake arm and take off the return spring.

3. Lift the shoes off by pulling them away from the anchor plate at their outside edges.

4. Remove the brake arm and pull out the cam spindle.

5. Inspect and service parts as in Steps 5-9 in prior section.

6. Hook the 2 return springs in the brake shoes (with the hooked spring ends up) and place the shoes over the pivot and cam.

7. Hold the shoes in a V-position until they are in place over the pivot and cam and then snap them down into their correct position. Leading and trailing shoes are not interchangeable and should be installed as shown in **Figure 5**.

BRAKE ADJUSTMENT
Front Brake
(Prior to Serial No. DU.66246)

1. Adjust the front brake fulcrum pin so the brake shoes make full contact on the drums.

2. Loosen the hex nut just behind the fork tube on the anchor plate.

3. Lock up the front brake at the handlebar lever and tighten the nut while the brake is fully applied.

Front Brake
(After Serial No. DU.66246)

1. Adjust the front twin leading shoe brakes during reassembly as detailed above.

2. Adjust free play at the handlebar control lever (with the adjusting nut) so that there is between 1/16 in. (1.6mm) and ⅛ in. (3.2mm) slack in the inner cable.

DISC BRAKE
Adjustment

The hydraulic disc brake system requires no adjustment. Wear of the brake pads is compensated for by displacement of hydraulic fluid in the system. See **Figures 6 and 7**.

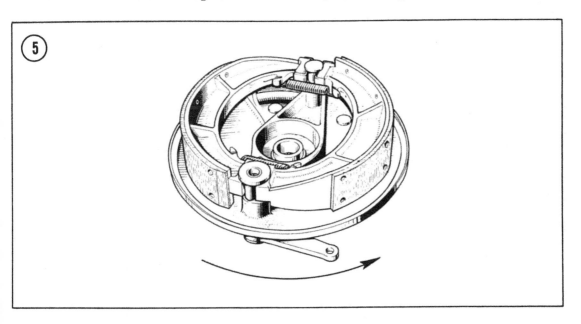

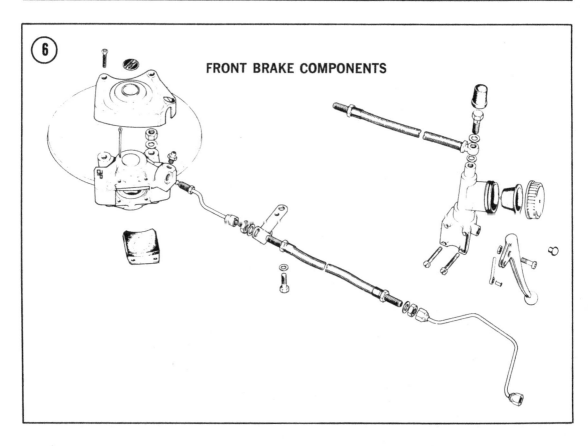

FRONT BRAKE COMPONENTS

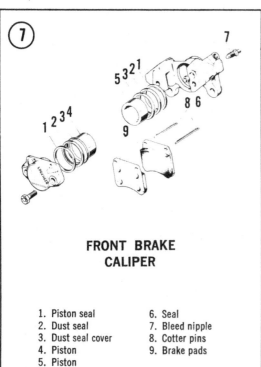

**FRONT BRAKE
CALIPER**

1. Piston seal
2. Dust seal
3. Dust seal cover
4. Piston
5. Piston
6. Seal
7. Bleed nipple
8. Cotter pins
9. Brake pads

Disc Replacement

1. The only service required of the brake disc is replacement if the disc is damaged or excessively scored. Check the runout of the disc with a dial indicator; it should not exceed 0.0035 in. (0.089mm).

2. Remove the front wheel as described in *Front Drum Brakes* section of this chapter.

3. Unscrew the 4 nuts which hold the disc in place. Remove the disc and install a new one.

4. Tighten the nuts diagonally to the torque given in Appendix I.

5. Reinstall the wheel in the fork and check the runout of the disc. If the runout is excessive, it may be corrected by rotating the disc relative to the wheel to find a better tolerance combination.

PAD REPLACEMENT

The brake pad lining should be periodically checked for wear and the pads replaced when lining wears down to about 1/16 in. (1.6mm).

To remove and reinstall the pads, remove the aluminum cover from the caliper and pull out both cotter pins. The pads can now be drawn out. Do not actuate the brake lever with the pads removed. Install new pads and cotter pins and reinstall the cover plate.

BRAKE CALIPER

Removal/Installation

It is recommended that service on the brake caliper be limited to removal and replacement of the unit. Special tools are required to satisfactorily rebuild the caliper and it is assembled with critical torque loadings.

1. To remove the caliper from the fork leg, first remove the cover plate.

2. Attach one end of a piece of tubing to the bleed screw and place other end in a container.

3. Open the bleed screw and pump out the fluid by actuating the brake lever several times.

4. Disconnect the brake line from the caliper.

5. Remove the 2 nuts which hold the caliper on the fork leg and pull the caliper off the mounting studs.

6. Reverse these steps to reinstall the caliper and fill and bleed the system as detailed next.

BLEEDING

The brake system must be bled and the main cylinder filled to the level mark anytime the brake line is disconnected. Any fluid drained from the system as a result of servicing should not be reused.

1. Connect a rubber tube to the bleed nipple. The tube should be long enough to loop upward from the nipple. This provides a fluid head that will prevent air from being drawn back into the system (**Figure 8**).

2. Place the opposite end of the tube in a glass jar containing about ½ in. (12mm) of clean brake fluid.

3. Unscrew the cap from the brake fluid reservoir and remove the rubber diaphragm. Check the level in the reservoir and add fluid if necessary.

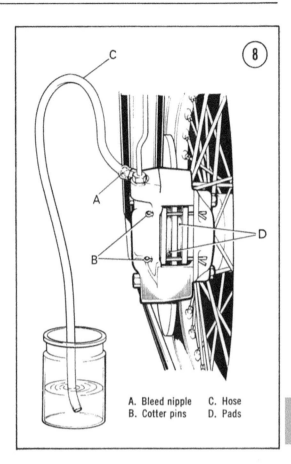

A. Bleed nipple C. Hose
B. Cotter pins D. Pads

4. Turn the bleed valve counterclockwise and pull the brake lever all the way to the handlebar and hold it there for a few seconds. Air being expelled from the system should be visible as bubbles rising in the brake fluid in the jar. Release the lever and pull it in again, repeating the procedure until air bubbles no longer appear in the jar. It will probably be necessary to add fluid to the reservoir during this operation.

5. When air bubbles have ceased to flow from the bleeder tube, hold the brake lever all the way on and tighten the bleed valve. Check and correct the level in the reservoir if necessary. Fold the diaphragm as shown in **Figure 9** and reinstall it in the reservoir. Put the fiber washer in place and firmly screw on the cap.

MASTER CYLINDER

The master cylinder (**Figure 10**) may be satisfactorily rebuilt without the use of special tools, providing that the bore of the cylinder is not scored.

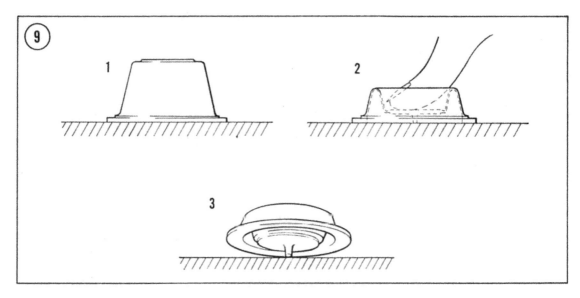

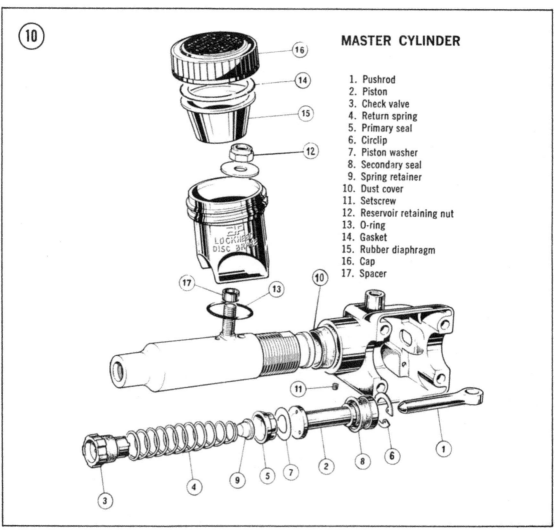

MASTER CYLINDER

1. Pushrod
2. Piston
3. Check valve
4. Return spring
5. Primary seal
6. Circlip
7. Piston washer
8. Secondary seal
9. Spring retainer
10. Dust cover
11. Setscrew
12. Reservoir retaining nut
13. O-ring
14. Gasket
15. Rubber diaphragm
16. Cap
17. Spacer

1. Drain the system and disconnect the flexible hose from the caliper. If the hose is cracked, chafed, or shows signs of deterioration, it should be replaced.

2. Replace the copper sealing washers in the unions and be careful not to overtighten the unions and strip the threads.

3. Unscrew the brake lever pivot bolt and remove the lever and pushrod.

4. Remove the switch console and master cylinder from the handlebar.

5. Unscrew the nut located in the bottom of the reservoir and remove the reservoir from the cylinder.

6. Note the locations of the washer, spacer, and O-ring for reference during reassembly.

7. Unscrew the setscrew that locks the cylinder in position with the switch housing and unscrew the cylinder.

8. Remove the rubber boot from the end of the cylinder. Press the piston into the cylinder with the pushrod to relieve the spring load and remove the circlip.

9. Remove the piston, washer, primary seal, return spring, and check valve from the cylinder. Pull the secondary seal off over the piston flange.

10. Check the bore of the cylinder for scoring and replace if necessary.

11. Clean all of the parts in brake fluid. Do not use gasoline or any other cleaning solvents.

12. Install a new secondary seal on the piston with the lip facing toward the piston face. Make sure the seal is completely seated in the groove.

13. Install the spring retainer on the small end of the spring and check valve on the large end.

14. Install the spring assembly in the cylinder, large end first.

15. Install a new primary seal in the cylinder with the seal lip facing inward.

16. Install the piston washer in the cylinder with its dished side facing inward.

17. Install the piston and push it in with the pushrod until the circlip can be fitted into its groove.

18. Reinstall the rubber boot and attach the reservoir with the O-ring, spacer, and washer correctly positioned.

19. Fill the reservoir with fluid and push the piston in and allow it to return by itself. After doing this several times, fluid should flow out of the outlet. Install cylinder in the switch housing.

20. Remove the reservoir from the cylinder and install the brake lever and pushrod. Hold the brake lever all the way in and screw in the cylinder as far as it will go. Cover the main feed port and blow through the cylinder outlet. No air should escape through the breather port.

21. Unscrew cylinder slowly until air escapes through the breather port. Unscrew the cylinder one complete turn and set cylinder angle at 10 degrees as in **Figure 11**. When the setscrew in the switch housing is tightened, it should contact the flat on the threads of the cylinder.

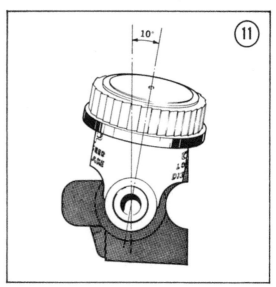

22. Install the switch housing and master cylinder on the handlebar, reconnect the brake line, and fill and bleed the system.

WHEEL BEARINGS

Wheel bearing lubrication and service is covered in Chapter Two.

WHEELS

Inspection

1. Check runout and wobble of the wheel rim.

2. Check the final driven sprocket for excessive wear. Compare with **Figure 12**.

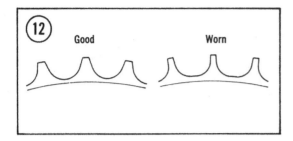

3. Check the final drive chain for wear and stretch. See Chapter Two for routine maintenance and repair.

Spokes

The spokes support the weight of the motorcycle and rider and transmit tractive and braking forces.

Check the spokes periodically for looseness or binding. A bent or otherwise faulty spoke will adversely affect neighboring spokes, and should therefore be replaced immediately. To remove the spoke, completely unscrew the threaded portion, then remove the bend end from the hub.

Spokes tend to loosen as the bike is ridden. Retighten each spoke one turn, beginning with those on one side of the hub, then those on the other side. Tighten the spokes on a new bike after the first 50 miles of operation, then at 50-mile intervals until they no longer loosen.

If the machine is subjected to particularly severe service, as in off-road or competition riding, check the spokes frequently.

Balance

An unbalanced wheel results in unsafe riding conditions. Depending on the degree of unbalance and the speed of the motorcycle, the rider may experience anything from a mild vibration to a violent shimmy which may result in loss of control. Balance weights are applied to the spokes on the light side of the wheel to correct this condition.

Before you attempt to balance the wheel, check to be sure that the wheel bearings are in good condition and properly lubricated and that the brakes do not drag so that the wheel rotates freely.

1. Mount the wheel on a fixture such as the one in **Figure 13** so it can rotate freely.

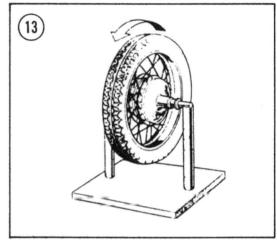

2. Give the wheel a spin and let it coast to a stop. Mark the tire at the lowest point.

3. Spin the wheel several more times. If the wheel keeps coming to a rest at the same point, it is out of balance.

4. Attach a weight to the upper, or light, side of the wheel at the spoke (**Figure 14**). Weights come in 4 sizes: 5, 10, 15, and 20 grams.

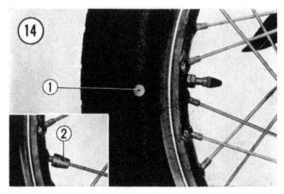

1. Balance marking   2. Balance weight

5. Experiment with different weights until the wheel, when spun, comes to rest at a different position each time.

TIRE

Removal

1. Place the wheel on a soft surface to prevent damage to the hub.

2. If a puncture is suspected, inspect the tread for sharp objects.

3. Remove the valve core and the valve stem retaining nut.

4. Break the bead free of the rim by stepping on it from both sides.

5. Insert 2 small tire irons 4-6 inches apart between the rim and the tire bead at the valve location.

6. Pry in and down with the irons, moving only one iron at a time until the bead is free of the rim all the way around.

7. Pull the inner tube out of the tire casing.

Installation

1. Inflate the new inner tube slightly, leaving the valve core in place.

2. Make sure the inner strip that protects the tube is in good condition and is centered over the spoke nipples.

3. Insert the tube into the tire casing with the valve stem and the tire balance mark aligned with the hole in the rim.

4. Insert the valve stem through the hole in the rim and partially tighten the retaining nut. Then remove the valve core.

5. Coat the bead surfaces and the edge of the rim with mounting solution (liquid detergent is a good substitute).

6. Push the tire into place with your feet. Start on the far side of the rim from the valve and work in opposite directions around the wheel with your heels in order to seat the bead in place.

7. Force the last bit of bead into place with a soft-headed mallet. Do not use tire irons or screwdrivers which could damage the tube.

8. Insert the valve core. Then inflate the tire to 10 psi *over* the recommended pressure to seat the bead against the rim. Deflate completely and reinflate to the standard pressure. Check for leaks.

FRONT FORKS

Three types of front forks have been used on unit construction Triumphs. Determine the frame number and follow the appropriate directions below for fork disassembly.

Removal, 750/650cc
(All Except Disc Brake Models)

1. Refer to **Figures 15-21** and drain the fork oil. Remove the drain plug at the bottom of each fork tube, lock the front brake, and pump the front end up and down several times to expel all the oil.

2. Support the motorcycle on a stand so that the front wheel is approximately 6 inches off the ground.

3. Remove the front wheel and front fender as outlined under the heading, *Brakes.*

4. Remove the headlight unit as detailed in Chapter Seven. Disconnect the throttle and the choke cables.

5. If an ignition lock is fitted, it must also be removed.

6. Remove the handlebars by unscrewing the locknuts on the eyebolts under the top triple clamp. Refer to **Figures 22 and 23**.

7. If there is a steering damper fitted, remove the damper plate pivot bolt.

8. Loosen the triple clamp pinch bolt.

9. Unscrew the sleeve nut on the steering stem.

10. Remove the fork cap nuts and guide tube assemblies if fitted.

11. Hold the forks and tap the top triple clamp until it comes free from the top fork tubes. Carefully lower the fork tubes and lower triple clamp so that the top ball race is not disturbed and the lower race ball bearings are caught when the assembly is lowered far enough.

12. The fork tubes can be removed individually, but special tools must be used for the removal and then the replacement.

13. Remove the fork cap nuts and loosen the lower triple clamp pinch bolt.

14. Remove the 2 oil filler plugs from the fork tubes (if fitted). Drive out the fork tubes. See **Figure 24**.

15. Refer to **Figure 25** and note that 3 different types of eyebolt securing methods are used. Methods marked USA Condition and Standard Condition require that the radiused washers be fitted only as shown. The rigid condition method washers cannot be fitted incorrectly.

8

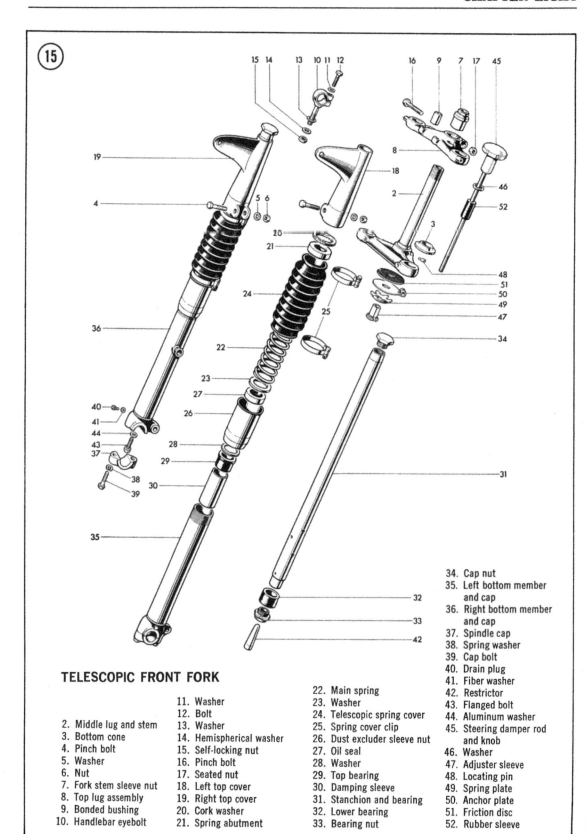

TELESCOPIC FRONT FORK

11. Washer
12. Bolt
13. Washer
14. Hemispherical washer
15. Self-locking nut
16. Pinch bolt
17. Seated nut
18. Left top cover
19. Right top cover
20. Cork washer
21. Spring abutment

2. Middle lug and stem
3. Bottom cone
4. Pinch bolt
5. Washer
6. Nut
7. Fork stem sleeve nut
8. Top lug assembly
9. Bonded bushing
10. Handlebar eyebolt

22. Main spring
23. Washer
24. Telescopic spring cover
25. Spring cover clip
26. Dust excluder sleeve nut
27. Oil seal
28. Washer
29. Top bearing
30. Damping sleeve
31. Stanchion and bearing
32. Lower bearing
33. Bearing nut

34. Cap nut
35. Left bottom member
    and cap
36. Right bottom member
    and cap
37. Spindle cap
38. Spring washer
39. Cap bolt
40. Drain plug
41. Fiber washer
42. Restrictor
43. Flanged bolt
44. Aluminum washer
45. Steering damper rod
    and knob
46. Washer
47. Adjuster sleeve
48. Locating pin
49. Spring plate
50. Anchor plate
51. Friction disc
52. Rubber sleeve

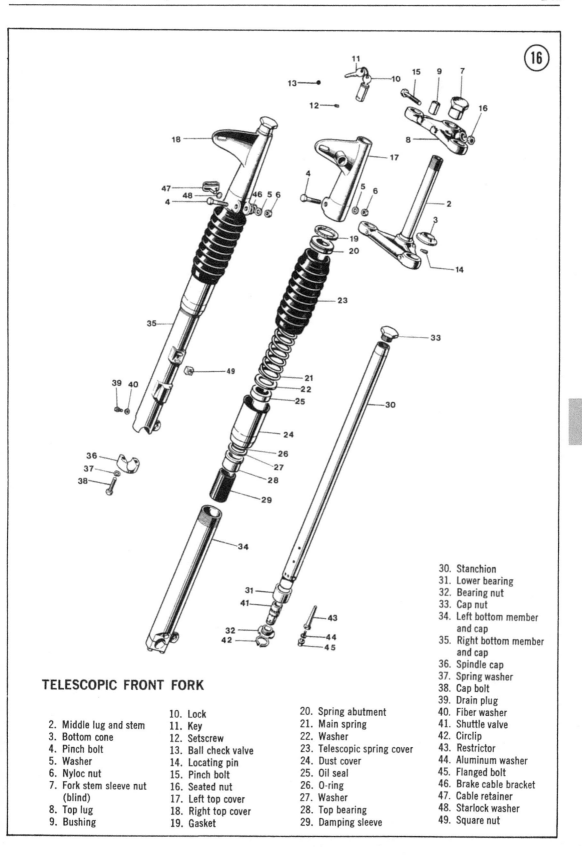

TELESCOPIC FRONT FORK

2. Middle lug and stem
3. Bottom cone
4. Pinch bolt
5. Washer
6. Nyloc nut
7. Fork stem sleeve nut (blind)
8. Top lug
9. Bushing
10. Lock
11. Key
12. Setscrew
13. Ball check valve
14. Locating pin
15. Pinch bolt
16. Seated nut
17. Left top cover
18. Right top cover
19. Gasket
20. Spring abutment
21. Main spring
22. Washer
23. Telescopic spring cover
24. Dust cover
25. Oil seal
26. O-ring
27. Washer
28. Top bearing
29. Damping sleeve
30. Stanchion
31. Lower bearing
32. Bearing nut
33. Cap nut
34. Left bottom member and cap
35. Right bottom member and cap
36. Spindle cap
37. Spring washer
38. Cap bolt
39. Drain plug
40. Fiber washer
41. Shuttle valve
42. Circlip
43. Restrictor
44. Aluminum washer
45. Flanged bolt
46. Brake cable bracket
47. Cable retainer
48. Starlock washer
49. Square nut

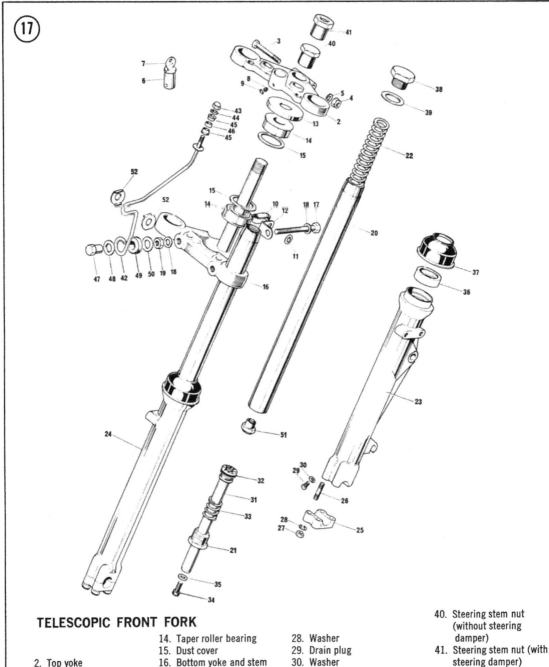

TELESCOPIC FRONT FORK

2. Top yoke
3. Pinch bolt
4. Nut
5. Washer
6. Lock
7. Key
8. Grub screw
9. Sealing washer
10. Brake cable retainer
11. Starlock washer
12. Retaining bracket
13. Abutment ring

14. Taper roller bearing
15. Dust cover
16. Bottom yoke and stem
17. Pinch bolt
18. Washer
19. Nut
20. Stanchion
21. End plug
22. Main spring
23. Outer member, R.H.
24. Outer member, L.H.
25. Cap wheel spindle
26. Stud
27. Nut

28. Washer
29. Drain plug
30. Washer
31. Damper tube and valve assembly
32. Damper ring O-ring
33. Recoil spring
34. Cap screw-damper tube
35. Cap screw seal
36. Oil seal outer member
37. Scraper sleeve outer member
38. Top cap nut
39. Washer

40. Steering stem nut (without steering damper)
41. Steering stem nut (with steering damper)
42. Headlamp brackets
43. Nut
44. Washer
45. Bush
46. Spacer
47. Sleeve nut
48. Washer
49. Grommet
50. Washer
51. Damper valve nut
52. Clamp washer

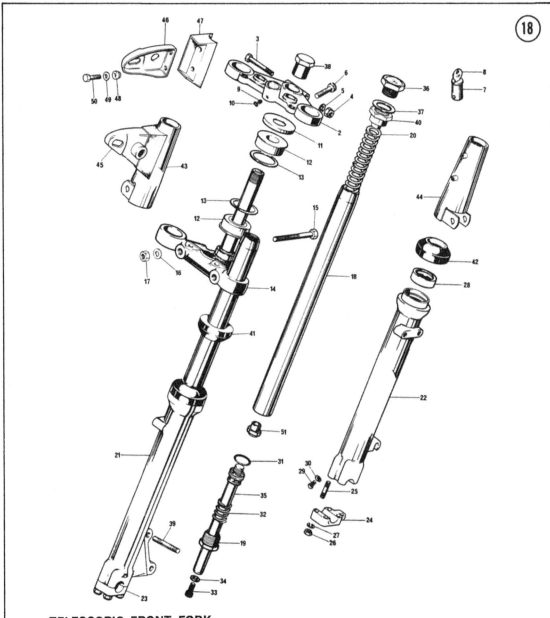

TELESCOPIC FRONT FORK

2. Top yoke
3. Pinch bolt
4. Nut
5. Washer
6. Cap screw
7. Lock
8. Key
9. Setscrew
10. Sealing washer
11. Abutment ring
12. Tapered roller bearing
13. Dust cover
14. Bottom yoke and stem
15. Pinch bolt
16. Washer
17. Nut
18. Stanchion
19. End plug
20. Main spring
21. Outer member
22. Outer member
23. Wheel spindle cap
24. Wheel spindle cap
25. Stud
26. Nut
27. Washer
28. Oil seal
29. Drain plug
30. Washer
31. O-ring
32. Recoil spring
33. Cap screw
34. Cap screw seal
35. Damper tube and valve assembly
36. Top cap nut
37. Washer
38. Steering stem nut
39. Caliper to outer member stud
40. Top cap nut cap screw
41. Outer cover rubber ring
42. Scraper sleeve
43. Outer cover
44. Outer cover
45. Headlight bracket
46. Headlight bracket
47. Rubber mounting
48. Rubber buffer
49. Backing washer
50. Attachment bolt
51. Damper valve nut

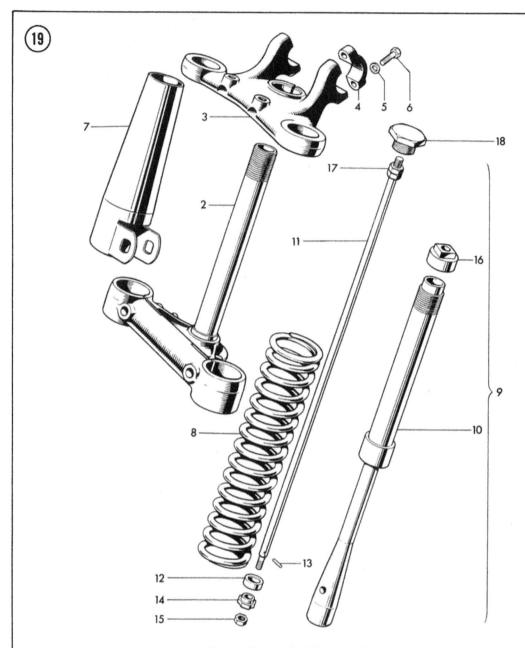

**TELESCOPIC FRONT FORK,
TR6C AND T120TT**

	7. Top cover	13. Pin
2. Middle lug and stem	8. Main spring (yellow/green)	14. Oil restrictor
3. Top lug and caps	9. Damper assembly	15. Nut
4. Cap	10. Restrictor body	16. Restrictor body cap
5. Washer	11. Rod	17. Nut
6. Bolt	12. Cup	18. Cap nut

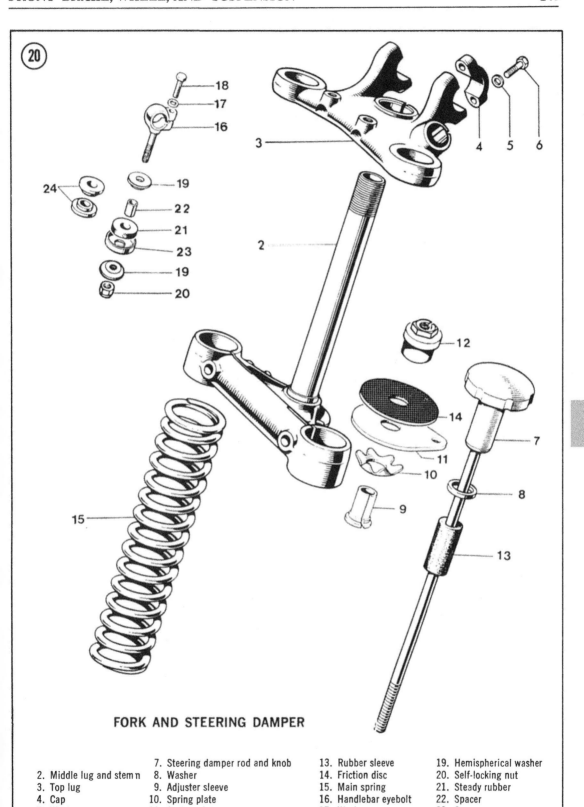

FORK AND STEERING DAMPER

2. Middle lug and stem n	7. Steering damper rod and knob	13. Rubber sleeve	19. Hemispherical washer
3. Top lug	8. Washer	14. Friction disc	20. Self-locking nut
4. Cap	9. Adjuster sleeve	15. Main spring	21. Steady rubber
5. Washer	10. Spring plate	16. Handlebar eyebolt	22. Spacer
6. Bolt	11. Anchor plate	17. Washer	23. Cup
	12. Fork stem sleeve nut	18. Bolt	24. Windshield bushings

8

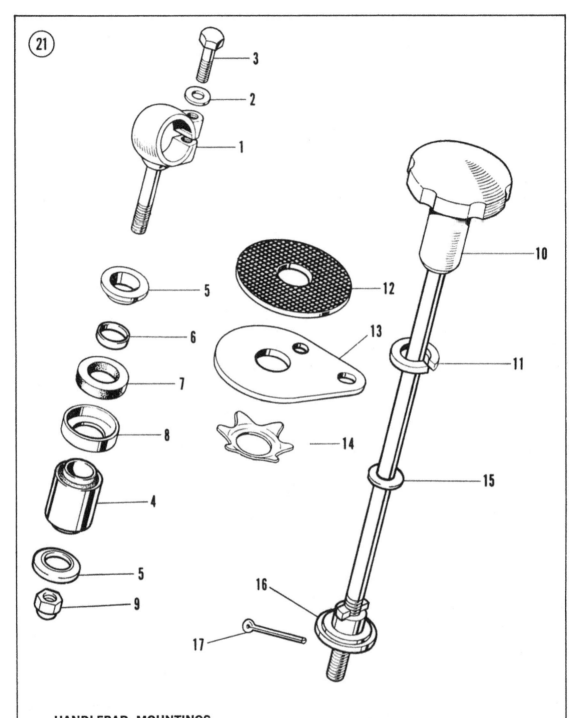

**HANDLEBAR MOUNTINGS
AND STEERING DAMPER**

1. Handlebar eyebolt
2. Washer
3. Bolt
4. Metal and rubber bushing
5. Hemispherical washer
6. Spacer
7. Rubber washer
8. Cup
9. Nut
10. Steering damper rod
11. Spring washer
12. Friction disc
13. Anchor plate
14. Spring plate
15. Washer
16. Bottom nut
17. Cotter pin

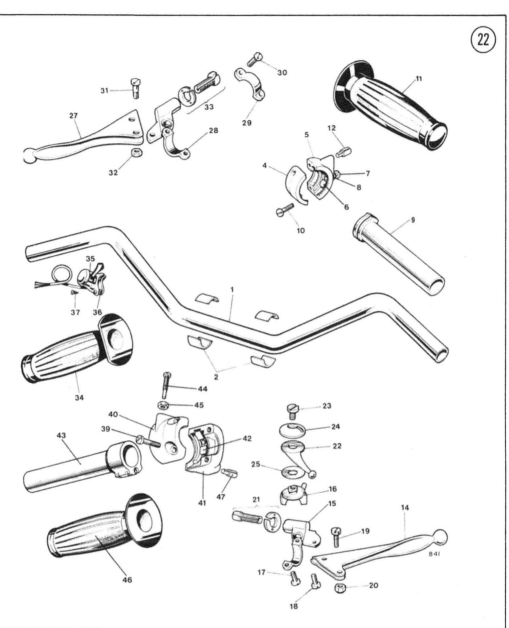

HANDLEBARS AND CONTROL LEVERS

1. Handlebar
2. Gasket
3. Twistgrip
4. Body top half
5. Body bottom half
6. Friction spring
7. Friction spring screw
8. Locknut
9. Rotor
10. Screw
11. Grip
12. Cable stop
13. Front brake and air lever assembly
14. Brake lever (ball end)
15. Lever bracket
16. Control body
17. Screw
18. Screw
19. Fulcrum screw
20. Self-locking nut
21. Cable adjuster and nut
22. Lever
23. Bolt
24. Cap
25. Spring washer
26. Clutch lever assembly
27. Clutch lever (ball end)
28. Clutch lever bracket
29. Clamp
30. Screw
31. Fulcrum screw
32. Self-locking nut
33. Cable adjuster and nut
34. Handlebar grip
35. Horn push and dip switch
36. Backing rubber
37. Self-tapping screw
38. Twin rotor twist grip
39. Screw
40. Body top half
41. Body bottom half
42. Friction spring
43. Rotor
44. Rotor stop
45. Nut
46. Grip
47. Cable stop

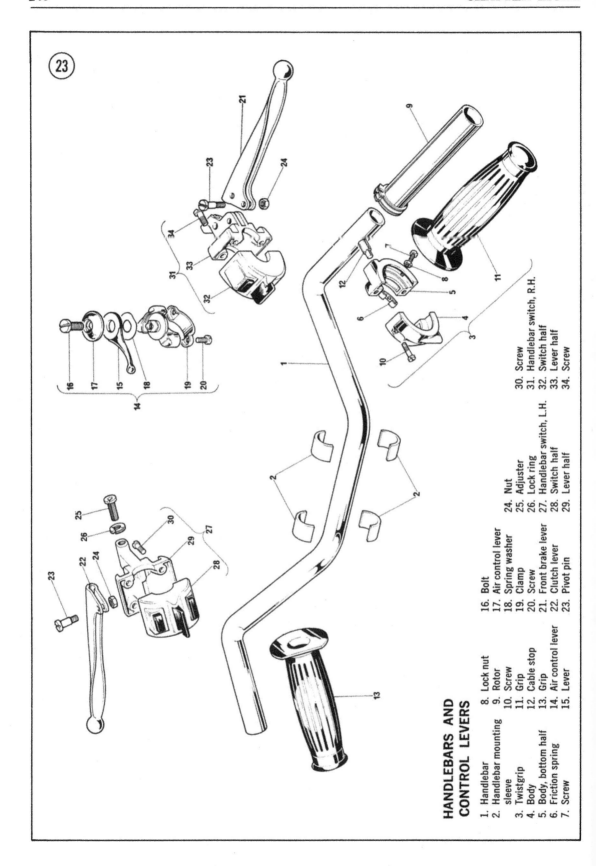

HANDLEBARS AND CONTROL LEVERS

1. Handlebar
2. Handlebar mounting sleeve
3. Twistgrip
4. Body
5. Body, bottom half
6. Friction spring
7. Screw
8. Lock nut
9. Rotor
10. Screw
11. Grip
12. Cable stop
13. Grip
14. Air control lever
15. Lever
16. Bolt
17. Air control lever
18. Spring washer
19. Clamp
20. Screw
21. Front brake lever
22. Clutch lever
23. Pivot pin
24. Nut
25. Adjuster
26. Lock ring
27. Handlebar switch, L.H.
28. Switch half
29. Lever half
30. Screw
31. Handlebar switch, R.H.
32. Switch half
33. Lever half
34. Screw

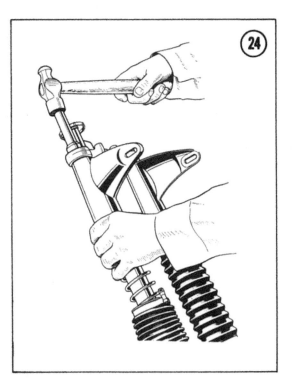

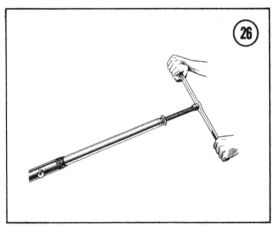

6. Remove the top tube, top bushing, and damping sleeve from the bottom tube. It may be necessary to use force to free the bushing. If the top tube will not come free it will be necessary to use a special service tool as shown in **Figure 26**.

7. Remove the hex head bolt in the axle lug and then pull out the oil restrictor rod assembly. Do not lose the aluminum washer on the bolt.

8. Remove the slotted nut which holds the bottom fork bearing bushing to the fork tube.

Disassembly
(Frame Nos. DU.5825-DU.66245)

1. Remove the front forks from the frame as detailed earlier.

2. Hold the triple clamp stem in a vise and remove the fork tubes as detailed before.

3. Remove the fork boots and nacelle shrouds if fitted.

4. Remove the spring covers, springs, and top and bottom washers.

5. Remove the fork top shrouds. Replace the felt sealing washers when forks are reassembled.

6. Hold the fork tubes in a vise at the wheel axle lug and remove the dust cover nut. It may be necessary to tap the wrench with a mallet to free the nut.

7. Check **Figure 27** and if the hydraulic damping unit shown is used, it will have to be removed before the top and bottom tubes can be separated. Unscrew the hex head bolt located in the axle lug. Separate the top and bottom tubes.

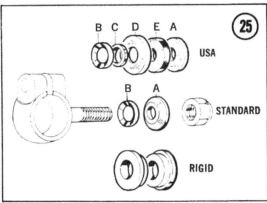

Disassembly
(Frame Nos. DU.101-DU.5824)

1. Loosen the triple clamp pinch bolts. Replace the fork bolts and use a rawhide mallet to tap the tubes out of the triple clamp.

2. Remove the nacelle bottom covers and sealing washers.

3. If boots are fitted, loosen their clips and remove them.

4. Place the assembly in a vise at the axle lug.

5. Remove the dust cover. It may be necessary to tap the wrench with a mallet to loosen the nut.

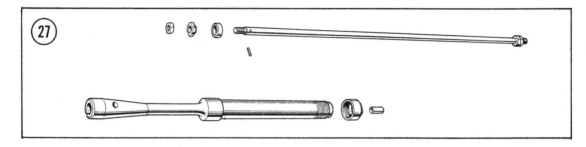

**Disassembly (Frame No. DU.66246 Up,
Except Disc Brake Models)**

1. Remove the front forks from the frame as detailed previously.

2. Hold the triple clamp stem in a vise and unscrew the 2 triple clamp pinch bolts and the boot clips.

3. Drive the fork tubes out of the triple clamp. An old top cap can be used as a tool.

4. Remove the spring abutments, springs, boots, and clips.

5. Remove the fork top shrouds. Replace the felt sealing washers when forks are reassembled.

6. Mount the axle lug in a vise and remove the dust cover nut. It may be necessary to tap the wrench with a mallet to free the nut.

7. Pull the top tube, bushing, and shuttle valve free from the bottom tube. The cone shaped restrictor can be removed from the bottom tube if necessary by removing the securing bolt. Do not lose the aluminum sealing washer. **Figure 28** shows details of the shuttle valves at the bottom end of the fork tubes.

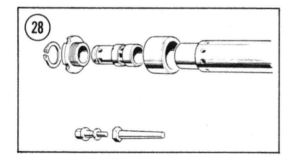

**Inspection, 750/650cc
(All Except Disc Brake Models)**

1. Checking fork component alignment requires special equipment and should be done by a Triumph dealer.

2. Check the fork tubes for bending by rolling them on a surface plate or pane of sheet glass. Slight bends less than 5/32 in. (3.97mm) may be corrected by using a press.

3. Check the top and bottom bushings for wear by measuring the inside diameter of the top bushing and the outside diameter of the bottom bushing. Check the data given in the tables in Appendix I. The top bushing can be placed on the top tube about 8 in. from the bottom and measured for diametrical clearance. The bottom bushing should be placed on the tube and then lowered into the bottom tube to a depth of 8 in. and checked for excessive clearance. Note that bronze bushings have replaced the old type iron ones. If iron ones are found, they should be replaced regardless of apparent wear.

4. Measure the main springs; the old type (long) should be within ½ in. (12.7mm) of the data given in Appendix I. The new type (short) should be within ¼ in. (6.4mm). Both springs should also be very close to same length. Check for signs of cracks and bending. See **Table 1**.

5. Check the steering stem cones and cups for pitting or small indentations. Replace if any damage or wear is apparent. The cups can be driven out using a long narrow drift. Drive in new cups after the frame heading has been thoroughly cleaned. Use a wood block to protect the cups and be sure the cups are driven in straight. Remove the bottom cone by prying it out. Any burrs on the stem can be filed off and a little grease applied to ease installation. Use a 1-1/16 in. (2.7 cm) inside diameter tube approximately 9 in. long. Always replace the steel balls if new cups and cones are used. There are 20 balls ¼ in. (6.4mm) in diameter in both the top and bottom races. The cups are a tight fit in the head lug and the bottom cone a tight fit on the triple clamp stem. If a new cone is

Table 1 FORK SPRING SPECIFICATIONS

Model	Spring Rate lb./in.	Load at Fitted Length lbs.	Color Code
All models after DU.13374			
Solo	26½	22	Yellow/Blue
Sidecar	32½	26½	Yellow/Green
All models DU.5825 to DU.3374			
Solo	30	50	Unpainted
Sidecar	37	60	Yellow/White
Engine No. DU.101 to DU.5824			
6T/T120 Solo	32	85	Black/Green
6T/T120 Sidecar	37	98	Red/White
TR6 Solo	30	46	Black/White
TR6 Sidecar	37	56	Black/Red

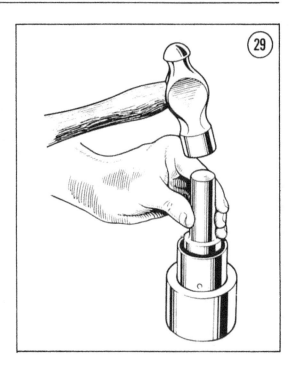

not a tight fit on the stem, the stem and clamp assembly may have to be replaced.

6. On late models, the fork oil seals are pressed into the dust excluder (dust cover) sleeve nuts. They are driven out with a drift against the seals. Press the new seals in with the lip and spring side toward the threaded end of the sleeve. Be sure the seals are seated properly. Oil seal removal on earlier models is accomplished by using a special service tool as shown in **Figure 29**. Place the dust excluder nut as shown and drive the seal out with the drift. Fit the new oil seal with its spring and lip side toward the threaded end of the bore. Press the sleeve in until it is flush with the oil seal's rear face. Replace the O-ring fitted into the threads of the chrome dust excluder (dust cover) on late models each time the forks are dismantled. See Figure 29.

Assembly/Installation (Frame Nos. DU.101-5824)

1. Slide the top tube and bottom bushing assembly into the bottom tube. Replace the damper sleeve and top bushing. Lightly oil the dust excluder sleeve (dust cover) and oil seal assembly and slide them over the top tube. Tighten the sleeve nut. Push the top tube down all the way and match the top tube to the restric-

tor rod. Replace the restrictor rod securing bolt (with its aluminum washer) until the bolt is nearly tight. Work the restrictor rod around until the slot in its base is level with the location plug hole, then replace the plug and tighten the securing bolts. There is a fiber washer under each of the locating plugs.

2. Clean and grease the steering head race cups and cones and place 20 balls in the top and bottom cups. Use heavy grease to keep them from falling out. Move the lower triple clamp and stem up into position and lower the top cone and dust cover into place. Replace the top triple clamp over the stem and tighten the sleeve nut until all slack is taken up.

3. Tighten the sleeve nut pinch bolt finger-tight and align the top and bottom triple clamps. Replace the 2 bottom nacelle covers and insert the pinch bolts, leaving them loose at this time.

4. The right fork tube can be identified by its brake anchor locating boss. Insert it into the lower triple clamp and tap the fork leg up into the top triple clamp by using the bottom tube as an impact driver. Tighten the pinch bolt when the tube is into the top clamp, and the oil filler plug hole is accessible through the headlight opening. Do the same for the other fork leg and then fill legs with ¼ pint (150cc) of fork oil.

8

5. Replace the main springs and screw in the cap nuts and guide tube assemblies until they begin to engage in the threads. Loosen the lower triple clamp pinch bolts and tighten the cap nuts securely. At this point adjust the steering head races as detailed later in this chapter. Fully tighten the bottom triple clamp pinch bolts and the sleeve nut pinch bolts. Replace the oil filler plugs (if used on your machine). Replace the handlebars and continue assembly in the reverse order of disassembly.

Assembly/Installation
(Frame Nos. DU.5825-66245)

1. Assemble the fork tube to the bottom tube and assemble the damper sleeve and top bushing. Screw on the dust excluder (dust cover) sleeve nut and oil seal assembly. Use gasket cement on the outer member threads and tighten the sleeve nut while the bottom tube is held in a vise at the wheel axle lug. Note that from engine No. DU.13375 the damper sleeve is fitted with its thicker end downward. Later forks also have an O-ring installed between the dust excluder (dust cover) and the bottom members. They may be fitted to any external sleeve forks.

2. Slide the oil restrictor tube down inside the upper tube and use a piece of ½ in. (12.7mm) inside diameter tubing 2 feet long to hold the rod in place while a few threads of the securing bolt are threaded onto the rod. New fiber washers should be used under the 2 locating plugs. Screw in the small locating plug and turn the restrictor rod with the tubing until the locating slot is lined up with the plug and tighten the securing bolt. Repeat the process with the other fork leg and then place the plain thrust washers, main springs, covers, and felt washers over the legs. A pair of plain steel washers are fitted under the felt washers on models equipped with a nacelle type headlight.

3. Replace and clamp the fork tube rubbers on the thrust washer and sleeve nut. Line up the top and bottom triple clamps and position the lower nacelle covers. Screw in the bottom triple clamp pinch bolts finger-tight. Note that the right fork tube has a brake anchor locating boss. Insert it as far as possible into the bottom triple clamp. Screw the adapter into the top tube and pull

the fork leg up into position. Tighten the pinch bolt and remove the tool. Screw the cap bolt loosely in place and assemble the other fork leg similarly.

4. Remove both cap nuts and fill each fork leg with ⅓ pint (190cc) of fork oil. Replace the cap nuts but do not fully tighten until the bottom triple clamp pinch bolt has been loosened. If a nacelle is fitted to the fork, turn the fork legs so the oil filler plug holes can be seen through the headlight aperture. Adjust the steering head bearings as detailed in this chapter and tighten the sleeve nut pinch bolts and the lower triple clamp pinch bolts. Reassemble the remaining components in the reverse order of disassembly.

Assembly/Installation
(Frame No. DU.66246 Up, Except
Disc Brake Models)

1. Early models (engine No. DU.66246 through DU.68362) in this series use C.E.I. threads on certain bolts, the upper pipes, cap nuts, and bottom bearing nuts. After DU.68363, Unified threads were used. The 2 systems cannot be interchanged unless all components involved are replaced as a unit.

2. Assemble the bottom bushing on the upper tube, replace the shuttle valve with its larger diameter upward, and secure it with the bearing retainer nut. Replace the circlip which keeps the valve from sliding into the tube. Replace the cone shaped restrictor into the bottom tube and hold it in place with the upper tube while the hex headed securing bolt and its aluminum washer are replaced. Fit the upper fork tube into the lower tube and replace the top bushing. Lightly oil the top tube and install the dust cover sleeve and O-ring. Tighten the dust cover.

3. Assemble the spring, rubber boot and clips, the top spring abutment, and cork washer. Tighten the boot clips over the top abutment and the dust cover. Align the triple clamps, install the top shrouds, and insert the bottom clamp pinch bolts (do not tighten at this time). Note that the right fork leg has the front brake anchor plate boss. Install the fork leg into the bottom clamp as far as possible. Screw the plug adapter into the top tube and pull the fork legs into position.

4. Tighten the pinch bolts and replace the fork caps. Remove both caps after the legs are in place and pour ⅓ pint (190cc) of fork oil in each leg. Replace the caps but do not tighten until the lower triple clamp pinch bolt has been loosened. Adjust the steering head bearing races as detailed later and tighten the sleeve nut pinch bolt and the lower triple clamp pinch bolts. Reassemble the remaining components in reverse order of disassembly.

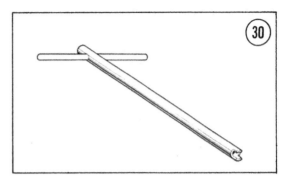

Removal (750/650cc Models With Disc Brakes)

1. Refer to Figure 18 and remove the plugs from the bottom of the fork legs to drain the oil. Lock the front brake and pump the front end up and down to expel the oil. Support the front of the motorcycle on a stand so that the wheel is about 6 inches off the ground. Remove the front wheel and fender. Remove the complete handlebar assembly and the fork cap nuts. Disconnect and remove the drive cables and wires from speedometer and tachometer.

2. Disconnect the brake line at the bottom fork clamp and the lower fork leg. Remove the brake caliper. Unscrew the pinch bolts at the rear of the top fork clamp and remove the aluminum cap screws. Pull out the fork springs.

3. Unscrew the Allen screw from the bottom of the lower leg. It may be necessary to hold the valve inside the fork tube with a special tool similar to the one shown in **Figure 30** while unscrewing the Allen screw. Pull the lower fork legs off the tubes. Loosen the pinch bolts in the bottom fork clamp and pull out the tubes. When reinstalling, the pinch bolts should be torqued to 18-20 ft.-lb. (2.49-2.77 mkg). Pull the dust covers off the fork legs.

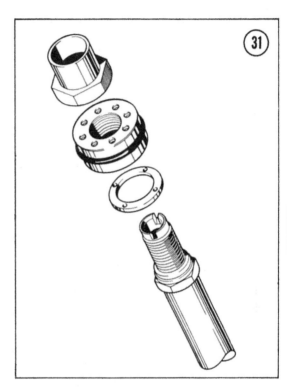

Disassembly (750/650cc Models With Disc Brakes)

1. The damper valve (**Figure 31**) need not be disassembled unless the bleed valve has been damaged. Remove the oil seal from the bleed valve and install a new one. If you use a screwdriver to lift the seal out of its groove, be careful not to damage the soft alloy bearing surface of the valve.

2. Remove the oil seals from the top of each fork leg with a tool like that shown in **Figure 32**. The tool can be made from a piece of mild steel approximately 12 in. long, 1 in. wide, and ⅛ in. thick. Its dimensions are not critical; however, it must be made so that the tool does not come in contact with the fork leg—only the seal. Work the seal loose all around the periphery before attempting to remove it. Thoroughly clean the seal recess and remove any burrs that may be found around the upper edge.

3. To install a new seal, place the tube in the fork leg and place a polyethylene bag over the top. This is necessary to protect the seal from being damaged by the sharp edge of the tube.

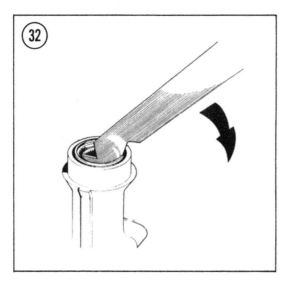

Push the seal onto the tube and into position at the top of the fork leg. The seal must be pressed into the fork leg with a drift like the one shown in **Figure 33**. A drift of this type can be made from the outer tube of an early fork leg with a shoulder welded to one end of the tube to contact the seal. The seal must be pressed in with the tube in place to ensure its concentricity in the fork leg.

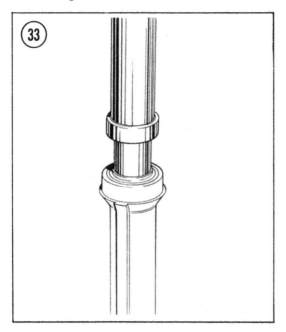

4. Remove the tube, clean all of the components in solvent and wipe dry. Be sure to clean the fork tube bores.

Assembly/Installation
(750/650cc Models With Disc Brakes)

1. Make sure the small sealing washer is installed in the bore in the bottom of the fork leg. If it is worn or damaged, replace it. Apply Loctite to the threads of the fork tube end nut and install the damper assembly in the bottom of the stanchion. Tighten the nut to 25 ft.-lb.

2. Install the rubber dust cover on the leg and install the fork tube. Push the tube into the leg, guiding the bottom of the valve tube into the bore in the bottom of the leg. It may be necessary to guide the valve. When the valve tube is in position, reinstall and tighten the Allen screw.

3. Reinstall the assembled forks in the fork clamps, reversing the order of disassembly. The tops of the tubes should be flush with the upper surface of the top fork clamp. Tighten the top pinch bolts to 20 ft.-lb. Reinstall the springs and fill each fork leg with 5 ounces (190cc) of automatic transmission fluid. Apply gasket cement to the threads of the cap nuts and install them in the tops of the stanchions. Tighten the nuts to 40 ft.-lb. Continue reassembling the rest of components in reverse order of disassembly.

4. When reassembly is complete, check the alignment of the forks and correct it if necessary. Tighten the spindle cap nuts on the right leg and slightly loosen those on the left. Loosen the cap nuts and the pinch bolts in both fork clamps.

5. Lock the front brake, and pump the front end up and down several times to align the forks. Tighten the bolts from the bottom up, beginning with the nuts on the left side wheel spindle cap, then the pinch bolts in the bottom fork clamp, the pinch bolts in the top clamp, the cap nuts, and the fork stem pinch bolt.

Removal (500cc Models)

1. Service of the front forks (**Figures 34 and 35**) is basically the same as that detailed for the 650cc. Minor differences exist for removal of the forks from the motorcycle. In addition to the instructions for the 650cc, disconnect the speedometer and tachometer cables and remove the instrument bulbs and sockets. Note the position of the red ground lead for reassembly. On early models, a steering damper was fitted; re-

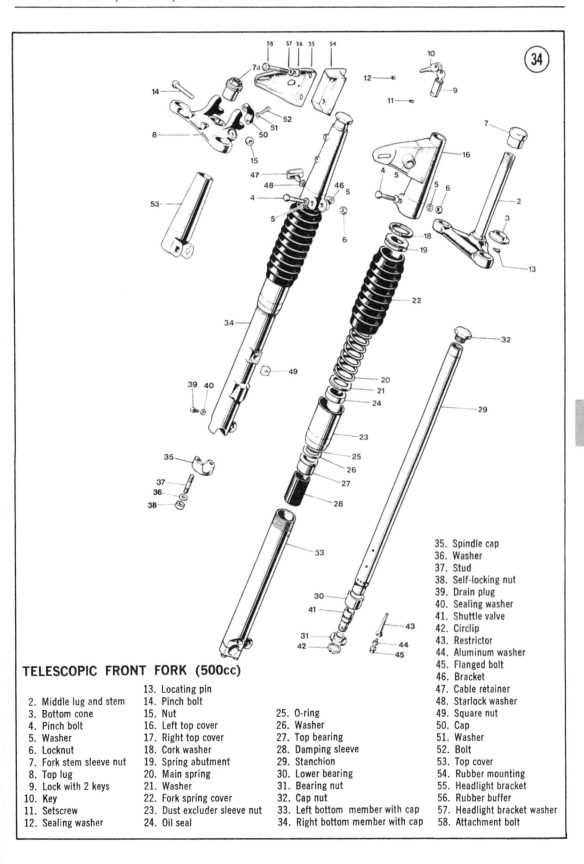

TELESCOPIC FRONT FORK (500cc)

2. Middle lug and stem
3. Bottom cone
4. Pinch bolt
5. Washer
6. Locknut
7. Fork stem sleeve nut
8. Top lug
9. Lock with 2 keys
10. Key
11. Setscrew
12. Sealing washer
13. Locating pin
14. Pinch bolt
15. Nut
16. Left top cover
17. Right top cover
18. Cork washer
19. Spring abutment
20. Main spring
21. Washer
22. Fork spring cover
23. Dust excluder sleeve nut
24. Oil seal
25. O-ring
26. Washer
27. Top bearing
28. Damping sleeve
29. Stanchion
30. Lower bearing
31. Bearing nut
32. Cap nut
33. Left bottom member with cap
34. Right bottom member with cap
35. Spindle cap
36. Washer
37. Stud
38. Self-locking nut
39. Drain plug
40. Sealing washer
41. Shuttle valve
42. Circlip
43. Restrictor
44. Aluminum washer
45. Flanged bolt
46. Bracket
47. Cable retainer
48. Starlock washer
49. Square nut
50. Cap
51. Washer
52. Bolt
53. Top cover
54. Rubber mounting
55. Headlight bracket
56. Rubber buffer
57. Headlight bracket washer
58. Attachment bolt

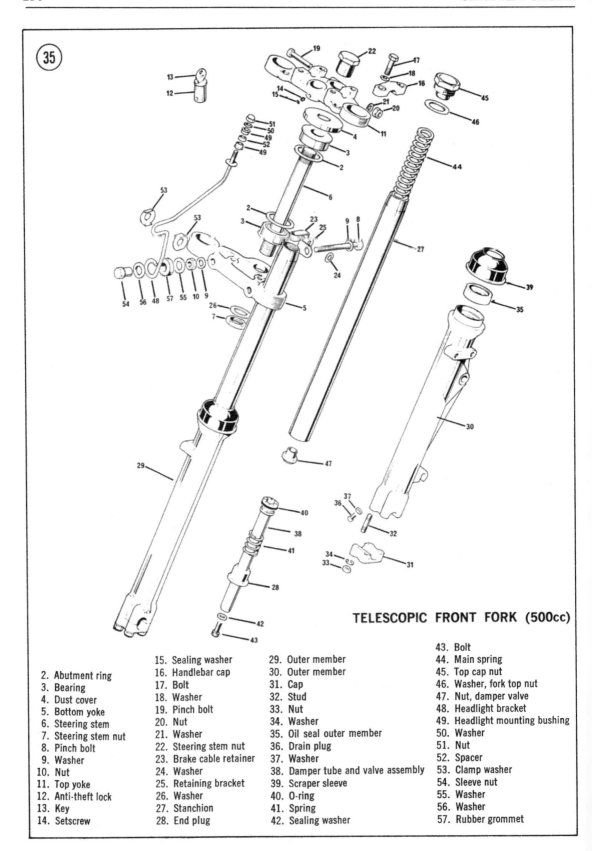

TELESCOPIC FRONT FORK (500cc)

2. Abutment ring
3. Bearing
4. Dust cover
5. Bottom yoke
6. Steering stem
7. Steering stem nut
8. Pinch bolt
9. Washer
10. Nut
11. Top yoke
12. Anti-theft lock
13. Key
14. Setscrew

15. Sealing washer
16. Handlebar cap
17. Bolt
18. Washer
19. Pinch bolt
20. Nut
21. Washer
22. Steering stem nut
23. Brake cable retainer
24. Washer
25. Retaining bracket
26. Washer
27. Stanchion
28. End plug

29. Outer member
30. Outer member
31. Cap
32. Stud
33. Nut
34. Washer
35. Oil seal outer member
36. Drain plug
37. Washer
38. Damper tube and valve assembly
39. Scraper sleeve
40. O-ring
41. Spring
42. Sealing washer

43. Bolt
44. Main spring
45. Top cap nut
46. Washer, fork top nut
47. Nut, damper valve
48. Headlight bracket
49. Headlight mounting bushing
50. Washer
51. Nut
52. Spacer
53. Clamp washer
54. Sleeve nut
55. Washer
56. Washer
57. Rubber grommet

move it by unscrewing. On later models, the leads must be disconnected from the ignition switch.

2. Drive the fork tubes from the triple clamps as shown in **Figure 36**.

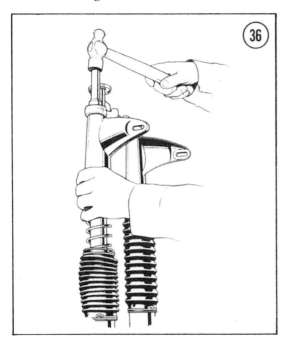

Disassembly (500cc Models)

1. Disassemble the forks by referring to the disassembly section given previously for frame Nos. DU.5825-DU.66245.

2. Service the steering head races as described previously, but note that 48 balls, 3/16 in. (4.76mm) diameter are used, 24 in each race.

Assembly/Installation (500cc Models)

1. When replacing the fork seals, follow the procedure given previously. See **Figure 37**.

2. Replace the oil restrictor. Slide the restrictor down inside the fork tube and use a piece of tubing approximately 2 feet long and ½ in. inside diameter to hold the restrictor while several threads of the securing bolt are engaged. Replace the aluminum sealing washer under the securing bolt. If hydraulic damper units are to be fitted instead of the restrictor, leave this procedure until later.

3. Screw in the location plug and drain plug and rotate the restrictor with the tubing until the

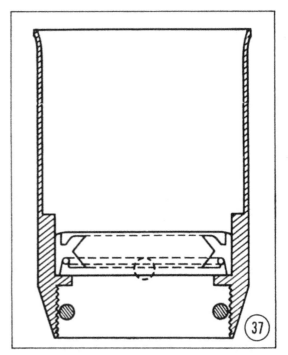

slot is aligned with the plug. Tighten the securing bolt. Use new washers under the location plugs. Fit the upper tube to the bottom member and assemble the damping sleeve, top bushing, and plain washer. Screw on the dust sleeve nut and oil seal assembly, using gasket cement on the outer threads. Tighten the sleeve nut while the bottom tube is held in a vise at the axle lug.

4. On T90 and T100 models, fit the plain thrust washer, main spring, fork boot, spring abutment, and cork washer on each pipe. On 3TA and 5TA models, fit the plain washer, main spring, top cover, plain washer, and cork washer. Align the triple clamps and position the left and right covers (nacelle covers on 3TA and 5TA) and insert the bottom triple clamp pinch bolts. Screw the nuts on finger-tight. Note that the right fork tube has the boss for locating the front brake anchor plate and fit the tubes up into the triple clamps as far as possible.

5. Pull the tubes all the way up into the triple clamps. Tighten the pinch bolts temporarily, remove the tool, and replace it with a cap bolt partially screwed in. When both tubes are in place, remove the cap nuts.

6. On models prior to Serial No. H57083, fit the hydraulic damper units. Place each unit in

the fork leg and fit the location and drain plug into the base of the bottom tube. Engage the securing bolt (using a new aluminum washer) into the base of the damper unit. Rotate the damper unit until the slot engages in the location and drain plug. Tighten the securing bolt and repeat the procedure for the other fork tube.

7. On later models, a shuttle valve type damping unit was used. Note that oil heavier than SAE 20 should never be used. Thread the damper unit into the cap nut and lock into position with the locknut. Pour ⅓ pint (200cc) of fork oil into each tube and replace the cap nuts until several threads are engaged in the tubes. Loosen the pinch bolts, then tighten the cap nuts securely. With nacelles fitted, the fork tubes will have to be turned so the oil fill holes can be seen through the headlight opening. Tighten the cap nuts after this has been done. Adjust the steering head bearings as detailed in this chapter. Tighten the sleeve nut pinch bolt and the 2 bottom triple clamp pinch bolts to the specifications given in Appendix I. Continue reassembly in the reverse order of disassembly.

8. Further details of the damping unit are given in this chapter under *Hydraulic Damper*. Details of damping unit fitted after Serial No. H57083 are shown in **Figure 38**.

HYDRAULIC DAMPER

Some bikes may be fitted with the type of damping unit shown in Figure 27. Refer to the figure for dismantling sequence and note that the top of the damper rod screws into the cap nut. During assembly do not forget to attach the rod to the cap nut or the rod will fall down into the fork tube.

STEERING HEAD BEARINGS
Adjustment

1. Steering head races must be adjusted before front forks are completely reassembled and whenever too much play is found.

2. Raise the machine so its front wheel is off the ground and check that the forks will turn to either side from a central position with a slight starting assist. If they move freely, the adjustment is not too tight. Try rocking the forks back and forth, holding the fender in one hand and the top triple clamp in the other. Any slack must be removed. Loosen the stem sleeve pinch bolt and tighten the stem sleeve nut ¼ turn at a time until all looseness is gone. Check again that the forks will turn readily from lock-to-lock and retighten the stem sleeve pinch bolt.

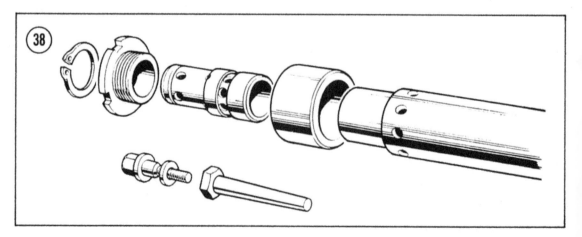

NOTE: If you own a 1978 or later model, first check the Supplement at the back of the book for any new service information.

CHAPTER NINE

REAR BRAKE, WHEEL,
AND SUSPENSION

REAR DRUM BRAKES

Three types of rear wheel arrangements are used on Triumph motorcycles.

Removal/Installation (Early Type)

1. Unscrew the rear brake adjuster and disconnect the drive chain at the master link.

2. Loosen the rear chainguard bolt and swing the chainguard up and out of the way.

3. Remove the rear torque arm securing bolt at the anchor plate.

4. Loosen the axle nuts and pull the wheel off.

5. Service the brake following procedure given under *Front Drum Brakes* in Chapter Eight.

Removal/Installation (Quick Detach Type)

1. Unscrew the axle from the right side and pull out the spacer between the wheel and swing arm end.

2. Pull the wheel into the space vacated by the spacer so it clears the splines and brake drum. The wheel will then come free.

3. Service the brake following procedure given under *Front Drum Brakes* in Chapter Eight.

Removal/Installation (Latest Type Wheel)

1. Disconnect the chain at the master link and remove it from the rear sprocket only.

2. Unscrew the brake anchor nut on the backing plate and loosen the bolt at the forward end of the anchor arm.

3. Loosen the bottom bolt on the left shock and lift the chainguard.

4. Unscrew the axle nut and pull the axle out of the wheel and swing arm.

5. Pull the wheel back and out of the swing arm.

6. Service the brake following procedure given under *Front Drum Brakes* in Chapter Eight.

Installation

To install the wheel, reverse the steps used for removal. Refer to Chapter Two, *Drive Chain Adjustment,* to adjust the chain. Refer to this chapter for brake adjustment procedure.

When a rear wheel is replaced, it may feel somewhat loose as it is fitted into the splines on the hub. Find the position at which it fits tightest and mark the hub and brake drum so it can be more easily positioned in the future. Check the rubber ring which fits over the splines on the wheel and replace it if worn or deteriorated.

9

BRAKE ADJUSTMENT

Front Brake
(Prior to Serial No. DU.66246)

1. Adjust the front brake fulcrum pin so the brake shoes make full contact on the drums.

2. Loosen the hex nut just behind the fork tube on the anchor plate.

3. Lock up the front brake at the handlebar lever and tighten the nut while the brake is fully applied.

Front Brake
(After Serial No. DU.66246)

1. Adjust the front twin leading shoe brakes during reassembly as detailed above.

2. Adjust free play at the handlebar control lever (with the adjusting nut) so that there is between 1/16 in. (1.6mm) and ⅛ in. (3.2mm) slack in the inner cable.

Rear Brake (Typical)

Adjust the wing nut on the rear brake rod so that there is approximately ½ in. (13mm) free play before the brake is engaged.

REAR DISC BRAKES

The 1976 Triumph is equipped with rear disc brakes. The rear caliper is identical to the front caliper (see *Disc Brakes,* Chapter Eight) and has the same part number. **Figure 1** shows the components of the rear disc system, and **Figure 2** shows component locations. Note that the master cylinder reservoir is located under the seat, while the cylinder itself is located inside the left side of the swing arm, near the caliper. Replacement of brake pads, removal/installation of calipers, bleeding, and other service can be accomplished by following the procedures given for front disc brakes in Chapter Eight.

1. Caliper
2. Disc
3. Master cylinder reservoir
4. Master cylinder (behind swing arm)
5. Stoplight switch
6. Brake pedal

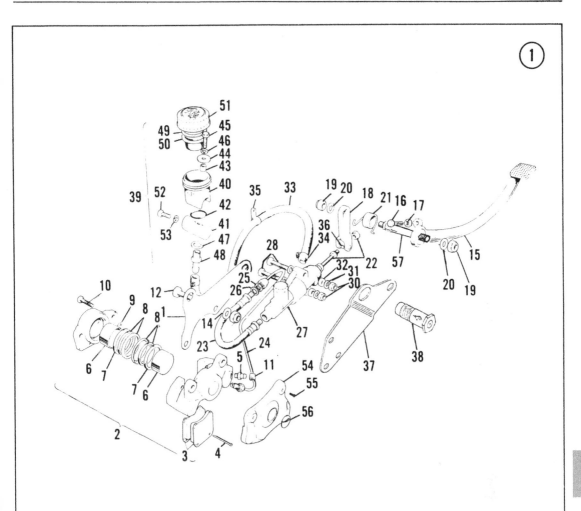

REAR DISC BRAKE

1. Anchor plate
2. Caliper
3. Brake pad
4. Retaining pin
5. Bleed nipple
6. Piston
7. Fluid seal
8. Retainer
9. Fluid seal
10. Bolt
11. Bleed nipple cover
12. Caliper
13. Bolt
14. Washer
15. Brake pedal
16. Stoplight adjuster screw
17. Adjuster screw nut
18. Brake lever
19. Brake lever/pedal nut
20. Washer
21. Brake pedal return spring
22. Pushrod adjuster nut
23. Master cylinder to pipe hose
24. Hose to caliper type
25. Hose to eyebolt nut
26. Serrated washer
27. Master cylinder
28. Bolt
30. Nut
31. Washer
32. Spacer
33. Hose
34. Hose clamp
35. Hose clamp
36. Trunnion
37. Master cylinder mounting plate
38. Pivot bolt
39. Reservoir assembly
40. Reservoir body
41. Reservoir mounting
42. O-ring
43. Spacer
44. Washer
45. Bolt
46. Spring washer
47. Sealing washer
48. Outlet stub
49. Paper gasket
50. Diaphragm
51. Filter cap
52. Mounting bolt
53. Washer
54. Caliper cover
55. Mounting screw
56. Decal
57. Brake pedal pivot arm

WHEELS

Servicing of rear wheels (inspection, spokes, balance, etc.) is identical to that for front wheels. Follow the procedures given in Chapter Eight. Instructions for packing wheel bearings are given in Chapter Two.

REAR SHOCKS
Removal/Repair/Installation

1. Remove rear shocks (**Figure 3**) with the rear wheel raised off the ground. On earlier 6T models, the rear wheel and fender must be removed to gain access to the bolts.

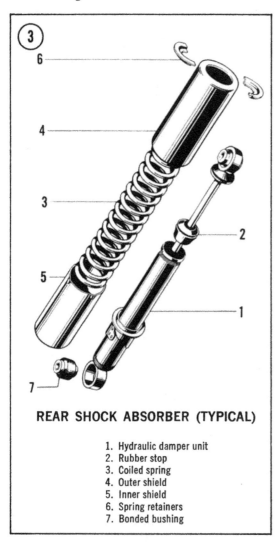

REAR SHOCK ABSORBER (TYPICAL)

1. Hydraulic damper unit
2. Rubber stop
3. Coiled spring
4. Outer shield
5. Inner shield
6. Spring retainers
7. Bonded bushing

2. Compress the outer spring so the 2 large locking collets may be removed.

3. Check the spring against a new one. Check the damper unit for perceptible damping when operated without the spring in place, and also for leaks. Resistance during an extension stroke should be noticeably greater than for a compression stroke, and the resistance should remain constant throughout a stroke. If the rod is bent it may be straightened by a machinist, but replacement is advised. Press out the rubber bushings in the mounts and replace with new ones, if necessary.

4. Squeaking rear shocks may be caused by springs rubbing against the bottom shield. Check for the cause.

5. Replace the components in the order shown in Figure 3 and compress the spring, making sure the shock adjuster is in the light load position. Replace the locking collets. Use a large drift to align the mounting holes.

SWING ARM
Removal/Installation
(750/650cc Models)

1. Refer to **Figures 4, 5, and 6.** Remove the rear enclosure panels, if fitted. Disconnect the rear chain at the master link.

2. Remove the front torque arm nut and brake rod adjusting nut, then loosen the rear axle nuts. Remove the rear wheel.

3. Remove the left and right engine mounting plates at the rear of the engine (early models).

4. Loosen the rear bolt, remove the front bolt, disconnect the stoplight switch leads, then remove the chainguard.

5. Remove the bottom shock absorber mounting bolts.

6. If the swing arm axle nut is on the right side of the machine, remove the oil scavenge hose from the oil tank to gain access to the nut.

7. Remove the locknut and unscrew the axle and pull it out.

8. Remove the end plates, outer sleeves, spacer, and dust covers.

9. If the swing arm bolt has a brazed-on head, it may be replaced with the newer one-piece type.

10. Clean all components and check for excessive wear. Check that the outer sleeves on the

SWING ARM AND SUSPENSION UNITS

1. Swing arm
2. Pivot bushing
3. Grease nipple
4. Fiber washer
5. Sleeve
6. Spacer
7. Flanged washer
8. Bolt
9. Tab washer
10. Nut
11. Brake torque stay
12. Bolt
13. Spring washer
14. Nut
15. Suspension unit
16. Bonded bushing
17. Spring retainer
18. Outer dirt shield
19. Inner dirt shield
20. Spring
21. Unit fixing bolt
22. Washer
23. Spring washer
24. Washer
25. Nut
26. Chainguard
27. Grommet
28. Bolt
29. Rubber tube
30. Washer
31. Self-locking nut
32. Stepped bolt
33. Washer
34. Nut
35. Clip
36. Nut
37. Bolt
38. Spring washer
39. Brake operating rod
40. Split pin
41. Adjuster nut
42. Stoplight switch
43. Screw
44. Self-locking nut
45. Switch lever
46. D-washer
47. Spring
48. O-ring

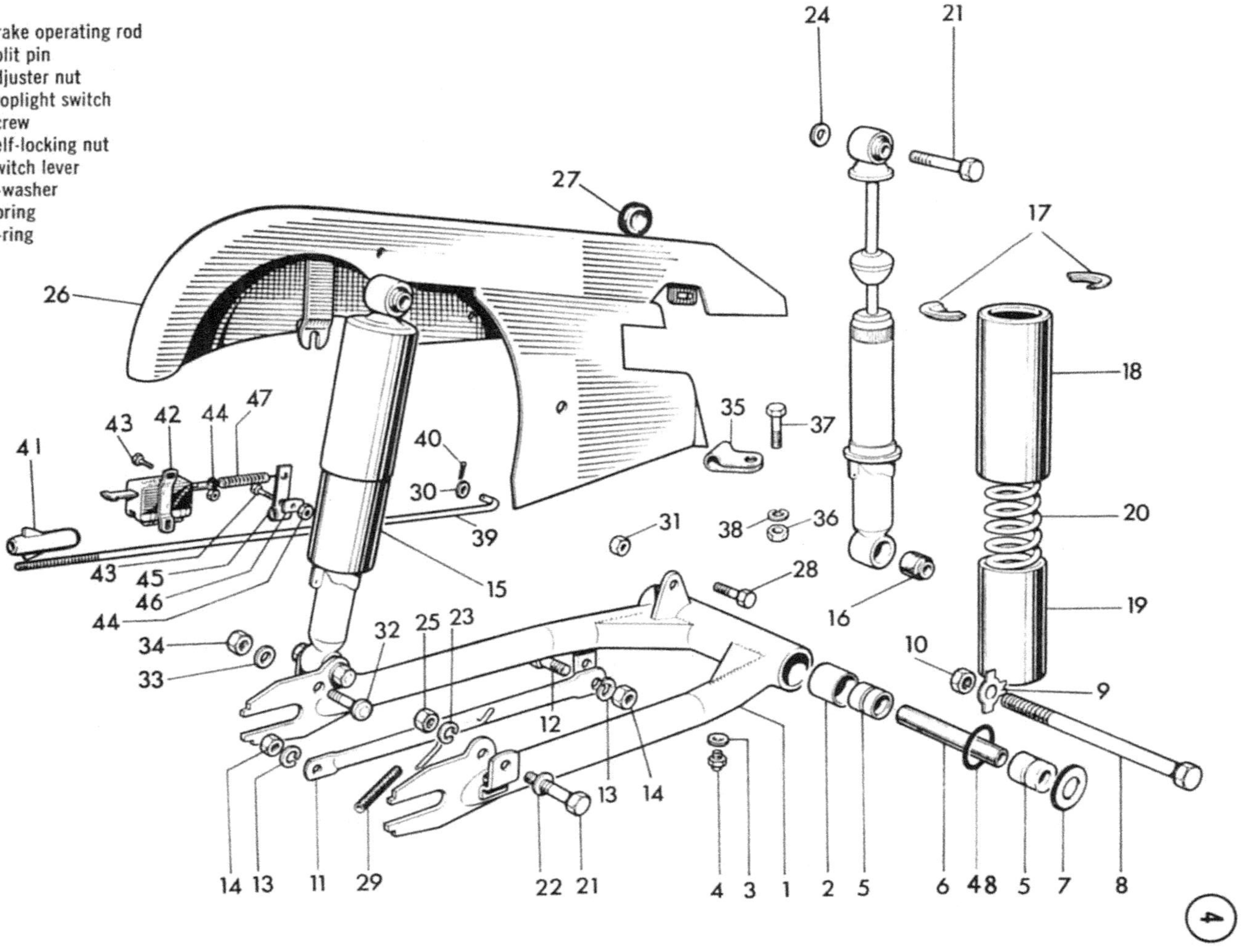

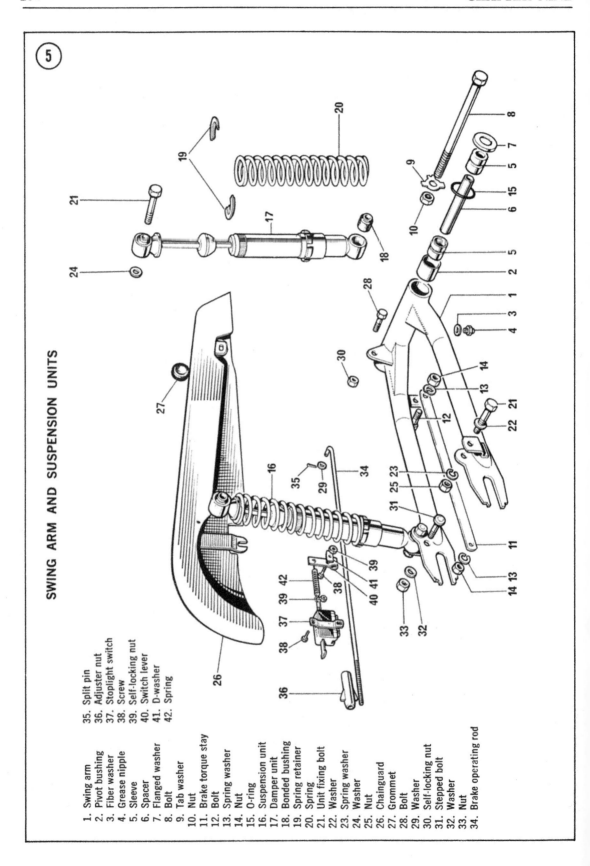

⑤

SWING ARM AND SUSPENSION UNITS

1. Swing arm
2. Pivot bushing
3. Fiber washer
4. Grease nipple
5. Sleeve
6. Spacer
7. Flanged washer
8. Bolt
9. Tab washer
10. Nut
11. Brake torque stay
12. Bolt
13. Spring washer
14. Nut
15. O-ring
16. Suspension unit
17. Damper unit
18. Bonded bushing
19. Spring retainer
20. Spring
21. Unit fixing bolt
22. Washer
23. Spring washer
24. Washer
25. Nut
26. Chainguard
27. Grommet
28. Bolt
29. Washer
30. Self-locking nut
31. Stepped bolt
32. Washer
33. Nut
34. Brake operating rod
35. Split pin
36. Adjuster nut
37. Stoplight switch
38. Screw
39. Self-locking nut
40. Switch lever
41. D-washer
42. Spring

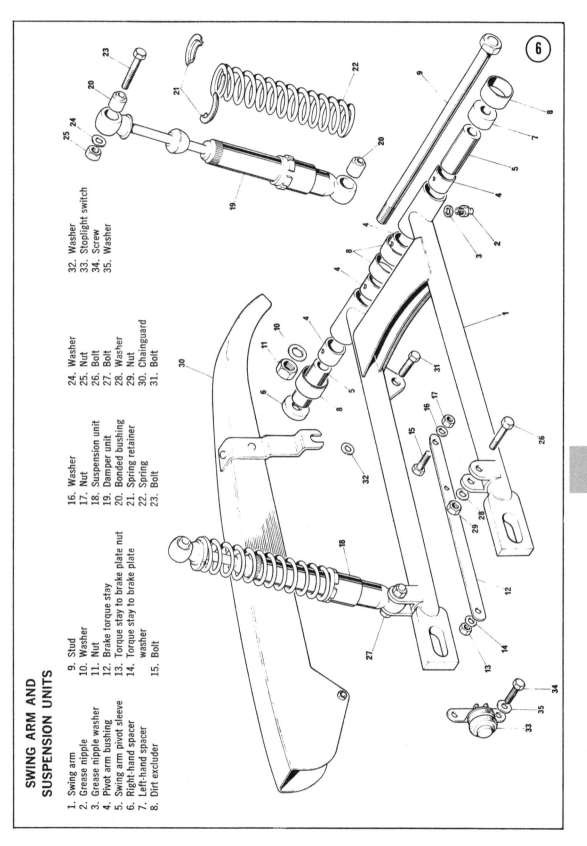

SWING ARM AND
SUSPENSION UNITS

1. Swing arm
2. Grease nipple
3. Grease nipple washer
4. Pivot arm bushing
5. Swing arm pivot sleeve
6. Right-hand spacer
7. Left-hand spacer
8. Dirt excluder

9. Stud
10. Washer
11. Nut
12. Brake torque stay
13. Torque stay to brake plate nut
14. Torque stay to brake plate washer
15. Bolt

16. Washer
17. Nut
18. Suspension unit
19. Damper unit
20. Bonded bushing
21. Spring retainer
22. Spring
23. Bolt

24. Washer
25. Nut
26. Bolt
27. Bolt
28. Washer
29. Nut
30. Chainguard
31. Bolt

32. Washer
33. Stoplight switch
34. Screw
35. Washer

swing arm bushings do not have excessive clearance (Appendix I). Drive out the old bushings and drive in new ones if necessary.

11. Grease all parts liberally before installing. Fill the space around the spacer tube with grease.

12. Install the components and tighten the bolt until the swing arm is just free enough to move up and down easily. If there is excessive side play and the bushings are in good shape, the spacer can be filed slightly to take out the side play. Thicker side plates are available for earlier models.

13. Tighten the locknut and bend over the locking washer tabs; install the remaining components. If a new bolt has to be used, check that the threads on the new one are the same as on the old one. A new system of Unified threads has been implemented and components are sometimes replaced with the new system.

Removal/Installation
(500cc Models, Except TR5T)

1. From serial No. H49833 up (**Figure 7, 8, and 9**), block the bike up so the frame is supported. Disconnect the centerstand spring. Remove the rear brake adjuster from the brake rod. Disconnect the drive chain at its master link.

2. Remove the torque arm anchor nut from the brake anchor plate and take the anchor arm off the stud. Loosen the nut at the rear of the left shock absorber bottom mounting and raise the rear of the chainguard. Disconnect the speedometer cable from the rear wheel and remove the wheel as detailed in this chapter. Remove the speedometer cable clips from the swing arm.

3. Remove the rubber tube from the chain oiler pipe on the torque arm. Remove the exhaust pipes and mufflers. Disconnect the stoplight switch spring and the connectors. Remove the shocks, the rear brake pedal, and rod.

4. Remove the passenger footpegs and mounting bolts. Remove the front chainguard and front lower switch panel. Remove the oil tank lower mount and tap the studs back through the lug. Remember to replace the spacer over the top stud between the oil tank bottom bracket and frame lug during installation. Remove the fender forward bottom securing bolt. Loosen

the top, and remove the bottom nuts and bolts holding the front and rear frame sections together.

5. Tap the swing arm spindle tab washers clear and remove the spindle end bolts. Remove the tab washers. While installing, note that the right end bolt has a grease nipple and the tabs on the washers fit into the rear frame side plates. Pivot the rear frame on the top stud and support it so the swing arm spindle is accessible. Remove the left and right spacers. When reassembling, the thicker one fits on the chain side. The ribbed side must be next to the rear frame side plate on the right side. Tap out the spindle with a drift and hammer.

6. Fit the correct shims between the swing arm pivot lug and the frame lug. Use the spacing washer and sufficient shims against the right frame lug so the swing arm will just move under its own weight.

7. Grease and then tap the spindle in from the left side. Fill the inside with grease and replace the end caps, retaining rod, and nut. Fill until grease comes out of the swing arm bushings.

Old swing arm bushings can be driven out and new ones pressed in, if necessary. A drift made from a 31/32 in. diameter bar turned down to 7/8 in. for one inch at the end can be used. Ream the bushings to the size given in Appendix I. Continue assembly in the reverse order of disassembly.

Removal/Installation
(500cc Models, Prior to H49832)

1. To remove the swing arm (Figure 8), place the bike on its centerstand.

2. Remove the bolt at the front of the chainguard and disconnect the stoplight switch wiring.

3. Remove the stoplight switch operating clip from the brake rod.

4. Remove the chainguard from the swing arm.

5. Remove the shock absorber bottom mounting bolts.

6. Remove the spindle retaining rod and caps. Use an extractor made as suggested in **Figure 10** to pull the spindle out from the right side.

7. Disconnect the drive chain at its master link

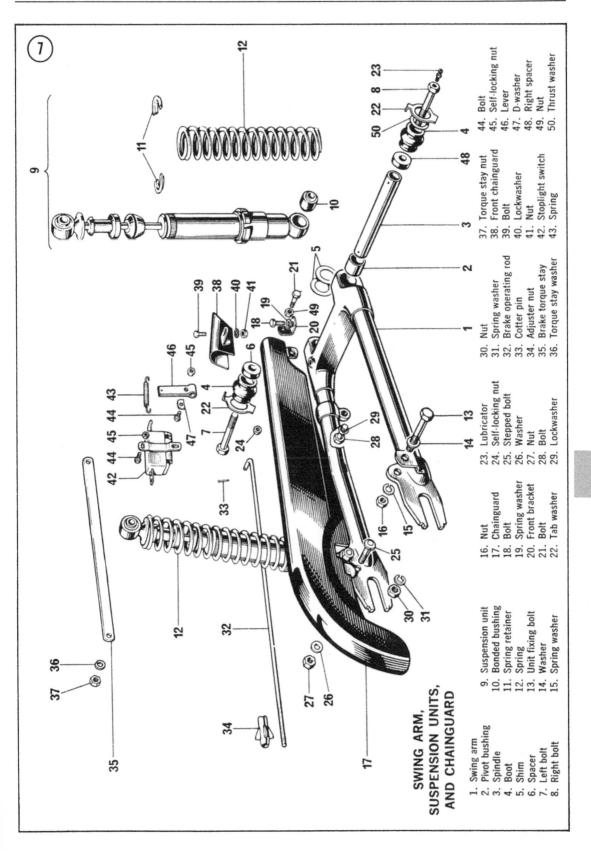

SWING ARM, SUSPENSION UNITS, AND CHAINGUARD

1. Swing arm
2. Pivot bushing
3. Spindle
4. Boot
5. Shim
6. Spacer
7. Left bolt
8. Right bolt
9. Suspension unit
10. Bonded bushing
11. Spring retainer
12. Spring
13. Unit fixing bolt
14. Washer
15. Spring washer
16. Nut
17. Chainguard
18. Bolt
19. Spring washer
20. Front bracket
21. Bolt
22. Tab washer
23. Lubricator
24. Self-locking nut
25. Stepped bolt
26. Washer
27. Nut
28. Bolt
29. Lockwasher
30. Nut
31. Spring washer
32. Brake operating rod
33. Cotter pin
34. Adjuster nut
35. Brake torque stay
36. Torque stay washer
37. Torque stay nut
38. Front chainguard
39. Bolt
40. Lockwasher
41. Nut
42. Stoplight switch
43. Spring
44. Bolt
45. Self-locking nut
46. Lever
47. D-washer
48. Right spacer
49. Nut
50. Thrust washer

9

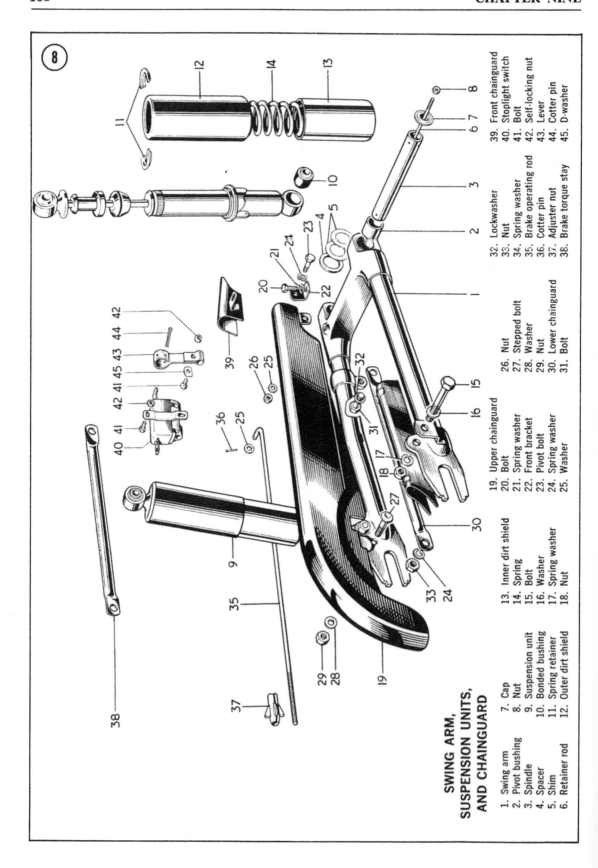

**SWING ARM,
SUSPENSION UNITS,
AND CHAINGUARD**

1. Swing arm
2. Pivot bushing
3. Spindle
4. Spacer
5. Shim
6. Retainer rod
7. Cap
8. Nut
9. Suspension unit
10. Bonded bushing
11. Spring retainer
12. Outer dirt shield
13. Inner dirt shield
14. Spring
15. Bolt
16. Washer
17. Spring washer
18. Nut
19. Upper chainguard
20. Bolt
21. Spring washer
22. Front bracket
23. Pivot bolt
24. Spring washer
25. Washer
26. Nut
27. Stepped bolt
28. Washer
29. Nut
30. Lower chainguard
31. Bolt
32. Lockwasher
33. Nut
34. Spring washer
35. Brake operating rod
36. Cotter pin
37. Adjuster nut
38. Brake torque stay
39. Front chainguard
40. Stoplight switch
41. Bolt
42. Self-locking nut
43. Lever
44. Cotter pin
45. D-washer

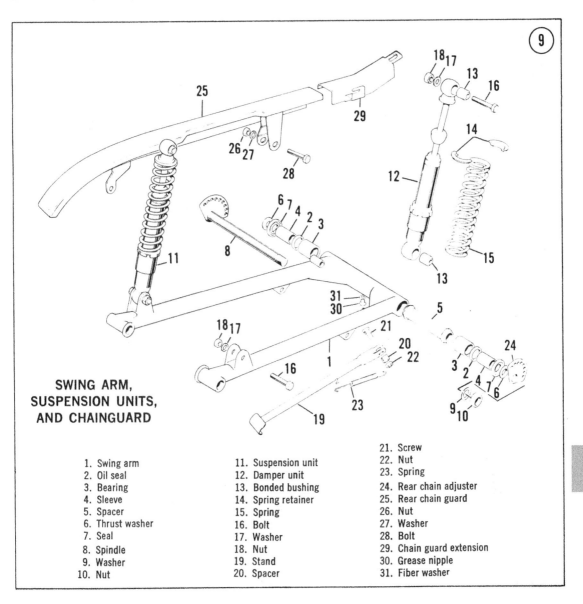

⑨

**SWING ARM,
SUSPENSION UNITS,
AND CHAINGUARD**

⑨

1. Swing arm	11. Suspension unit	21. Screw
2. Oil seal	12. Damper unit	22. Nut
3. Bearing	13. Bonded bushing	23. Spring
4. Sleeve	14. Spring retainer	24. Rear chain adjuster
5. Spacer	15. Spring	25. Rear chain guard
6. Thrust washer	16. Bolt	26. Nut
7. Seal	17. Washer	27. Washer
8. Spindle	18. Nut	28. Bolt
9. Washer	19. Stand	29. Chain guard extension
10. Nut	20. Spacer	30. Grease nipple
		31. Fiber washer

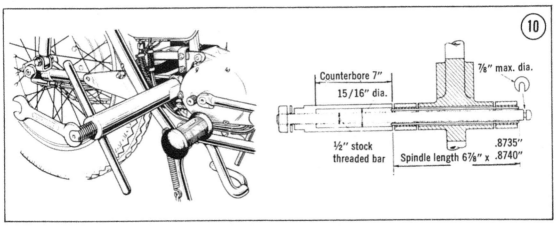

⑩

Counterbore 7"

15/16" dia.

⅞" max. dia.

½" stock
threaded bar

Spindle length 6⅞" x

.8735"
.8740"

and remove the bolt holding the anchor arm to the backing plate at the rear wheel.

8. Remove the brake rod adjusting nut.

9. Loosen the bolt at the rear of the chainguard and move the chainguard up so the rear wheel can be removed once the speedometer cable is removed and the axle nuts loosened.

10. Remove the swing arm.

11. Clean and inspect all parts.

12. Fit shims to the swing arm as described in Step 6 of the prior section.

13. Grease the swing arm bushings and spindle and drive the spindle in with the extractor previously used. Fill the inside with grease.

14. Replace end caps, retaining rod, and nut.

15. Grease the fitting (**Figure 11**) until grease comes out the swing arm bushings.Continue reassembly in the reverse order of disassembly.

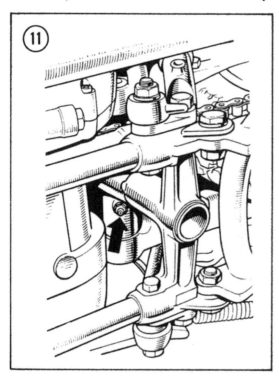

Removal/Installation (TR5T Model)

1. Refer to Figure 9 and block up the rear of the motorcycle so the wheel is a foot above the ground.

2. Disconnect the rear chain and remove it from the wheel sprocket.

3. Unscrew the brake rod adjuster and push the rod out of the brake arm.

4. Unscrew the speedometer cable from the drive on the right side of the wheel and remove the left passanger footrest to release the brake anchor.

5. Unscrew the axle nut, support the rear wheel, pull out the axle, and remove the wheel.

6. Remove the front bolt on the chainguard and loosen the bolt on the bottom of the rear shock to remove the chainguard.

7. Remove the rear shocks.

8. Unscrew the pivot bolt for the brake pedal and remove the pedal and rod.

9. Remove the rear bolt from the stoplight switch so the switch is permitted to swing down and away from the adjuster cam. Note the location of the spacer and washers for reassembly.

10. Unscrew the nut from the right end of the swing arm pivot bolt. Support the swing arm and pull the pivot bolt out of the frame.

11. Disconnect the electrical harness at the top of the left shock.

12. Remove the 5 bolts which hold the fender in place and remove the fender. Remove the swing arm from the frame.

13. Clean all of the parts for the swing arm pivot with solvent or kerosene and dry thoroughly (**Figure 12**).

14. Check the condition of the bearings which should turn freely and smoothly, without any radial play. If replacement is necessary, drive the spacer tube out through one of the bearings.

15. Remove the spacer tube and both bearings.

16. Grease the new bearings, install one, making sure it is square with the bore before pressing in. Install a new spacer tube from the other end and install the opposite bearing.

17. Install the swing arm in the frame, referring to Figure 12 for location of the seals, cups, and washers.

18. Assemble the remaining components, reversing the order for disassembly, and grease the swing arm through the fitting located on the bottom of the pivot tube. Adjust the free play in the rear chain.

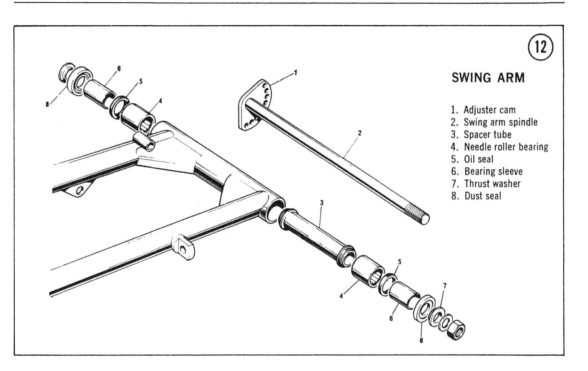

⑫

SWING ARM

1. Adjuster cam
2. Swing arm spindle
3. Spacer tube
4. Needle roller bearing
5. Oil seal
6. Bearing sleeve
7. Thrust washer
8. Dust seal

9

CHAPTER TEN

FRAME

Figures 1, 2, and 3 are exploded views of early and late frames for 750/650cc models.

Triumph 500cc frame component service is similar and in most cases identical to the procedures for servicing the 750/650cc models. Diagrams showing the differences between the models are included whenever available. Refer to **Figures 4, 5, 6, and 7** for Triumph 500cc frame construction details.

Note that U.N.F. (Unified) threads have been introduced on frame fittings. All nuts, bolts, and threaded components should be checked for proper thread type before they are replaced.

OIL TANK

Removal/Installation

1. On early models, drain the oil tank. It may be necessary to remove a body panel on some models to gain access to the oil tank. See **Figure 8** for oil tank illustration.

2. Remove the tool tray if one is fitted.

3. Loosen or remove the engine oil feed hose clip. Unscrew the return pipe union nut and oil filter.

4. Remove the rear chain oiler tube (if fitted).

5. Disconnect the rocker feed pipe.

6. Refer to the previous section and remove the battery carrier as detailed. Remove the bottom bracket bolt and remove the bracket from the bottom grommet.

7. Secure the mounting pegs and remove the nuts. Push the pegs back through their sleeves.

8. Raise the oil tank and move its top to clear the frame brackets and then remove the tank. On earlier models, it may be necessary to lower the bottom bracket into the space behind the gearbox after the lower mounting bolt is removed in order to remove the tank.

9. Remove the oil tank filter body.

10. Clean the filter and oil tank with solvent.

11. To install, attach the seat check strap and ground wire to the front and rear mounting bolts. Use care while slipping the rubber hoses in place. Continue installation in the reverse order of removal. Fill with the proper grade and quantity of oil as given in Chapter Two.

BATTERY CARRIER

Refer to **Figure 9** for 750/650cc models (typical) and to **Figure 10** for 500cc models (typical).

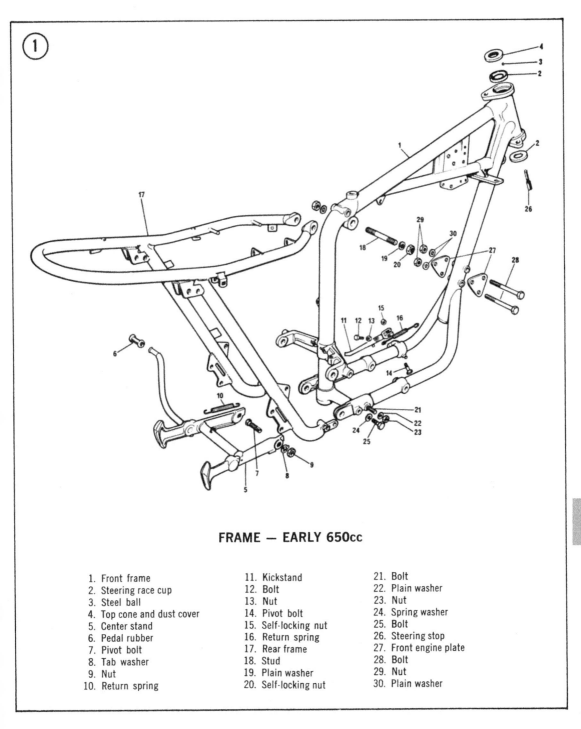

FRAME — EARLY 650cc

1. Front frame	11. Kickstand	21. Bolt
2. Steering race cup	12. Bolt	22. Plain washer
3. Steel ball	13. Nut	23. Nut
4. Top cone and dust cover	14. Pivot bolt	24. Spring washer
5. Center stand	15. Self-locking nut	25. Bolt
6. Pedal rubber	16. Return spring	26. Steering stop
7. Pivot bolt	17. Rear frame	27. Front engine plate
8. Tab washer	18. Stud	28. Bolt
9. Nut	19. Plain washer	29. Nut
10. Return spring	20. Self-locking nut	30. Plain washer

1. Lift the seat and disconnect the negative and positive battery leads. Remove the rubber battery strap and lift the battery out. For some models, the battery vent tube is routed under the motorcycle and behind the swing arm lug during replacement.

2. Remove the nut securing the ground lead and the rectifier to the carrier. Loosen the nuts on the cross straps and lift the carrier out. On earlier models, the carrier should be lifted after the battery is removed so the zener diode heat sink bolts can be removed.

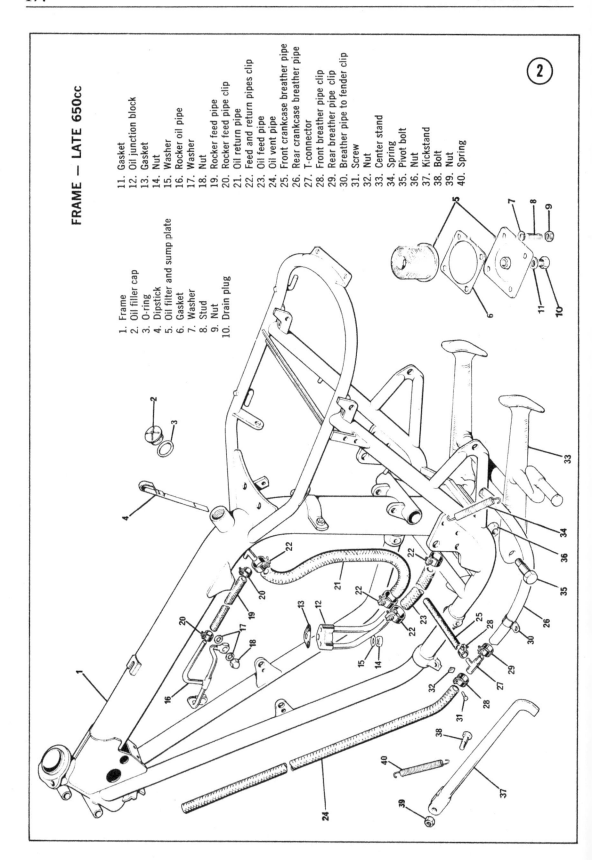

FRAME — LATE 650cc

1. Frame
2. Oil filler cap
3. O-ring
4. Dipstick
5. Oil filter and sump plate
6. Gasket
7. Washer
8. Stud
9. Nut
10. Drain plug
11. Gasket
12. Oil junction block
13. Gasket
14. Nut
15. Washer
16. Rocker oil pipe
17. Washer
18. Nut
19. Rocker feed pipe
20. Rocker feed pipe clip
21. Oil return pipe
22. Feed and return pipes clip
23. Oil feed pipe
24. Oil vent pipe
25. Front crankcase breather pipe
26. Rear crankcase breather pipe
27. T-connector
28. Front breather pipe clip
29. Rear breather pipe clip
30. Breather pipe to fender clip
31. Screw
32. Nut
33. Center stand
34. Spring
35. Pivot bolt
36. Nut
37. Kickstand
38. Bolt
39. Nut
40. Spring

FRAME — 750cc

1. Frame
2. Oil filler cap
3. O-ring
4. Oil filter
5. Sump plate
6. Gasket
7. Washer
8. Stud
9. Nut
10. Drain plug
11. Gasket
12. Oil junction block
13. Gasket
14. Nut
15. Washer
16. Rocker oil pipe
17. Washer
18. Nut
19. Rocker feed pipe
20. Rocker feed pipe clip
21. Oil return pipe
22. Return pipe clip
23. Feed pipe clip
24. Oil feed pipe
25. Oil vent pipe
26. Front crankcase breather pipe
27. Rear crankcase breather pipe
28. T-connector
29. Front breather pipe clip
30. Rear breather pipe clip
31. Breather pipe to fender clip
32. Screw
33. Nut
34. Center stand
35. Spring
36. Pivot bolt
37. Nut
38. Kickstand
39. Bolt
40. Nut
41. Spring

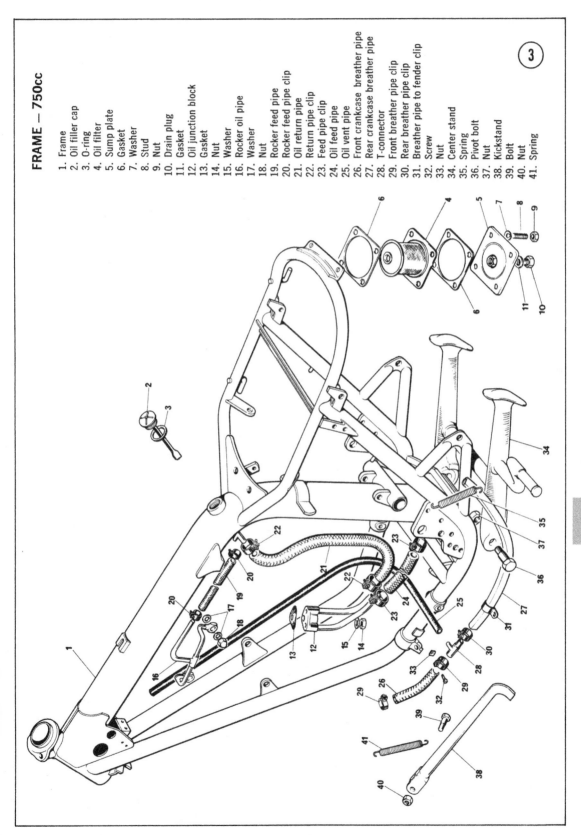

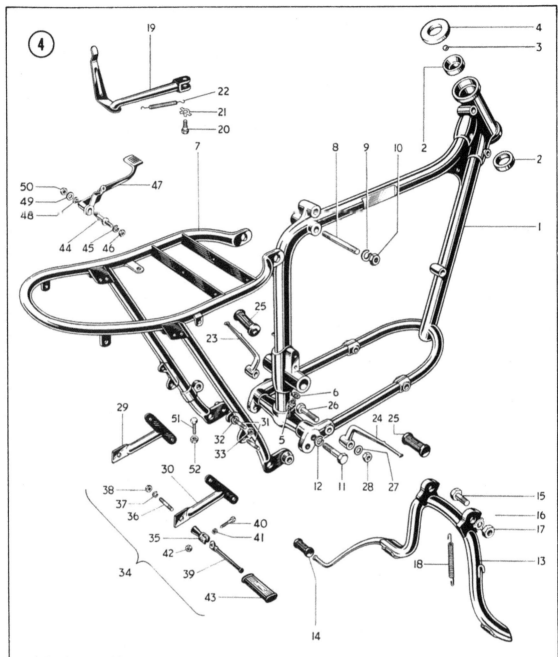

FRAME — 500cc

1. Front frame	11. Bolt	22. Return spring	31. Stud	42. Nut
2. Steering race cup	12. Spring washer	23. Left footrest	32. Spring washer	43. Pedal rubber
3. Steel ball	13. Center stand	24. Right footrest	33. Nut	44. Brake pedal spindle
4. Top cone and dust cover	14. Pedal rubber	25. Pedal rubber	34. Pillion footrest	45. Spring washer
5. Grease nipple	15. Pivot bolt	26. Bolt	35. Footrest hanger	46. Nut
6. Fiber washer	16. Tab washer	27. Plain washer	36. Stud	47. Brake pedal
7. Rear frame	17. Nut	28. Nut	37. Spring washer	48. Spring washer
8. Stud	18. Return spring	29. Left pillion footrest	38. Nut	49. Plain washer
9. Spring washer	19. Prop stand	support	39. Pedal	50. Nut
10. Nut	20. Pivot bolt	30. Right pillion footrest	40. Pivot bolt	51. Pedal stop screw
	21. Tab washer	support	41. Spring washer	52. Nut

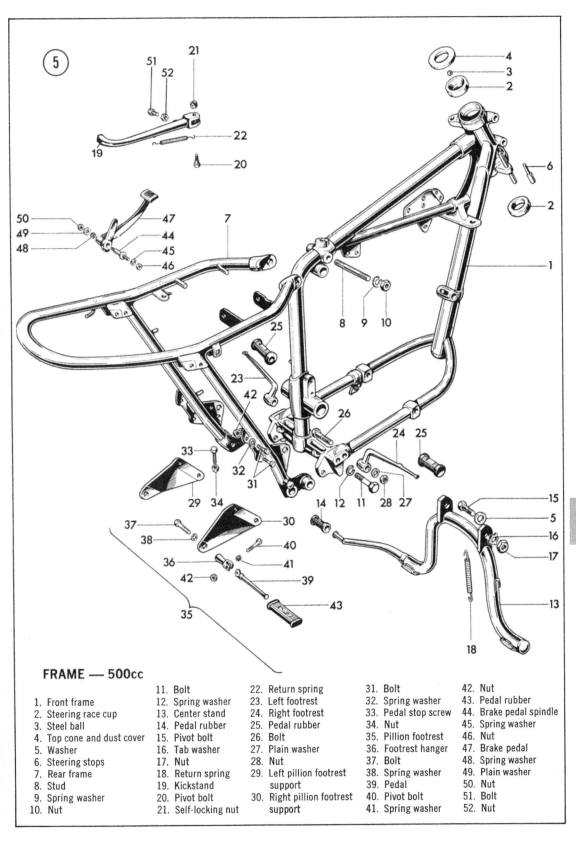

FRAME — 500cc

1. Front frame
2. Steering race cup
3. Steel ball
4. Top cone and dust cover
5. Washer
6. Steering stops
7. Rear frame
8. Stud
9. Spring washer
10. Nut
11. Bolt
12. Spring washer
13. Center stand
14. Pedal rubber
15. Pivot bolt
16. Tab washer
17. Nut
18. Return spring
19. Kickstand
20. Pivot bolt
21. Self-locking nut
22. Return spring
23. Left footrest
24. Right footrest
25. Pedal rubber
26. Bolt
27. Plain washer
28. Nut
29. Left pillion footrest support
30. Right pillion footrest support
31. Bolt
32. Spring washer
33. Pedal stop screw
34. Nut
35. Pillion footrest
36. Footrest hanger
37. Bolt
38. Spring washer
39. Pedal
40. Pivot bolt
41. Spring washer
42. Nut
43. Pedal rubber
44. Brake pedal spindle
45. Spring washer
46. Nut
47. Brake pedal
48. Spring washer
49. Plain washer
50. Nut
51. Bolt
52. Nut

10

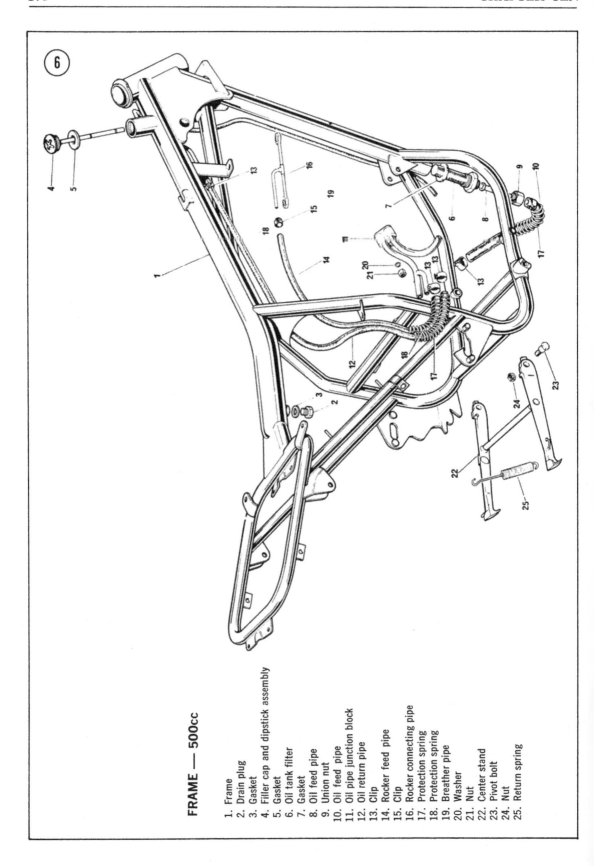

⑥

FRAME — 500cc

1. Frame
2. Drain plug
3. Gasket
4. Filler cap and dipstick assembly
5. Gasket
6. Oil tank filter
7. Gasket
8. Oil feed pipe
9. Union nut
10. Oil feed pipe
11. Oil pipe junction block
12. Oil return pipe
13. Clip
14. Rocker feed pipe
15. Clip
16. Rocker connecting pipe
17. Protection spring
18. Protection spring
19. Breather pipe
20. Washer
21. Nut
22. Center stand
23. Pivot bolt
24. Nut
25. Return spring

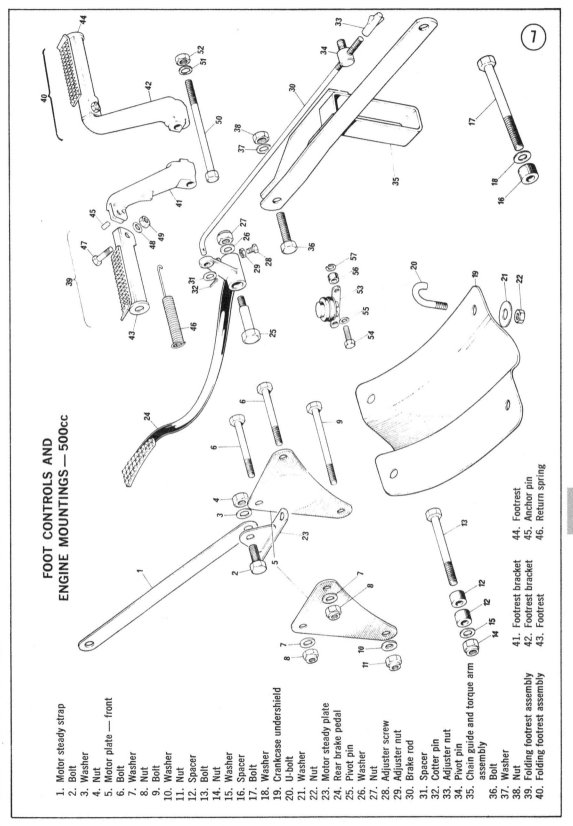

**FOOT CONTROLS AND
ENGINE MOUNTINGS — 500cc**

1. Motor steady strap
2. Bolt
3. Washer
4. Nut
5. Motor plate — front
6. Bolt
7. Washer
8. Nut
9. Bolt
10. Washer
11. Nut
12. Spacer
13. Bolt
14. Nut
15. Washer
16. Spacer
17. Bolt
18. Washer
19. Crankcase undershield
20. U-bolt
21. Washer
22. Nut
23. Motor steady plate
24. Rear brake pedal
25. Pivot pin
26. Washer
27. Nut
28. Adjuster screw
29. Adjuster nut
30. Brake rod
31. Spacer
32. Cotter pin
33. Adjuster nut
34. Pivot pin
35. Chain guide and torque arm
    assembly
36. Bolt
37. Washer
38. Nut
39. Folding footrest assembly
40. Folding footrest assembly
41. Footrest bracket
42. Footrest bracket
43. Footrest
44. Footrest
45. Anchor pin
46. Return spring

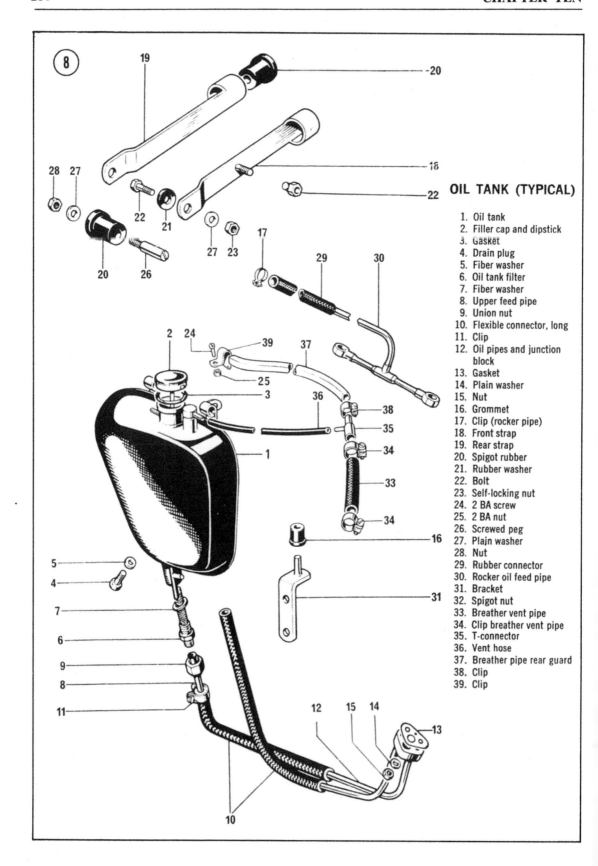

OIL TANK (TYPICAL)

1. Oil tank
2. Filler cap and dipstick
3. Gasket
4. Drain plug
5. Fiber washer
6. Oil tank filter
7. Fiber washer
8. Upper feed pipe
9. Union nut
10. Flexible connector, long
11. Clip
12. Oil pipes and junction
    block
13. Gasket
14. Plain washer
15. Nut
16. Grommet
17. Clip (rocker pipe)
18. Front strap
19. Rear strap
20. Spigot rubber
21. Rubber washer
22. Bolt
23. Self-locking nut
24. 2 BA screw
25. 2 BA nut
26. Screwed peg
27. Plain washer
28. Nut
29. Rubber connector
30. Rocker oil feed pipe
31. Bracket
32. Spigot nut
33. Breather vent pipe
34. Clip breather vent pipe
35. T-connector
36. Vent hose
37. Breather pipe rear guard
38. Clip
39. Clip

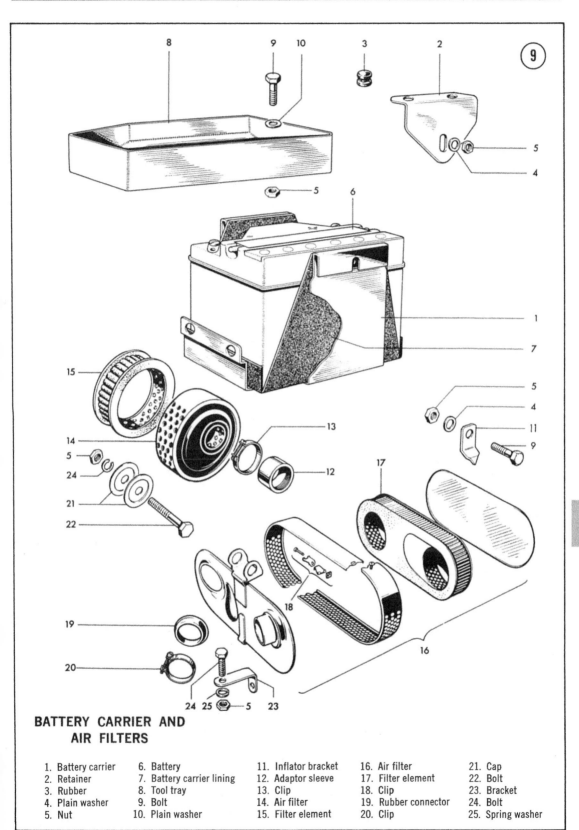

BATTERY CARRIER AND AIR FILTERS

1. Battery carrier	6. Battery	11. Inflator bracket	16. Air filter	21. Cap
2. Retainer	7. Battery carrier lining	12. Adaptor sleeve	17. Filter element	22. Bolt
3. Rubber	8. Tool tray	13. Clip	18. Clip	23. Bracket
4. Plain washer	9. Bolt	14. Air filter	19. Rubber connector	24. Bolt
5. Nut	10. Plain washer	15. Filter element	20. Clip	25. Spring washer

⑩

EXHAUST SYSTEM, SIDE COVERS, AND BATTERY

1. Exhaust pipe
2. Coupling pipe
3. Clip
4. Clip
5. Bolt
6. D-washer
7. Washer
8. Nut
9. Spring
10. Muffler
11. Clip
12. Tailpipe
13. Side cover
14. Side cover
15. Dzus fastener
16. Rubber washer
17. Front fixing bracket
18. Bolt
19. Washer
20. Nut
21. Transfer
22. Battery carrier
23. Locating rubber
24. Support tray rubber
25. Bolt
26. Washer
27. Nut
28. Rubber
29. Battery strap hook
30. Battery strap
31. Battery
32. Vent tube
33. Clip
34. Tool tray
35. Washer
36. Bolt
37. Tab washer
38. Spring plate
39. Support strap
40. Bolt
41. Washer
42. Washer
43. Nut

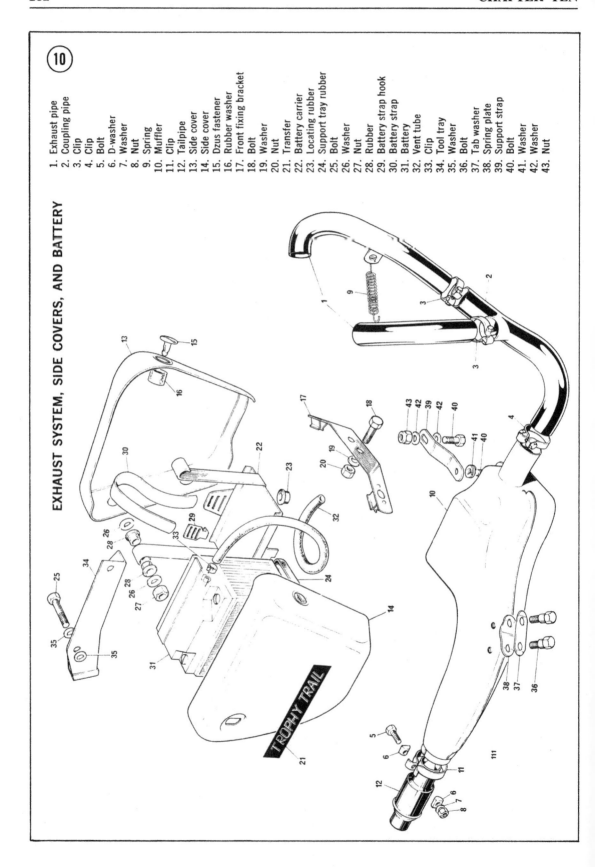

TROPHY TRAIL

APPENDIX

SPECIFICATIONS

Triumph is gradually switching to the Unified thread standard. For older parts, the table below will help determine the thread count for replacements.

THREADS PER INCH

		UNIFIED		WHITWORTH	
Size	Cycle Engineers Institute (C.E.I.)	Unified Fine (UNF)	Unified Coarse (UNC)	British Standard Fine (BSF)	British Standard Whitworth (Coarse) (BSW)
1/4"	26	28	20	26	20
5/16"	26	24	18	22	18
3/8"	26	24	16	20	16
7/16"	26	20	14	18	14
1/2"	20	20	13	16	12
9/16"	20	18	12	16	12
5/8"	20	18	11	14	11

SPECIFICATIONS — 750cc MODELS (T140V and TR7V)
Specifications for all models otherwise stated

LUBRICATION SYSTEM

OIL PUMP

Bore diameter: Feed	.40675/.40625 in. (10.33/10.317mm)
Scavenge	.4877/.4872 in. (12.388/12.375mm)
Plunger diameter: Feed	.40615/.40585 in. (10.315/10.307mm)
Scavenge	.4872/.4869 in. (12.375/12.377mm)
Valve spring length	½ in. (12.70mm)
Ball diameter	7/32 in. (5.556mm)
Aluminum crosshead width	.497/.498 in. (12.624/12.639mm)
Working clearance in plunger heads	.0015/.0045 in. (.038/.124mm)

OIL PRESSURE RELEASE VALVE

Piston diameter	.5605/.5610 in. (14.237/14.249mm)
Working clearance	.001/.002 in. (.0254/.0508mm)
Pressure release operates	60 lb./sq. in. (4.22 kg./sq. cm.)
Spring length	1 17/32 in. (38.894mm)
Load at 1 3/16 in.	12/12½ lbs.
Rate	37 lb./in.

OIL PRESSURE

Normal running	65/80 lb./sq. in. (4.575/.62 kg./sq./sq. cm.)
Idling	20/25 lb./sq. in. (1.406/1.76 kg./sq. cm.)

OIL PRESSURE SWITCH

Operating pressure	7/11 lb./sq. in. (0.492/0.773 kg./sq. cm.)

ENGINE

BASIC DETAILS

Bore and stroke	2.953 in. × 3.228 in. (75 × 82mm)
Cubic capacity	44 cu. in. (724cc)
Compression ratio	8.6 : 1
Bore and stroke (after Engine No. T140V XH 22019 and TR7V AH 24040)	2.992 in. × 3.228 in. (76 × 82mm)
Capacity (after Engine No. T140V XH 22019 and TR7V AH 24040)	45 cu. in. (747cc)

CRANKSHAFT

Crankshaft type	Forged two throw crank with bolt-on flywheel
Main bearing (drive side) size and type	2 13/16 × 1⅛ × 13/16 in. Single lipped roller bearing (71.5 × 28.6 × 20.6mm)
Main bearing (timing side) size and type	2.837 × 1.182 × 0.749. Ball race (72mm × 30mm × 19mm)
Main bearing journal diameter (timing side)	1.1812/1.1808 in. (30/29.99mm)
Main bearing journal diameter (drive side)	1.1247/1.1250 in. (28.567/28.575mm)
Main bearing journal diameter	1.1247/1.1250 in. (28.567/28.575mm)
Big end journal diameter	2.8095/2.8110 in. (71.361/71.399mm)
Minimum regrind diameter	1.6235/1.6240 in. (41.237/41.25mm)
Crankshaft end float	1.6035/1.6040 in. (40.73/40.742mm)
	.003/.017 in. (.0762/.432mm)

CONNECTING RODS

Length (centers)	6.001/5.999 in. (152.4254/152.3746mm)
Bearing side clearance	.012/.016 in. (.305/.406mm)
Bearing diametric clearance	.0005/.0020 in. (.0127/.0508mm)

CYLINDER BLOCK

Bore	See Engine Section
Maximum oversize	+ .040 in. (1.016mm)
Tappet guide block housing diameter	.9990/.9985 in. (25.37/25.362mm)

CYLINDER HEAD

Inlet port size	1.12 in. (28.5mm)
Exhaust port size	1⅜ in. dia. (34.9mm)

VALVES

Stem diameter: Inlet	.3095/.3100 in. (7.86/7.874mm)
Exhaust	.3090/.3095 in. (7.849/7.861mm)
Head diameter: Inlet	1.592/1.596 in. (40.437/40.538mm)
Exhaust	1.434/1.440 in. (36.388mm/36.576mm)

SPECIFICATIONS — 750cc MODELS (T140V and TR7V) — Continued
Specifications for all models otherwise stated

VALVE GUIDES

Bore diameter (inlet and exhaust)	.3127/.3137 in. (7.943/7.958mm)
Outside diameter (inlet and exhaust)	.5005/.5010 in. (12.713/12.725mm)
Length: Inlet	1 31/32 in. (50mm)
Exhaust	2 11/64 in. (55.2mm)

VALVE SPRINGS (RED SPOT INNER)
(GREEN SPOT OUTER)

	Outer	Inner
Free length	1½ in. (38mm)	1 17/32 in. (38.9mm)
Total number of coils	5½	7¼
Total fitted load:	Inlet	Exhaust
Valve open	143 lbs. (64.9 kg.)	155 lbs. (70.3 kg.)
Valve closed	75 lbs. (34 kg.)	87 lbs. (39.5 kg.)
Fitted length (valve closed):		
Inner	1 3/16 in. (30.2mm)	1⅛ in. (28.6mm)
Outer	1 7/32 in. (31mm)	1 5/32 in. (29.4mm)

VALVE TIMING

Checked at nil tappet clearance	
Inlet opens	0.190 in. (4.85mm) at TDC
Exhaust closes	0.130 in. (3.27mm) at TDC

ROCKERS

Bore diameter	.5002/.5012 in. (12.705/12.731mm)
Rocker spindle diameter	.4990/.4995 in. (12.675/12.687mm)
Tappet clearance (cold): Inlet	.008 in. (0.203mm)
Exhaust	.006 in. (0.15mm)

CAMSHAFTS

Journal diameter: Left	.8100/.8105 in. (20.564/20.577mm)
Right	.8730/.8735 in. (22.174/22.187mm)
Diametric clearance: Left	.0010/.0025 in. (.0254/.0635mm)
Right	.0005/.0020 in. (.0127/.0508mm)
End float	.013/.020 in. (.331/.508mm)
Cam lift: Inlet	.347 in. (8.814mm)
Exhaust	.305 in. (7.75mm)
Base circle diameter	.812 in. (20.625mm)

TAPPETS

Tip radius: Inlet	.75 in. (19.1mm)
Exhaust	1.125 in. (28.56mm)
Tappet diameter	.3110/.3115 in. (7.9/7.913mm)
Clearance in guide block	.0005/.0015 in. (.0127/.267mm)

TAPPET GUIDE BLOCK

Diameter of bores	.3120/.3125 in. (7.925/7.938mm)
Outside diameter	1.0000/.9995 in. (25.4/25.387mm)
Interference fit in cylinder block	.0005/.0015 in. (.0127/.267mm)

CAMSHAFT BEARING BUSHINGS

Bore diameter (fitted): Left	.8125/.8135 in. (20.648/20.663mm)
Right	.874/.875 in. (22.1/22.1mm)
Outside diameter: Left	1.0010/1.0015 in. (25.425/25.438mm)
Right	1.126/1.127 in. (28.601/28.628mm)
Length: Left inlet	1.104/1.114 in. (28.042/28.296mm)
Left exhaust	.932/.942 in. (23.637/23.927mm)
Right inlet and exhaust	1.010/1.020 in. (25.025/25.508mm)
Interference fit in crankcase: Left	.001/.002 in. (.025/.051mm)
Right	.0010/.0025 in. (.025/.064mm)

TIMING GEARS

Inlet and exhaust camshaft pinions:	
No. of teeth	50
Interference fit on camshaft	.000/.001 in. (.0254mm)
Intermediate timing gear:	
No. of teeth	47
Bore diameter	.5618/.5625 in. (14.27/14.288mm)
Intermediate timing gear bushing:	
Outside diameter	.5635/.5640 in. (14.313/14.326mm)
Bore diameter	.4990/.4995 in. (12.675/12.687mm)
Length	.6775/.6825 in. (17.209/17.336mm)
Working clearance on spindle	.0005/.0015 in. (.0127/.267mm)
Intermediate wheel spindle:	
Diameter	.4980/.4985 in. (12.649/12.662mm)
Interference fit in crankcase	.0005/.0015 in. (.0127/.267mm)
Crankshaft pinion:	
No. of teeth	25
Fit on crankshaft	.0003/—.0005 in. (.0076/—.0127mm)

11

SPECIFICATIONS — 750cc MODELS (T140V and TR7V) — Continued
Specifications for all models otherwise stated

IGNITION TIMING

Crankshaft position (BTDC)
Static timing .... 14°
Fully advanced .... 38°

Piston position (BTDC)
Static timing .... .060 in. (1.5mm)
Fully advanced .... .415 in. (10.4mm)

Advance range:
Contact breaker .... 12°
Crankshaft .... 24°

CONTACT BREAKER

Gap setting .... .014/.016 in. (.35/.40mm)
Fully advanced at .... 2,000 rpm

SPARK PLUG

Type .... Champion N3
Gap .... .025 in. (.635mm)
Thread size .... 14mm × ¾ in. reach

PISTONS

Clearance: Top of skirt .... See Engine Section
          Bottom of skirt .... See Engine Section
Piston pin hole diameter .... .7502/.7504 in. (19.011/19.162mm)

PISTON RINGS

Compression rings (tapered):
Width .... .121/.113 in. (3.073/2.896mm)
Thickness .... .0625/.0615 in. (1.589/1.563mm)
Fitted gap .... .008/.013 in. (.203/.330mm)
Clearance in groove .... .0035/.0015 in. (.089/.038mm)

Oil control ring:
Thickness .... .125 in. (3.18mm)
Width .... .121 in. (3.073mm)
Fitted gap .... .010/.040 in. (.254/1.016mm)
Clearance groove .... .0015/.0025 in. (.038/.0635mm)

FUEL SYSTEM

	Twin Carburetors	Single Carburetors
Amal type	L930/92 R930/93	R930/89
Main jet size	190	280
Needle jet size	.106	.106
Needle type	STD	STD
Needle position	1	2
Throttle valve:		
Cutaway	3	3½
Carburetor nominal bore size	30mm	30mm
Air cleaner type	Surgical gauze and metal gauze	Surgical gauze and metal gauze

TRANSMISSION

CLUTCH DETAILS

Type .... Multiplate with integral shock absorber
No. of plates: Driving (bonded) 6
              Driven (plain) 6

Pressure springs:
Number 3
Free length .... 1.75 in. (43.5mm)
No. of working coils 7½
Spring rate .... 169 lb./in.
Approximate fitted load .... 83 lb. (37.65 kg.)

Bearing rollers:
Number 20
Diameter .... .2495/.2500 in. (6.337/6.35mm)
Length .... .231/.236 in. (5.831/5.958mm)

Clutch hub bearing diameter .... 1.3733/1.3743 in. (33.882/34.907mm)
Clutch sprocket bore diameter .... 1.8745/1.8755 in. (47.612/47.638mm)
Thrust washer thickness .... .052/.054 in. (1.312/1.372mm)
Engine sprocket teeth 29
Clutch sprocket teeth 58
Chain details .... Triplex endless—⅜ in. pitch × 84 links

CLUTCH OPERATING MECHANISM

Conical spring:
Number of working coils 2
Free length .... 13/32 in. (10.3mm)

SPECIFICATIONS — 750cc MODELS (T140V and TR7V) — Continued

Specifications for all models otherwise stated

Diameter of balls ... $3/8$ in. (9.525mm)
Clutch operating rod:
Diameter of rod ... $7/32$ in. (5.6mm)
Length of rod ... 11.822/11.812 in. (300.279/300.025mm)

KICKSTARTER MECHANISM

Bushing bore diameter ... .751/.752 in. (19.085/19.11mm)
Spindle working clearance in bushing ... .003/.005 in. (.076/.127mm)
Ratchet spring free length ... $1/2$ in. (12.7mm)

GEARSHIFT MECHANISM

Plungers:
Outer diameter ... .4315/.4320 in. (10.92/10.937mm)
Working clearance in bore ... .0005/.0015 in. (.0127/.038mm)
Plunger springs:
No. of working coils ... 12
Free length ... $1\,1/4$ in. (31.75mm)
Inner bushing bore diameter ... .6245/.6255 in. (15.86/15.888mm)
Clearance on shaft ... .0007/.0032 in. (.0178/.081mm)
Outer bushing bore diameter ... .7495/.7505 in. (19.047/19.073mm)
Clearance on shaft ... .0005/.0025 in. (.0127/.064mm)
Quadrant return springs:
No. of working coils ... $9\,1/2$
Free length ... $1\,3/4$ in. (44.5mm)

GEARBOX (5-SPEED)

RATIOS

Primary ratios: 5th ... 1.00 : 1
4th ... 1.19 : 1
3rd ... 1.40 : 1
2nd ... 1.837 : 1
1st ... 2.585 : 1
Overall ratios: 5th ... 4.70
4th ... 5.59
3rd ... 6.58
2nd ... 8.63
1st ... 12.25
Engine rpm @ 10 mph in 5th gear ... 627
Main sprocket teeth ... 20

GEAR DETAILS

Main shaft high gear:
Bearing type ... Needle roller (torrington B1314)
Bearing length ... .875/.865 in. (22.23/21.97mm)
Spigot diameter (high gear) ... 1.5077/1.5072 in. (38.36/38.28mm)

GEARBOX SHAFTS

Main shaft:
Left end diameter ... .8103/.8098 in. (20.58/20.57mm)
Right end diameter ... .7494/.7498 in. (19.044/19.054mm)
Length ... 11.23 in. (285.2mm)
Countershaft:
Left end diameter ... .6875/.6870 in. (17.46/17.404mm)
Right end diameter ... .6875/.6870 in. (17.46/17.404mm)
Length ... 6.47 in. (164.33mm)

BEARINGS

Main shaft bearing (left) ... $1\,1/2 \times 2\,1/2 \times 5/8$ in. Roller bearing (38.1 $\times$ 63.5 $\times$ 15.9mm)
Main shaft bearing (right) ... $3/4 \times 1\,7/8 \times 9/16$ in. Ball journal (19 $\times$ 47.5 $\times$ 14.3mm)
Countershaft bearing (left) ... $11/16 \times 7/8 \times 3/4$ in. Needle roller (17.5 $\times$ 22.23 $\times$ 19mm)
Countershaft bearing (right) ... $11/16 \times 7/8 \times 3/4$ in. Needle roller (17.5 $\times$ 22.23 $\times$ 19mm)
Countershaft 1st gear bushing:
Bore diameter ... .800/.795 in. (20.32/20.203mm)
Shaft diameter ... .8075/.8070 in. (20.511/20.498mm)
Countershaft 2nd gear bushing:
Bore diameter ... .800/.795 in. (20.32/20.203mm)
Shaft diameter ... .8075/.8070 in. (20.511/20.498mm)

FRAME AND ATTACHMENT DETAILS

STEERING HEAD BEARINGS

Type ... Timken taper roller bearing
Bore size ... .7508/.7500 in. (19.08/19.06mm)
Outer diameter ... 1.7820/1.7810 in. (45.27/45.24mm)

11

SPECIFICATIONS – 750cc MODELS (T140V and TR7V) – Continued
Specifications for all models otherwise stated

SWINGING ARM

Bushing bore diameter	1 in. nominal (25.4mm)
Sleeve diameter	.9984/.9972 in. (25.35/25.32mm)
Distance between fork ends	8.018 in. (203.653mm)

REAR SUSPENSION

Type	Swinging fork controlled by combined coil spring/hydraulic damper units
Fitted length	8 in. (203.2mm) at mid-position
Free length	9.5 in. (241.3mm)
Spring rate	88 lbs./in.
Mean coil diameter	1.98 in. (50.29mm)

WHEELS, BRAKES, AND TIRES

WHEELS

Rim size: Front	WM2-19
Rear	WM3-18

REAR WHEEL

Bolt size for detachable sprocket	¼ dia. × 2 in. U.H. (6.35 × 51mm)
Number of bolts	5

REAR WHEEL DRIVE

Gearbox sprocket	See Gearbox
Rear wheel sprocket	47 teeth
Chain details: No. of links	106 links
Pitch	⅝ in. (15.875mm)
Width	⅜ in. (9.525mm)
Speedometer drive gearbox ratio	1.25/1
Speedometer cable length (outer)	66 in. (1676.4mm)
Speedometer cable length (inner)	67.63 in. (minimum) (1717.7mm)

BRAKES

Front: Type	Hydraulically operated disc
Disc diameter	10 in. (254mm)
Friction pad type	Mintex M64
Lining thickness	.25 in. (6.35mm)
Rear: Type	Internal expanding single leading shoe
Lining thickness	.187/.197 in. (4.75/5mm)
Drum diameter	7 in. (+ 0.002 in.) 177.8mm (+ 0.0508mm)

TIRES

Size: Front	3.25 × 19 in. (82.5 × 484.6mm)
Rear	4.00 × 18 in. (101.6 × 457.2mm)
Tire pressure: Front	24 lb./sq. in. (1.685 kg./sq. cm.)
Rear	24 lb./sq. in. (1.685 kg./sq. cm.)

FRONT FORK

TELESCOPIC FORK

Type	Telescopic (hydraulic damping)
Spring details: Free length	19.1 in. (485mm)
Compressed length	11.4 in. (289.5mm)
Fitted length	18.5 in. (217mm)
Spring rate	25 lbs./in.
Maximum load	194 lbs. (88 kg.)
Color identification	Orange
Stanchion diameter: (Top)	1.355/1.350 in. (34.4/34.3mm)
(Bottom)	1.3610/1.3605 in. (33.04/33.03mm)
Outer member bore diameter	1.365/1.363 in. (33.15/33.1mm)

ELECTRICAL SYSTEM

ELECTRICAL EQUIPMENT

Battery type (12 volt)	Lucas PUZ 5A
Rectifier type	Lucas 2DS 506
Alternator type	Lucas RM21
Horn type	Lucas 6H
Zener diode type	Lucas 2D715
Ignition coil type	Lucas 17M12
Ignition switch type	Lucas S45
Handlebar switch type	Lucas 169SA
Flasher unit type	Lucas 8FL
Contact breaker type	Lucas 6CA
Condenser	Lucas 54441582

SPECIFICATIONS — 750cc MODELS (T140V and TR7V) — Continued
Specifications for all models otherwise stated

BULBS (original equipment)

	No.	Type
Headlight	Lucas 370	45/35W pre-focus
Parking light	Lucas 989	6W M.cc
Stop and taillight	Lucas 380	5/21W
High beam, ignition and direction indicator warning lights	Lucas 281	2W (WL.15)
Speedometer, tachometer	Lucas 987	3W MES
Direction indicators	Lucas 382	21W

GENERAL

CAPACITIES

Fuel tank	2 gals. (2½ U.S. gals.)
Oil tank	4 pints (4.8 U.S. pints, 2.27 liters)
Gearbox	⅞ pint (500cc)
Primary chaincase (initial fill only)	⅝ pint (350cc)
Telescopic fork legs	1/3 pint (200cc)

BASIC DIMENSIONS

Wheelbase	55 in. (140cm)
Overall length	87.5 in. (222cm)
Overall width	33 in. (84cm)
Seat height	31½ in. (80cm)

WEIGHTS

Unladen weight	408 lb. (185 kg.) 402 lb. (182 kg.)
Engine unit (dry)	139 lb. (67 kg.) 137 lb. (63 kg.)

TIGHTENING TORQUES (DRY)

Flywheel bolts	33 ft. lb. (4.6 kgm)
Connecting rod bolts	22 ft. lb. (3.9 kgm)
Crankcase junction bolts	13 ft. lb. (1.8 kgm)
Crankcase junction studs	20 ft. lb. (2.8 kgm)
Rocker box bolts—inner (5/16 in. dia.)	10 ft. lb. (1.38 kgm)
Cylinder head bolts—outer (⅜ in. dia.)	18 ft. lb. (2.49 kgm)
Cylinder head bolt—center (5/16 in. dia.)	16 ft. lb. (2.07 kgm)
Cylinder head bolt—inner (⅜ in. dia.)	18 ft. lb. (2.49 kgm)
Rocker box nuts	5 ft. lb. (.7 kgm)
Rocker box bolts (¼ in. diameter)	5 ft. lb. (.7 kgm)
Rocker spindle domed nuts	22 ft. lb. (3.0 kgm)
Oil pump nuts	5 ft. lb. (.7 kgm)
Kickstarter ratchet pinion nut	45 ft. lb. (6.3 kgm)
Clutch center nut	70 ft. lb. (9.7 kgm)
Rotor fixing nut	40 ft. lb. (4.1 kgm)
Stator fixing nuts	20 ft. lb. (2.8 kgm)
Primary cover domed nuts	10 ft. lb. (1.4 kgm)
Headlight pivot bolts	10 ft. lb. (1.4 kgm)
Head race sleeve nut pinch bolt	15 ft. lb. (2.1 kgm)
Stanchion pinch bolts	25 ft. lb. (3.5 kgm)
Front wheel spindle cap bolts	25 ft. lb. (3.5 kgm)
Rear brake drum to hub bolts	15 ft. lb. (2.1 kgm)
Brake cam spindle nuts	20 ft. lb. (2.8 kgm)
Zener diode fixing nut	1.5 ft. lb. (.21 kgm)
Fork cap nut	80 ft. lb. (11.1 kgm)
Brake disc retaining bolts	20 ft. lb. (2.8 kgm)

11

SPECIFICATIONS 650cc MODELS

LUBRICATION SYSTEM

OIL PUMP

Body material	Brass
Bore diameter: Feed	.40625/.40675 in.
Scavenge	.4877/.4872 in.
Plunger diameter: Feed	.40620/.40623 in.
Scavenge	.4872/.4869 in.
Valve spring length	7/32 in.
Ball diameter	½ in.
Aluminum crosshead width	.497/.498 in.
Working clearance in plunger heads	.0015/.0045 in.

OIL PRESSURE RELEASE VALVE

Piston diameter	.5605/.5610 in.
Working clearance	.001/.002 in.
Pressure release operates	60 lb./sq. in. (4.22 kg/sq. cm.)
Spring length	1 17/32 in.
Load at 1 3/16 in.	12/12½ lb.
Rate	37 lb./in.

OIL PRESSURE

Normal running	65/80 lb./sq. in.
Idling	20/25 lb./sq. in.

OIL PRESSURE SWITCH

Operating pressure	7-11 lb./sq. in.

ENGINE

BASIC DETAILS

Bore and stroke	2.795 × 3.228 in. (71 × 82mm)
Cubic capacity	40 cu. in. (649cc)
Compression ratio	9:1
Power output (bhp @ rpm)	47 @ 6,700

CRANKSHAFT

Crankshaft type	Forged two-throw crank with bolt-on flywheel. Located by the timing side main bearing
Main bearing (drive side) size and type	2 13/16 × 1⅛ × 13/16 in. Single lipped roller bearing
Main bearing (timing side) size and type	2 13/16 × 1⅛ × 13/16 in.
Main bearing journal diameter	1.1247/1.1250 in.
Main bearing housing diameter	2.8095/2.8110 in.
Big end journal diameter	1.6235/1.6240 in.
Minimum regrind diameter	1.6035/1.6040 in.
Crankshaft end float	.003/.017 in.

TIMING SIDE MAIN BEARING (After Engine No. GE27209)

Main bearing, size and type	72×30×19mm
Main bearing, journal diameter	1.1812/1.1808 in. (30/29.99mm)

CONNECTING RODS

Length (centers)	6.499/6.501 in.
Big end bearings—type	Steel backed white metal
Bearing side clearance	.012/.016 in.
Bearing diametric clearance	.0005/.0020 in.

PISTON PIN

Material	High tensile steel
Fit in small end bushing	.0005/.0012 in. clearance
Diameter	.6882/.6885 in.
Length	2.151/2.156 in.

SMALL END BUSHING

Material	Phosphor bronze
Outer diameter	.8140/.8145 in.
Length	1.030/1.031 in.
Finished bore diameter	.6890/.6894 in.

CYLINDER BLOCK

Material	Cast iron
Bore size	2.7948/2.7953 in.
Maximum oversize	2.8348/2.8353 in.
Tappet guide block housing diameter	.999/.9985 in.

CYLINDER HEAD

Material	D.T.D. 424 Aluminum
Intake port size	1 3/16 in. dia. tapering to 1⅛ in.
Exhaust port size	1⅜ in. diameter
Valve seatings:	
Type	Cast-in
Material	Cast iron

VALVES

Stem diameter: Intake	.3095/.3100 in.
Exhaust	.3090/.3095 in.
Head diameter: Intake	1.592/1.596 in.
Exhaust	1.434/1.440 in.
Exhaust valve material	21/4NS

SPECIFICATIONS 650cc MODELS (continued)

VALVE GUIDES

Material	Aluminum—Bronze
Bore diameter (intake and exhaust)	.3127/.3137 in.
Outside diameter (intake and exhaust)	.5005/.5010 in.
Length: Intake	1 31/32 in.
Exhaust	2 11/64 in.

VALVE SPRINGS (RED SPOT INNER)
(GREEN SPOT OUTER)

	Outer	Inner
Free length	1½ in.	1 17/32 in.
Total number of coils	5¾	7¼

	Intake	Exhaust
Total fitted load:		
Valve open	143 lbs.	155 lbs.
Valve closed	75 lbs.	87 lbs.
Fitted length (valve closed):		
Inner	1 3/16 in.	1⅛ in.
Outer	17/32 in.	1 5/32 in.

VALVE TIMING

Intake opens	34° B.T.D.C.
Intake closes	55° A.B.D.C.
Exhaust opens	55° B.B.D.C.
Exhaust closes	34° A.T.D.C.

Set all tappet clearances @ .002 in. (.5mm) for checking

ROCKERS

Material	High tensile steel forging
Bore diameter	.5002/.5012 in.
Rocker spindle diameter	.4990/.4995 in.
Tappet clearance (cold): Intake	.002 in. (.05mm)
Exhaust	.004 in. (.10mm)

CAMSHAFTS

Journal diameter: Left	.8100/.8105 in.
Right	.8730/.8735 in.
Diametric clearance: Left	.0010/.0025 in.
Right	.0005/.0020 in.
End float	.013/.020 in.
Cam lift: Intake and exhaust	.314 in.
Base circle diameter	.812 in.

TAPPETS

Material	High tensile steel body—Stellite tip
Tip radius	1.125 in.
Tappet diameter	.3110/.3115 in.
Clearance in guide block	.0005/.0015 in.

Exhaust tappet has flat, 3/32 in. tall, 1.884 in./1.881 in. below top of tappet, flat is .032 in. deep, and oil hole is .047 in. diameter.

TAPPET GUIDE BLOCK

Diameter of bores	.3120/.3125 in.
Outside diameter	1.0000/.9995 in.
Interference fit in cylinder block	.0005/.0015 in.

CAMSHAFT BEARING BUSHINGS

Material	High density sintered bronze
Bore diameter (fitted): Left	.8125/.8135 in.
Right	.874/.875 in.
Outside diameter: Left	1.0010/1.0015 in.
Right	1.126/1.127 in.
Length: Left intake	1.104/1.114 in.
Left exhaust	.932/.942 in.
Right intake and exhaust	1.010/1.020 in.
Interference fit in crankcase: Left	.001/.002 in.
Right	.0010/.0025 in.

TIMING GEARS

Intake and exhaust camshaft pinions:	
No. of teeth	50
Interference fit on camshaft	.000/.001 in.
Intermediate timing gear:	
No. of teeth	47
Bore diameter	.5618/.5625 in.
Intermediate timing gear bushing:	
Material	Phosphor bronze
Outside diameter	.5635/.5640 in.
Bore diameter	.4990/.4995 in.
Length	.6775/.6825 in.
Working clearance on spindle	.0005/.0015 in.
Intermediate wheel spindle:	
Diameter	.4980/.4985 in.
Interference fit in crankcase	.0005/.0015 in.
Crankshaft pinion:	
No. of teeth	25
Fit on crankshaft	+.0003/−.0005 in.

11

SPECIFICATIONS 650cc MODELS (continued)

IGNITION TIMING

Crankshaft position (B.T.D.C.):
Static timing .... 14°
Fully advanced .... 38°
Piston position (B.T.D.C.):
Static timing .... .060 in. (1.5mm)
Fully advanced .... .415 in. (10.4mm)
Advance range:
Contact breaker .... 12°
Crankshaft .... 24°

POINTS

Gap setting .... .014-.016 in. (.35-.40mm)
Fully advanced at .... 2,000 rpm

SPARK PLUG

Type .... Champion N3
Gap setting .... .025 in. (.635mm)
Thread size .... 14mm × ¾ in. reach

PISTONS

Material .... Aluminum Alloy—diecasting
Clearance: Top of skirt .... .0106/.0085 in.
                    Bottom of skirt .... .0061/.0046 in.
Piston pin hole dia. .... .6882/.6886 in.

PISTON RINGS

Material .... Cast iron
Compression rings (tapered):
Width .... .0615/.0625 in.
Thickness .... .092/.100 in.
Fitted gap .... .010/.014 in.
Clearance in groove .... .001/.003 in.
Oil control ring:
Width .... .092/.100 in.
Thickness .... .124/.125 in.
Fitted gap .... .010/.014 in.
Clearance in groove .... .0005/.0025 in.

FUEL SYSTEM

Twin carburetors

Amal type	Concentric float R930/9 and L930/10	Monobloc
Main jet size	220	206
Pilot jet size	—	25
Needle jet size	.106	.106
Needle type	STD	D
Needle position	2	3
Throttle valve: Type	2½	389/95
Return spring free length	2½ in.	2½ in.
Carburetor nominal bore size	30mm	1⅛ in. dia.
Air cleaner type	Not fitted	Not fitted

TRANSMISSION

CLUTCH DETAILS

Type .... Multiplate with integral shock absorber
No. of plates: Driving (bonded) 6
                       Driven (plain) 6
Pressure springs:
Number 3
Free length .... 1 13/16 in.
No. of working coils 9½
Spring rate .... 113 lbs./in.
Approximate fitted load .... 62 lbs.
Bearing rollers:
Number 20
Diameter .... .2495/.2500 in.
Length .... .231/.236 in.
Clutch hub bearing diameter .... 1.3733/1.3743 in.
Clutch sprocket bore diameter .... 1.8745/1.8755 in.
Thrust washer thickness .... .052/.054 in.
Engine sprocket teeth 29
Clutch sprocket teeth 58
Chain details .... Duplex endless—⅜ in. pitch × 84 links

CLUTCH OPERATING MECHANISM

Conical spring:
Number of working coils 2
Free length .... 13/32 in.
Diameter of balls .... ⅜ in.
Clutch operating rod:
Diameter of rod .... 7/32 in.
Length of rod .... 11.822/11.812 in.

SPECIFICATIONS 650cc MODELS (continued)

GEARBOX

RATIOS

Internal ratios (Std.) 4th (Top) .... 1.00 : 1
3rd .... 1.19 : 1
2nd .... 1.69 : 1
1st (Bottom) .... 2.44 : 1

	Solo	Sidecar
Overall ratios: 4th (Top)	4.84	5.41
3rd	5.76	6.44
2nd	8.17	9.15
1st (Bottom)	11.8	13.4
Engine rpm @ 10 mph in 4th (Top) gear	648	725
Gearbox sprocket teeth	19	17

GEAR DETAILS

Mainshaft high gear:
Bore diameter (bushing fitted) .... .8135/.8145 in.
Working clearance on shaft .... .0032/.0047 in.
Bushing length .... 2 19/32 in.
Bushing protrusion length .... 7/16 in. None after DU66246

Layshaft low gear:
Bore diameter (bushing fitted) .... .8135/.8145 in.
Working clearance on shaft .... .0025/.0045 in.

GEARBOX SHAFTS

Mainshaft:
Left end diameter .... .8098/.8103 in.
Right end diameter .... .7494/.7498 in.
Length .... 10 63/64 in. (up to DU48143) 11 19/64 in. (DU48144 onwards)

Layshaft:
Left end diameter .... .6845/.6850 in.
Right end diameter .... .6845/.6850 in.
Length .... 6 31/64 in.
Camplate plunger spring:
Free length .... 2½ in.
No. of working coils 22
Spring rate .... 5.6 lb./in.

BEARINGS

High gear bearing .... 1¼ × 2½ × ⅝ in. Ball Journal
Mainshaft bearing .... ¾ × 1⅞ × 9/16 in. Ball Journal
Layshaft bearing (left) .... 11/16 × ⅞ × ¾ in. Needle Roller
Layshaft bearing (right) .... 11/16 × ⅞ × ¾ in. Needle Roller

KICKSTART OPERATING MECHANISM

Bushing bore diameter .... .751/.752 in.
Spindle working clearance in bushing .... .003/.005 in.
Ratchet spring free length .... ½ in.

GEARCHANGE MECHANISM

Plungers:
Outer diameter .... .4315/.4320 in.
Working clearance in bore .... .0005/.0015 in.
Plunger springs:
No. of working coils 12
Free length .... 1¼ in.
Inner bushing bore diameter .... .6245/.6255 in.
Clearance on shaft .... .0007/.0032 in.
Outer bushing bore diameter .... .7495/.7505 in.
Clearance on shaft .... .0005/.0025 in.
Quadrant return springs:
No. of working coils 9½
Free length .... 1¾ ins.

FRAME AND ATTACHMENT DETAILS

HEAD RACES

No. of balls: Top 20
Bottom 20
Ball diameter .... ¼ in.

SWINGING FORK

Bushing type .... Pre-sized, steel-backed— phosphor bronze
Bushing bore diameter .... 1.4460/1.4470 in.
Sleeve diameter .... 1.4445/1.4450 in.
Distance between fork ends .... 7½ in.

11

SPECIFICATIONS 650cc MODELS (continued)

REAR SUSPENSION

Type .... Swinging fork controlled by combined coil spring/hydraulic damper units

Spring details:
Fitted length .... 8 in.
Free length .... 8 3/16 in.
Mean coil diameter .... 1 3/4 in.
Spring rate .... 145 lbs./in.
Color code .... Blue/yellow
Load at fitted length .... 38 lb.

WHEELS, BRAKES AND TIRES

WHEELS
Rim size: Front and rear ...... WM2-19 front, WM2-18 rear
Type: Front .... Spoke—single cross lacing
Rear .... Spoke—double cross lacing
Spoke details: Front: Left side ... 20 off 8/10 SWG butted 5⅝ in. U.H. straight
Right side .... 10 off 8/10 SWG butted 4 25/32 in. U.H. 78° head
Right side .... 10 off 8/10 SWG butted 4⅞ in. U.H. 100° head
Rear: Left side .... 20 off 8/10 SWG butted 7 9/16 in. U.H. 90° head
Right side .... 20 off 8/10 SWG butted 7⅞ in. U.H. 90° head

WHEEL BEARINGS
Front and rear, dimensions and type ... 20 × 47 × 14mm—Ball Journal
Front and rear, spindle dia. (at bearing journals) ... .7862/.7867 in.

STANDARD REAR WHEEL
Bolt size for detachable sprocket ... ¼ in. dia. × 13/16 in. U.H. × 26 C.E.I.
Number of bolts 8

Q.D. REAR WHEEL
Bearing type .... ¾ × 1⅞ × 9/16 in. Ball Journal
Bearing sleeve: journal diameter ... .7500/.7495 in.
Brake drum bearing .... ⅞ × 2 × 9/16 in. Ball Journal
Bearing sleeve: journal diameter ... .8745/.8740 in.
Bearing housing: internal diameter ... 1.9890/1.9980 in.

REAR WHEEL DRIVE
Gearbox sprocket .... See "Gearbox"
Rear wheel sprocket teeth 46
Chain details:
No. of links 104
Pitch .... ⅝ in.
Width .... ⅜ in.
Speedometer drive gearbox ratio ... 2:1
Speedometer cable length ..... 65 ins.

BRAKES
Type .... Internal expanding twin leading shoes
Drum Diameter: Front .... 8 in. } ± .002 in.
Rear .... 7 in. }
Lining thickness: Front .... .183/.193 in.
Rear .... .177/.187 in.
Lining area: Front .... 24.4 sq. in.
Rear .... 14.6 sq. in.

TIRES
Size: Front .... 3.25 × 19 in.
Rear .... 3.50 × 18 in.
Tire pressure: Front .... 24 lb./sq. in. (1.685 Kg/sq. cm.)
Rear .... 24 lb./sq. in. (1.685 Kg/sq. cm.)

FRONT FORKS

TELESCOPIC FORK
Type .... Telescopic—Shuttle valve damping

Spring details:

	Solo	Sidecar
Free length	9¾ in.	9¾ in.
No. working coils	12½	15½
Spring rate	26½ lb. in.	32½ lb. in.
Gauge	6 SWG	5 SWG
Color code	Yellow/blue	Yellow/green

Damper sleeve
Length .... 2⅛ in.
Internal diameter .... 1.387–1.393 in.
Material .... Black polypropylene

Bushing details:

	Top bush	Bottom bush
Length	.870/.875 in.	
Outer diameter	1.498/1.499 in.	1.4935/1.4945 in.
Inner diameter	1.3065/1.3075 in.	1.2485/1.2495 in.

SPECIFICATIONS 650cc MODELS (continued)

TELESCOPIC FORK (continued)

Stanchion diameter	1.3025/1.3030 in.
Working clearance in top bushing	.0035/.0050 in.
Fork leg bore diameter	1.498/1.500 in.
Working clearance of bottom bushing	.0035/.0065 in.
Shuttle valve:	
Outer diameter (large)	1.018/1.016 in.
Outer diameter (small)	0.875/0.874 in.

ELECTRICAL SYSTEM

ELECTRICAL EQUIPMENT

Battery type (12v.)	PUZ 5A
Rectifier type	2DS 506
Alternator type	RM.19
Horn type (12v.)	6H

Bulbs:

	No.	Type
Headlight (L/H dip)	414	50/40 watts—pre-focus
Parking light	989	6 watts—MCC
Stop and tail light	380	6/21 watts—offset pin
Speedometer light	987	3 watts—MES
Ignition warning light	281	2 watts (BA 7S)
High beam indicator light	281	2 watts (BA 7S)

Zener diode type	ZD 715
Coil type (2 off)	MA12 (12v.) 2 off
Contact breaker type	6CA
Fuse rating	35 amp.

GENERAL

CAPACITIES

Fuel tank	4 gal. (4.8 U.S. gals., 18 liters)
Oil tank	6 pint (7¼ U.S. pints, 3 liters)
Gearbox	⅞ pint (500cc)
Primary chaincase	⅝ pint (350cc)
Telescopic fork legs	⅓ pint (200cc)

BASIC DIMENSIONS

Wheel base	55 in. (140 cm)
Overall length	84 in. (214 cm)
Overall width	27½ in. (70 cm)
Overall height	38 in. (97 cm)
Ground clearance	5 in. (13 cm)

WEIGHTS

Empty weight	365 lb. (166 kgm)
Engine unit (dry)	130 lb. (59 kgm)

TORQUE WRENCH SETTINGS (DRY)

Flywheel bolts	33 lb. ft. (4.6 kgm)
Conn. rod bolts	28 lb. ft. (3.9 kgm)
Crankcase junction bolts	13 lb. ft. (1.8 kgm)
Crankcase junction studs	20 lb. ft. (2.8 kgm)
Cylinder block nuts	35 lb. ft. (4.8 kgm)
Cylinder head bolts (⅜ in. dia.)	18 lb. ft. (2.49 kgm)
Cylinder head bolt (5/16 in. dia.)	15 lb. ft. (2.1 kgm)
Rocker box nuts	5 lb. ft. (.7 kgm)
Rocker box bolts	5 lb. ft. (.7 kgm)
Rocker spindle domed nuts	22 lb. ft. (3.0 kgm)
Oil pump nuts	5 lb. ft. (.7 kgm)
Kickstart ratchet pinion nut	45 lb. ft. (6.3 kgm)
Clutch center nut	50 lb. ft. (.7 kgm)
Rotor fixing nut	30 lb. ft. (4.1 kgm)
Stator fixing nuts	20 lb. ft. (2.8 kgm)
Primary cover domed nuts	10 lb. ft. (1.4 kgm)
Headlamp pivot bolts	10 lb. ft. (1.4 kgm)
Headrace sleeve nut pinch bolt	15 lb. ft. (2.1 kgm)
Stanchion pinch bolts	25 lb. ft. (3.5 kgm)
Front wheel spindle cap bolts	25 lb. ft. (3.5 kgm)
Brake cam spindle nuts	20 lb. ft. (2.8 kgm)
Zener diode fixing nut	1.5 lb. ft. (.21 kgm)
Fork cap nut	80 lb. ft. (11.1 kgm)

N.B.—18 lb. ft. replaces earlier recommendation of 25 lb. ft. for ⅜ in. dia. cylinder head bolts for all unit construction 650cc models.

11

MISCELLANEOUS INFORMATION PRIOR TO ENGINE NUMBER DU.66246

LUBRICATION SYSTEM

OIL PUMP (Before DU.44394 all models)

Scavenge bore diameter .... .4372/.4377 in.
Scavenge plunger diameter .... .4369/.4372 in.

OIL PUMP (DU.44394 to DU.66245)

Scavenge bore diameter .... .4877/.4872 in.
Scavenge plunger diameter .... .4872/.4869 in.
Feed bore diameter .... .3748/.3753 in.
Feed plunger diameter .... .3747/.3744 in.

OIL PRESSURE RELEASE VALVE (with indicator button prior to DU.13375)

Indicator spring length (free) .... 9/32 in.
Release spring length (free) .... 31/32 in.

BASIC ENGINE DETAILS

	Compression ratio	Combustion chamber capacity	Power output (bhp @ rpm)
6T up to DU.44393	7.5 : 1	50cc	37 @ 6,700
TR6 up to DU.44393	8.5 : 1	43.3cc	40 @ 6,500
T120 up to DU.24874	8.5 : 1	43.3cc	46 @ 6,500

MAIN BEARING (All models up to DU.24875)

Both ball journal size ... 2 13/16 × 1 1/8 × 13/16 in. ball journal

CRANKSHAFT LOCATION

All models DU.101 to DU.13374 .... Located to timing side
All models DU.13375 to DU.24874 ..... Located to drive side
All models DU.24875 onwards .... Located to timing side

FLYWHEEL (prior to DU.24875)

Weight .... 8½ lbs.

PISTON CLEARANCES (Triumph pistons) before DU.44394

	6T (DU.101-DU.5824)	6T (DU.5825-DU.44393)	TR6
Top of skirt	.0088/.0098 in.	.0046/.0057 in.	.0088/.0098 in.
Bottom of skirt	.0033/.0043 in.	.0016/.0027 in.	.0033/.0043 in.

	T120	T120TT (11 : 1 CR)
Top of skirt	.0093/.0103 in.	.0093/.0103 in.
Bottom of skirt	.0038/.0048 in.	.0073/.0083 in.

VALVES

Sizes

	6T & TR6 DU.101-DU.44393	T120 DU.101-DU.5824	T120 DU.5824 onwards
Intake	1½ in.	1½ in.	1 19/32 in.
Exhaust	1 11/32 in.	1 11/32 in.	1 7/16 in.

VALVE SPRINGS

6T DU.101 onwards .... Red spot
TR6 DU.101 onwards .... Red spot
T120 DU.101 to DU.24874 .... White spot
T120 DU.24875 to DU.44394 .... Red spot

VALVE SPRINGS (RED SPOT)

	Outer	Inner
Free length	1⅝ in.	1 17/32 in.
Total number of coils	5½	7¼
Total fitted load valve open	125 lbs.	
Total fitted load valve closed	50 lbs.	

VALVE SPRINGS (WHITE SPOT)

	Outer	Inner
Free length	2 1/32 in.	1⅝ in.
Total number of coils	6½	7
Total fitted load valve open	130 lbs.	
Total fitted load valve closed	77 lbs.	

VALVE TIMING

NOTE:
Set all tappets to 0.020 in. for checking

	6T (DU.101 onwards)	TR6 (DU.101-DU.44393)	T120 (DU.101-DU.24874)	T120 (certain later machines series DU.24874 and DU.44393)
Intake opens	25° BTDC	34° BTDC	34° BTDC	34° BTDC
Intake closes	52° ABDC	55° ABDC	55° ABDC	55° ABDC
Exhaust opens	60° BBDC	48° BBDC	48° BBDC	55° BBDC
Exhaust closes	17° ABDC	27° ABDC	27° ABDC	34° ABDC

MISCELLANEOUS INFORMATION PRIOR TO ENGINE NUMBER DU.66246 (continued)

TAPPETS (T120 prior to DU.24875)

Tip radius .... .750 in.

CAMSHAFT

	6T	TR6	T120
Cam lift	.305 in.	.314 in. Intake / .296 in Exhaust	.314 in. Intake / .296 in Exhaust

Certain later T120 between DU.24875 and DU.44393 have .314 in. both Intake and Exhaust.

IGNITION TIMING A.C. Magneto (E.T.) Ignition equipment

Crankshaft position (BTDC)
- Static timing .... 29°
- Fully advanced .... 39°

Piston position (BTDC)
- Static timing .... 1/4 in. (6.3mm)
- Fully advanced .... 7/16 in. (11.5mm)

Advance range:
- Contact breaker .... 5°
- Crankshaft .... 10°

IGNITION TIMING

	6T	TR6	T120
Static Crankshaft position DU.101-DU.5824	7°	10°	10°
Static Crankshaft position DU.5825-onwards	11°	14°	15°

FULLY ADVANCED

	6T	TR6	T120
Static Piston position DU.101-DU.5824	35°	39°	39°
Static Piston position DU.5825-onwards	1/64 in. / 1/32 in.	1/32 in. / 1/16 in.	1/32 in. / 1/16 in.

Advance range:

	6T	TR6	T120
Contact breaker Up to DU.5824	14°	14°	14°
Crankshaft	28°	28°	28°
Contact breaker From DU.5825	12°	12°	12°
Crankshaft	24°	24°	24°

CARBURETOR

	6T DU.101 onwards	TR6 DU.101-DU.5824	TR6 DU.5825-DU.44393	T120 DU.101-DU.5824	T120 DU.5825-DU.66246
Type	376	376	389	376	389
Main jet	230	250	310	240	260
Needle jet	.106	.106	.106	.106	.106
Needle type	C	C	D	C	D
Needle position	3	3	1	2	3
Throttle valve	376/4	376/3½	389/3½	376/3½	389/3
Pilot jet	25	25	25	25	25
Bore	1 1/16 in.	1 1/16 in	1 1/8 in.	1 1/16 in.	1 1/8 in.

GEARBOX

Mainshaft length (prior to DU.44394) .... 10 63/64 in.

Gearbox sprocket teeth number:
- 6T up to DU.44393 .... 20 teeth solo, 18 teeth sidecar
- T120 and TR6 DU.101 onwards .... 19 teeth solo, 17 teeth sidecar

Speedometer gear ratio (prior to DU.24875) .... solo 1.50:1, sidecar 1-67:1

FRONT BRAKE (prior to DU.24875)

Lining area .... 16.2 sq. in.

REAR CHAIN

All models before DU.44394 .... 103 links
USA T120TT only .... 102 links

WHEELS

Front wheel 6T and T120 DU.101-DU.44393
- Size .... 3.25 × 18
- Spoke details: Left .... 20 off 8/10 SWG × 5 1/16 in. straight
  - Right .... 10 off 8/10 SWG × 4 5/16 in. 78° head
  - Right .... 10 off 8/10 SWG × 4 5/16 in. 100° head

Q.D. Rear wheel prior to DU.24875 rim size .... WM3 × 18

Q.D. REAR WHEEL (Before Eng. No. DU.13375)

Bearing type .... .75 × 1.8504 × .566 in. Taper Roller

Bearing sleeve: journal diameter .... .7500/.7495 in.

Brake drum bearing .... 7/8 × 2 × 9/16 in. Ball Journal

Bearing sleve: journal diameter .... 8745/8740 in.

Bearing housing: internal diameter .... 1.9990/1.9980 in.

TIRES

Size: Front 6T and T120 .... 3.25 × 18 in.
- Rear TR6 only .... 4.00 × 18 in.

Tire pressures: Front .... 20 lb./sq. in. (1.4 kg./sq. cm.)
- Rear .... 18 lb./sq. in. (1.3 kg./sq. cm.)

11

MISCELLANEOUS INFORMATION PRIOR TO ENGINE NUMBER DU.66246   (continued)

6-VOLT MODELS (Eng. No. DU.101-DU.5824)

Battery type	6v.—MLZ9E or ML9E
Rectifier type	Lucas 2DS506
Alternator type	Lucas RM.19
Horn	8H (6v.)

Bulbs (6v.)

	No.	Type
Headlight	Lucas 373	30/24 watts, pre-focus
Parking light	Lucas 988	6 watts—MCC
Stop and tail light	Lucas 384	6/18 watts, offset
Speedometer light	Smiths P52305	2 watts special
Coil type		Lucas MA6 (6v.)
Contact breaker type		Lucas 4CA (14°)

12-VOLT MODELS (Eng. No. DU.5825 to DU.24874)

Battery type	MK9E, 6v.—2 in series
Rectifier type	Lucas 2DS506
Alternator type	Lucas RM.19
Horn	8H, 12v.

Bulbs

	No.	Type
Headlight	Lucas 414	50/40 watts, pre-focus
Parking light	Lucas 222	4 watts—MCC
Stop and tail light	Lucas 380	6/21 watts, offset pin
Speedometer light	Lucas 987	2 watts MES
Zener diode type	ZD.715	
Coil type		Lucas MA.12 (12v.) 2 off
Contact breaker type		Lucas 4CA (12°)
Fuse rating		25 amp.

FRONT FORKS

Engine Numbers and Model

	DU.101 - DU.5824			DU.5825 - DU.13374		DU.13375 - up to DU.66245	
	6T	TR6	6T	6T	TR6 & T120	6T	TR6 & T120
STANCHIONS							
Part Number	H1123	H1299	H1595	H1695	H1649	H1890	H1889
Length	20⅞ in.	22 3/16 in.	20⅞ in.	22 in.	22 in.	22 in.	22 in.
One 3/16 in. Filler Hole	Yes	No	No	Yes	No	Yes	No
Damping Holes:— Level from Bottom—5⅝ in.							
—4½ in.	2 No. 42 holes	2 No. 42 holes	2 No. 42 holes	2 3/16 in.	2 3/16 in.	—	—
—3 1/16 in.	—	—	—	2 3/16 in.	2 3/16 in.	2 3/16 in.	2 3/16 in.
—2 5/16 in.	4 3/16 in.	4 3/16 in.	4 3/16 in.	2 3/16 in.	2 3/16 in.	2 3/16 in.	2 3/16 in.
—1⅞ in.	2 No. 42	2 No. 42	2 No. 42	2 No. 42	2 No. 42	2 No. 42	2 No. 42
	To use 6T type stanchion where appropriate for TR6 or T120, solder up the filler hole. **Note:**—No. 42 hole is 0.0935 in.						
DAMPER SLEEVE							
Length	4 5/16 in.	3 11/16 in.	4 5/16 in.	3⅛ in.	3⅛ in.	2⅛ in.	2⅛ in.
Internal Diameter	1.387/1.393 ins.						

Variations among models sold in the U.S.

FRONT FORKS

TELESCOPIC FORK

Type: Telescopic—Oil Damping

	Eng. No. DU.101 to DU.5824	Eng. No. DU.5825 to DU.66245	Eng. No. DU.13375 onwards
Spring details: Solo			
Free length	17 1/16 in.	8¾ in.	9¾ in.
No. of working coils	52	13	12½
Spring rate	32 lb./in.	30 lb./in.	26½ lb./in.
Color code	Black/Green	Unpainted	Yellow/blue

Bushing details: Material .... Sintered bronze

	Top bushing	Bottom bushing
Length	1 in.	.870/.875 in.
Outer diameter	1.498/1.499 in.	1.4935/1.4945 in.
Inner diameter	1.3065/1.3075 in.	1.2485/1.2495 in.
Stanchion diameter	1.3025/1.3030 in.	
Working clearance in top bushing	.0035/.0050 in.	
Fork leg bore diameter	1.498/1.500 in.	
Working clearance of bottom bushing	.0035/.0065 in.	

CAPACITIES

Fork legs: DU.101-DU.5824 .... ¼ pt. (150cc)
DU.5825-DU.66245 .... 1/3 pt. (190cc)
Oil tank: DU.101-DU.24874 .... 5 pt. (6 U.S. pints; 3 liters)

ENGINE

Compression ratio T120TT .... 11 : 1
Power output T120TT .... 54 @ 6,500

CYLINDER HEAD T120TT

Intake port size .... 1 3/16 dia. tapering to 1⅛ in.
Exhaust port size .... 1⅜ in. dia.
Carburetor adaptors .... 1 3/16 in. dia.

CARBURETORS

	T120TT
Choke size	1 3/16 in.
Type	389/95
Main jet size	330
Pilot jet size	25
Needle jet size	.106
Needle type	D
Needle position	2
Throttle valve	389/4

AIR CLEANER

T120R and T120TT .... Coarse felt
TR6R and TR6C .... Cloth

SPARK PLUGS

T120TT .... Champion N58R

IGNITION TIMING (A.C. MAGNETO) TR6C AND T120TT (Up to DU.66245)

Crankshaft position (BTDC)
Static timing .... 29°
Fully advanced .... 39°
Piston position (BTDC)
Static timing .... ¼ in. (6.3mm)
Fully advanced .... 7/16 in. (11.5mm)
Advance range:
Contact breaker .... 5°
Crankshaft .... 10°

CONTACT BREAKER (A.C. MAGNETO) TR6C and T120TT (Up to DU.66245)

Gap setting .... .014/.016 in. (.35/.40mm)
Advance range .... 5°
Fully advanced at .... 2,000 rpm

TRANSMISSION

REAR CHAIN

T120TT .... 103 links

GEARBOX SPROCKET

TR6R .... 19 teeth
TR6C .... 18 teeth
T120R .... 19 teeth
T120TT .... 17 teeth

11

Variations among models sold in the U.S. (continued)

FRAME

FUEL TANK

TR6R	3½ gals. (4.117 U.S. gals.)
TR6C, T120R, T120TT, TR6R	2½ gals. (2.912 U.S. gals.)

SUSPENSION UNITS TR6C, TR6R, T120R, T120TT

Spring details:

Fitted length	8⅜ in.
Free length	8⅝ in.
Mean coil diameter	1¾ in.
Spring rate	100 lb./in.
Color code	Green/Green
Load at fitted length	28 lb.

WHEELS

REAR WHEEL NON-Q.D.

Rim size	WM3-18
Tire size	4.00 × 18 in.
Security bolt	WM3

FORKS

TR6C and T120TT only (Up to DU.66245)

Oil restrictor assembly:

Rod diameter	.3115/.3125 in.
Cup outer diameter	.5845/.5855 in.
Body bore diameter	.590/.592 in.

N.B. With this hydraulic damping unit it is essential that SAE 20 oil only is used.

ELECTRICAL

TR6C and T120TT (Up to DU.66245)

Alternator type	RM19E.T.
Horn, type	Clear hooter, A.C. 585, S.G.
Coil type	3 E.T.
Condensers (Capacitors)	54441582
Contact breaker type	Lucas 4CA

Lighting system:

	No.	Type
Bulbs (6v.):		
Headlight	Lucas 166	24/24 watts—Pre-focus
Stop and tail light	Lucas 384	6/18 watts—Offset pin
Tail lamp type		L679
Kill button type		151.SA
		SS5 up to DU.66245

GEARBOX (4-SPEED)

RATIOS

Primary ratios (Std.)		
4th		1.00 : 1
3rd		1.258
2nd		1.71
1st		2.47

Overall ratios:		
4th		4.95
3rd		6.15
2nd		8.36
1st		12.10

Engine rpm @ 10 mph in 4th gear	660
Gearbox sprocket teeth	19 (18 for TR6C only)

VARIATIONS – 1972 MODELS

FUEL SYSTEM

	Twin Carburetors	Single Carburetor	Single Carburetor (Early models)
Amal type	R930/66, L930/67	R930/60	930/65
Main jet size	180	230	210
Needle jet size	.106	.106	.160
Needle type	STD	STD	STD
Needle position	1	2	2
Throttle valve:			
Type	3	3½	3½
Return spring free length	2½ in. (63.5mm)	2½ in. (63.5mm)	2½ in. (63.5mm)
Carburetor nominal bore size	30mm	30mm	30mm
Air cleaner type (where fitted)	Surgical gauze and metal gauze	Surgical gauze and metal gauze	Surgical gauze and metal gauze

VARIATIONS – 1972 MODELS – continued

GEARBOX (5-SPEED)

RATIOS

Primary ratios:		
5th	....	1.00 : 1
4th	....	1.19 : 1
3rd	....	1.40 : 1
2nd	....	1.837 : 1
1st	....	2.585 : 1

Overall ratios:		
5th	....	4.95
4th	....	5.89
3rd	....	6.92
2nd	....	9.10
1st	....	12.78

Engine rpm @ 10 mph in 5th gear	....	659
Gearbox sprocket teeth	....	19

GEAR DETAILS

Main shaft high gear:
Bearing type	....	Needle roller (torrington B1314)
Bearing length	....	.875/.865 in. (22.23/21.97mm)
Spigot diameter (high gear)	....	1.5077/1.5072 in. (38.36/38.28mm)

GEARBOX SHAFTS

Main shaft:
Left end diameter	....	.8103/.8098 in. (20.58/20.57mm)
Right end diameter	....	.7494/.7498 in. (19.044/19.054mm)
Length	....	11.23 in. (285.2mm)

Countershaft:
Left end diameter	....	.6875/.6870 in. (17.46/17.404mm)
Right end diameter	....	.6875/.6870 in. (17.46/17.404mm)
Length	....	6.47 in. (164.33mm)

BEARINGS

Main shaft bearing (left)	....	1½ × 2½ × ⅝ in. Roller bearing (38.1 × 63.5 × 15.9mm)
Main shaft bearing (right)	....	¾ × 1⅞ × 9/16 in. Ball journal (19 × 47.5 × 14.3mm)
Countershaft bearing (left)	....	11/16 × ⅞ × ¾ in. Needle roller (17.5 × 22.23 × 19mm)
Countershaft bearing (right)	....	11/16 × ⅞ × ¾ in. Needle roller (17.5 × 22.23 × 19mm)

Countershaft 1st gear bushing:
Bore diameter	....	.800/.795 in. (20.32/20.203mm)
Shaft diameter	....	.8075/.8070 in. (20.511/20.498mm)

Countershaft 2nd gear bushing:
Bore diameter	....	.800/.795 in. (20.32/20.203mm)
Shaft diameter	....	.8075/.8070 in. (20.511/20.498mm)

FRAME AND ATTACHMENT DETAILS

STEERING HEAD BEARINGS

Type	....	Timken taper roller bearing
Bore size	....	.7508/.7500 in. (19.08/19.06mm)
Outer diameter	....	1.7820/1.7810 in. (45.27/45.24mm)

SWINGING ARM

Bushing bore diameter	....	1 in. nominal (25.4mm)
Sleeve diameter	....	.9984/.9972 in. (25.35/35.32mm)
Distance between fork ends	....	8.018 in. (203.653mm)

11

VARIATIONS – 1972 MODELS – continued

REAR SUSPENSION

Type .... Swinging fork controlled by combined coil spring/hydraulic damper units

Fitted length .... 8 in. (203.2mm) at mid-position

Free length .... 9.5 in. (241.3mm)
Spring rate .... 88 lbs./in.
Mean coil diameter .... 1.98 in. (50.29mm)

WHEELS, BRAKES AND TIRES

WHEELS

Rim size: Front WM2-19
     Rear WM3-18

REAR WHEEL

Bolt size for detachable sprocket ... ¼ dia. × 2 in. U.H. (6.35 × 51mm)

Number of bolts 5

REAR WHEEL DRIVE

Gearbox sprocket .... See Gearbox
Rear wheel sprocket .... 47 teeth
Chain details: No. of links ... 106 links
     Pitch .... ⅝ in. (15.875mm)
     Width .... ⅜ in. (9.525mm)
Speedometer drive gearbox ratio ... 1.25/1
Speedometer cable length (outer) ... 66 in. (1676.4mm)
Speedometer cable length (inner) ... 67.63 in. (minimum) (1717.7mm)

BRAKES

Type .... Internal expanding twin leading shoes
Drum diameter: Front .... 8 in. ± .002 in. (203.2mm ± .0508mm)
     Rear .... 7 in. ± .002 in. (177.8mm ± .0508mm)
Lining front .... Lockheed 4523-166
Lining rear: thickness .... .187/.197 in. (4.75/5mm)
Area (total) .... 14.2 sq. in.

TIRES

Size: Front .... 3.25 × 19 in. (82.5 × 484.6mm)
   Rear .... 4.00 × 18 in. (101.6 × 457.2mm)
Tire pressure: Front .... 24 lb./sq. in. (1.685 kg./sq. cm.)
     Rear .... 24 lb./sq. in. (1.685 kg./sq. cm.)

FRONT FORK

TELESCOPIC FORK

Type .... Telescopic (hydraulic damping)
Spring details: Free length .... 19.1 in. (485mm)
    Compressed length ... 11.4 in. (289.5mm)
    Fitted length ...... 18.5 in. (217mm)
    Spring rate .... 25 lbs./in.
    Maximum load .... 194 lbs. (88 kg.)
    Color identification ... Orange
Stanchion diameter: (Top) ..... 1.355/1.350 in. (34.4/34.3mm)
     (Bottom) ..... 1.3610/1.3605 in. (33.04/33.03mm)
Outer member bore diameter .... 1.365/1.363 in. (33.15/33.1mm)

ELECTRICAL SYSTEM

ELECTRICAL EQUIPMENT

Battery type (12 volt) .... Lucas PUZ 5A
Rectifier type .... Lucas 2DS 506
Alternator type .... Lucas RM21
Horn type .... Lucas 6H
Zener diode type .... Lucas 2D715
Ignition coil type .... Lucas 17M12
Ignition switch type .... Lucas 149SA
Condenser pack type .... Lucas 2CP
Handlebar switch type .... Lucas 169SA
Flasher unit type .... Lucas 8FL
Contact breaker type .... Lucas 6CA

VARIATIONS — 1972 MODELS — continued

GENERAL

CAPACITIES

Fuel tank	3 gals. (3½ U.S. gals.) (4¾ U.S. gals. optional)
Oil tank	4 pints (4.8 U.S. pints, 2.27 liters)
Gearbox	⅞ pint (500cc)
Primary chaincase (initial fill only)	⅝ pint (350cc)
Telescopic fork legs	1/3 pint (200cc)

BASIC DIMENSIONS

Wheelbase	55 in. (140cm)
Overall length	87.5 in. (222cm)
Overall width	33 in. (84cm)
Seat height	32 in. (18cm)

WEIGHTS

Unladen weight	382 lb. (173 kg.)
Engine unit (dry)	130 lb. (59 kgm.)

TORQUE WRENCH SETTINGS (DRY)

Flywheel bolts	33 ft. lb. (4.6 kgm)
Connecting rod bolts	28 ft. lb. (3.9 kgm)
Crankcase junction bolts	13 ft. lb. (1.8 kgm)
Crankcase junction studs	20 ft. lb. (2.8 kgm)
Cylinder block nuts	35 ft. lb. (4.8 kgm)
Rocker box bolts—inner (5/16 in. dia.)	10 ft. lb. (1.38 kgm)
Cylinder head bolts—outer (⅜ in. dia.)	18 ft. lb. (2.49 kgm)
Cylinder head bolt—center (5/16 in. dia.)	15 ft. lb. (2.07 kgm)
Cylinder head bolt—inner (⅜ in. dia.)	18 ft. lb. (2.49 kgm)
Rocker box nuts	5 ft. lb. (.7 kgm)
Rocker box bolts (¼ in. dia.)	5 ft. lb. (.7 kgm)
Rocker spindle domed nuts	22 ft. lb. (3.0 kgm)
Oil pump nuts	5 ft. lb. (.7 kgm)
Kickstarter ratchet pinion nut	45 ft. lb. (6.3 kgm)
Clutch center nut	50 ft. lb. (7.0 kgm)
Rotor fixing nut	30 ft. lb. (4.1 kgm)
Stator fixing nuts	20 ft. lb. (2.8 kgm)
Primary cover domed nuts	10 ft. lb. (1.4 kgm)
Headlamp pivot bolts	10 ft. lb. (1.4 kgm)
Head race sleeve nut pinch bolt	15 ft. lb. (2.1 kgm)
Stanchion pinch bolts	25 ft. lb. (3.5 kgm)
Front wheel spindle cap bolts	25 ft. lb. (3.5 kgm)
Rear brake drum to hub bolts	15 ft. lb. (2.1 kgm)
Brake cam spindle nuts	20 ft. lb. (2.8 kgm)
Zener diode fixing nut	1.5 ft. lb. (.21 kgm)
Fork cap nut	80 ft. lb. (11.1 kgm)

SPECIFICATIONS 500cc MODELS — MODEL T100C Model T100R

For data not given here refer to General Data — Model T100R

ENGINE

VALVE GUIDES

Material	Cast iron

CARBURETOR

Main jet	170
Needle jet	.106
Throttle valve	4
Needle position	2

GEARBOX SPROCKET — 18T

WHEELS, BRAKES AND TIRES

BRAKES

Type—front	Internal expanding twin leading shoe 7 in. (177.8mm)

TIRES (EAST COAST MODELS ONLY)

Front	3.25 × 19 Trials Universal
Rear	4.00 × 19 Trials Universal

FRONT FORKS

Main spring—load	32 lb. in. (.4 kgm)
Color identification	Yellow/green

REAR CHAIN — 102 pitches

11

SPECIFICATIONS   500cc   MODELS — MODEL T100S
For data not given here refer to General Data — Model T100R

CRANKSHAFT

Left main bearing, size and type .... 72 × 30 × 19mm
Ball Journal

Right crankshaft main bearing journal dia. .... 1.4375/1.4380 in. (36.513/36.525mm)

Right main bearing bore, size and type .... 1.4390/1.4385 in. (36.55/36.564mm) Steel backed copper lead lined bush. Under sizes available: −0.010 in., −0.020 in., −0.030 in.

Right main bearing housing dia. .... 1.8135/1.8140 in. (46.06/46.07mm)

PISTON PIN

Fit in small end bushing .... .0005/.0012 in. (.0127/.0305mm)

SMALL END BUSHING

Material .... Phosphor Bronze
Outer dia. .... .782/.783 in. (19.863/19.888mm)
Length .... .890/.910 in. (22.61/23.1mm)
Finished bore dia. .... .6905/.6910 in. (17.54/17.55mm)

VALVE TIMING

Set all tappet clearances @ .020 in. (.5mm) for checking ....
Intake opens 34° BTDC
Intake closes 55° ABDC
Exhaust opens 48° BBDC
Exhaust closes 27° ATDC

TAPPETS

Tip radius .... ¾ in. (19.05mm)

CAMSHAFTS

Cam lift (exhaust) .... .296 in. (7.518mm)

CYLINDER HEAD

Intake port size .... 1 in. (25.4mm)

GEAR DETAILS

Layshaft low gear: bushing protrusion .... ¾ in. (before H.57083) (9.525mm)

FUEL SYSTEM

Carburetor:	Before H.57083	After H.57083
Amal type	376/273	628/8
Main jet size	190	180
Pilot jet size	25	—
Needle jet size	.106	.106
Needle type	C	Two scribed rings above needle clip grooves*
Needle position	3	2
Throttle valve: Type	376.32	4
Carburetor nominal bore size	1 in.	26mm
Air cleaner type	Felt or paper element	Felt

*An alternative needle is available for use with alcohol, marked Y

WHEELS,   BRAKES   AND   TIRES

WHEELS

Rim size: Front and rear WM2–18
Type: Front .... Spoke—single cross lacing
Rear .... Spoke—double cross lacing
Spoke details: Front .... 40 off 8/10 SWG butted 5 17/32 in. U.H. Straight
Rear: Left side .... 20 off 8/10 SWG butted 7 9/16 in. U.H. 90°
Right side .... 20 off 8/10 SWG butted 7⅞ in. U.H. 90°

BRAKES

Type .... Internal Expanding
Drum diameter: Front .... 7 in.
Rear .... 7 in. ± .002 in. $\left(\left. \begin{matrix} 177.8 \\ 177.8 \end{matrix} \right\} \pm .0508mm \right)$
Lining thickness: Front and rear .... .187/.197 in. (4.75/5.0mm)
Lining area: Front and rear .... 14.6 sq. in. (93.7 cm²)

SPECIFICATIONS   500cc MODELS – MODEL T100S (continued)

TIRES

Size: Front .... 3.25 × 18 in. Dunlop ribbed (82.5 × 457.2mm)

Rear .... 3.50 × 18 in. Dunlop K70 (88.9 × 457.2mm) (U.S.A. T100C, T100R 4.0018 in.) (101.6 × 457.2mm)

Tire pressure: Front .... 24 lb./sq. in. (1.7 kg/sq. cm)

Rear .... 25 lb./sq. in. (1.7 kg/sq. cm)

ELECTRICAL

	No.	Type
Bulbs:		
Headlight	Lucas 414	50/40 watts pre-focus
Parking light	Lucas 989	6 watts MCC
Stop and tail light	Lucas 380	6/21 watts offset pins
Speedometer light	Lucas 987	2 watt MES
Main beam indicator light (where fitted)	Lucas 281	2 watt (BA7S)
Ignition warning light	Lucas 281	2 watt (BA7S)
Coil type	Lucas 17M12 (12v.) 2 off	
Contact breaker type	Lucas 4CA (12° range) After H.57083 Lucas 6CA (12° range)	

MODEL T100T

For data not given here refer to General Data — Model T100R

ENGINE

BASIC DETAILS

Compression ratio .... 9 : 1

Power output (bhp @ rpm) .... 39 @ 7,400

CAMSHAFTS

Cam lift: Intake .... .314 in. (8.98mm)

Exhaust .... .314 in. (8.98mm)

VALVES

Head diameter intake .... 1 17/32 in. (38.89mm)

Head diameter exhaust .... 1 5/16 in. (33.33mm)

TIMING GEARS

Intake and Exhaust camwheels .... 3 Keyway

CYLINDER HEAD

Intake port size .... 1 1/16 in. (26.98mm)

PISTONS

Clearance: Top of skirt .... .0050/.0072 in. (.127/.183mm)

Bottom of skirt .... .0030/.0045 in. (.076/.114mm)

FUEL SYSTEM

Carburetors (two)

	Before H.57083	After H.57083
Amal type	376/324 and 325	626/9 and 10
Main jet size	200	140
Pilot jet size	25	—
Needle jet size	.106	.106
Needle type	C	Two rings above needle clip grooves*
Needle position	3	2
Throttle valve:		
Type	376/3½	3
Nominal bore size	1 1/16 in. (27mm)	26mm
Air cleaner type	Coarse felt	Coarse cloth

*An alternative needle is available for use with alcohol, marked Y

VALVE TIMING

Set all tappet clearances @ .020 in. (.5mm) for checking:

Intake opens 40° BTDC
Intake closes 52° ABDC
Exhaust opens 61° BBDC
Exhaust closes 31° ATDC

TAPPETS

Tip radius .... 1⅛ in. (28.57mm)

11

SPECIFICATIONS   500cc MODELS — MODEL T100T (continued)

WHEELS AND BRAKES

Drum diameter: Front	8 in. (203.2mm)
Lining thickness: Front	.183 in. (4.65mm)
Lining area: Front	23.4 sq. in. (150.9 cm²)

REAR WHEEL DRIVE

Rear chain: No. of links	102

MODEL 5TA

(DISCONTINUED AFTER ENGINE NUMBER H.49833)

FOR DATA NOT GIVEN HERE REFER TO GENERAL DATA — MODEL T100R

ENGINE

BASIC DETAILS

Bore and stroke	69 × 65.5mm
Bore and stroke	2.7165 × 2.58 in.
Cubic capacity	490cc (30 cu. in.)
Compression ratio	7 : 1
Capacity of combustion chamber	35cc (2.14 cu. in.)
Power output (bhp @ rpm)	27 @ 6,500

PISTONS

Material	Aluminum Alloy die casting
Clearance: Top of skirt	.0065/.0075 in. (.165/.190mm)
Bottom of skirt	.001/.002 in. (.0254/.0508mm)
Piston pin hole diameter	.6882/.6886 in. (17.45/17.46mm)

VALVE TIMING

See all tappet clearances @ .020 in. (.50mm) for checking	Intake opens 34° BTDC Intake closes 55° ABDC Exhaust opens 48° BBDC Exhaust closes 27° ATDC

VALVES

Seat angle (included)	90°
Head diameter: Intake	1 7/16 in. (36.51mm)
Exhaust	1 5/16 in. (33.34mm)

GENERAL

BASIC DIMENSIONS

Weight	341 lb. (154.7 kg)
Tire size: Front	3.25 × 18 Dunlop island
Rear	3.50 × 18 Dunlop K70

VALVE GUIDES

Material	Cast iron
Bore diameter (intake and exhaust)	.3130/.3120 in. (7.91/7.88mm)
Outside diameter (intake and exhaust)	.5005/.5010 in. (12.71/12.73mm)
Length: Intake	1¾ in. (44.45mm)
Exhaust	1¾ in. (44.45mm)

IGNITION TIMING

Crankshaft position (BTDC)	
Static timing	12°
Fully advanced	36°
Piston position (BTDC)	
Static timing	0.035 in. (0.90mm)
Fully advanced	0.310 in. (7.93mm)
Advance range:	
Contact breaker	12°
Crankshaft	24°

FUEL SYSTEM

Carburetor:	
Amal type	376/273
Main jet size	190
Pilot jet size	25
Needle jet size	.106
Needle type	C
Needle position	3
Throttle valve:	
Type	376/3½
Return spring free length	2½ in. (63.5mm)
Carburetor nominal bore size	1 in. (25.4mm)
Air cleaner type	Felt or paper element

SPECIFICATIONS 500cc MODELS — MODEL 5TA (continued)

GEARBOX

RATIOS

Internal ratios: 4th (Top)	1.00 : 1
3rd	1.22 : 1
2nd	1.61 : 1
1st (Bottom)	2.47 : 1
Overall ratios: 4th (Top)	5.40
3rd	6.59
2nd	8.69
1st (Bottom)	13.34
Engine rpm @ 10 mph in 4th (Top) gear	720
Gearbox sprocket teeth	20
Layshaft bushes, material	Bronze
Bore size L/H	.6865/.6885 in. (17.44/17.487mm)
Bore size R/H	.690/.689 in. (17.53/17.56mm)
Interference fit in casing R/H	.0005/.0015 in. (.0127/.038mm)
Interference fit in Kickstart assembly L/H	.0005/.0015 in. (.0127/.038mm)

FRAME AND ATTACHMENT DETAILS

REAR SUSPENSION

Spring details:	
Fitted length	8 in. (203.2mm)
Free length	8 3/16 in. (207.96mm)
Mean coil diameter	1 3/4 in. (44.45mm)
Spring rate	145 lb. in. (1.66 kgm)
Color code	Blue/yellow
Load at fitted length	38 lb. (17.24 kg)

WHEELS, BRAKES AND TIRES

WHEELS

Rim size: Front	WM2-18
Spoke details: Front	40 off 8/10 SWG butted 5 17/32 in. U.H. Straight
Rim size: Rear	WM2-18

REAR WHEEL DRIVE

Gearbox sprocket teeth	See "Gearbox"
Rear wheel sprocket teeth	46
Chain details:	
No. of links	103
Pitch	5/8 in. (15.87mm)
Width	3/8 in. (9.53mm)
Speedometer gearbox drive ratio	2 : 1

TIRES

Size: Front	3.25 × 18 in. Avon Speedmaster
Rear	3.50 × 18 in. Avon Speedmaster
Tire pressures: Front	24 lb./sq. in. (1.7 kg/sq. cm)
Rear	24 lb./sq. in. (1.7 kg/sq. cm)

FRONT FORKS

Spring details:	
Free length	9 3/4 in. (247.65mm)
No. of working coils	12 1/2
Spring rate	261 1/2 lb. in. (0.3 kgm)
Color code	Yellow/blue

ELECTRICAL SYSTEM

ELECTRICAL EQUIPMENT

Battery type	Lucas 12-volt (type PUZ5A) or alternatively two Lucas 6-volt (type MKZ9E) connected in series	
Rectifier type	Lucas 2DS506	
Alternator type	Lucas RM19	
Horn	Lucas 8H (12 volt)	
Bulbs: (6-volt)	No.	Type
Headlight	Lucas 414	50/40 watts pre-focus
Parking light	Lucas 989	6 watts MCC
Stop and tail light	Lucas 380	6/21 watts offset pins
Speedometer light	Lucas 987	2 watt MES
Coil type	Lucas MA12 (12 volt)	
Contact breaker type	Lucas 4CA	
Fuse rating	35 amp.	
Spark plugs:		
Type	Champion N4	
Plug gap setting	.020 in. (.5mm)	
Thread size	14mm	

11

SPECIFICATIONS 500cc MODELS — MODEL 5TA (continued)

GENERAL

BASIC DIMENSIONS

Wheelbase	53½ in. (136 cm)
Overall length	83¼ in. (211.5 cm)
Overall width	26½ in. (67.3 cm)
Overall height	38 in. (96.5 cm)
Ground clearance	7½ in. (19 cm)

VALVE TIMING

Set all tappet clearances @ .020 in.
(.5mm) for checking.

Inlet opens	34° before top center
Inlet closes	55° after bottom center
Exhaust opens	48° before bottom center
Exhaust closes	27° after top center

TAPPETS

Tip radius	¾ in. (19.05mm)
Tappet diameter	.3110/.3115 in. (7.89/7.912mm)
Clearance in guide block	.0005/.0015 in. (.0127/.038mm)

FUEL SYSTEM

Carburetor:

Amal type	930/87
Main jet size	200
Needle jet size	.106
Needle type	Std.
Needle position	MID
Throttle valve type	3
Carburetor nominal bore size	30mm
Air cleaner type	Paper element

WEIGHTS

Empty weight	340 lb. (155 kgm)
Engine unit (Dry)	106 lb. (48 kgm)

SPECIFICATIONS — MODEL TR5T

FRAME AND FITTINGS

FRONT FORKS

Type	Coil-spring (hydraulically damped)
Springs: Free length	19.1 in. (48.5mm)
Compressed length	11.4 in. (289.5mm)
Fitted length	18.5 in. (217mm)
Spring rate	25 lb./in.
Maximum lead	194 lbs. (45.5 kg.)
Color identification	Orange
Stanchion diameter top	1.355/1.350 in. (34.4/34.3mm)
Stanchion diameter bottom	1.3610/1.3605 (34.6/34.55mm)
Stanchion internal bore	1.095/1.089 in. (27.8/27.66mm)
Outer member bore size	1.365/1.363 in. (27.05/27mm)

REAR SHOCK ABSORBERS

Type	Coil-spring (hydraulically damped)
Springs: Free length	8.68 in. (220.5mm)
Spring rate	100 lb./in.
Color identification	Green/pink (applies both to chrome or black springs)

SPECIFICATIONS – MODEL TR5T – continued

SWINGING ARM

Bearing type	Torrington needle roller bearings B1616 2 off
Spindle diameter	.801/.800 in. (20.35/20.32mm)
Bearing sleeve diameter	1.0000/.9995 in. (25.4/25.388mm)
Thrust washer depth	.199/.197 in. (5.05/4.99mm)

GEARBOX

RATIOS

Primary ratios (Std.) 4th	1.00 : 1
3rd	1.22 : 1
2nd	1.76 : 1
1st	2.84 : 1
Overall ratios: 4th	6.57
3rd	8.04
2nd	11.6
1st	18.7
Engine rpm @ 10 mph in 4th gear	857
Gearbox sprocket teeth	18

WHEELS, BRAKES, AND TIRES

WHEELS

Rim size and type (front)	WM1-21
Rim size and type (rear)	WM3-18

WHEEL BEARINGS

Front (left and right-hand)	20 × 47 × 14mm Ball journal
Rear (left and right-hand)	20 × 47 × 14mm Ball journal
Rear brake drum	20 × 47 × 14mm Ball journal
Spindle diameter (front)	.8740/.8745 in. (22.199/22.212mm)
Spindle diameter (rear, left-hand)	.8745/.8750 in. (22.212/22.225mm)
Spindle diameter (rear, right-hand)	.685/.686 in. (17.399/17.424mm)

BRAKES

Front (diameter) single leading shoe	6 in. (152mm)
Front (width) single leading shoe	.875 in. (22.2mm)
Rear (diameter)	7 in. (177.8mm)
Rear (width)	1.125 in. (28.5mm)
Lining thickness (front)	.145/.156 in. (3.66/3.94mm)
Lining thickness (rear)	.197/.217 in. (4.98/5.5mm)

TIRES

Size (front)	3.00 × 21 in.
Size (rear)	4.00 × 18 in.
Pressure (front)	25 psi (1.75 kg./sq. cm.)
Pressure (rear)	25 psi (1.75 kg./sq. cm.)
Speedometer drive ratio	2 : 1

12 VOLT ELECTRICAL EQUIPMENT

Battery	Lucas PU25A
Coil	Lucas MA.12
Contact breaker unit	Lucas 6CA
Generator	Lucas RM.21
Generator output	115 watt
Horn	Lucas 6H
Rectifier	Lucas 2DS.506
Zener diode	Lucas ZD.715
Bulbs: Headlamp (main)	40/50 watt Lucas 380
Headlamp (pilot)	6 watt Lucas 989
Warning lamps	2 watt
Stop tail lamp	6/21 watt
Condenser	Lucas 54441582
Capacitor	Lucas 2MC
Flasher unit	Lucas 8FL
Headlamp	Lucas MCH66
Ignition switch	Lucas S45
Rear stop switch	Lucas 1185A

11

SPECIFICATIONS — MODEL TR5T — continued

CAPACITIES

CAPACITIES

Fuel tank	2½ U.S. gals. (14.774 liters)
Oil reservoir	4 pints/4.8 U.S. (2.273 liters)
Gearbox	375cc
Primary chaincase (initial fill)	150cc
Front fork (each leg)	190cc

BASIC DIMENSIONS

BASIC DIMENSIONS

Overall length	85 in. (216cm)
Handlebar width	32 in. (81.28cm)
Seat height	32 in. (81.28cm)
Ground clearance	7 in. (18cm)
Wheelbase	54 in. (137cm)

WEIGHTS

WEIGHTS

Unladen weight	322 lbs.
Engine/gearbox unit (less carburetor)	85 lbs. (39 kg.)

TORQUE SETTINGS

TIGHTENING TORQUE (DRY)

Flywheel bolts	33 ft. lb. (4.24 kgm)
Connecting rod bolts	18 ft. lb. (2.483 kgm)
Crankcase junction bolts	15 ft. lb. (2.074 kgm)
Crankcase junction studs	20 ft. lb. (2.765 kgm)
Cylinder block nuts	35 ft. lb. (4.835 kgm)
Cylinder head bolts (⅜ in. dia.)	18 ft. lb. (2.489 kgm)
Rocker box/head bolts (5/16 in. dia.)	16 ft. lb. (2.2128 kgm)
Rocker box nuts	5 ft. lb. (0.69 kgm)
Rocker box bolts	5 ft. lb. (0.69 kgm)
Rocker spindle domed nuts	25 ft. lb. (3.318 kgm)
Oil pump nuts	6 ft. lb. (0.83 kgm)
Kickstarter ratchet pinion nut	40 ft. lb. (5.53 kgm)
Clutch center nut	70 ft. lb. (6.9 kgm)
Rotor fixing nut	30 ft. lb. (4.148 kgm)
Stator fixing nuts	20 ft. lb. (2.765 kgm)
Headlamp pivot bolts	10 ft. lb. (1.38 kgm)
Head race sleeve nut pinch bolt	15 ft. lb. (2.074 kgm)
Stanchion pinch bolts	18-20 ft. lb. (2.765 kgm)
Front wheel spindle cap bolts	25 ft. lb. (3.318 kgm)
Brake cam spindle nuts	20 ft. lb. (2.765 kgm)
Zener diode fixing nut	1½ ft. lb. (0.207 kgm)
Twin carburetor manifold socket screws	10 ft. lb. (1.38 kgm)
Fork stanchion end plug	25 ft. lbs. (3.318 kgm)

SPECIFICATIONS — MODEL T100R

LUBRICATION SYSTEM

OIL PUMP

Body material	Brass
Bore diameter: Feed	.3748/.3753 in. (9.529/9.532mm)
Scavenge	.4372/.4377 in. (11.105/11.118mm)
Scavenge (Before H.49833)	.4877/.4872 in. (12.375/12.388mm)
Plunger diameter: Feed	.3744/.3747 in. (9.5098/9.5174mm)
Scavenge	.4369/.4372 in. (11.097/11.105mm)
Scavenge (Before H.49833)	.4872/.4869 in. (12.375/12.367mm)
Valve spring length	1/2 in. (12.7mm)
Ball diameter	7/32 in. (5.537mm)
Aluminum crosshead width	.497/.498 in. (12.624/12.649mm)
Working clearance in plunger heads	.0015/.0045 in. (0.038/0.11mm)

OIL PRESSURE RELEASE VALVE

Piston diameter	.5605/.5610 in. (14.236/14.239mm)
Pressure release operates	60 lb./sq. in. (4.22 kg/sq. cm.)
Springs length (Free)	1 3/8 in. (39.926mm)
Load at 1 3/16 in.	8 lb. (3.629 kg)
Rate	42.3 lb. (19.18 kg)

OIL PRESSURE

Normal running	60 lb./sq. in. (4.218 kg/cm²)
Idling	20/25 lb./sq. in. (1.406/1.687 kg/cm²)

ENGINE

BASIC DETAILS

Bore and stroke	69 × 65.5mm
Bore and stroke	2.7165 × 2.58 in.
Cubic capacity	490cc (30 cu. in.)
Compression ratio	9.0 : 1
Capacity of combustion chamber	31cc

CRANKSHAFT

Type	Forged two-throw crank with bolt on flywheel
Left main bearing: Size and type	72 × 30 × 90mm Roller bearing
Left crankshaft main bearing journal diameter	1.1805-1.1808 in. (29.985/29.992mm)
Left bearing housing diameter	2.8336-2.8321 in. (71.973/71.935mm)
Right main bearing: Size and type	72 × 35 × 17mm Ball bearing
Right crankshaft main bearing journal diameter	1.3774-1.3777 in. (34.986/34.994mm)
Right bearing housing diameter	2.8336-2.8321 in. (71.973/71.935mm)
Big-end journal diameter	1.4375-1.4380 in. (36.513/36.525mm)
Minimum regrind diameter	1.4075-1.4080 in. (35.751/35.763mm)
Crankshaft and float	0.0008-0.017 in. (.2032/.4318mm)

11

SPECIFICATIONS   500cc   MODELS — MODEL T100R   (continued)

CONNECTING RODS
Material .... Alloy "H" Section RR.56
Length (center) .... 5.311/5.313 in.
       (134.899/134.95mm)
Big-end bearings type .... Steel backed white metal
   Bearing side clearance .... .013/.017 in.
       (.3302/.4018mm)
Bearing diametric clearance ..... .0005/.0020 in. min.
       (.0127/.0508mm)

PISTON PIN
Material .... High tensile steel
Diameter .... .6882/.6885 in.
       (17.48/17.49mm)
Length .... 2.151/2.156 in.
       (54.635/54.76mm)

SMALL END (No Bushing)
Diameter .... 0.689-0.6894 in.
       (17.5/17.508mm)

PISTONS
Material .... Aluminum Alloy die casting

	From H.49833	Before H.49833
Clearance: Top of skirt	.0030/.0045 in. (.1270/.183mm)	.0075/.0085 in. (.1905/.2159mm)
Bottom of skirt	.0030/.0045 in. (.0762/.1143mm)	.002/.003 in. (.0508/.0762mm)
Piston pin hole dia.	.6882/.6886 in. (17.48/17.49mm)	.6882/.6886 in. (17.48/17.49mm)

PISTON RINGS
Material .... Cast iron
Compression rings (taper faced):
   Width .... .0615/.0625 in.
       (1.562/1.587mm)
   Thickness .... .092/.100 in.
       (2.34/2.54mm)
   Fitted gap .... .010/.014 in.
       (.254/.356mm)
   Clearance in groove .... .001/.003 in.
       (.0254/.076mm)
Oil control ring:
   Width .... .124/.125 in.
       (3.15/3.18mm)
   Thickness .... .092/.100 in.
       (2.34/2.54mm)

PISTON RINGS (continued)
   Fitted gap .... .010/.014 in.
       (.254/.356mm)
   Clearance in groove .... .0005/.0025 in.
       (.01270/.0635mm)

VALVES
Seat angle (included) .... 90°
Head diameter: Intake .... 1 17/32 in. (38.89mm)
   Intake (before H.49833) .... 1 7/16 in. (36.5mm)
   Exhaust .... 1 5/16 in. (33.34mm)
Stem diameter: Intake .... .3095/.3100 in.
       (7.86/7.87mm)
   Exhaust .... .3090/.3095 in.
       (7.849/7.86mm)

VALVE GUIDES
Material .... Hidural
Bore diameter (intake and exhaust) ... .312/.313 in.
       (7.925/7.95mm)
Outside diameter (intake and exhaust) ... .5005/.5010 in.
       (12.713/12.75mm)
Length: Intake and exhaust .... 1.760/1.770 in.
       (19.3/19.56mm)

VALVE SPRINGS (Inner—Yellow, Outer—L/Blue Spot)

	Outer	Inner
Free length	1½ in. (38.1mm)	1 19/32 in. (40.48mm)
Total number of coils	6 (152.4mm)	8¼ (209.55mm)
Total fitted load:		
Valve open	136 lb. (60.8 kg)	
Valve closed	63 lb. (28.1 kg)	

VALVE TIMING
Set all tappet clearances @ .020 in. (.5mm)
   for checking ....
       Intake opens   40°   BTDC
       Intake closes   52°   ABDC
       Exhaust opens   61°   BBDC
       Exhaust closes   31°   ATDC

ROCKERS
Material .... High tensile steel forging
Bore dia. .... .4375/.4380 in.
       (11.113/11.125mm)
Rocker spindle diameter .... .4355/.4360 in.
       (11.06/11.07mm)
Tappet clearance (cold): Intake ... .002 in. (.05mm)
   Exhaust .... .004 in. (.10mm)

SPECIFICATIONS 500cc MODELS — MODEL T100R (continued)

TAPPETS
Material .... High tensile steel forging —Stellite Tip
Tip radius .... 1 1/8 in. (28.57mm)
Tappet diameter .... .3110/.3115 in. (7.89/7.912mm)
Clearance in guide block .... .0005/.0015 in. (.0127/.038mm)

TAPPET GUIDE BLOCK
Diameter of bores .... .3120/.3125 in. (7.925/7.938mm)
Outside diameter .... 1.000/.9995 in. (25.4/25.27mm)
Interference fit in cylinder block .... .0005/.0015 in. (.0127/.038mm)

CAMSHAFTS
Journal diameter: Left .... .8100/.8105 in. (20.6/20.58mm)
Diametric clearance: Left .... .0010/.0025 in. (.0254/.0635mm)
End float .... .005/.008 in. (.127/.2032mm)
Cam lift: Intake .... .314 in. (7.98mm)
　　　　 Exhaust .... .314 in. (7.98mm)
Base circle diameter: Intake and exhaust .... .812 in. (20.62mm)

CAMSHAFT BEARING BUSHINGS
Material .... Steel backed bronze
Bore diameter (fitted): Left .... .8125/.8135 in. (20.64/20.663mm)
Outside diameter: Left .... .906/.907 in. (23.01/23.05mm)
Length: Left intake .... 1.114/1.094 in. (28.29/27.78mm)
　　　　 Left exhaust .... .922/.942 in. (23.42/23.93mm)
Interference fit in crankcase: Left .... .002/.003 in. (.05/.076mm)

TIMING GEARS
Intake and exhaust camshaft pinions:
No. of teeth 50
Interference fit on camshaft .... .000/.001 in. (.000/.0254mm)
Intermediate timing gear:
No. of teeth 42
Bore diameter .... .5618/.5625 in. (14.27/14.29mm)

TIMING GEARS (continued)
Intermediate timing gear bushing:
Material .... Phosphor bronze
Outside diameter .... .5635/.5640 in. (14.31/14.33mm)
Bore diameter .... .4990/.4995 in. (12.685/12.687mm)
Length .... .6775/.6825 in. (17.31/17.34mm)
Working clearance on spindle .... .0005/.0015 in. (.0127/.038mm)

Intermediate wheel spindle:
Diameter .... .4980/.4985 in. (12.65/12.66mm)
Interference fit in crankcase .... .0005/.0015 in. (.0127/.038mm)

Crankcase pinion:
No. of teeth 25
Fit on crankcase .... +.0003 in. (.00762mm)
　　　　　　　　　　　　　　　　　　 −.0005 in. (.0127mm)

IGNITION TIMING
Crankshaft position: (BTDC)
Static timing 14
Fully advanced 38
Piston position: (BTDC)
Static timing .... .052 (1.32mm)
Fully advanced .... .330 (8.38mm)
Advance range:
Contact breaker 12
Crankshaft 24

POINTS
Gap setting .... .014-.016 in. (35-40mm)
Advance range .... 12° (24° crankshaft)
Fully advanced at .... 2,000 rpm

CYLINDER BLOCK
Material .... Cast iron
Bore size .... 2.7160/2.7165 in. (68.98/68.99mm)
　　　　　　　　　　　　　　　　　　 2.7360/2.7365 in. (69.49/69.5mm)
Maximum oversize ....
Tappet guide block housing diameter .... .9985/.9990 in. (25.36/25.37mm)

11

SPECIFICATIONS 500cc MODELS — MODEL T100R (continued)

CYLINDER HEAD

Material	DTD 424 Aluminum Alloy
Intake port size	1 1/16 in. dia. (26.99mm)
Exhaust port size	1 1/4 in. dia. (31.75mm)
Valve seatings:	
Type	Cast-in
Material	Cast iron

FUEL SYSTEM

	U.S.A.	Before H.57083	After H.57083
Carburetor:			
Amal type	626/36.38	376/324 and 325	626/9 and 10
Main jet size	170	200	140
Pilot jet size	—	25	—
Needle jet size	.106	.106	.106
Needle type	—	C	Two rings above needle clip grooves*
Needle position	1	3	2
Throttle valve:			
Type	3 1/2	3 1/2	3
Carburetor nominal bore size	26mm	1 1/16 in.	26mm
Air cleaner type	—	Coarse felt	Coarse cloth

*An alternative needle is available for use with alcohol, marked Y

CLUTCH

CLUTCH DETAILS

Type	Multiplate with integral shock absorber
No. of plates: Driving (bonded)	6
Driven (plain)	6
Pressure springs:	
Number	3
Free length	1 31/32 in. (50mm)
No. of working coils	9 1/2
Spring rate	58 1/2 lb./in. (4.1 kg/cm²)
Approximate fitted load	42 lb. (2.95 kg/cm²)
Bearing rollers:	
Number	20
Diameter	.2495/.2500 in. (6.34/6.35mm)
Length	.231/.236 in. (5.87/5.98mm)
Clutch hub bearing diameter	1.3733/1.3743 in. (9.48/9.5mm)

CLUTCH DETAILS (continued)

Clutch sprocket bore diameter	1.0745/1.0755 in. (25.4189/25.942mm)
Thrust washer thickness	.052/.054 in. (.002/.0021mm)
Engine sprocket teeth	26
Clutch sprocket teeth	58
Chain details	Duplex endless-3/8 in. pitch × 78 links

CLUTCH OPERATING MECHANISM

Conical spring:	
Number of working coils	2
Free length	13/32 in. (10.32mm)
Diameter of balls	3/8 in. dia. (9.53mm)
Clutch operating rod:	
Diameter of rod	3/16 in. dia. (4.76mm)
Length of rod	9.562/9.567 in. (13.26/13.36mm)

GEARBOX

RATIOS

Internal ratios (Std.) 4th (Top)	1.00 : 1
3rd	1.22 : 1
2nd	1.61 : 1
1st (Bottom)	2.47 : 1
Overall ratios: 4th (Top)	5.70
3rd	6.95
2nd	9.18
1st (Bottom)	14.09
Engine rpm @ 10 mph in 4th (Top) gear	763
Gearbox sprocket teeth	18

GEAR DETAILS

Mainshaft high gear:	
Bore diameter (bushing fitted)	.7520/.7530 in. (18.9/18.93mm)
Working clearance on shaft	.0020/.0035 in. (.0508/.0889mm)
Layshaft low gear:	
Bushing length	2 19/32 in. (40.48mm)
Bore diameter (bushing fitted)	.689/.690 in. (17.5/17.53mm)
Working clearance on shaft	.0015/.003 in. (.038/.0762mm)

SPECIFICATIONS 500cc MODELS — MODEL T100R (continued)

GEARBOX SHAFTS
Mainshaft:
Left end diameter .... .7495/.7500 in. (19.03/19.05mm)
Right end diameter .... .6685/.6689 in. (16.98/16.99mm)
Length .... 9 1/64 in. (229mm)
Length (before H.49833) .... 8 51/64 in. (223.4mm)

GEARBOX SHAFTS (continued)
Layshaft:
Left end diameter .... .6845/.6850 in. (17.386/17.399mm)
Right end diameter .... .6870/.6875 in. (17.45/17.5mm)
Length .... 5 3/8 in. (136.53mm)
Camplate plunger spring:
Free length .... 2 1/2 in. (63.5mm)
No. of working coils 22
Spring rate .... 5-6 lb./in. (.35/.42 kg/cm²)

BEARINGS
High gear bearing .... 30 × 62 × 16mm Ball journal
Mainshaft bearing .... 17 × 47 × 14mm Ball journal
Layshaft bearing (left) .... 11/16 × 7/8 × 3/4 in. Needle roller (17.46 × 22.23 × 19.05mm)
Layshaft bearing (right) .... 5/8 × 13/16 × 3/4 in. Needle roller (15.87 × 20.64 × 19.05mm)

KICKSTART OPERATING MECHANISM
Ratchet spring free length .... 1/2 in. (12.7mm)

GEARCHANGE MECHANISM
Plungers:
Outer diameter .... .3402/.3412 in. (8.64/8.66mm)
Working clearance in bore .... .0015/.0035 in. (.038/.0889mm)
Plunger springs:
No. of working coils 16
Free length .... 1 1/16 in. (26.99mm)
Outer bush bore diameter .... .623/.624 in. (15.82/15.85mm)

GEARCHANGE MECHANISM (continued)
Clearance on shaft .... .001/.003 in. (.0254/.0762mm)
Quadrant return springs:
No. of working coils 18
Free length .... 1 7/8 in. (47.63mm)

FRAME AND ATTACHMENT DETAILS

HEAD RACES
No. of balls: Top 24
Bottom 24
Ball diameter .... 3/16 in. (4.763mm)

SWINGING FORK
Bushing type .... Phosphor bronze strip
Bushing bore diameter .... .8745/.8750 in. (21.91/21.93mm)
Spindle diameter .... .8735/.8740 in. (22.19/22.2mm)
Distance between fork ends .... 7 7/16 in. (189mm)

REAR SUSPENSION
Type .... Swinging fork controlled by combined spring/hydraulic damper units. (Bolted up after H.49833)
Spring details:
Fitted length .... 8 in. (203.2mm)
Free length .... 8 3/16 in. (207.96mm)
Mean coil diameter .... 1 3/4 dia. (44.45mm)
Spring rate .... 145 lb./in. (10.12 kg/cm²)
Color code .... Blue/Yellow
Load at fitted length .... 38 lb. (2.67 kg/cm²)

WHEELS, BRAKES, AND TIRES

WHEELS
Rim size: Front/rear WM2-19/WM3-18
Type: Front .... Spoke—Single cross lacing
Rear .... Spoke—Double cross lacing
Spoke details: Front: Left side .... 20 off 10 SWG 5 5/8" U.H. (142.86mm)
Right side .... 20 off 10 SWG 4 11/16" U.H. (116.1mm)
Rear: Left side .... 20 off 10 SWG 7 9/16" U.H. (102.1mm)
Right side .... 20 off 10 SWG 7 7/8" U.H. (199mm)

11

SPECIFICATIONS   500cc MODELS — MODEL T100R (continued)

WHEEL BEARINGS

Front and rear, dimensions and type	20 × 47 × 14mm
	Ball journal
Front spindle diameter (at bearing journals)	.7868/.7873 in.
	(19.98/19.99mm)
Rear spindle diameter (at bearing journals)	.7862/.7867 in.
	(19.97/19.98mm)

Q.D. REAR WHEEL

Bearing type	¾ × 1⅞ × 9/16 in.
	Ball journal
	(19 × 47.6 × 14.3mm)
Bearing sleeve: journal diameter	.7490/.7495 in.
	(18.92/18.94mm)
Brake drum bearing	⅞ × 2 9/16 in.
	(22.23 × 50.8 ×
	14.3mm)
Bearing sleeve: journal diameter	.8740/.8745 in.
	(22.199/22.21mm)
Bearing housing: internal diameter	1.9980/1.9990 in.
	(25.35/25.37mm)

REAR WHEEL DRIVE

Gearbox sprocket	See "Gearbox"
Rear wheel sprocket teeth	46
Chain details:	
No. of links	102
Pitch	⅝ in. (15.87mm)
Width	⅜ in. (9.52mm)
Speedometer drive gearbox ratio	19 : 10

BRAKES

Type: Front	Internal expanding, twin
	leading shoe
Rear	Internal expanding
Diameter front and rear	7 in. (177.8mm)
Lining thickness	.179/.190 in.
	(4.54/4.83mm)

TIRES

Size: Front	3.25 × 19 in. Dunlop K70
Rear	4.00 × 18 in. Dunlop K70
Tire pressure: Front	24 lb./sq. in.
	(1.7 kg/sq. cm)
Rear	25 lb./sq. in.
	(1.7 kg/sq. cm)

FRONT FORKS

TELESCOPIC FORK

Type	Telescopic with oil damping
	Shuttle valve
Spring details:	
Free length	9¾ in. (247.65mm)
No. of working coils	12½
Spring rate	26½ lb./in. (1.79 kg/cm²)
Color code	Yellow/blue

Bushing details:	Top bushing	Bottom bushing
Length	1 in. (25.4mm)	.870/.875 in.
		(22.098/22.23mm)
Outer diameter	.498/1.499 in.	1.4935/1.4945 in.
	(38.04/38.06mm)	(37.94/37.96mm)
Inner diameter	.3065/1.3075 in.	1.2485/1.2495 in.
	(33.19/33.21mm)	(31.71/31.74mm)
Stanchion diameter		1.3025/1.3030 in.
		(33.08/33.096mm)
Working clearance in top bushing		.0035/.0050 in.
		(.0889/.127mm)
Fork leg bore diameter		1.498/1.500 in.
		(38.04/38.1mm)
Working clearance of bottom bushing		.0035/.0065 in.
		(.0889/.165mm)
Shuttle valve: outer diameter (large)		1.018/1.106 in.
		(24.046/28.09mm)
outer diameter (small)		.875/0.874 in.
		(22.27/22.205mm)

12 VOLT ELECTRICAL SYSTEM

Battery	1 Lucas 12-volt battery
	PUZ5A or earlier 2 Lucas
	6-volt batteries connected
	in series (MKZ9E)
Rectifier type	Lucas 2DS506
Alternator type	Lucas RM19
Horn	Clear hooter 27899 12-volt

SUPPLEMENT

1978 AND LATER SERVICE INFORMATION

The following supplement provides procedures unique to the 1978 and later Triumph twins. All other service procedures are identical to earlier models.

The headings in this supplement correspond to those in the main portion of this book. If a change is not included in the supplement, there are no changes affecting the 1978 and later models.

Use the data in this supplement in conjunction with the procedures outlined in the various chapters in the main body of this manual, for these late model motorcycles.

CHAPTER TWO

LUBRICATION AND MAINTENANCE

ENGINE TUNE-UP

Valve Clearance

The procedure for adjusting valve clearance is identical to earlier models (refer to Chapter Two, *Engine Tune-up* section under *Valve Clearance (750/650cc Models)* in the front of this book). **Figure 1** shows the configuration of the 1977-1978 rocker box cover and retaining nuts. Remove this cover to adjust valves.

Ignition Timing

A new breakerless ignition is used on the 750 model covered in this supplement. Refer to Chapter Seven of this supplement for adjustment procedures.

CHAPTER FOUR

ENGINE

ENGINE
REMOVAL/INSTALLATION

The engine removal/installation procedure for the 750cc twins covered in this supplement is identical to earlier models (refer to Chapter Four, *Engine Removal/Installation* section under *750/650cc Models* in the front of this book). However, for clarification, refer to **Figures 2 through 6** when performing this procedure.

1. Torque stay-to-cylinder head mounting bolts   2. Torque stay-to-chassis mounting bolts

1. Front engine mounting bolt   2. Front engine support

1. Footrest mounting nut   3. Rear engine
2. Swing arm pivot bolt/nut   mounting bolts/nuts

1. Kickstarter retaining bolt
2. Kickstarter lever

CHAPTER SIX

FUEL SYSTEM

12

CARBURETOR

Removal, disassembly, reassembly, and installation procedures for the Amal Mk II concentric carburetors used on the 1978-1979 Triumph twins are given in this section.

Removal

1. Turn the fuel shutoff valve to the OFF position (**Figure 7**).

2. Remove the side covers.

3. Loosen the carburetor-to-cylinder head flexible coupling hose clamps, then disconnect the upper fuel feed hoses on each side of the engine (**Figures 8 and 9**).

4. Disconnect the lower fuel feed hose on each side of the engine (**Figure 10**).

5. Remove the linkage pin retainers (refer to **Figures 8 and 9**), then push the pins out in the direction shown in **Figure 11**.

6. Pull carburetors off their mounts and lift off of engine (**Figure 12**).

7. To install, reverse the preceding steps.

1. Carburetor-to-cylinder head flexible
   coupling (left side)
2. Clamp screws

1. Carburetor-to-cylinder head flexible
   coupling (right side)
2. Clamp screws

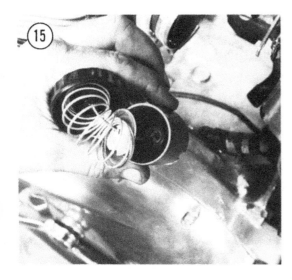

Disassembly/Reassembly

1. Unscrew the mixing chamber cap and lift out the entire mechanism as shown in **Figure 13**.

2. The control wire installation is shown in **Figures 14 and 15**. Simply push the retainer up and out of the throttle valve as shown in **Figure 16**, then pull the control wire out (**Figure 17**).

3. Slide the plate off the end of the control wire (**Figure 18**), then slip the mixing chamber cap off of the control wire (**Figure 19**). The protective boot can be left on the control cable (**Figure 20**).

4. **Figure 21** shows the components removed in the preceding steps.

5. Unscrew the jet needle mechanism (**Figure 22**), then disconnect the hoses shown in **Figure 23**.

6. Remove the banjo bolt and housing (**Figure 24**), then remove the filter screen (**Figure 25**).

7. Unscrew the jet holder (**Figure 26**).

8. Remove the float bowl retaining screws (**Figure 27**) and lift the float bowl off (**Figure 28**).

9. Remove the gasket and neoprene cap (**Figures 29 and 30**).

10. Unscrew the jet holder (**Figure 31**), then unscrew the main jet (**Figure 32**).

11. Unscrew the needle jet (**Figure 33**).

12. Lift the float out of the float bowl, then slide the float pivot pin out of the float (**Figure 34**).

13. To reassemble, reverse the preceding steps.

12

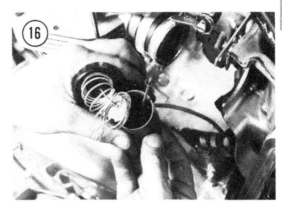

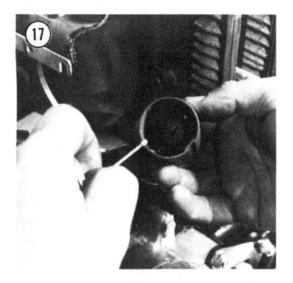

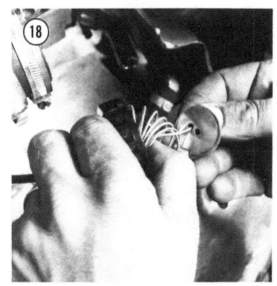

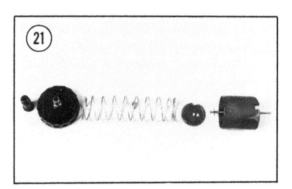

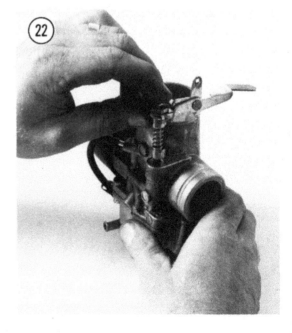

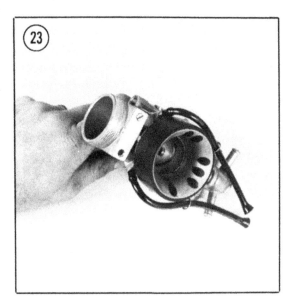

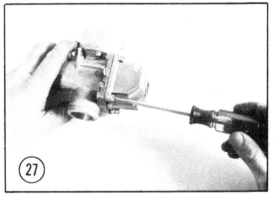

12

1. Neoprene cap   2. Gasket

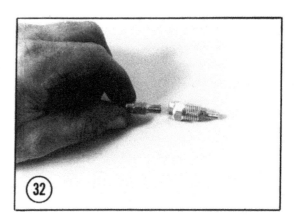

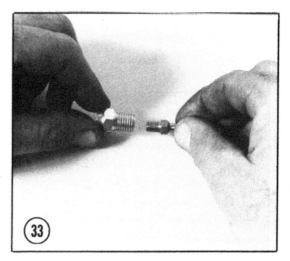

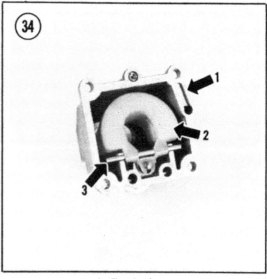

1. Float bowl
2. Float
3. Float pivot pin

CHAPTER SEVEN

ELECTRICAL SYSTEM

The models covered in this supplement are fitted with a new electronic (breakerless) ignition, and an 11 amp alternator which offers improved slow speed output.

ALTERNATOR

The RM24 alternator used on all models covered in this supplement is an 11 amp version of the basic 3-phase RM21 and RM23 units used in previous years. The physical dimensions and mounting holes are identical to previous models, in fact, which has been accomplished by retaining the same inside and outside stator diameters and 3 hole mounting positions. However, the lamination stack thickness is increased by 0.030 in., and the RM23 hot staked rotor is fitted as standard equipment. Slow speed output has been considerably improved due to the stator's 9 poles (rather than 6 as previously fitted). A 3-phase output is the result (the rotor has 6 poles).

Stator Specifications

The AC output between any 2 leads is 4.5 volts minimum @ 1,000 rpm and 6.5-7.0 volts minimum @ 5,000 rpm. Resistance between any 2 leads should be between 0.80-0.95 ohms. Insulation resistance must not test less than 100 meg. @ 500 volts DC, and must be able to withstand a 500 volt AC flash test when tested between any one of the cable leads and the lamination assembly.

DC Output Regulation

A single zener diode is installed to regulate the DC output between 14.7 and 15.8 volts.

Rectification

A 6-diode rectifier pack handles rectification.

CAUTION
The rectifier pack is polarity and surge conscious. Therefore, DC circuits must never be disconnected while engine is running.

ELECTRONIC IGNITION

The electronic ignition consists of an AB11 amplifier, 5PU pickup, and a reluctor. The amplifier (a remotely mounted electronic switching system) is installed in a cast aluminum box. The pickup consists of a magnetic base plate and encapsulated winding mounted in the crankcase, around the reluctor. (The reluctor is a ground steel timing device situated on the end of the camshaft.)

Operation

When the ignition switch closes, current flows through the primary windings of the two 6-volt coils in series, then through the conducting amplifier to ground. The pickup's magnetic base plate generates a magnetic field between the pickup's poles when the rotating reluctor's arms approach these poles, and the changing field strength generates a pulse in the winding. This pulse is transmitted to the amplifier, causing it to switch off. The primary winding field collapses in the ignition coils, which generates the high tension voltage for the spark.

Ignition Timing

1. Remove the ignition cover retaining screws (**Figure 35**).

12

2. Loosen base plate retaining nuts (**Figure 36**) slightly and move base plate to center of adjustment slot.

3. Tighten base plate retaining screws lightly (refer to **Figure 36**), then start engine.

4. With a stroboscope (connect according to the manufacturer's instructions), adjust base plate until timing mark on rotor appears stationary under the fixed timing mark (engine at 3,500 rpm).

> NOTE: *The timing mark is located behind the forward Phillips head cover in the left crankcase cover about the word "Triumph".*

5. Install the ignition cover and retaining screws (refer to **Figure 35**).

CHAPTER EIGHT

FRONT BRAKE, WHEEL, AND SUSPENSION

INSTRUMENTS AND OPERATING CONTROLS

The latest Triumph twins feature a new instrument layout (**Figure 37**), and handlebar switches (**Figures 38 and 39**).

BRAKE CALIPER

Refer to Chapter Eight in the front of this book under *Brake Caliper, Removal/Installa-* tion, and refer to **Figure 40** for bleeder screw location and brake line connection.

FRONT FORKS

Refer to Chapter Eight in the front of this book under *Front Forks* section, and refer to **Figures 41 through 45** for location of the various hydraulic hoses, fork retaining bolts, etc.

1. Hydraulic brake line-to-caliper connection
2. Bleeder screw (with protective cap in place)

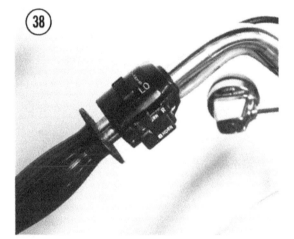

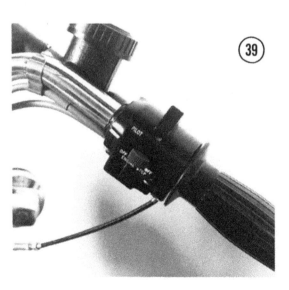

1. Upper fork retaining screws
2. Hydraulic brake line connections
3. Hydraulic brake line-to-fork connecting nuts
4. Hydraulic valve-to-front fork connecting nut

12

1. Handlebar clamp nuts
2. Steering stem nut
3. Steering stem clamp nut
4. Upper fork retaining screws

1. Front fork bleeder screw (left side)
2. Front fork cap retaining nuts (left side)

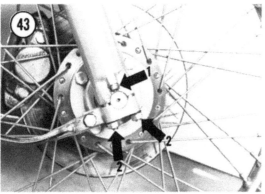

1. Front fork bleeder screw (right side)
2. Front fork cap retaining nuts (right side)

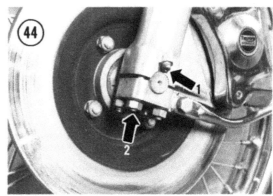

1. Hydraulic brake line-to-fork connecting nut
2. Hydraulic brake line connection

CHAPTER NINE

REAR BRAKE, WHEEL, AND SUSPENSION

REAR SHOCKS

Refer to Chapter Nine in the front of this book under *Rear Shocks, Removal/Repair/ Installation*, and refer to **Figures 46 and 47**.

SWING ARM

Refer to Chapter Nine in the front of this book under *Swing Arm* section, and refer to **Figure 48** for mounting bolt and rear stoplight switch connections.

A. Lower left-hand connection

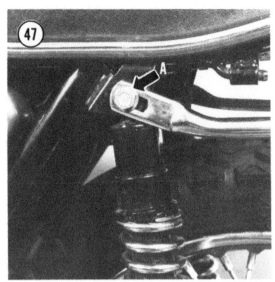

A. Upper left-hand connection

1. Swing arm pivot bolt/nut
2. Rear brake retaining bolt/nut
3. Rear brake light adjusting nuts
4. Rear engine mounting bolts/nuts

12

INDEX

13

13

500cc WIRING DIAGRAM — U.S.A.

12 volt coil ignition, separate headlamp
From Eng. Nos. H49833 to H57082

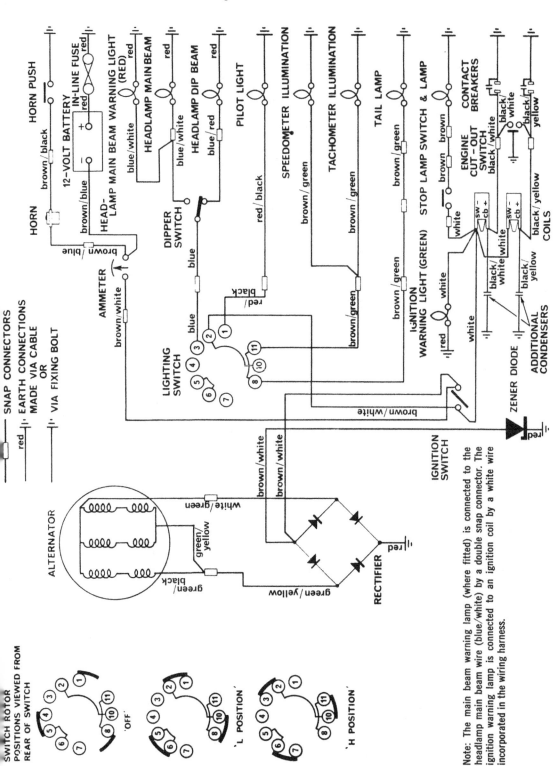

Note: The main beam warning lamp (where fitted) is connected to the headlamp main beam wire (blue/white) by a double snap connector. The ignition warning lamp is connected to an ignition coil by a white wire incorporated in the wiring harness.

14

500cc WIRING DIAGRAM — U.S.A.

12-volt coil ignition, separate headlamp
From Eng. No. H57083

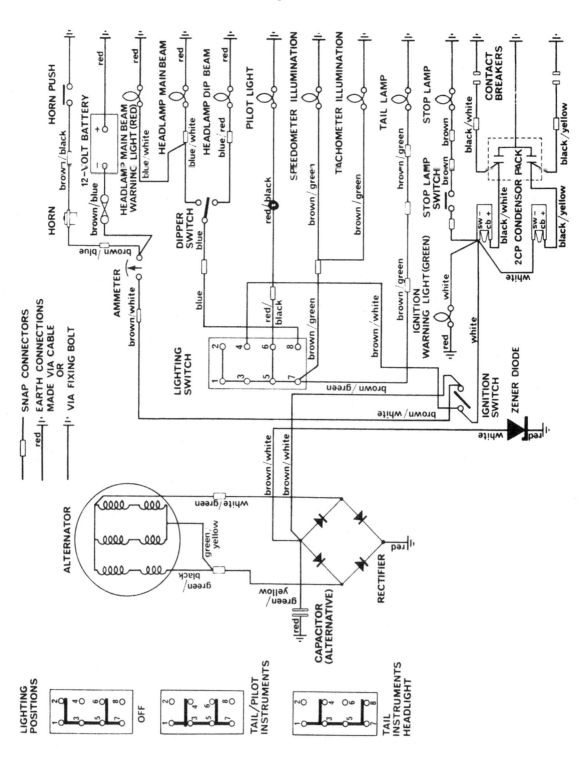

500cc WIRING DIAGRAM — U.S.A.
12-Volt Coil Ignition
From Engine No. H65573

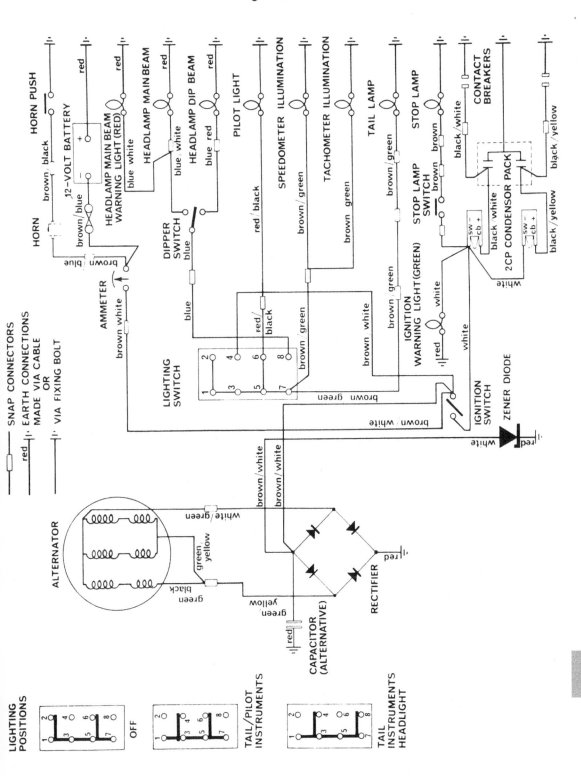

14

500cc WIRING DIAGRAM — U.S.A.
12-volt coil ignition
From Eng. No. KD27850

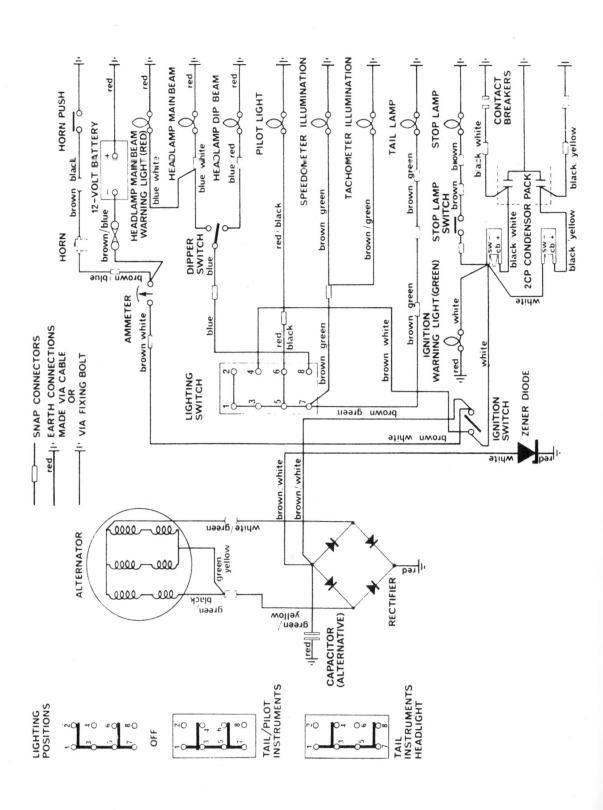

500cc WIRING DIAGRAM — U.S.A.
12-volt coil ignition
From Eng. No. KE00001

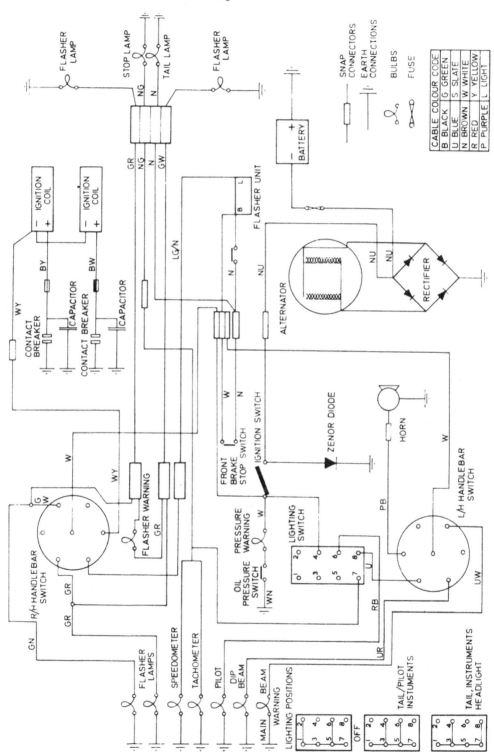

14

500cc WIRING DIAGRAM — TR5T

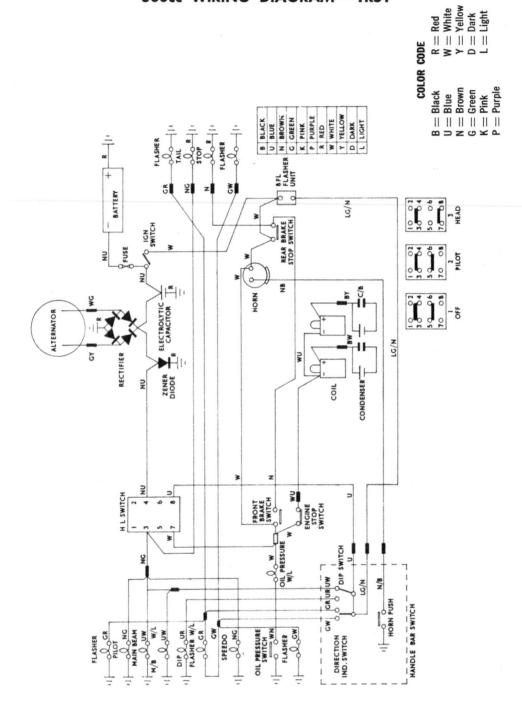

500cc WIRING DIAGRAM — United Kingdom

12-volt coil ignition, separate headlamp
From Eng. No. H57083

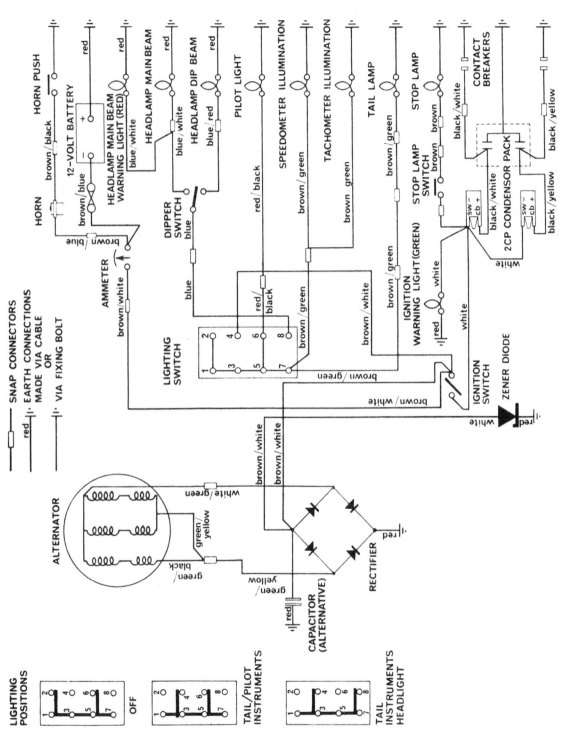

14

500cc WIRING DIAGRAM — United Kingdom

12-volt coil ignition
From Eng. No. KE00001

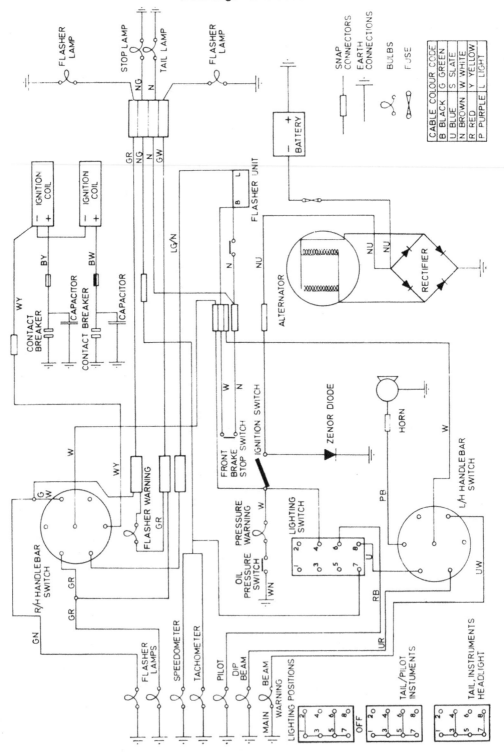

500cc WIRING DIAGRAM — United Kingdom & General Export
12-volt coil ignition, without nacelle
Before Eng. No. H49832

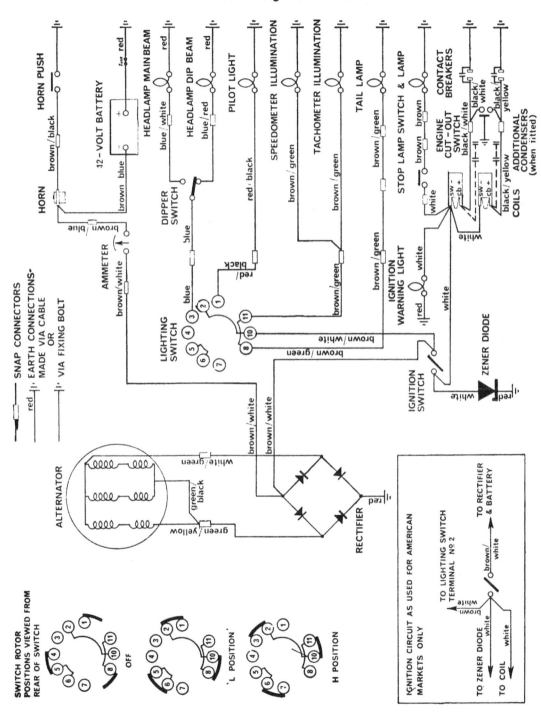

Note: The main beam warning lamp (where fitted) is connected to the headlamp main beam wire (blue/white) by a double snap connector. The ignition warning lamp is connected to an ignition coil by a white wire incorporated in the wiring harness.

500cc **WIRING DIAGRAM** — United Kingdom

12-Volt Coil Ignition, Separate Headlamp
Engine Nos. H49833-H57082

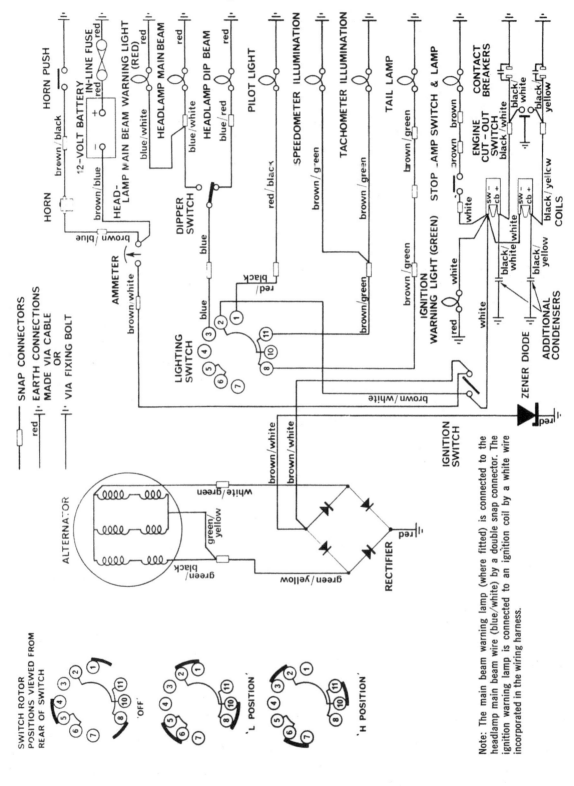

Note: The main beam warning lamp (where fitted) is connected to the headlamp main beam wire (blue/white) by a double snap connector. The ignition warning lamp is connected to an ignition coil by a white wire incorporated in the wiring harness.

500 & 650cc WIRING DIAGRAM
12-Volt Coil Ignition, With Nacelle

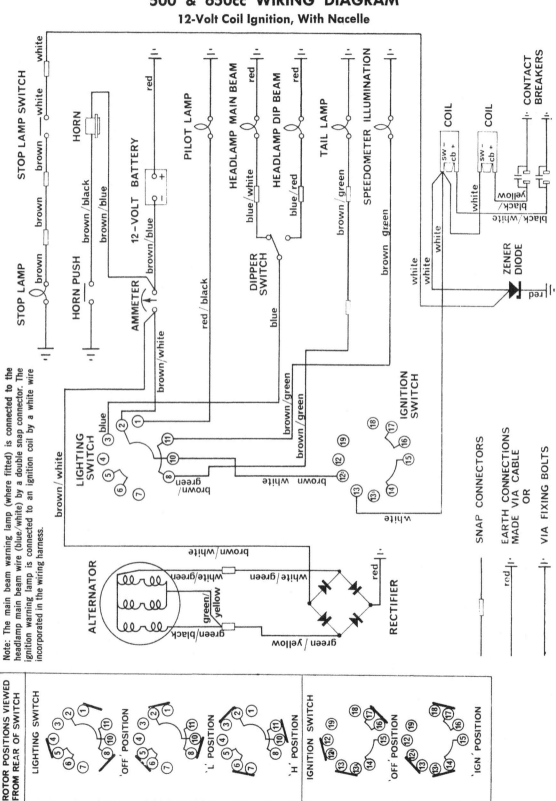

Note: The main beam warning lamp (where fitted) is connected to the headlamp main beam wire (blue/white) by a double snap connector. The ignition warning lamp is connected to an ignition coil by a white wire incorporated in the wiring harness.

STOP LAMP SWITCH

STOP LAMP

HORN

HORN PUSH

AMMETER

12-VOLT BATTERY

PILOT LAMP

HEADLAMP MAIN BEAM

HEADLAMP DIP BEAM

DIPPER SWITCH

TAIL LAMP

SPEEDOMETER ILLUMINATION

IGNITION SWITCH

LIGHTING SWITCH

ALTERNATOR

RECTIFIER

COIL

COIL

CONTACT BREAKERS

ZENER DIODE

SNAP CONNECTORS

EARTH CONNECTIONS MADE VIA CABLE OR

VIA FIXING BOLTS

ROTOR POSITIONS VIEWED FROM REAR OF SWITCH

LIGHTING SWITCH — 'OFF' POSITION — 'L' POSITION — 'H' POSITION

IGNITION SWITCH — 'OFF' POSITION — 'IGN' POSITION

14

500 & 650cc WIRING DIAGRAM
With AC Magneto (ET Ignition)

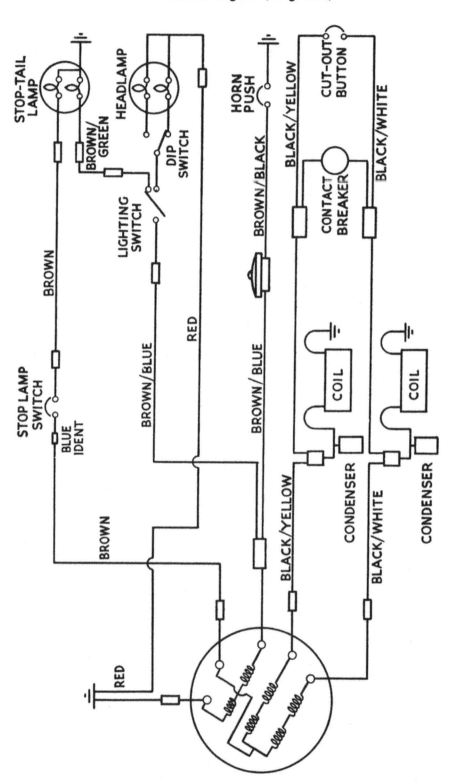

Note: The main beam warning lamp (where fitted) is connected to the headlamp main beam wire by a double snap connector. No ignition warning lamp is fitted.

500 & 650cc WIRING DIAGRAM
6-Volt Coil Ignition

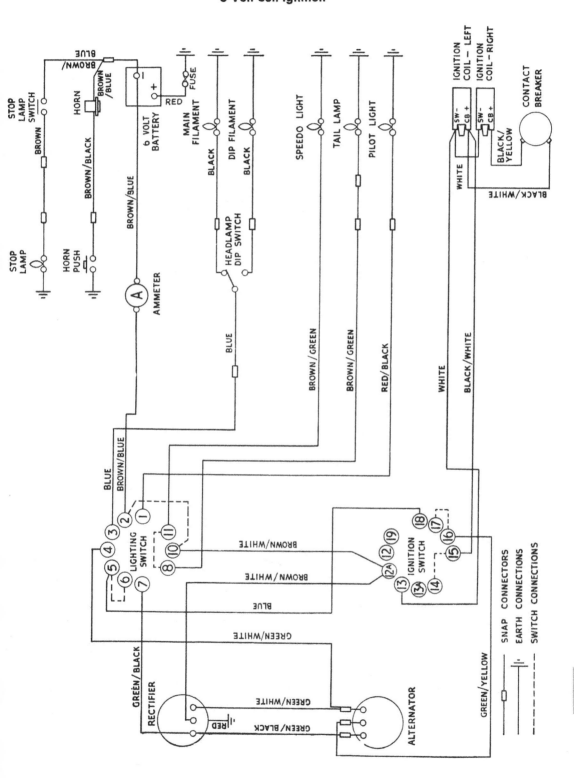

500 & 650cc WIRING DIAGRAM — Police Models
6-Volt Coil Ignition, With Boost Switch

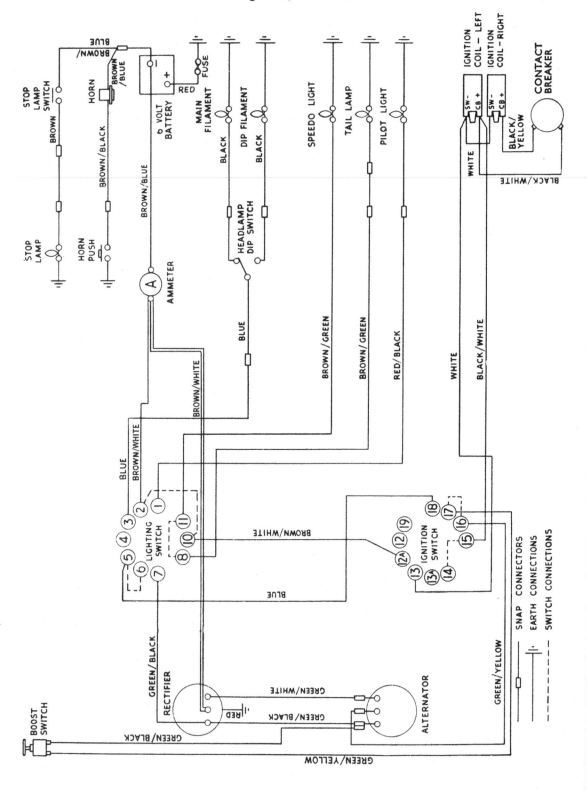

650cc WIRING DIAGRAM

From engine number HG 30870

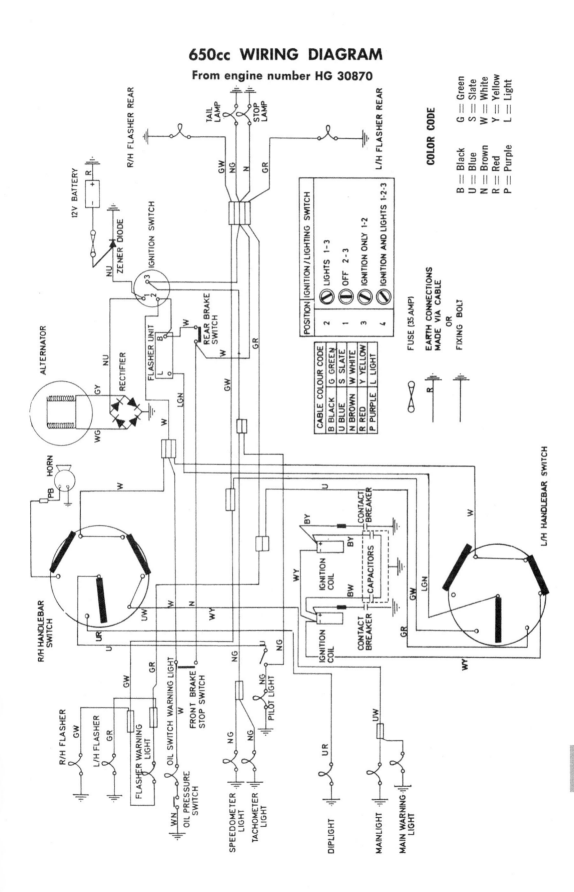

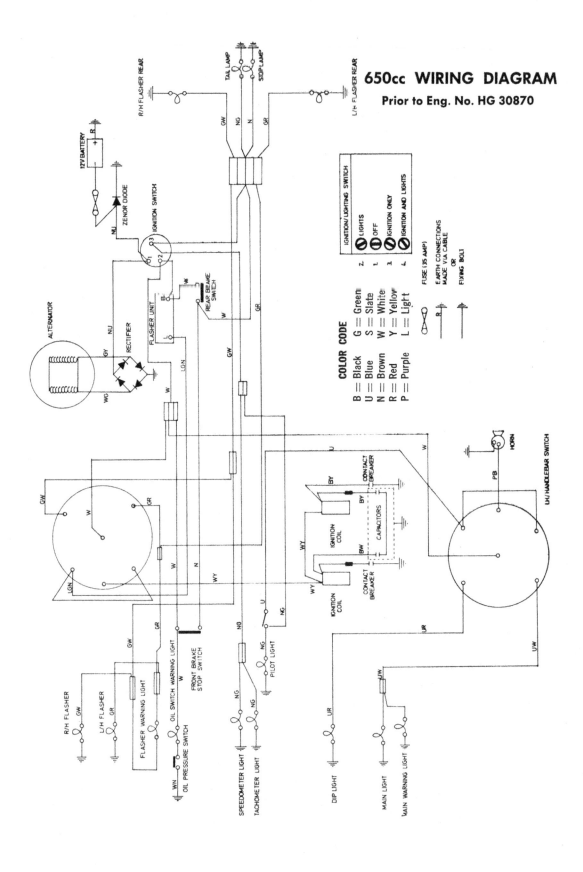

650cc WIRING DIAGRAM
Prior to Eng. No. HG 30870

650cc WIRING DIAGRAM

12-volt coil ignition with Zener diode and
2 rate charge control — Before Eng. No. DU. 24875

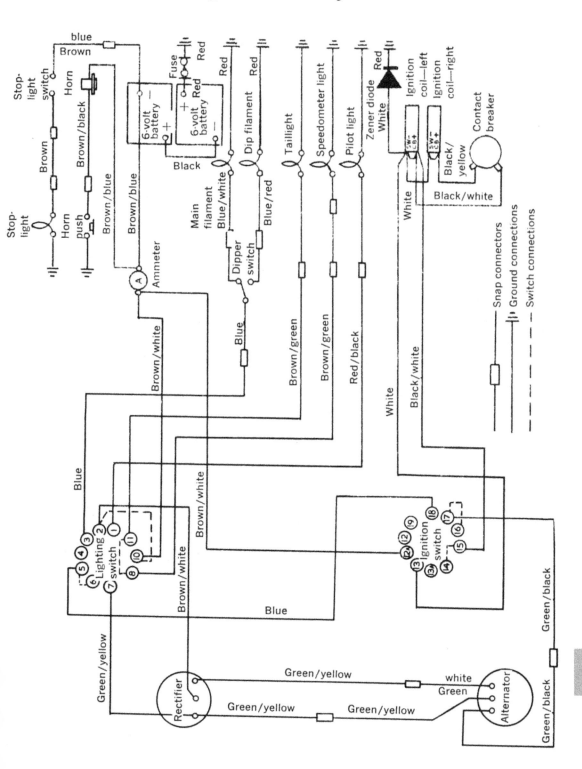

650cc WIRING DIAGRAM — U.S.A.

12-volt coil ignition, separate headlamp
From Eng. Nos. DU. 44394 to DU. 66245

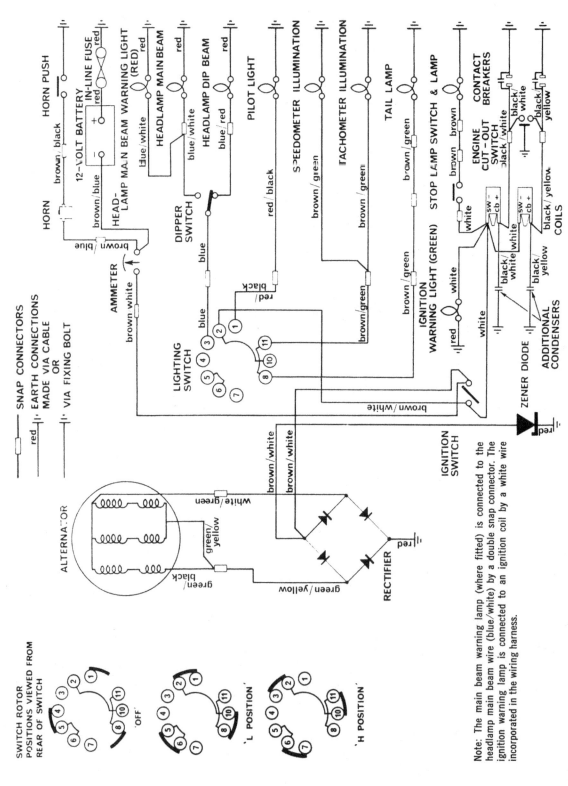

Note: The main beam warning lamp (where fitted) is connected to the headlamp main beam wire (blue/white) by a double snap connector. The ignition warning lamp is connected to an ignition coil by a white wire incorporated in the wiring harness.

650cc WIRING DIAGRAM — United Kingdom

12-volt coil ignition, separate headlamp
From Eng. No. DU. 24875

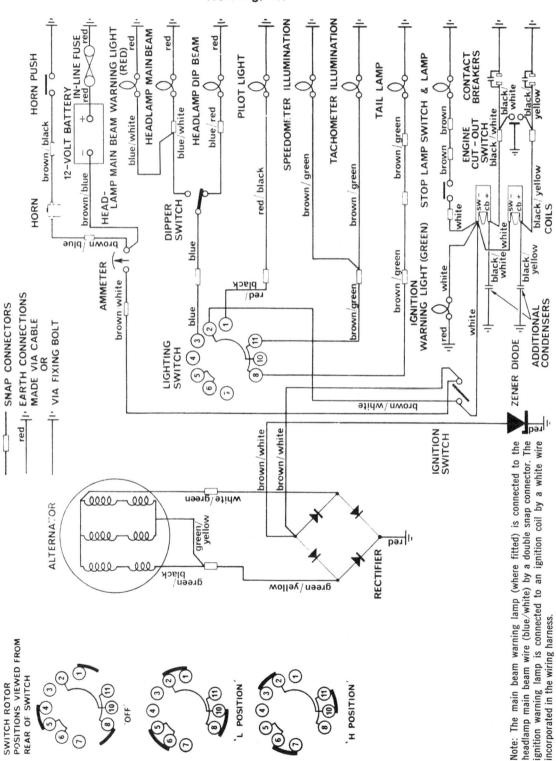

Note: The main beam warning lamp (where fitted) is connected to the headlamp main beam wire (blue/white) by a double snap connector. The ignition warning lamp is connected to an ignition coil by a white wire incorporated in the wiring harness.

14

650cc WIRING DIAGRAM — United Kingdom
Eng. No. DU. 66246 and subsequent

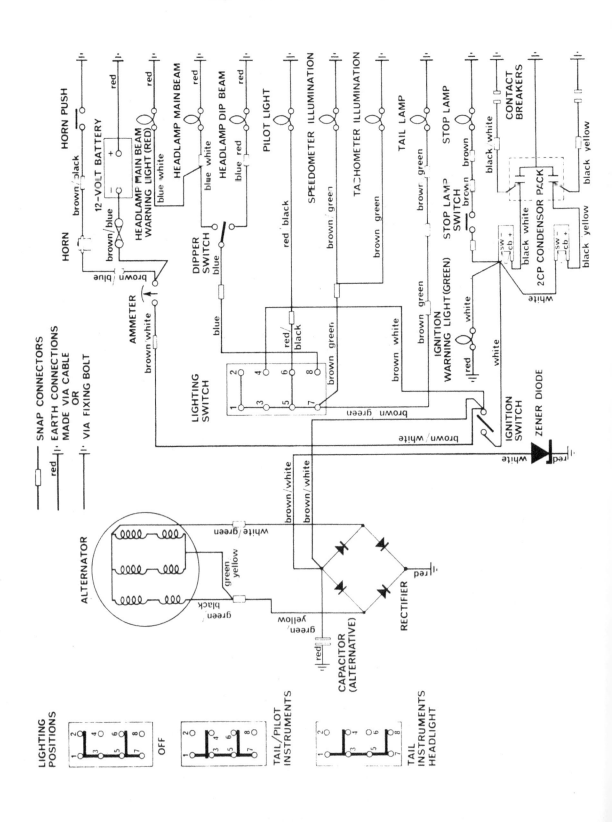

650cc WIRING DIAGRAM — General Export

From Eng. No. DU. 66246

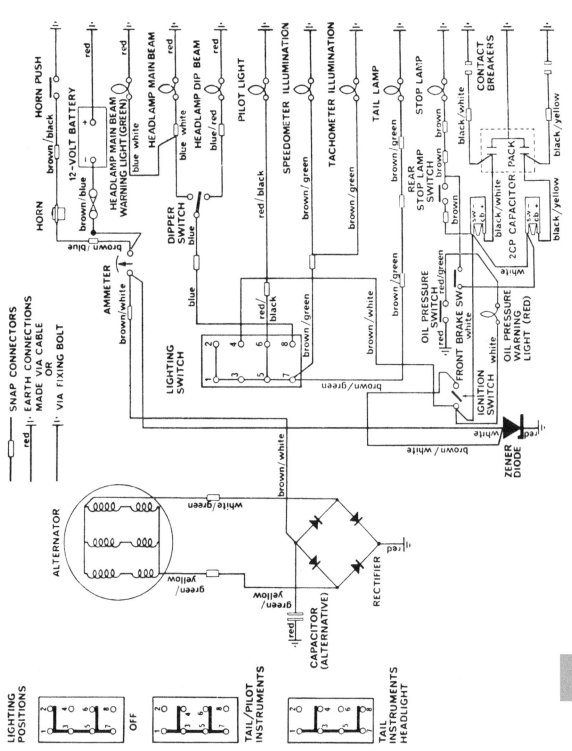

14

750cc WIRING DIAGRAM

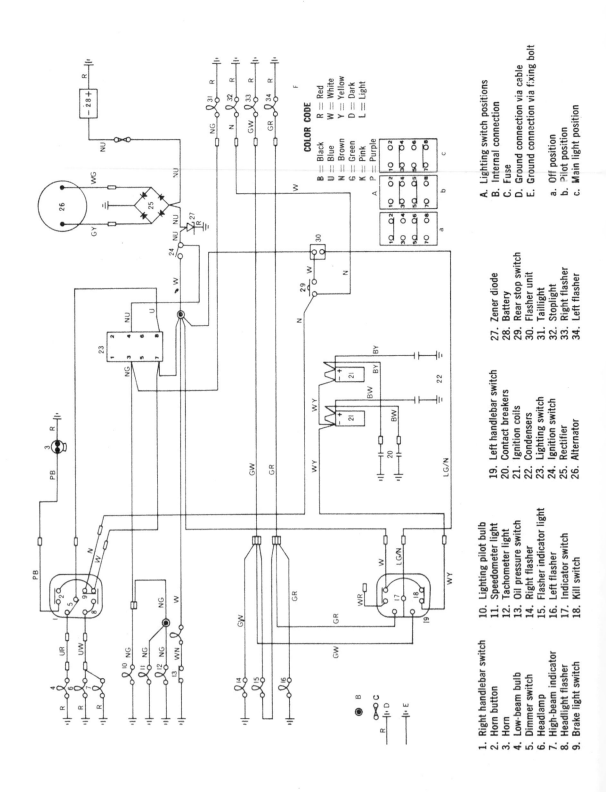

COLOR CODE

B = Black   R = Red
U = Blue   W = White
N = Brown   Y = Yellow
G = Green   D = Dark
K = Pink   L = Light
P = Purple

A. Lighting switch positions
B. Internal connection
C. Fuse
D. Ground connection via cable
E. Ground connection via fixing bolt

a. Off position
b. Pilot position
c. Main light position

1. Right handlebar switch
2. Horn button
3. Horn
4. Low-beam bulb
5. Dimmer switch
6. Headlamp
7. High-beam indicator
8. Headlight flasher
9. Brake light switch
10. Lighting pilot bulb
11. Speedometer light
12. Tachometer light
13. Oil pressure switch
14. Right flasher
15. Flasher indicator light
16. Left flasher
17. Indicator switch
18. Kill switch
19. Left handlebar switch
20. Contact breakers
21. Ignition coils
22. Condensers
23. Lighting switch
24. Ignition switch
25. Rectifier
26. Alternator
27. Zener diode
28. Battery
29. Rear stop switch
30. Flasher unit
31. Taillight
32. Stoplight
33. Right flasher
34. Left flasher

NOTES

NOTES

NOTES

NOTES

NOTES

NOTES

NOTES

NOTES

MAINTENANCE LOG

Date	Miles	Type of Service

Check out *clymer.com* for our full line of powersport repair manuals.

BMW

M308	500 & 600 CC Twins, 55-69
M309	F650, 1994-2000
M500-3	BMW K-Series, 85-97
M501	K1200RS, GT & LT, 98-05
M502-3	BMW R50/5-R100 GSPD, 70-96
M503-3	R850, R1100, R1150 and R1200C, 93-05

HARLEY-DAVIDSON

M419	Sportsters, 59-85
M428	Sportster Evolution, 86-90
M429-4	XL/XLH Sportster, 91-03
M427-1	Sportster, 04-06
M418	Panheads, 48-65
M420	Shovelheads, 66-84
M421-3	FLS/FXS Evolution, 84-99
M423-2	FLS/FXS Twin Cam, 00-05
M422-3	FLH/FLT/FXR Evolution, 84-99
M430-4	FLH/FLT Twin Cam, 99-05
M424-2	FXD Evolution, 91-98
M425-3	FXD Twin Cam, 99-05

HONDA

ATVs

M316	Odyssey FL250, 77-84
M311	ATC, TRX & Fourtrax 70-125, 70-87
M433	Fourtrax 90 ATV, 93-00
M326	ATC185 & 200, 80-86
M347	ATC200X & Fourtrax 200SX, 86-88
M455	ATC250 & Fourtrax 200/250, 84-87
M342	ATC250R, 81-84
M348	TRX250R/Fourtrax 250R & ATC250R, 85-89
M456-3	TRX250X 87-92; TRX300EX 93-04
M215	TRX250EX, 01-05
M446-2	TRX250 Recon & ES, 97-04
M346-3	TRX300/Fourtrax 300 & TRX300FW/Fourtrax 4x4, 88-00
M200-2	TRX350 Rancher, 00-06
M459-3	TRX400 Foreman 95-03
M454-3	TRX400EX 99-05
M205	TRX450 Foreman, 98-04
M210	TRX500 Rubicon, 98-04

Singles

M310-13	50-110cc OHC Singles, 65-99
M319-2	XR50R, CRF50F, XR70R & CRF70F, 97-05
M315	100-350cc OHC, 69-82
M317	Elsinore, 125-250cc, 73-80
M442	CR60-125R Pro-Link, 81-88
M431-2	CR80R, 89-95, CR125R, 89-91
M435	CR80, 96-02
M457-2	CR125R & CR250R, 92-97
M464	CR125R, 1998-2002
M443	CR250R-500R Pro-Link, 81-87
M432-3	CR250R, 88-91 & CR500R, 88-01
M437	CR250R, 97-01
M352	CRF250, CRF250X & CRF450R, CRF450X, 02-05
M312-13	XL/XR75-100, 75-03
M318-4	XL/XR/TLR 125-200, 79-03
M328-4	XL/XR250, 78-00; XL/XR350R 83-85; XR200R, 84-85; XR250L, 91-96
M320-2	XR400R, 96-04
M339-7	XL/XR 500-650, 79-03

Twins

M321	125-200cc Twins, 65-78
M322	250-350cc Twins, 64-74
M323	250-360cc Twins, 74-77
M324-5	Twinstar, Rebel 250 & Nighthawk 250, 78-03
M334	400-450cc Twins, 78-87
M333	450 & 500cc Twins, 65-76
M335	CX & GL500/650 Twins, 78-83
M344	VT500, 83-88
M313	VT700 & 750, 83-87
M314-3	VT750 Shadow (chain drive), 98-06
M440	VT1100C Shadow, 85-96
M460-3	VT1100C Series, 95-04

Fours

M332	CB350-550cc, SOHC, 71-78
M345	CB & 650, 03-05
M336	CB650, 79-82
M341	CB750 SOHC, 69-78
M337	CB750 DOHC, 79-82
M436	CB750 Nighthawk, 91-93 & 95-99
M325	CB900, 1000 & 1100, 80-83
M439	Hurricane 600, 87-90
M441-2	CBR600F2 & F3, 91-98
M445-2	CBR600F4, 99-06
M220	CBR600RR, 03-06
M434-2	CBR900RR Fireblade, 93-99
M329	500cc V-Fours, 84-86
M438	Honda VFR800, 98-00
M349	700-1000 Interceptor, 83-85
M458-2	VFR700F-750F, 86-97
M327	700-1100cc V-Fours, 82-88
M340	GL1000 & 1100, 75-83
M504	GL1200, 84-87
M508	ST1100/PAN European, 90-02

Sixes

M505	GL1500 Gold Wing, 88-92
M506-2	GL1500 Gold Wing, 93-00
M507-2	GL1800 Gold Wing, 01-05
M462-2	GL1500C Valkyrie, 97-03

KAWASAKI

ATVs

M465-2	KLF220 & KLF250 Bayou, 88-03
M466-4	KLF300 Bayou, 86-04
M467	KLF400 Bayou, 93-99
M470	KEF300 Lakota, 95-99
M385	KSF250 Mojave, 87-00

Singles

M350-9	Rotary Valve 80-350cc, 66-01
M444-2	KX60, 83-02; KX80 83-90
M448	KX80/85/100, 89-03
M351	KDX200, 83-88
M447-3	KX125 & KX250, 82-91 KX500, 83-04
M472-2	KX125, 92-00
M473-2	KX250, 92-00
M474-2	KLR650, 87-06

Twins

M355	KZ400, KZ/Z440, EN450 & EN500, 74-95
M360-3	EX500, GPZ500S, Ninja R, 87-02
M356-5	Vulcan 700 & 750, 85-06
M354-2	Vulcan 800 & Vulcan 800 Classic, 95-04
M357-2	Vulcan 1500, 87-99
M471-2	Vulcan Classic 1500, 96-04

Fours

M449	KZ500/550 & ZX550, 79-85
M450	KZ, Z & ZX750, 80-85
M358	KZ650, 77-83
M359-3	900-1000cc Fours, 73-81
M451-3	1000 &1100cc Fours, 81-02
M452-3	ZX500 & 600 Ninja, 85-97
M453-3	Ninja ZX900-1100 84-01
M468-2	Ninja ZX-6, 90-04
M469	ZX-7 Ninja, 91-98
M453-3	Ninja ZX900, ZX1000 & ZX1100, 84-01
M409	Concours, 86-04

POLARIS

ATVs

M496	Polaris ATV, 85-95
M362	Polaris Magnum ATV, 96-08
M363	Scrambler 500, 4X4 97-00
M365-2	Sportsman/Xplorer, 96-03

SUZUKI

ATVs

M381	ALT/LT 125 & 185, 83-87
M475	LT230 & LT250, 85-90
M380-2	LT250R Quad Racer, 85-92
M270	LT-Z400, 03-07
M343	LTF500F Quadrunner, 98-00
M483-2	Suzuki King Quad/ Quad Runner 250, 87-98

Singles

M371	RM50-400 Twin Shock, 75-81
M369	125-400cc 64-81
M379	RM125-500 Single Shock, 81-88
M476	DR250-350, 90-94
M477	DR-Z400, 00-06
M384-3	LS650 Savage, 86-04
M386	RM80-250, 89-95
M400	RM125, 96-00
M401	RM250, 96-02

Twins

M372	GS400-450 Twins, 77-87
M481-4	VS700-800 Intruder, 85-04
M260	Volusia/C50, 01-06
M482-2	VS1400 Intruder, 87-01
M484-3	GS500E Twins, 89-02
M361	SV650, 1999-2002

Triple

M368	380-750cc, 72-77

Fours

M373	GS550, 77-86
M364	GS650, 81-83
M370	GS750 Fours, 77-82
M376	GS850-1100 Shaft Drive, 79-84
M378	GS1100 Chain Drive, 80-81
M383-3	Katana 600, 88-96 GSX-R750-1100, 86-87
M331	GSX-R600, 97-00
M264	GSX-R600, 01-05
M478-2	GSX-R750, 88-92 GSX750F Katana, 89-96
M485	GSX-R750, 96-99
M377	GSX-R1000, 01-04
M338	GSF600 Bandit, 95-00
M353	GSF1200 Bandit, 96-03

YAMAHA

ATVs

M499	YFM80 Badger, 85-01
M394	YTM/YFM200 & 225, 83-86
M488-5	Blaster, 88-05
M489-2	Timberwolf, 89-00
M487-5	Warrior, 87-04
M486-6	Banshee, 87-06
M490-3	Moto-4 & Big Bear, 87-04
M493	YFM400FW Kodiak, 93-98
M280-2	Raptor 660R, 01-05

Singles

M492-2	PW50 & PW80, BW80 Big Wheel 80, 81-02
M410	80-175 Piston Port, 68-76
M415	250-400cc Piston Port, 68-76
M412	DT & MX 100-400, 77-83
M414	IT125-490, 76-86
M393	YZ50-80 Monoshock, 78-90
M413	YZ100-490 Monoshock, 76-84
M390	YZ125-250, 85-87 YZ490, 85-90
M391	YZ125-250, 88-93 WR250Z, 91-93
M497-2	YZ125, 94-01
M498	YZ250, 94-98 and WR250Z, 94-97
M406	YZ250F & WR250F, 01-03
M491-2	YZ400F, YZ426F, WR400F WR426F, 98-02
M417	XT125-250, 80-84
M480-3	XT/TT 350, 85-00
M405	XT500 & TT500, 76-81
M416	XT/TT 600, 83-89

Twins

M403	650cc, 70-82
M395-10	XV535-1100 Virago, 81-03
M495-4	V-Star 650, 98-05
M281-2	V-Star 1100, 99-05
M282	Road Star, 99-05

Triple

M404	XS750 & 850, 77-81

Fours

M387	XJ550, XJ600 & FJ600, 81-92
M494	XJ600 Seca II, 92-98
M388	YX600 Radian & FZ600, 86-90
M396	FZR600, 89-93
M392	FZ700-750 & Fazer, 85-87
M411	XS1100 Fours, 78-81
M397	FJ1100 & 1200, 84-93
M375	V-Max, 85-03
M374	Royal Star, 96-03
M461	YZF-R6, 99-04
M398	YZF-R1, 98-03
M399	FZ1, 01-05

VINTAGE MOTORCYCLES

Clymer® Collection Series

M330	Vintage British Street Bikes, BSA, 500-650cc Unit Twins; Norton, 750 & 850cc Commandos; Triumph, 500-750cc Twins
M300	Vintage Dirt Bikes, V. 1 Bultaco, 125-370cc Singles; Montesa, 123-360cc Singles; Ossa, 125-250cc Singles
M301	Vintage Dirt Bikes, V. 2 CZ, 125-400cc Singles; Husqvarna, 125-450cc Singles; Maico, 250-501cc Singles; Hodaka, 90-125cc Singles
M305	Vintage Japanese Street Bikes Honda, 250 & 305cc Twins; Kawasaki, 250-750cc Triples; Kawasaki, 900 & 1000cc Fours